AF290404

Food Hydrocolloids
Structures, Properties, and Functions

Food Hydrocolloids

Structures, Properties, and Functions

Edited by

Katsuyoshi Nishinari

Osaka City University
Osaka, Japan

and

Etsushiro Doi

Kyoto University
Kyoto, Japan

Springer Science+Business Media, LLC

Library of Congress Cataloging-in-Publication Data

Food hydrocolloids : structures, properties, and functions / edited by
  Katsuyoshi Nishinari and Etsushiro Doi.
        p.   cm.
    "Proceedings of an International Conference and Industrial
  Exhibition on Food Hydrocolloids, held November 16-20, 1992, in
  Tsukuba, Japan"--T.p. verso.
    Includes bibliographical references and index.
    ISBN 978-1-4613-6059-9   ISBN 978-1-4615-2486-1 (eBook)
    DOI 10.1007/978-1-4615-2486-1
    1. Colloids--Congresses. 2. Food--Analysis--Congresses.
  I. Nishinari, Katsuyoshi. II. Doi, Etsushiro. III. International
  Conference and Industrial Exhibition on Food Hydrocolloids (1992 :
  Tsukuba-shi, Japan)
    [DNLM: 1. Colloids--congresses. 2. Food Analysis--congresses.
  QU 133 F6864 1992]
  TP453.C65F72   1993
  664'.06--dc20
  DNLM/DLC
  for Library of Congress   93-44786
                                                          CIP

Proceedings of an International Conference and Industrial Exhibition on Food Hydrocolloids, held November 16–20, 1992, in Tsukuba, Japan

ISBN 978-1-4613-6059-9

© 1993 Springer Science+Business Media New York
Originally published by Plenum Press, New York in 1993
Softcover reprint of the hardcover 1st edition 1993

PREFACE

It is now well recognised that the texture of foods is an important factor when consumers select particular foods. Food hydrocolloids have been widely used for controlling in various food products their viscoelasticity, emulsification, gelation, dispersion, thickening and many other functions. An international journal, FOOD HYDROCOLLOIDS, launched in 1986 has published a number of stimulating papers, and established an active forum for promoting the interaction between academics and industrialists and for combining basic scientific research with industrial development.

Although there have been various research groups in many food processing areas in Japan, such as fish paste (kamaboko, surimi), soybean curd (tofu), agar jelly dessert, kuzu starch jelly, kimizu (Japanese style mayonnaise), their activities have been conducted in isolation of one another. The interaction between the various research groups operating in the various sectors has been weak. Symposia on food hydrocolloids have been organised on several occasions in Japan since 1985. Professor Glyn O. Phillips, the Chief Executive Editor of FOOD HYDROCOLLOIDS, suggested to us that we should organise an international conference on food hydrocolloids. We discussed it on many occasions, and eventually decided to organise such a meeting, and extended the scope to include recent development in proteinaceous hydrocolloids, and their nutritional aspects, in addition to polysaccharides and emulsions.

The international conference in Tsukuba was organised by the Japanese Society of Food Hydrocolloids under the auspices of many relevant societies and organisations. It was an epoch-making-event for Japanese researchers as well as overseas participants working in the field of food hydrocolloids. Many of the participants who know each other through their publications, could benefit from face-to-face discussions during tea time, lunch time, poster sessions and industrial exhibitions. The social and scientific exchange was so fruitful that all the participants felt very pleased to be there. We are particularly grateful for the overwhelming response to the success of the meeting. As a result, it has been decided to organise the 2nd International Conference on Food Hydrocolloids at Ohio State University in USA during the first week of September 1994.

This book we trust will convey the essence and enthusiasm of the conference. Papers included in this book are classified into the following six groups:
* *Polysaccharides*
* *Functional Properties*
* *Proteins*
* *Emulsions*
* *Interactions*
* *Nutrition and Physiology*

The information collected in this book supplies valuable scientific information and provides suggestions for further developments which will further assist our understanding and utilisation of food hydrocolloids.

Katsuyoshi Nishinari
Etsushiro Doi

CONTENTS

FUNCTIONAL PROPERTIES

PROTEINS

EMULSION

INTERACTION

PHYSIOLOGY AND NUTRITION

STRUCTURAL ASPECTS OF SOME FUNCTIONAL POLYSACCHARIDES

Akira Misaki

Department of Food and Nutrition
Osaka City University
Sumiyoshi, Osaka 558, Japan

INTRODUCTION

Carbohydrate polymers are widespread in nature, and play important roles in all life-stages of the organisms. Biological, physiological and other functions of natural polysaccharides are summarized in Table 1. Although in a nutritional point of view, the most important function of carbohydrate must provide biological fuel, which is reserved in the form of starch or glycogen, some undigestible polysaccharides, either plant hemicelluloses or microbial polysaccharides, have recently attracted attention because of the physiological effect as dietary fibers. In the field of food science and technology, pectic substances in plant cell wall matrix, and a variety of gum exudates, have long been used as emulsion stabilizers, water absorption agents, gelling and thickening agents. In addition, some microbial polysaccharides having unique properties, elaborated extracellularly in abundant amounts, have recently been utilized as a gel-forming agent or a food-packing film. In Table 2 natural polysaccahrides as food hydrocolloids are classified according to their biological sources. Although functionality of individual polysaccharides are closely related to their own molecular shapes, we must first look into the primary structure of the building units, *i. e.* the carbohydrate components, mode of linkages and their sequential arrangements. In this symposium structural correlations with functionality of some plant and microbial polysaccharides, important food hydrocolloids, are discussed.

Food Hydrocolloids: Structures, Properties, and Functions
Edited by K. Nishinari and E. Doi, Plenum Press, New York, 1994

Table 1. Functionality and utilization of polysaccharides.

Biological functions
   a) Cell wall architecture (plants and microorganisms)
   b) Hydration or gelation of intracellular fluids
   c) Cell-cell recognition (cell surface)
   d) Stimulation of host defense
Physiological functionality
   a) Nutrition (biological fuel)
   b) Dietary fiber
   c) Source of functional oligosaccharides
Functionality as food hydrocolloid
   a) Stabilizer
   b) Thickener
   c) Gelling agent
   d) Water-holding agent

GENERAL MOLECULAR STRUCTURE OF NATURAL POLYSACCHARIDES AND METHODS FOR STRUCTURAL ANALYSIS

The rheological characteristics of natural polysaccharides are principally related to their molecular conformations, which may be regulated by the sequences of carbohydrate chains. Therefore, it must be necessitated to know their chemical structure, especially, the mode of glycosidic linkages in the building units. Even in homoglycans, there are wide varieties in modes of glycosidic linkages, arrangements of glycosyl residues, and molecular shape, either a linear or branched molecule. For example, there are a variety of glucose polymers occurring in nature as classfied:

	Glucosidic linkage	Name	Occurrence
α-D-GLUCANS;	$(1{\to}4)$ linear	amylose	plant
	$(1{\to}4)$ 1,6 branch.	amylopectin	plant
	$(1{\to}4)$ 1,6 branch.	glycogen	mammal
	$(1{\to}4)_3(1{\to}3)$	elsinan	*Elsinoe* sp.
	$(1{\to}6)$ 1,3 branch.	dextran	*L. mesenteroides*
	$(1{\to}3)$ 1,6 branch.	(mutan)	*Str. mutans*
	$(1{\to}3)G(1{\to}4)$	nigeran	*Asp. niger*
β-D-GLUCANS;	$(1{\to}4)$ linear	cellulose	plant cell wall
	$(1{\to}4)_nG(1{\to}3)$	cereal gum	endosperm
	$(1{\to}4)_2G(1{\to}3)$	lichenan	lichen
	$(1{\to}3)$ linear, DP 30	laminaran	seaweed
	$(1{\to}3)$ linear	curdlan	*Alc. faecalis*
	$(1{\to}3)$ less branch.		yeast cell wall
	$(1{\to}3)$ 1,6 branch.		fungi, mushroom
	$(1{\to}2)$ cyclic		*Arg. tumefaciens*

These plant and microbial glucans of either α- or β-configuration, with different linkages, exhibit their specific properties and functionality. For instance, amylose and cellulose, both linear molecules build up with $(1{\to}4)$-linked D-glucose residues, but with different anomeric linkages (α-, or β-), giving different physical properties and enzymic

Table 2. Polysaccharides as food hydrocolloid.

PLANT POLYSACCHARIDES
  Starch and its modified products
  Seed gums
      Cereal gums $(1,3/1,4-\beta$-Glucan)
      Galactomannans (Guaran, Locust bean, Tara bean etc.)
      Xyloglucan (Tamarind)
  Tuber gums
      Glucomannan (Konjac), Inulin
  Cell wall constituents
      Cellulose (chemical modified)
      Fucoxyloglucan
      Arabinoxylan (monocot), Arabinogalactan (dicot)
      Pectinic substances (Rhamnogalacturonan, Galacturonan etc.)
  Exudate mucilages
      Gum arabic, Gum tragacanth, Karaya gum and others
SEAWEED EXTRACTS
      Agar (Gal., 3,6-AnhydroGal)
      Carrageenan (sulfated Gal), Fucoidan
      Alginic acid (Guluro-mannuronide)
      Laminaran
MICROBIAL POLYSACCHARIDES
  Neutral polymers
      Dextran $(\alpha-1,6$ glucan)
      Pullulan $(\alpha-1,6/1,4)$; low viscosity, film sheet
      Elsinan $(\alpha-1,3/1,4)$; high viscosity, film sheet
      Curdlan $(\beta-1,3)$; heat-gelation
      Screloglucan (6-O-branchedβ-1,3)
      Levan (2,6-fructan)
  Acidic polymers
      Xanthan (GlA, Pyruvate, Man, 4Glc$_\beta$-); synergistic gelation effect
      Gellan ($\rightarrow$3Glc$_\beta\rightarrow$, $\rightarrow$4GlA$_\beta\rightarrow$, $\rightarrow$4Rha$_\alpha\rightarrow$, $\rightarrow$4Glc$_\beta\rightarrow$); gel forming
  Others having potential utilization
      Welan gum (Kelco)
      Rhamsan gum (Kelco)
      B. polymyxa acidic Ps
      PS 1004 from methnol (Ga, Man, Al, GlA)
      T. fuciformis acidic PS (GlA, $\rightarrow$3Man$_\alpha\rightarrow$)
      Acetobacter sp. acidic PS (GlA, Gal, $\rightarrow$4Glc$_\beta\rightarrow$)

susceptibilities. In a similar way, $(1\rightarrow3)$-linked α- and β-D-glucans possess significantly different conformations from each other, shown in Fig. 1. For instance, curdlan, a gel-forming bacterial glucan[1], and the backbone moiety of many antitumor, branched β-D-glucans, e.g., scleroglucan, lentinan, schizophyllan and other fungal glucans, consist of β-$(1\rightarrow3)$ glucosyl chains. They are mostly present as a triple helical strands, which are most likely responsible for their host-mediated antitumor activity[2]. In contrast, the $(1\rightarrow3)$-β-D-glucosyl chain, which forms a backbone of the dental cariogenic α-glucan ("mutan"), elaborated by *Strept. mutans* may have a ribbon-like single-

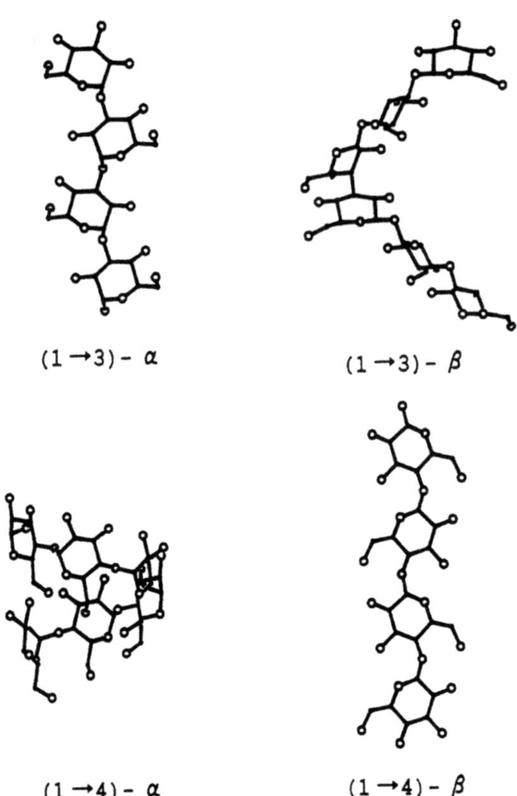

(1→3)- α   (1→3)- β

(1→4)- α   (1→4)- β

Fig. 1 Crystalline conformation of α - and β -D-glucans.

(Courtesy of Dr. K. Ogawa, RIAT, University of Osaka Prefecture)

chain conformation, which may give an insolubility characteristic; the adhesive characteristic may be exhibited by α -(1→6)-linked D-glucosyl side chains.

Many of natural polysaccharides, currently used as potent food hydrocolloids, such as, emulsion stabilizer, water-holding, thickening or gelling agents, are obtained from various tissues of plants, *i. e.* seeds, tubers, and exuded mucilages, and also from seaweed extracts (see, Table 2). Dextran was introduced into pharmaceutical industry as a blood plasma substitute after the second world war. Then, xanthan and pullulan, and recently gellan gum have also been put on commercial market, and some other microbial polymers have attracted attention because of their potential utilizations, as will be discussed later.

Most of these polysaccharides are heteropolysaccharides comprising different carbohydrate components. Some contain charged groups, *e. g.*, carboxyl, sulfonyl, or amino groups. Their unique rheological characteristics are basically attributable to the molecular shapes as well as localization of the charged groups. Fig. 2 shows viscosities of aqueous solutions of typical natural gums. For instance, galactomannans, *e. g.*, guar and locust bean gum, give high viscosity, most probably due to the characteristic features as the linear chain of the β -(1→4)-linked mannan and solubility property may be attributed to the presence of regularly arranged α -galactosyl branches. On the other hand, gum Arabic, a multiply branched acidic heteropolysaccharide, shows extremely

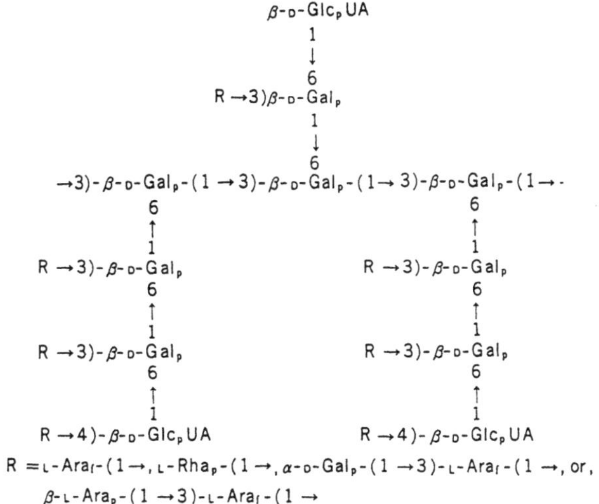

Gum arabic

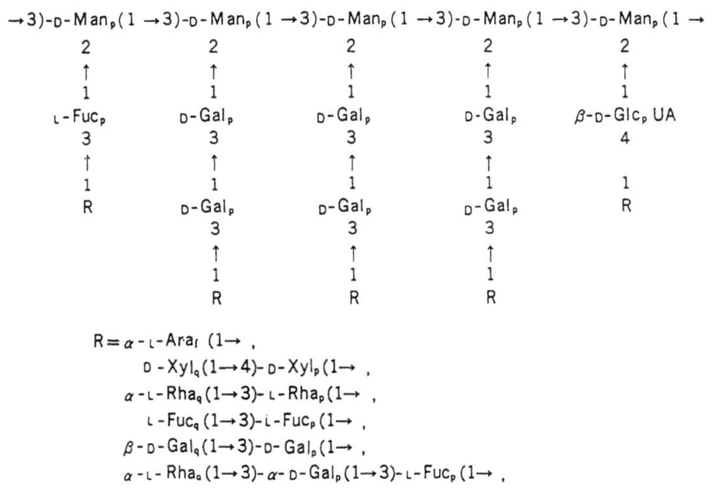

Polysaccharide of Junsai mucilage

low viscosity. In connection to this, it is interesting to note that a gelatinous mucilage which covers young leaves of "Junsai" (*Brasenia schreberi*) was characterized as a highly branched acidic polysaccharide[3], resembling that of gum arabic, and its aqueous solution gives a very low viscosity.

As mentioned above, elucidation of the molecular structure, both chemical structure and conformation, may be a key to open the functionality and application of a particular polysaccharide.

Information on chemical structure of a polysaccharide, either from plant or microbial source, can be provided, first, by analysis of the carbohydrate components and their molar ratios, and, second, the mode of their glycosidic linkages revealed by methylation analysis which includes gas-liquid chromatography-mass spectrometric analysis (GC-MS). The sequence of glycosyl residues may be elucidated by examination of the oligosaccharide fragments by partial hydrolysis with acid, acetolysis and Smith degradation etc. The use of purified endo-glycanases may also

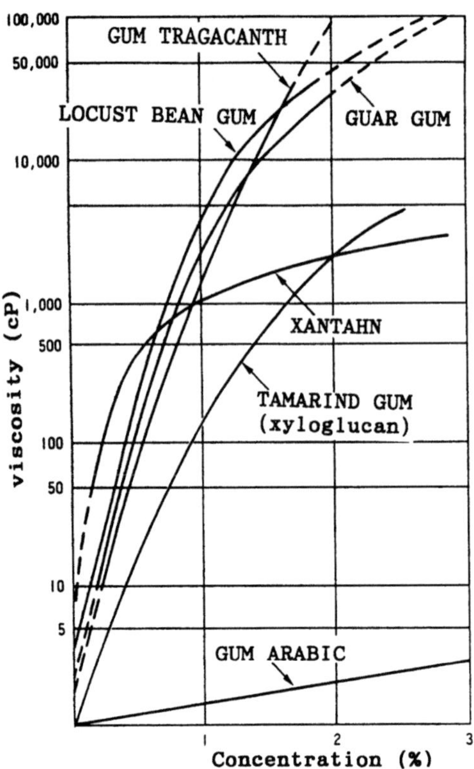

Fig. 2 Comparison in viscosities of natural gums in aqueous solutions. [Reproduced from a technical paper of Dainippon Pharmaceutical Ind. Co. Japan. Glyloid, tamarind seed polysaccharide]

provide valuable information on the sequential arrangements of the glycosyl residues. A general procedure for structural analyses may involve following items.

1. Constituent analysis
   Carbohydrate groups
   Other constituents (Amino acid, Acyl, Sulfate group, etc.)
2. Modes of glycosidic linkages
   Methylation , Smith degradation, Glycosidase action
   ^{1}H-NMR, ^{13}C-NMR
3. Linkage with non-carbohydrate residues
   Glycoprotein, Glycolipid
   Ester linkage (Acyl etc.)
4. Sequence of sugar residues
   Fragmentation analysis
        a) Partial acid hydrolysis
        b) Acetolysis
        c) Smith degradation
        d) Enzymic degradation
   HPLC, HPAEC(PA-derivative)
   FARB/MS

5. Biochemical analysis
    Immunochemical
    Lectins
6. Conformation analysis
    X-ray diffraction

In the past two decades, great advantages have been made in the precise instrumental analyses, such as, gas-liquid chromatography (GC) and high performance liquid chromatography (HPLC). Recently developed HPAEC, a new type of anion exchange LC, equipped with a pulse amperometric detector, may give a great help for precise fractionation and analysis of a series of oligosaccharides. For instance, a series of α-(1→4)-linked gluco-oligomers, released from amylopectin and glycogen by debranching enzymes, e. g., isoamylase, can be precisely fractionated, as shown in Fig. 3, and enabled to elucidate the fine structural

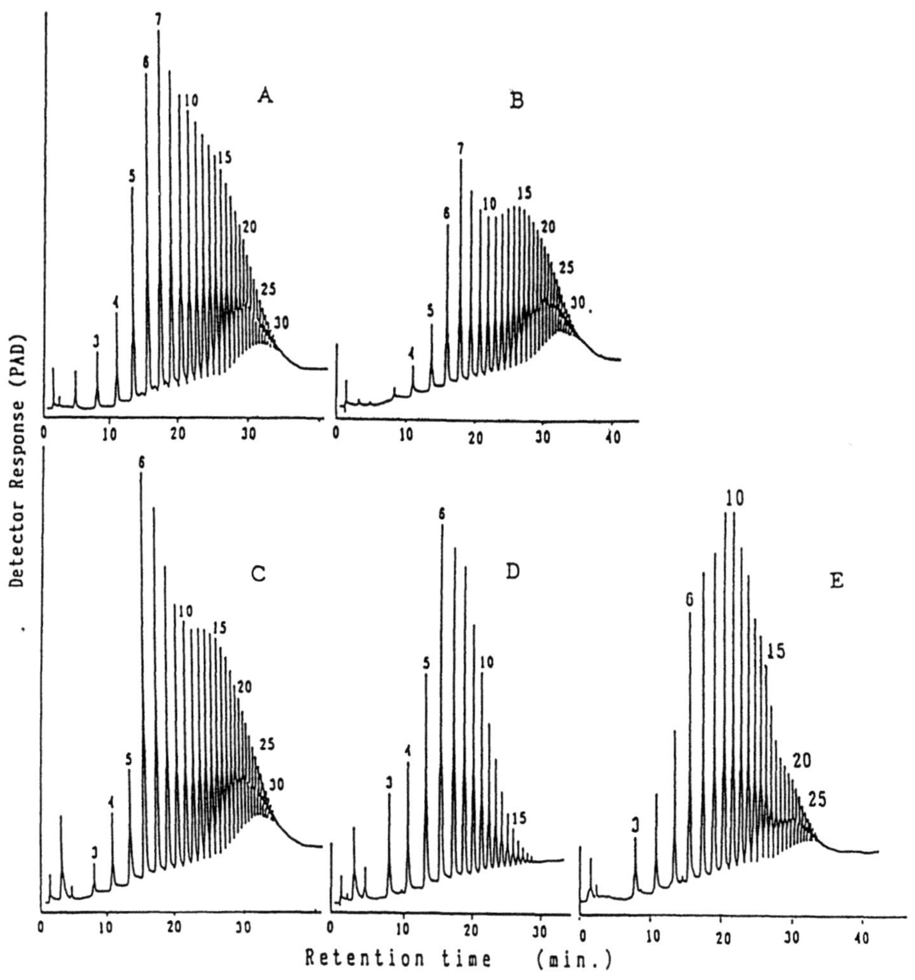

Fig. 3 HPAEC profiles of unit-chain distribution of glycogens.
A: Bovine liver, B: Rabbit liver, C: *Aspergillus fumigatus*, D: *Mycobacterium tuberculosis*

features[4]. For ultra-micro analysis of carbohydrate chains, HPLC of the pyridylamino-derivatives of the heterosaccharides has recently been developed by Dr. Hase *et al.* This method may especially be recommended for analysis of various types of glycopeptide chains. Carbohydrate sequences of the fragmentation products could also be determined by FARB/MS. Together with knowledge on the chemical structure, conformational analysis should be necessitated to know the steric molecular structure.

PLANT POLYSACCHARIDES

Food polysaccharides of plant origins may be classified, 1) components in the cell wall architecture, *i.e.* cellulose as the cell wall skeleton, hemicellulose (xyloglucan, arabinoxylan of monocotyledon), and pectic polysaccharides which constitute cell wall matrix of dicotyledon, and, 2) gums and exudates, *e.g.*, cereal gums, galactomannan, gum Arabic and Tragacanth etc. The seed gums such as guaran, a galactomannan containing single galactosyl branches every two mannose residues, are readily soluble to give a highly viscous solution, so, have frequently been used as a good food additive. It has been well known that xanthan gum produced by *Xanthomonas campestris* synergistically interacts with locust bean galactomannan, but not with guaran. This might be explained by intermolecular association between the unsubstituted segments of the β-(1→4)mannan backbone of locust bean gum and the β-(1→4)-linked glucosyl backbone chain of xanthan molecule (Fig. 4). However, Tako recently proposed an alternate mechanism, involved in the interaction of the side-chains of the single strand xanthan with the adjacent linear mannosyl segments of the galactomannan. Thus, one xanthan molecule may possibly combine with two or three molecules of xanthan (Fig. 5). In connection to possible utilization of the galactomannan oligosaccharides, we have purified endo-

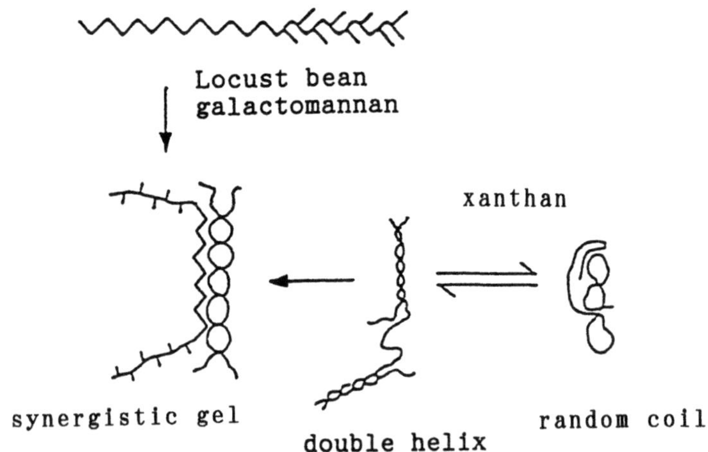

Fig. 4 Synergistic gel formation between xanthan and locust bean galactomannan. [I. C. M. Dea and E. R. Morris, "Extracellular Microbial Polysaccharides" p. 174 (ACS Symposium series 45, Ed. A. Laskins (1977))]

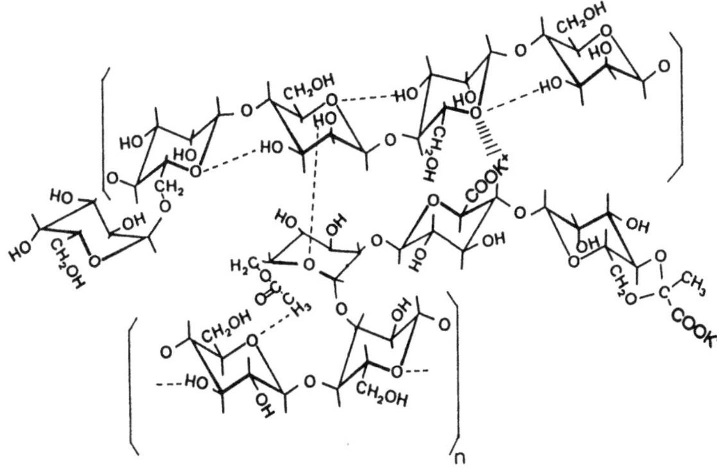

Fig. 5 Possible binding sites for D-mannose-specific interaction between xanthan and plant galactomannan. [Reproduced from M. Tako, *J. Carbohydr. Chem.* 16:239(1991)]

β-mannase from *Trametes sanguinear*. This enzyme, devoid of any galactosidase, hydrolyzes preferentially locust bean galactomannan and yields manno-triose, and -tetraose having the galactosyl side chain, beside O-β-(1→4)mannosyl mannose (Toyoda and Misaki, unpublished results).

Xyloglucan, usually contained, approximately, 20% in the primary cell walls of dicots, and 5% in those of monocots, has recently been recognized as a physiologically important polysaccharide for regulation of the plant cell growth. The seeds of tamarind (*Tamarindus indica*) contains exceptionally a high content of xyloglucan, approx. 50 % of the kernel. The tamarind xyloglucan is currently commercially available as a food stabilizer. Xyloglucan can be hydrolyzed by endo-β-(1→4)-glucanase, e. g., *Trichoderma viride*, which cleaves the sites of unsubstituted β-(1→4)-glucose residues of the backbone, to give hepta-, octa- and nonasaccharide. These oligosaccharides especially containing the terminal fucose residue may have an important role to regulate the elongation of the cell wall. They may also propose possible uses for dietary functionality. We recently prepared antibodies to these XG oligosaccharides and showed successful use for histochemical detection of the xyloglucan in the primary cell wall layers[5].

Pectic polysaccharides, which account for ca. 35% of dicotyl cell walls, are mostly located in the middle lammela. This type of polysaccharide consists mainly of the backbone of rhamnogalacturonan, some of the rhamnose residues are attached with side chains of arabinan, galactan or arabinogalactan (we have failed to isolate arabinogalactan from the soybean cotyledon). The most characteristic physical feature of the pectic polysaccharides must be gel-formation in the presence of large quantity of sugars which attract water molecules (Fig. 6). Another gel-formation of a pectic polysaccharide which is devoid essentially of side chains may be involved in formation of so-called "egg box" structure through ionic interactions between Ca^{++} and the adjacent α-(1→4)-linked galacturonic acid residues. We have recently found the latter case.

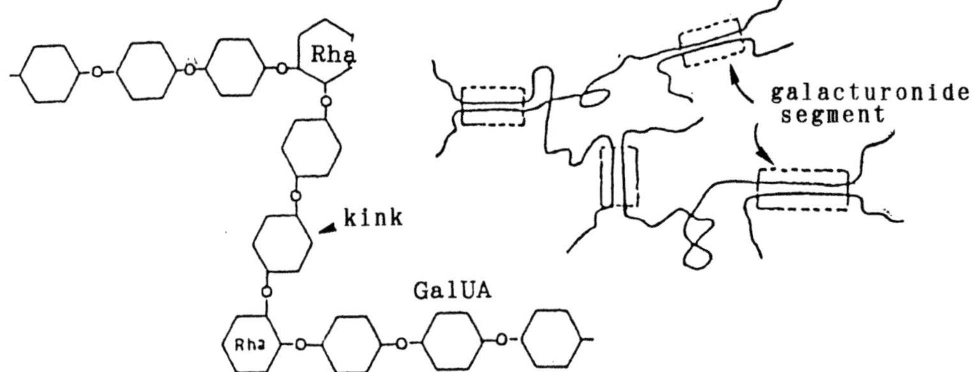

Fig. 6 Network formation of pectin by junction with galacturonide zones. [Reproduced from W. Pilnik and F. M. Rombouts, *Carbohydr. Res.*, 142:93(1985)]

A spontaneous gel-forming polysaccharides of "Ai-Geok"

"Ai-Geok" is an edible, jelly cake formed spontaneously during squeezing the seeds of *Ficus awkeotsang* with water. This plant is grown on mountainsides of Taiwan. We became aware that the spontaneous gelation of the water-extracted mucilage might be attributed to the unique chemical structure of the pectin-like substance which would be de-esterified by an intrinsic esterase of the seeds, whereby regular arrangements of the molecular chains respond to gel formation in the presence of Ca^{++}.

Before water-extraction, whole seeds were heated at 90°C for 30 min to inactivate any intrinsic enzyme. By fractionation of the cold water-extract of the pre-heated seeds, an acidic polysaccharide, which is localized on outer layer of the seeds, was isolated. The purified polysaccharide ($[\alpha]_D$ +239, and mol. wt. 3.4×10^5) was essentially unsubstituted linear molecule containing highly methoxylated α-(1→4)-linked D-galacturonic acid residues (DE, 69.3%); there was no L-rhamnose or any other neutral sugars detected. Incubation of its 0.5% aqueous solution with pectin esterase of lemon peel at 25°C and pH 7.7 in the presence of 1 mM Ca^{++}, afforded gradual formation of a rigid but somewhat brittle gel (breaking strength, 6.08×10^4 dyne/cm^2, after 60 min incubation), strongly suggesting formation of so-called egg-box structure containing Ca^{++}, as shown in Fig. 7[8]. The particular enzyme which should be responsible for spontaneous gelation of Ai-Geok mucilage was found in the red tepals attached to the pedicels of the seeds. The pectin esterase (EC 3.1.1.11) was purified by column chromatographies on QAE-Sephadex A-25, Q-Sepharose, hydroxyapatite and then by Sephacryl S-200. The purified methylesterase, which showed a single protein band on SDS-PAGE, had mol. wt. 42,000, at pH 4.4. The enzyme activity toward the highly methoxylated Ai-Geok galacturonan to lead to spontaneous gelation was stimulated by the presence of a low concentration of metal ions, $Ca^{++} >$ $Mg^{++} > K^+ > Na^+$. By using the purified enzyme, it was confirmed that the gelation mechanism of Ai-Geok must be involved in the enzyme-induced de-esterification of methoxyl groups followed by formation of the "egg-box structure" by situating Ca^{++} between the regularly arranged adjacent chains of α-(1→4)-linked D-galacturonic acid residues[8].

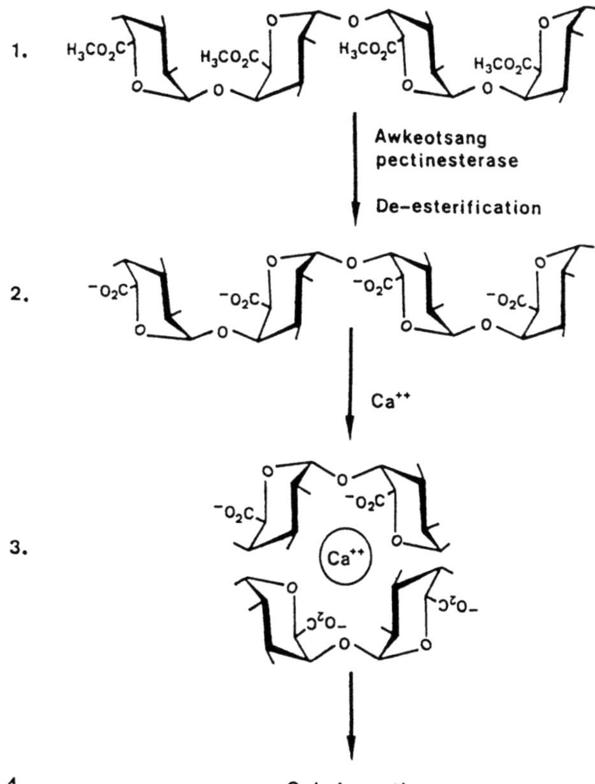

Fig. 7 Gel-formation mechanism of "Ai-Geok" galacturonan by "egg-box" struture through enzymic de-esterification.
1:Highly methoxylated polygalacturonide of the seed extract, 2:Enzymatically de-esterified polygalacturonide, 3:Formation of the "egg-box" structure.

We were also greatly interested in this pure galacturonan for preparation of a phytoalexin elicitor, which is capable of stimulating production of phytoalexin, an antimicrobial agent. Oligo-galacturonides(OLGAs) were obtained by the action of the purified endo-pectate lyase of *Erwinia carotovora*, and precise fractionation by QAE-Sephadex A-25 column chromatography, and were assayed for stimulation of production of glyceolin in soybean cotyledon. Among the purified OLGAs (DP 3-12), 4,5-unsaturated hexa-1,4-galacturonide was found to exhibit a potent elicitor activity[9], in addition to that of DP 10, previously reported by the group of Albersheim.

Structure and antitumor activity of *O*-6-branched fungal (1→3)-β-glucans

Most of fungal-D-glucans, which constitute fruiting body (mushroom) of *Basidiomycetes* are produced in cultural fluids, are (1→3)-linked β-D-glucans having D-glucosyl branches at *O*-6, but there may be a great structural diversity, with regard to branch distributions, degree of branching (db), 4/5 - 1/23, that may affect their immunomodulating activities. Since we first elucidated, in 1963, the structure of a branched (1→3)-β-glucan of *Screlotum glucanicum*[10], similar types of branched β-(1→3)-glucans attracted much attention because of their host mediated tumor-inhibiting activity (antitumor activity). Among them

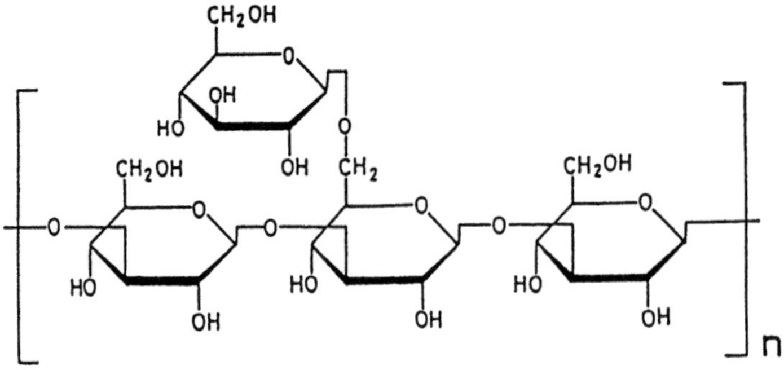

Fig. 8 Repeating units of antitumor-active schizophyllan and lentinan.

lentinan of Shiitake, fruiting body of *Lentinus edodes*, and schizophyllan (SPG), an extracellular product of *Schizophyllum commune*, have essentially the same structure with that of screloglucan (Fig. 8), and are currently used as an immunotherapic drug[2]. Although the mechanism of tumor-growth inhibition of these $(1{\rightarrow}3)$-β-glucans has not fully been understood, their immunomodulating actions appear to be closely related to the triple helix conformation of the backbone chains, as established by viscosity studies and X-ray analysis. The relationship between mol. wt. and antitumor activity of SPG on Sarcoma 180 of sonic-degraded low molecular weight fractions indicated that the triple helix conformation (mol. wt. $\geq$ 1×10^5) is essential[11].

Although schizophyllan and similar types of fungal glucans have moderate branched structure (db, 1/3), during studies on correlation of structure and antitumor activity of antitumor polysaccharides we could isolate a variety of 0-6-branched $(1{\rightarrow}3)$-β-D-glucans, having db in the range from 3/4 to 1/25, and different antitumor activities against Sarcoma 180 transplanted JCL mice[2]. Table 3 summarizes main repeating units of these glucans and their antitumor activities. It appeared that exhibition of the antitumor activity is closely related not only to the helical triple chains but also to the mode of branching, especially its distribution along the backbone chain. For instance, moderately branched glucans, db 1/3 - 1/5, like schizophyllan, Fukurotake (*Volvariella volvacea*) glucan (VVG), exhibit strong antitumor actions, whereas those having either extraordinary highly branched glucans, e. g., *Ganoderma* (Reishi) hot alkali-extracted glucan (db, 1/17-25) or *Auricularia* (Kikurage) alkali-insoluble glucan (db, 2/4) exhibit a very low activity[12]. Although such low activity might be, in part, due to physical property, i. e., insolubility, it must be recognized that conversion of the glucosyl branches into the corresponding polyhydroxyl groups, by periodate oxidation followed by borohydride reduction, gave remarkable enhancement of the original activity (see, Fig. 9)[13]. It is also true that, if the modified polyol groups were cleaved, the resulting linear $(1{\rightarrow}3)$-β-glucan molecules lost their original activity (Table 4).

Among these 0-6-branched antitumor $(1{\rightarrow}3)$-β-glucans, recently we obtained a highly antitumor-active polysaccharide by cold alkali extraction of an edible mushroom (*V. volvacea*), "tso ru" in Chinese, or "Fukurotake" in Japanese. This glucan, designated VVG, having db, 1/5, exhibited significantly high antitumor activity against allogeneic tumors, and also certain type of syngeneic tumors, such as Meth-A fibrosarcoma in

Table 3. Antitumor activities of glucans of V. volvacca and other fungal sources.

Glucan	db	Dose* (mg/kg)	Average tumor weight(g)		Tumor growth inhibition (%)	Complete regression
			Treated	Control		
Against Sarcoma 180 solid tumor in ICR-JCL mice						
V. volvacea						
Cold alkali-extracted	1/5	1	2.5	9.5	73.2	3/6
(VVG)		5	0.3	10.2	97.0	4/5
		10	2.0	10.9	81.7	1/5
Hot alkali-extracted	1/2	10	7.2	10.9	33.9	0/5
Schizophyllum commune						
Schizophyllan	1/3	1	0.3	4.1	95.1	6/10
A. auricula-judae						
Alkali-insoluble glucan	3/4	10	4.7	5.8	18.9	0/4
G. lucidum						
Water extracted	1/3	10	0.2	9.4	97.7	4/5
Cold alkali-extracted	1/17	10	3.8	9.4	59.6	2/6
Pestalotia sp. 815						
Pestalotan	2/3	5	4.2	9.8	57.3	0/6
Against Methylcholanthrene-induced fibrosarcoma in DBA/2 mice						
VVG		5	0	2.04	100	6/6

*Injected intraperitoneally daily for 10 days.

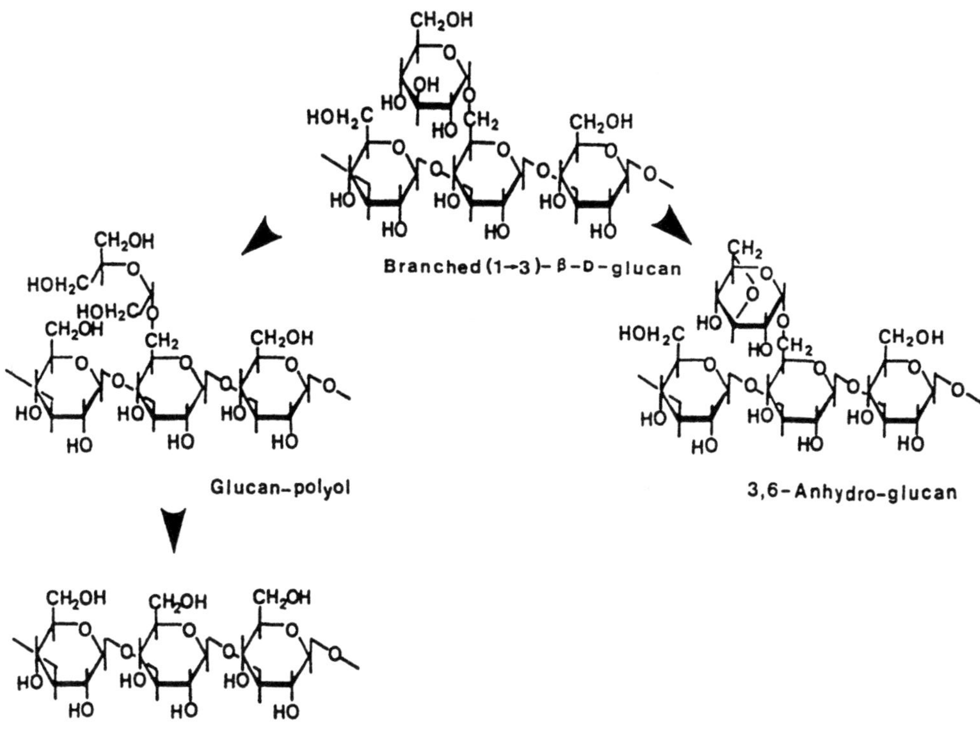

Fig. 9 Chemical modification of antitumor O-6-branched $(1{\rightarrow}3)$-β-D-glucan.

Table 4. Comparison of antitumor activities of branched (1→3)-β-D-glucans and their derivatives of fungal.

Polysaccharide	Dose* (mg/kg)	Average tumor weight(g)		Inhibition ratio (%)	Complete regression
		Treated	Control		
Against Sarcoma 180 solid tumor in ICR-JCL mice					
V. volvacea					
VVG-polyol	1	0	9.5	100	6/6
	5	0	9.5	100	6/6
	10	2.8	9.5	70.7	3/6
1.3-Linear-VVG	5	11.6	9.5	-22.5	0/6
3,6-Anhydro-VVG	5	15.6	9.5	-63.8	0/6
A. auricula-judae					
Glucan	10	8.9	11.0	18.9	0/6
Glucan-polyol	5	0.3	11.0	97.4	6/8
Pestalotia					
Pestalotan	5	4.2	9.8	57.3	0/6
Pestalotan-polyol	1	0	10.2	100	6/6
Against Meth-A fibrosaecoma in BALB/c mice					
VVG-polyol	5	0.1	3.5	97.1	3/5
*Against Methylcholanthrene-induced fibrosarcoma in DBA/2 mice***					
VVG-polyol	5	0.06	2.04	97.1	3/5

*Injected intraperitoneally daily for 10 days.
**Tested by Dr T.Sasaki, National Cancer Center Research Institute, Japan.

DBA/2 mice. In connection to the antitumor activity, VVG was purified by ion exchange column chromatography, and its fine structural feature was elucidated by chemical and enzymic methods. There may be molecular heterogeneity with regard to the arrangements in the glucosyl side chains, consisting less branched, moderately branched and highly branched segments[14].

MICROBIAL POLYSACCHARIDES

Many bacteria, fungi and yeasts produce abundant amounts of unique polysaccharides during growth under the appropriate cultural conditions. Since dextran, produced by *Leuconostoc mesenteroides*, had been utilized as blood plasma substitute, a few of extracellular polysaccharides, *i.e.*, pullulan, of *Aureobasidium pullulans*, xanthan of *Xanthomonas campestris*, curdlan of *Alcaligenes faecalis*, and gellan elaborated by *Pseudomonas elodea* are currently commercially produced and used in food industry. These polysaccharides, either neutral or acidic polymers, possess unique structures and properties, useful in food industry, as food additives, gelling agents, edible packing films.

ELSINAN: In the course of study on new polysaccharides, we found that a fungus on tea leaves, *Elsinoe leucospila*, produces a new type of α-D-glucan, designated *elsinan*, which contains (1→4)- and (1→3)-D-glucosidic linkages[15]. Later, we found that *E. fawcetti* and other *Elsinoe* species produce the same type of glucan. Elsinan (mol.wt., 2×

10^5) is a linear, but water soluble α-D-glucan containing consecutive three or four (1→4)-linked glucose residues which are flanked by a (1→3)-linkage, as shown by methylation, Smith degradation, and other chemical fragmentation analyses. Fig. 10 shows the repeating unit of elsinan, compared with that of pullulan. Although chemical analysis indicated only general sequences of the glucose units, the use of particular α-amylolytic enzymes could provide important information on the detailed glucose arrangements. When α-amylase of salivary, pancreatic and *B. subtilis* acted on elsinan 3^2-α-glucosyl maltose was produced together with a small amount of glucose which should have been released from 3-maltosyl-maltose. The action of another type of α-amylase from *Aspergillus oryzae* (Taka-amylase) gave tetrasaccharide, 3^2-Glucosyl-maltotriose [4-0-α-nigerosyl-maltose] and a series of novel oligosaccharides, hepta-, deca-, trideca- and hexadeca-saccharide, each containing α-(1→3)-linkage at the nonreducing site, by cleaving specifically maltotetraose unit(s) inserted between maltotriose units; a

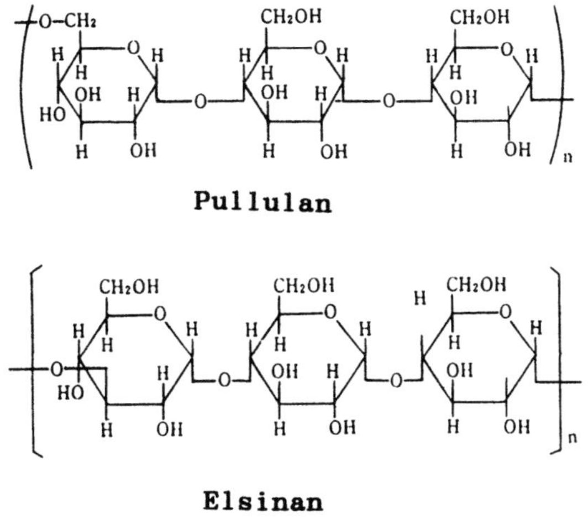

Pullulan

Elsinan

Fig. 10   Repeating units of elsinan and pullulan.

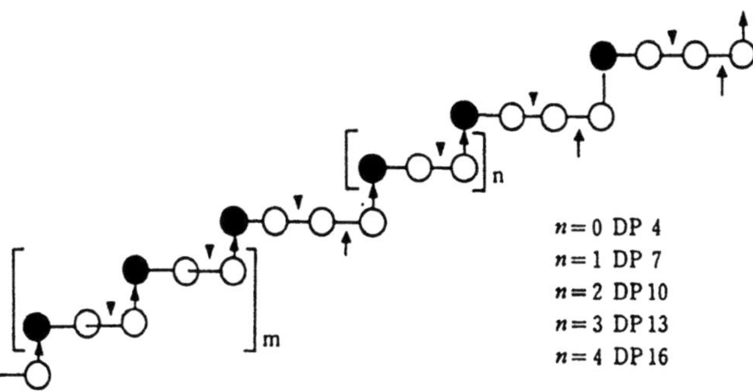

Fig. 11   Distribution of α-3-0-linked maltotriose and maltotetraose units and oligosaccharides released by salivary- and Taka-amylase.
(● : α-1, 3-, and ─○─ : 1, 4-glucosidic linkage)
▼ : salivary-type amylase, and ↑ : Taka-amylase

consecutive 3-0-linked maltotriose units in elsinan chain must be insusceptible to Taka-amylase, leaving high molecular products (yield, 33%)[16]. These results clearly indicated that maltotriose and maltotetraose units are rather randomly arranged, except for segments consisting of consecutive 3-0-linked maltotriose residues, as shown in Fig. 11.

As elsinan contains α-(1→3)-glucosidic linkages instead of α-(1→6)-linkages in pullulan, 1) it forms a weak acid-resistant, oxygen-impervious film, and 2) its aqueous solution gives high viscosity, ten times higher than those of pullulan, and forms a gel at conc. > 5%[17]. In chemical modification of elsinan, we attempted to introduce 3, 6-anhydro ring in the 0-4-substituted glucose residues through sulfation by treatment with DMSO-SO₃ followed by desulfation under alkali condition[18]. It was interesting that 3, 6-anhydro-elsinan showed unexpectedly very low viscosity as shown in Fig. 12. The 3, 6-anhydro elsinan, thus formed, was resistant to amylolytic enzymes, but by mild acid treatment it yielded novel oligosaccharides containing 3, 6-anhydro glucose group at the reducing end site, which may provide potential functionality. Although elsinan may be slowly digested to oligosaccharides in the digestive tracts, this glucan exhibits an interesting dietary fiber-effect, such as lowering cholesterol levels in blood in rats fed with high cholesterol diets. In connection to such hypocholesterolemic effect in rats was also

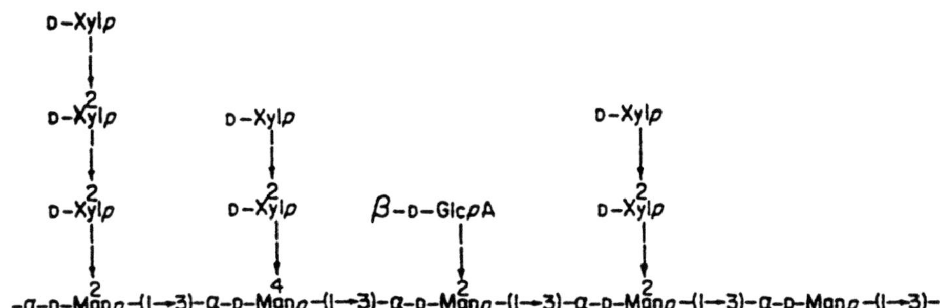

Glucuronoxylomannans

exhibited by some acidic polysaccharides mixed into the diet (1~2%), including an acidic polysaccharide produced by *B. polymyxa* S-4[19], and also glucuronoxylomannan from "Kikurage", and also as the extracellular product of *Tremella fuciformis*.

Other microbial polysaccharides

Recently, hyaluronic acid, originally isolated from mammalian cornea, has recently been produced by fermentation using non-pathogenic streptococcal strains for use in cosmetics as one of the best water-retaining agents. Other microbial biopolymers, which are proposed for potential utilizations in food industry and under commercial exploitation may include Welan produced by a strain of *Alcaligenes* species, and Rhamsan by another strain. (see, Table 2). In the course of searches on useful microbial polysaccahrides, we have recently discovered an interesting acidic polysaccharide, like a plant pectic substance, comprising (1→3)-linked L-rhamnose, D-glucose, and D-galacturonic acid residues (1:1:0. 8) (Y. Ohiso *et al*; Annual meeting of Jpn. Soc. Biotech.

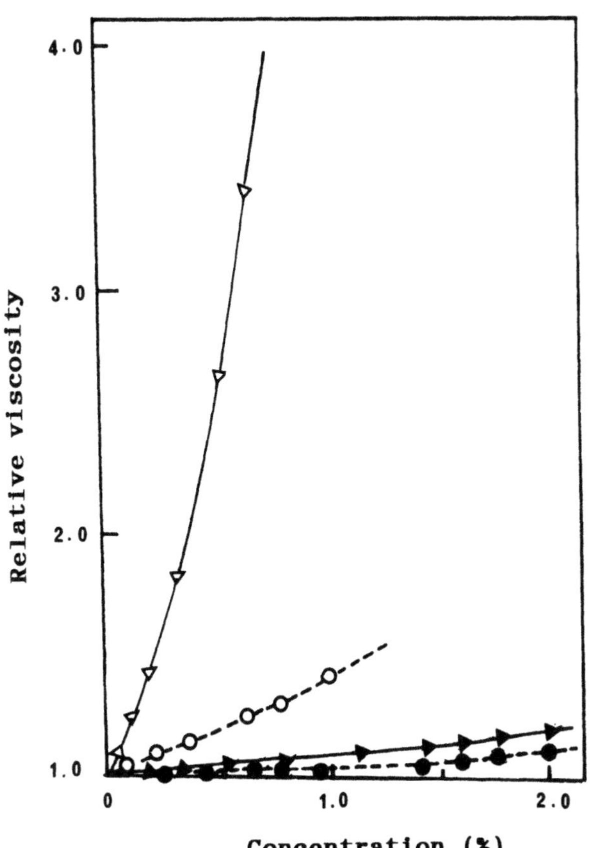

Fig. 12 Comparison of viscosities of aqueous solutions of elsinan, pullulan, and their 3,6-anhydro glucan derivatives. -▽- elsinan, -▶ anhydro-elsinan, -O-pullulan, -●-3,6-anhydro-pullulan

Biosci. Agrochem, 1993). Since this polysaccharide appears to possess a high water-retaining capability and other interesting properties, we are trying to elucidate its complete structure and mechanism of biosynthesis.

REFFERENCES

1. T. Harada, H. Saito and A. Misaki, Curdlan: a bacterial gel-forming β-1, 3-glucan, *Arch. Biochem. Biophys.*, 124:292(1968).
2. A. Misaki, E. Kishida, M. Kakuta and K. Tabata, Antitumor fungal $(1\rightarrow 3)$-β-D-glucans:structural diversity and effects of chemical modification, in: "Carbohydrates and Carbohydrate Polymers", ACS Symposium series, in press.
3. M. Kakuta and A. Misaki, Polysaccharide of "Junsai" mucilage, *Agric. Biol. Chem.*, 43:993, 1269(1979).
4. M. Matsui, M. Kakuta and A. Misaki, Fine structure of oyster glycogen, in: "Abstr. paper of XVIth International Carbohydrate Symposium, Paris", pp. 487(1992)
5. Y. Sone, J. Kuramae, S. Shibata and A. Misaki, Immunochemical specificities of antibodies to heptasaccharide unit of plant xyloglucan, *Agric. Biol. Chem.*, 53:2821(1989).
6. W. Pilnik and F. M. Rombouts, Polysaccharides and food processing, *Carbohydr. Res.* 142:93(1985).
7. K. Komae and A. Misaki, Isolation and characterization of the gel-forming polygalacturonide from seeds of *Ficus awkeotsang*, *Agric. Biol. Chem.*, 53:1237(1989).
8. K. Komae, Y. Sone, M. Kakuta and A. Misaki, Purification and characterization of pectineesterase from *Ficus awkeotsang*, *Agric. Biol. Chem.*, 54:1469(1990).
9. K. Komae, A. Komae and A. Misaki, A 4, 5-unsaturated low molecular oligosaccharide as a potent phytoalexin-elicitor isolated from polygalacturonide of *Ficus awkeotsang*, *Agric. Biol. Chem.*, 54:2155(1991).
10. J. Johnson, S. Kirkwood, A. Misaki, T. E. Nelson, J. V. Scaletti and F. Smith, Structure of a new glucan, *Chem. & Ind.* (London), 5:820(1963).
11. T. Yanaki, W. Ito, K. Tabata, T. Kojima, and T. Norisuye, Correlation between the antitumor activity of a polysaccharide schizophyllan and its triple-helical conformation in dilute aqueous solution, *Biophys. Chem.*, 17:337(1883).
12. E. Kishida, Y. Sone and A. Misaki, Effects of branch distribution and chemical modifications of antitumor $(1\rightarrow 3)$-β-D-glucans, *Carbohydr. Polymers*, 17:89(1993).
13. M. Kakuta, M. Sasaki, T. Tanaka, and H. Miyaji, Studies on interrelation of structure and antitumor effects of polysaccharides: antitumor action of periodate-modified, branched $(1\rightarrow 3)$-β-D-glucan of *Auricularia auricula-judae*, and other polysaccharides containing $(1\rightarrow 3)$-glycosidic linkages, *Carbohydr. Res.*, 92:115(1981).
14. E. Kishida, Y. Sone, and A. Misaki, Purification of antitumor-active, branched $(1\rightarrow 3)$-β-D-glucan from *Volvariella volvacea*, and elucidation of its fine structure, *Carbohydr. Res.*, 193:227(1989)
15. Y. Tsumuraya and A. Misaki, A new fungal α-D-glucan, Elsinan, elaborated by *Elsinoe leucospila*, *Carbohydr. Res.*, 66:491(1978).
16. A. Misaki, H. Nishio and Y. Tsumuraya, Degradation of elsinan by α-amy-

lases, and elucidation of its fine structure. *Carbohydr. Res.*, 109:207 (1982).

17. A. Misaki and Y. Tsumuraya, Structure and enzymic degradation of elsinan, a new α-D-glucan produced by *Elsinoe leucospila*, ACS symposium Series NO. 126, *Fungal Polysaccharides*, 109:197(1980).

18. Y. Ohe, K. Ohtani, Y. Sone and A. Misaki, Preparation and characterization of some β-D-glucans containing 3,6-anhydro-D-glucose, *Biosci. Biotech. Biochem.*, 57(in press) (1993).

19. H. Fukui, M. Tanaka and A. Misaki, Structure of physiologically active polysaccharide by *Bacillus polymyxa* S-4, *Agric. Biol. Chem.*, 49:2343 (1985).

ON THE RELATION STRUCTURE-PROPERTIES OF SOME
POLYSACCHARIDES USED IN THE FOOD INDUSTRY

M. Rinaudo

CERMAV-CNRS
B.P. 53 X
38041 Grenoble cedex - France
affiliated with the University Joseph Fourier of Grenoble

INTRODUCTION

Many polysaccharides are recognized as valuable food additives due to their original physicochemical properties. The polymers used for food applications are mainly thickening or gelling polysaccharides. Some are also used to stabilize emulsion or to suspend solid particles[1-5].

In this paper, one would give examples of specific properties of some polysaccharides in relation with their chemical structure and their molecular weight. The role of polysaccharides on the viscosity of aqueous solution is directly related to the stiffness of the molecules depending directly on the primary and secondary structures. The gelling ability depends mainly on cooperative intermolecular interactions controlled by the chemical structure and thermodynamic conditions.

Due to their stereoregularity and/or their ionic behaviour, secondary forces play a large role on intermolecular interactions (H- bond, divalent counterion linkage...) but also on the conformation of the molecules. In some cases, the mixture of two thickening polymers gives a gel as demonstrated with the system galactomannan-xanthan.

SOLUTION PROPERTIES

Many polysaccharides from natural sources were introduced since a long time in food applications. They are mainly water soluble polymers causing an increase of the viscosity of aqueous medium. Many of them are considered as dietary fibers and are non digestible.

Food Hydrocolloids: Structures, Properties, and Functions
Edited by K. Nishinari and E. Doi, Plenum Press, New York, 1994

Table 1. Water soluble polysaccharides used in food applications

Type	Origin	Function	Conditions
Pectin	Plant cell walls		
	DS > 70	Gelation	pH acid/sugar 65%
	DS < 50	Thickening	Na-form
		Gelation	Presence of divalent ions
Alginate	Brown seaweed	Stabilising	Na form
		Thickening	
		Emulsification	
		Suspending	
		Gelation	Presence of divalent ions
Agar	Red seaweed	Gelation	Firm gel under melting
		Suspending	temperature
Carrageenan	Red seaweed		
	λ	Thickening	
	i	Gelation	elastic gel in presence of K^+ ions
		Thickening	
	k	Gelation	Firm gel in presence of K^+ ions
Guar	Leguminous tree	Thickening	Galactomannan ratio galactose/mannose ~ 0.25
		Emulsification	
		Stabilising	
		Suspending	
Locust bean gum	Carob tree	Thickening	Galactomannan ratio galactose/mannose ~ 0.25
		Gelling	
		Stabilising	
		Emulsification	
Xanthan	Fermentation	Stabilising	
		Thickening	
		Emulsification	
		Pseudoplasticiser	
Gellan	Fermentation	Gelling	
		Thickening	
Curdlan	Fermentation	Gelling	Firm gel

Then, they are mostly used as additive to modify the functional properties for food applications by controlling the texture (Table 1).

Their role concerns the thickening of many systems. In some applications they also form a transitory network allowing suspension stabilization (i.e. reducing the rate of sedimentation of solid particles) or emulsification. Many sources of polysaccharides are industrially employed. Polysaccharides such as guar or locust bean gums are extracted from seeds ; some are extracted from algae such as carrageenans, agarose or alginates or from fruits such as pectins. Microbial polysaccharides are now developing rapidly with fungal sources for curdlan and scleroglucan or bacterial sources for gellan and xanthan. In this paper, cellulose, starch, hemicelluloses and their derivatives are not considered.

Solution properties and mainly rheological properties of polysaccharide solutions are controlled by the chemical structure and macromolecular characteristics of the solubilized polymers. Highly soluble and flexible polymers usually will be only thickeners. In that respect, depending slightly on temperature, galactomannans with high ratio galactose/mannose[6], alginate under sodium form[7], λ-carrageenan or i and k-carrageenan under sodium form, Na-pectins[8], and xanthan[9,10] when dissolved in water increase the viscosity.

This effect is directly related to the polymer concentration (C) and its molecular weight or its intrinsic viscosity following the relation :

$$\eta_{sp} = C[\eta] + k' \, (C[\eta])^2 + B \, (C[\eta])^n \qquad (1)$$

valid in the newtonian regime ($\dot{\gamma} \to 0$) ; in this relation, η_{sp} is the specific viscosity, $C[\eta]$ is the overlap parameter, $\dot{\gamma}$ is the shear rate, k' the Huggins constant and n is an exponent varying from 3 to 4 as observed for the behaviour of polymer in the melt. (Figure 1).

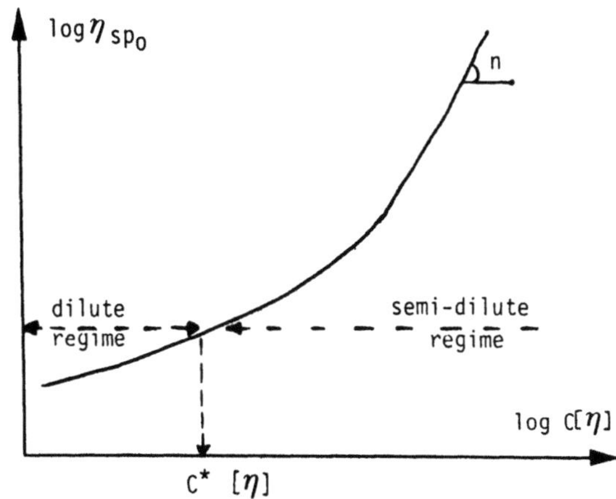

Figure 1. Dependence of the zero shear rate specific viscosity η_{spo} as a function of the overlap parameter $C[\eta]$. Critical overlap concentration C*.

All the polymers give the same type of curve only separated from each others due to the effect of k'. The linear dependence of the reduced viscosity $\frac{\eta_{sp}}{C}$ in the range of the dilute regime is reflected by the Huggins relation :

$$\frac{\eta_{sp}}{C} = [\eta] + k' \, C[\eta]^2 \qquad (1')$$

Deviation from Huggins behaviour appears for a critical value of $(C[\eta])^*$ corresponding to the critical concentration for overlap C* ; $(C[\eta])^*$ value is in the range of

1[10]. Over this polymer concentration, entanglements between chains develop in the solution which presents some elastic properties.

This general behaviour based on the C[η] also explains the role of the molecular structure and molar mass on the specific viscosity. It is well known that the intrinsic viscosity is directly related to the molecular weight M of a polymer and its stiffness by the Mark-Houwink relation :

$$[\eta]_{\dot{\gamma} \to 0} = KM^a \tag{2}$$

K and a are two parameters depending on the polymer-solvent system and temperature. The role of a is important and the relation between a and the flexibility of the polymer was recently discussed[11]. In the intermediate range of molecular weights ($10^4 < M < 10^6$), polysaccharides behave as wormlike chain molecules with partial draining effect even in θ-conditions. Then, the conformation adopted is that of a swollen coil ; the expansion of the coil is controlled by the persistence length (Lp) perturbed by electrostatic contribution for ionic polymers in low concentration of external salt[12]. Some data were also previously discussed by Robinson et al.[13]. (Table 2).

Table 2. Intrinsic persistence length of some polysaccharides

	Lp (Å)	a predicted[11]
Cellulose derivatives	25-50	
Pectins[14]	40-180	
Alginates M rich	40	0.62
M/G = 1.92		
G rich	90	0.78
M/G = 0.28		
Xanthan	430	0.73
Gellan double helix	710	

Then for a given molecular weight, [η] will increase when the persistence length increases ; one predicts that alginate under Na form and rich in guluronic acid will have better viscometric properties than a mannuronic rich sample[15]. This also explain the large viscosity obtained with xanthan (Lp ~ 400 Å) and the high "a" exponent obtained (a~1)[16].

In addition, external salt concentration often plays a large role on the viscosity of water soluble polysaccharide solution having polyelectrolyte properties. This effect is caused by the screening effect on long range electrostatic repulsions (figure 2). The decrease of the viscosity when salt is added is larger when the flexibility of the polymer is larger ; this is based on the contribution of electrostatic persistence length (Le) compared to the intrinsic value Lp. Le depends only on the ionic concentration then the larger is Lp the lower is the effect of salt on the viscosity. This explains the low sensitivity of xanthan solution viscosity to salt excess. The stiffness of the molecule depends directly on the conformation for stereoregular polymers ; bacterial polysaccharides such as xanthan adopt a locally ordered conformation in given thermodynamic conditions. A helix-coil transition is induced when temperature increases or when ionic concentration decreases causing a decrease of the viscosity.

The viscosity of solution also presents a pseudoplastic behaviour in many cases[9,10] (figure 3) ; the relative viscosity (η/η_0 with η_0 the viscosity of the solvent)

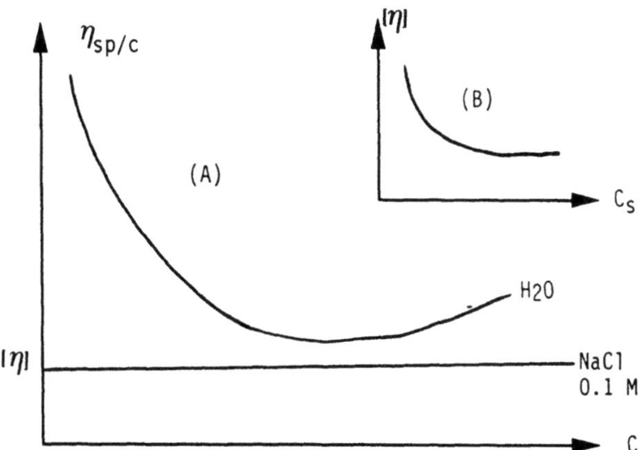

Figure 2. (A) Reduced viscosity as a function of polymer concentration for dilution with H2O or NaCl 0.1 M. (B) Influence of the salt concentration Cs on the intrinsic viscosity.

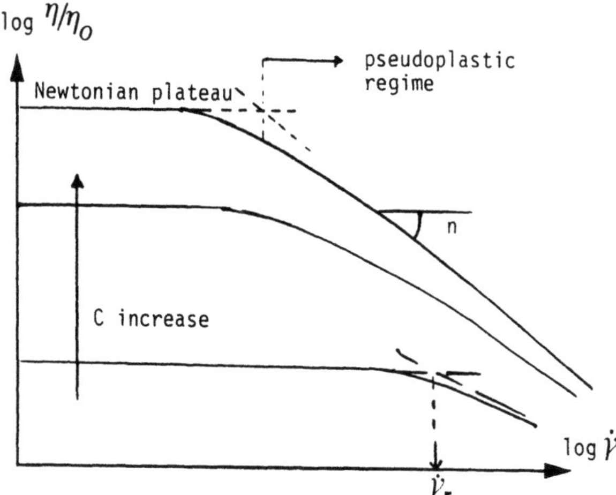

Figure 3. Steady shear relative viscosity (η/η_0) as a function of the shear rate ($\dot{\gamma}$). The transition between Newtonian and pseudoplastic regime ($\dot{\gamma}_r$) is displaced to lower values when polymer concentration increases.

plotted as a function of the shear rate ($\dot\gamma$) presents two parts separed by a critical value of $\dot\gamma$, ($\dot\gamma_r$) in the range of usual $\dot\gamma$ values tested. For low $\dot\gamma$, the newtonian plateau corresponds to the domain in which the viscosity is independant of the shear rate. This domain becomes very small when the polymer concentration increases or when interactions occur. $\dot\gamma_r$ corresponds to the inverse of the longer relaxation time in the solution ; it is independent on the polymer concentration when C<C*. The value of $\dot\gamma_r = \tau^{-1}$ is well estimated from the Rouse theory with :

$$\tau^{-1} = 6\eta_0\,[\eta]M/\pi^2 RT \tag{2}$$

with RT the Boltzmann term.

It follows that $\dot\gamma_r$ decreases when the stiffness of the molecules increases. Then, γ_r decreases when polymer concentration increases following the relation :

$$\tau^{-1} = 6\,(\eta_{\dot\gamma\to0}-\eta_0)\,M/\pi^2\,C\,RT \tag{2'}$$

The pseudoplastic behavior is reflected by the usual power law :

$$\eta = \kappa\,\dot\gamma^p \tag{3}$$

reflecting the shear-thinning behaviour. The slope $|p|$ varies also as a function of $C[\eta]$[10] ; it was demonstrated by Graessley that (p) goes asymptotically to -0.82 for large $C[\eta]$[16].

The thickener behaviour of many polysaccharides was examined previously specially guar[6], xanthan[10] and alginates[7].

At end, yield stress appears in some polysaccharide solution for high enough concentration ; it was found with xanthan solution for $C[\eta]>8$[10] (figure 4).

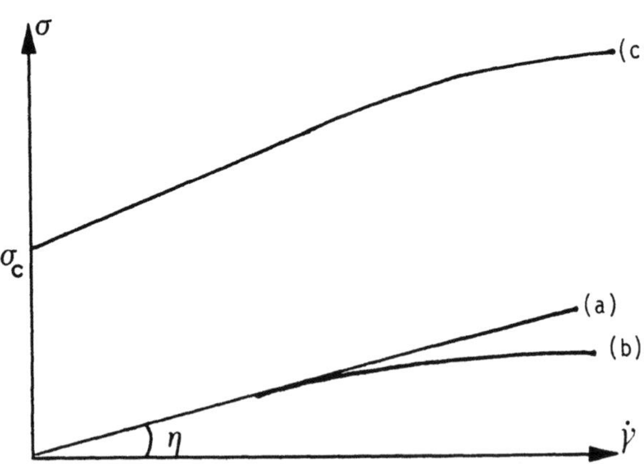

Figure 4. Exemples of flow curves for different polymeric solution. Shear stress versus shear rate for (a) newtonian behaviour (b) pseudoplastic behaviour (c) yield stress behaviour ; σ_c is the critical stress over which viscous flow occurs.

This characteristic is often invoked to explain suspension stabilization obtained with xanthan solution. In other cases the rate of sedimentation (V) is reduced enough by the level of viscosity in the absence of yield stress following the relation :

$$V = \frac{2}{9} \, r^2 \, \frac{(\rho - \rho_0)}{\eta} \tag{4}$$

in which r is the radius of the particles, ρ and ρ_0 the specific mass of the particles and of the fluid respectively[18].

To conclude, when no specific interchain interaction exists with water soluble polysaccharides the general solution properties explain their behaviour in dilute and semi-dilute solution.

Nevertheless, the question of solubilization of the polymers has to be considered ; specially for these OH rich polymers, cooperative and irreversible H bond crosslinkage may be caused depending on the conditions adopted for purification and dehydration[19].

GEL FORMATION - GEL PROPERTIES

This problem was recently discussed elsewere[20] ; non covalent gels specially in food applications are obtained with polysaccharides in given thermodynamic conditions. First obtained with a single polymers, now new textures result from the mixture of two gelling or non gelling polymers.

Ionic Crosslink

Pectins with low degree of esterification and alginates form gels in the presence of divalent counterions except Mg^{2+} ; usually, the role of Ca^{2+} is considered[21,22]. When calcium chloride is added to a polymeric solution a sol-gel phase transition is observed for a critical content of Ca^{2+} depending on the nature of the polymer and on the polymer concentration in particular (figure 5). The percolation mechanism was previously

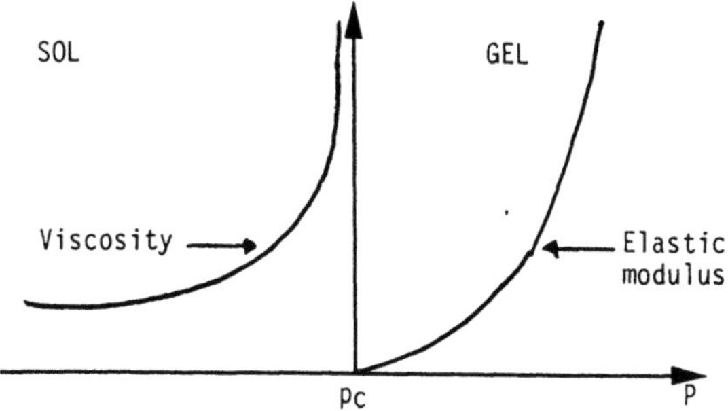

Figure 5. Sol-gel transition ; when the fraction of effective linkage (p) increases, a phase transition occurs for a critical value p_c for which viscosity diverges in liquid phase going to solid state behaviour characterized by an elastic modulus.

discussed[23,24] ; the crosslink was related for a long time to the cooperative interaction with Ca^{2+} involving the specific configuration of α-D-galacturonic unit in $^4C_1$ and α-L-guluronic units in $^1C_4$; the model adopted was that of an egg-box[25]. The elastic modulus of the gels determined by compression (Figure 6) depends on the polymer concentration

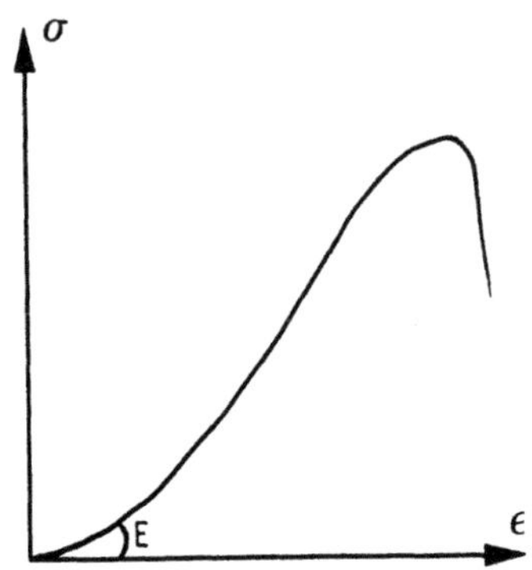

Figure 6. Typical stress-strain curve for mechanical characterization of a gel by compression.

$(E\alpha C^2)$ and on the chemical structure (degree of esterification or content in guluronic acid for alginates[26] ; other systems are crosslinked by divalent conterions such as k-carrageenans[27] and gellan[28] but these polymers are recognized to form more specific gels in presence of monovalent counterion (K^+).

Thermoreversibles Gels

Cooperative interaction between locally ordered polymers causes strong interchain interaction forming a 3D-network. Specially H-bonded systems in k-carrageenan, gellan, agarose, pectin with high degree of esterification are formed giving thermoreversible gels.

These gels are based on thermodynamic properties of the systems and controlled by the temperature. Specially for k-carrageenan and gellan under K^+ ionic form investigated in our laboratory, the mechanisms suggested was the following :

- in presence of K^+ counterion strong interactions exist with the polyelectrolytes forming ion pairs and decreasing the net charge and solubility[29].

- then, when temperature decreases, double helices are formed which associated laterally forming junction zones which connect[30]. The two steps mechanism was demonstrated for k-carrageenans and confirmed with gellan[31].

Kinetic of gelation can be investigated by dynamic measurements as shown on figure 7 ; this type of curves allows to determine the gel time when G' becomes larger than

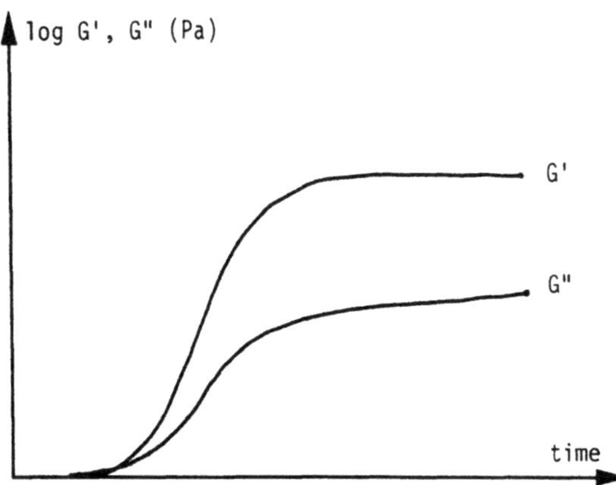

Figure 7. Typical plot of storage (G') and loss (G") components of the dynamic shear modulus as a function of time during gelation at a low frequency (i.e 1Hz).

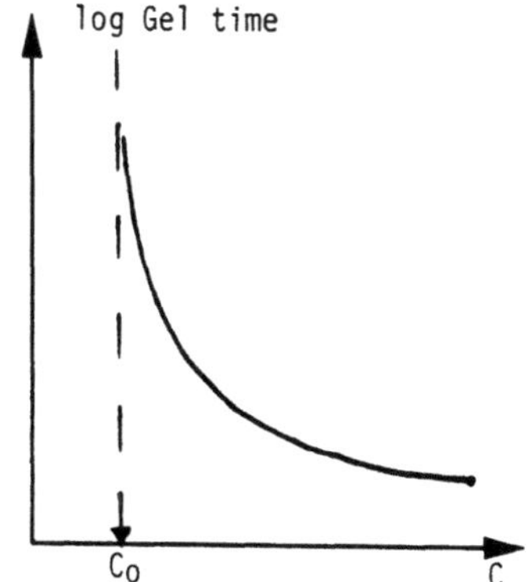

Figure 8. Logarithm of the gelation time as a function of polymer concentration. Gel time diverges for a critical concentration Co.

G". The gel time depends largely on the polymer concentration and temperature. A three dimensionnal network is formed over a critical concentration Co, which is of the same order of magnitude than C* (figure 8).

The stereoregularity of the polymers is a very important factor for cooperative interaction ; the ionic density also plays a role. It was demonstrated that the interchain interactions decrease with $-SO_3^-$ yield going from :

Agarose > k-carrageenan > i-carrageenan.

In the same order, the elastic modulus decreases, the junction zone stability decreases and the syneresis decreases. The stronger interaction corresponds to the lower solubility of the polymer[32]. On these systems, the elastic modulus of the gels are related to the polymer concentration $(E\alpha C^2)$ but also to the ionic strength[20,33] and to the nature of counterions[34].

Ionic selectivity is a very important property of polysaccharide systems as established previously[27, 28] ; in thermoreversible gel, the role of monovalent counterions is particularly significative.

A series of new thermoreversible gels were obtained when two polymers are mixed[35]

- xanthan-galactomannan (guar)[36,37]
- galactomannan - k-carrageenan[38]
- glucomannan (konjac) - xanthan[39]

Interaction between two polymers in semi-dilute solution is clearly demonstrated by viscosity when solution of two different polymers are mixed ; synergistic effect is reflected by an abnormal increase of the viscosity (Figure 9).

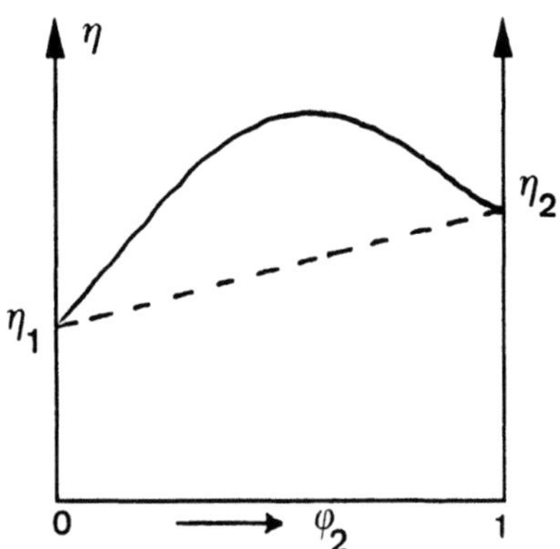

Figure 9. Synergy observed when two polymers are mixed in solution. φ_2 is the volume fraction of component 2, η_1 and η_2 the viscosity of components 1 and 2 solutions before mixing.

The properties depend directly on the ratio galactose/mannose when xanthan-galactomannan blends are considered. It was demonstrated clearly that the presence of acetyl group in xanthan modifies the interaction and that the xanthan interacts in the disordered conformation[36].

At end, it is important to point out that these thermoreversible gels are temperature sensitive due to their enthalpic nature ; increase of temperature decreases the stability of junction zones going to a melting of the gel and to the helix-coil transition (30). On opposite to rubber elasticity prediction, the elastic modulus usually decreases with temperature.

CONCLUSION

In this paper, some important properties of polysaccharides were discussed in relation with their applications in food industry. The tickening behaviour is directly related to the stiffness of the chain (depending on the primary structure and on the conformation) and to the molecular weight. The ionic content of the medium also control the electrostatic interaction and causes a decrease of the viscosity ; in that respect, divalent counterions are more effective than monovalent. The viscosity of a polymeric solution depends also on the conditions of applications or measurements.

In some conditions depending on their chemical structure, polysaccharides form a 3D-network with original properties : ionic selectivity, rigidity based on the structure of the junction zones, thermal stability related to the charge density of the polymer. Based on specific interchain interactions, two non gelling polysaccharides are able to form a gel in defined thermodynamic conditions. Then, new systems based on mixture of two polymers are developing to produce new textures.

REFERENCES

1. J.M.V. Blanshard and J.R. Mitchell, "Polysaccharides in Food", Butterworths London (1979).

2. P. Harris. "Foods Gels", Elsevier Applied Science, Barking (1990).

3. A. Larry Branen, P. Michael Davidson and Seppo Salminen "Food Additives", Dekker New-York (1990).

4. E. Dickinson, "Food Polymers, Gels and Colloids", Royal Society of Chemistry, Special publication n°82, Cambridge (1991).

5. G.O. Phillips, P.A. Williams, D.J. Wedlock,"Gums and Stabilisers for the Food Industry 6", IRL press, Oxford (1992).

6. J.L.M.S. Ganter, J.B.C. Correa, E. Reicher, M. Milas and M. Rinaudo, Study of solution properties of galactomannan from the seeds of Mimosa scabrella, *Carbohydr. Polym.* 17:171 (1992).

7. M. Rinaudo and D. Graebling, On the viscosity of sodium alginates in the presence of external salt, *Polym. Bull.* 15:253 (1986).

8. J.F. Thibault and M. Rinaudo, Interaction of mono and divalent counterions with alcali and enzyme deesterified pectins in salt free solutions, *Biopolymers*. 24:2131 (1989).

9. M. Milas, M. Rinaudo, and B. Tinland, The viscosity dependence on concentration, molecular weight and shear rate of xanthan solution, *Polym. Bull.* 14:157 (1985).

10. M. Milas, M. Rinaudo, M. Knipper, and J.L. Schuppiser, Flow and viscoelastic properties of xanthan gum solutions, *Macromolecules*. 23:2506 (1990).

11. M. Rinaudo, Wormlike chain behaviour of some bacterial polysaccharides (to be published) 32th IUPAC International symposium on macromolecules, Prague (1992).

12. E. Fouissac, M. Milas, M. Rinaudo, and R. Borsali, Influence of ionic strength on the dimension of Na-hyaluronate. *Macromolecules* (in press).

13. G. Robinson, C.E. Manning, and E.R. Morris, Conformation and physical properties of the bacterial polysaccharides gellan, welan and rhamsan, in "Food Polymers, Gels and Colloids", E. Dickinson, ed., Royal Society of Chemistry, Special publication n°82, Cambridge ; p. 22 (1991).

14. M.A.V. Axelos and J.F. Thibault, Influence of the substituents of the carboxyl groups and of the rhamnose content on the solution properties and flexibility of pectins. *Int. J. Biol. Macromol.* 13:77 (1991).

15. M. Rinaudo, The relation between the chemical structure of polysaccharides and their physical properties, in: "Gums and Stabilisers for the Food Industry 6", G.O. Phillips, P.A. Williams, D.J. Wedlock, ed., IRL Press, Oxford, p. 51 (1992).

16. M. Rinaudo, On the abnormal exponents a_η and a_D in Mark Houwink type equations for wormlike chain polysaccharides, *Polym. Bull.* 27:585 (1992).

17. W.W. Graessley, The entanglement concept in polymer rheology, *Adv. Polym. Sci.* 16:1 (1974).

18. H. Luyten, W. Kloek, and T. Van Vliet, The sedimentation of a single particle in xanthan-galactomannan gum mixtures, in "Gums and Stabilisers for the Food Industry 6", G.O. Phillips, P.A. Williams, D.J. Wedlock, ed, IRL Press, Oxford, p. 217 (1992).

19. A. Malovikova, M. Rinaudo, and M. Milas, On characterization of polygalacturonate salts in diluted solution, *Carbohydr. Polym.* (to be published).

20. M. Rinaudo, Gelation of polysaccharides, *The journal of Intelligent Material Systems and Structures* (in press).

21. M. Rinaudo, Gelation of ionic polysaccharides, in: "Gums and Stabilisers for the Food Industry 4", G.O. Philipps, D.J. Wedlock, P.A. Williams, ed., IRL Press, Oxford, p. 119 (1988).

22. J.F. Thibault and M. Rinaudo, Chain association of pectic molecules during calcium induced gelation, *Biopolymers* 25:455 (1986).

23. J.F. Thibault and M. Rinaudo, Gelation of pectinic acids in the presence of calcium counterions, *British Polym. J.* 17:181 (1985).

24. M.A.V. Axelos, and M. Kolb, Crosslinked biopolymers : experimental evidence for scalar percolation theory, *Phys. Review Letters*, 64:1457 (1990).

25. D. Thom, G.T. Grant, E.R. Morris, and D.A. Rees, Characterisation of cation binding and gelation of polyuronates by circular dichroïsm, *Carbohydr. Res.* 100:29 (1982).

26. C. Bouffar-Roupe, Structure et propriétés gélifiantes des alginates. Thesis Grenoble (1989).

27. C. Rochas, Etude de la transition sol-gel du kappa-carraghénane. Thesis Grenoble (1982)

28. X. Shi, Relation entre la conformation et les propriétés d'un polysaccharide bactérien, le gellane. Thesis Grenoble (1990).

29. M. Rinaudo, C. Rochas, and B. Michels, Etude par absorption ultrasonore de la fixation selective du potassium sur le k-carraghénane, *J. Chim. Phys.* 80:305 (1983).

30. C. Rochas and M. Rinaudo, On the mechanism of gel formation in kappa carrageenan, *Biopolymers.* 23:735 (1984).

31. M. Milas, X. Shi, and M. Rinaudo, On the physicochemical properties of gellan gum, *Biopolymers.* 30:451 (1990).

32. M. Rinaudo and S. Landry, On the volume change of non covalent gels in solvent-non solvent mixtures, *Polym. Bull*, 17:563 (1987).

33. C. Rochas, M. Rinaudo and S. Landry, Relation between the molecular structure and mechanical properties of carrageenan gels, *Carbohydr. Polym.* 10:115 (1989).

34. M. Rinaudo, Physico chemical properties of polysaccharides in relation to their molecular structure, in "Viscoelasticity of Biomaterials", W.G. Glasser and H. Hatakeyama, ed., ACS Symposium series 489:24, Washington (1992).

35. V.J. Morris, Designing polysaccharides for synergistic interactions, in: "Gums and Stabilisers for the Food Industry 6", G.O. Philipps, D.J. Wedlock, P.A. Williams, ed., IRL Press, Oxford, p. 161 (1992).

36. L. Lopez, C.T. Andrade, M. Milas, and M. Rinaudo, Role of conformation and acetylation of xanthan on xanthan-guar interactions, *Carbohydr. Polym.* 17:121 (1992).

37. R.C. Clark, Viscoelastic response of xanthan gum/guar gum blends, in: "Gums and Stabilisers for the Food Industry 4", G.O. Philipps, D.J. Wedlock, P.A. Williams, ed., IRL Press, Oxford, p. 165 (1988).

38. I. Tvaroska, C. Rochas, F.R. Taravel and T. Turquois, Kappa-carrageenan-mannan interaction : a theoretical approach in "Gums and Stabilisers for the Food Industry 6", G.O. Philipps, D.J. Wedlock, P.A. Williams, ed., IRL Press, Oxford, p. 231 (1992).

39. P.A. Williams, S. Clegg, M.J. Langdon, K. Nishinari and G.O. Philipps, Studies on the synergistic interaction of konjac mannan and locust bean gum with kappa carrageenan in "Gums and Stabilisers for the Food Industry 6", G.O. Philipps, D.J. Wedlock, P.A. Williams, ed., IRL Press, Oxford, p. 231 (1992).

GELATION OF (SOME) SEAWEED POLYSACCHARIDES

Lennart Piculell, Svante Nilsson, Christer Viebke and Wei Zhang

Physical Chemistry 1
Chemical Center
University of Lund
S-221 00 Lund, Sweden

ABSTRACT

The mechanism of gelation of carrageenans is discussed in the light of recent experimental and theoretical results. It is concluded that association of helices, rather than the double-helix formation in itself, provides the junctions of the gel network. Although the helical regions are typically much shorter than the carrageenan chains, the average number of helical regions on a chain is insufficient to generate crosslinking. The effects of gel-promoting cations and gel-impeding anions on the gelation (helix aggregation) is caused by the binding of these ions to the carrageenan helices, whereby the electrostatic helix-helix repulsion is affected through the decreased (for cation binding) or increased (for anion binding) charge density of the helix.

INTRODUCTION

Many noncovalent polymer gels form according to the scheme

$$coil \rightarrow helix \rightarrow gel, \qquad (I)$$

where the transitions to the right may be induced by lowering the temperature, by changing the solvent or by adding salt (for charged polymers). For convenience, we will call such gels *helical gels*. An increasing number of helical gels are being studied in ever greater detail, and currently known examples include "classical" gels like gelatin, agarose and carrageenans, newly discovered polysaccharides like gellan gum, and also polypeptides in organic solvents[1]. The mechanism of gelation of helical gels is thus of quite general interest. Nevertheless, satisfactory answers are lacking even to basic questions, such as:

Why does helix formation sometimes lead to gel formation, sometimes not?

Food Hydrocolloids: Structures, Properties, and Functions
Edited by K. Nishinari and E. Doi, Plenum Press, New York, 1994

-    Is indeed the molecular gelation mechanism similar for all helical gels, or is the similarity implied by scheme (I) only superficial?

A better understanding of gelation mechanisms is desirable also for other reasons. Helical gels are typically very sensitive to environmental factors such as the presence of cosolutes (salt, denaturants, other polymers). A spectacular example is the strong effect on the gelation of agarose and carrageenans by addition of certain galactomannans.[2] Clearly, we shall not expect to understand this effect without a good molecular understanding of the gelation in the absence of the added polymer. Moreover, it is not clear why structurally very similar molecules can give rise to gels with very different properties. In general, any intelligent design of a helical gel - by modification of the polymer itself or by modifying its environment - requires a better understanding of gelation mechanisms.

In the present contribution, the gelation of a classical family of helical gels, the carrageenans from red seaweeds,[3] will be discussed. Some basic observations regarding carrageenan gelation will be reviewed, with an emphasis on recent findings, and the implications of these observations on the issue of gelation mechanisms will be discussed.

GELATION OF CARRAGEENANS: PHENOMENOLOGY

The gelling ability of carrageenans[3] is restricted to helix-forming structures containing a repeating disaccharide pattern (with 0.6, 1 and 2 sulfate groups per disaccharide for furcellaran, κ- and ι-carrageenan, respectively), and a number of investigations confirm the general notion that helix formation is a prerequisite for carrageenan gelation.* There is good evidence that the carrageenan helix is a double helix, although this notion has been questioned repeatedly. (A detailed discussion of the arguments raised in favor of and against the double helix is given elsewhere.[3])

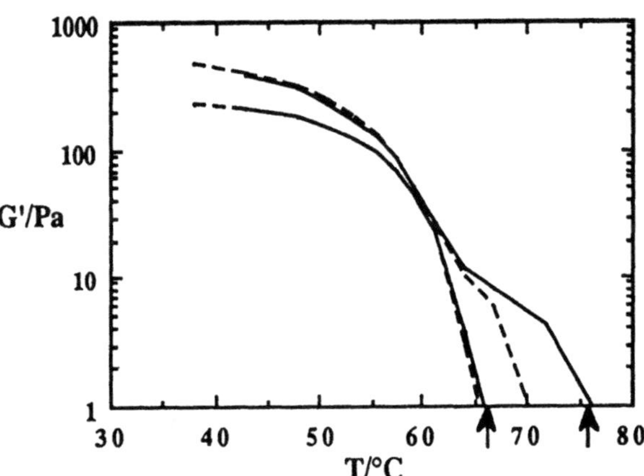

Fig. 1. Shear storage moduli obtained on cooling (dashed) and heating (solid) native (right) and purified (left) ι-carrageenan. Arrows indicate T_0 for the ι- (left) and κ- (right) carrageenan fractions. Data from ref. 5.

---

*   This view has recently been challenged by one group of researchers,[4] who claim that the gelation temperature, T_{gel}, may actually supercede the temperature of onset of helix formation, T_0, for κ-carrageenan. Space does not allow a detailed discussion of the results and arguments of these workers here. It may be noted however, that quantitatively, their T_{gel} values are in disagreement with results obtained by other investigators.[3] Qualitatively, on the other hand, their results confirm the strong correlation between T_0 and T_{gel}, which strongly suggests that the two transition temperatures reflect the same molecular event.

The gelation of the lower charged carrageenans, κ-carrageenan and furcellaran, is strongly ion specific. Certain cations (NH_4^+, K^+, Cs^+, Rb^+) promote both helix formation and gelation, whereas certain anions (I^-, SCN^-) promote helix formation, but impede gel formation. In the presence of gel-promoting cations, strongly aggregated gels are formed, as implied by gel turbidity and a thermal hysteresis of the transition. Both the gel strength and the thermal hysteresis increases with added salt, implying that a higher degree of association leads to an increased gel strength. Rochas and coworkers have shown[6,7] that, in excess salt, the elastic modulus of carrageenan gels increases with the square of the concentration over a large concentration range, and that the dependence of the gel strength on the molecular weight of κ-carrageenan levels off beyond a molecular weight of the order of 10^5.

For sufficiently pure ι-carrageenan, on the other hand, an absence of ion specificity has recently been demonstrated, although this fact is often obscured by the dramatic effects of small proportions of κ-carrageenan impurities.[3,5,8] Pure ι-carrageenan forms weak gels, with no thermal hysteresis (Fig. 1). Further notable differences between ι-carrageenan and the lower charged carrageenans is that the conformational transitions of the latter are typically sharper, and that synergistic interactions with galactomannans are only found[2] for the lower charged carrageenans. In mixed systems of ι- and κ-carrageenan, the gelation proceeds in two steps, corresponding to the ι- and κ-carrageenan transitions, respectively (Fig. 1).[5,8] Furthermore, by selecting the salt conditions, it is possible to make the κ-carrageenan step occur either at lower or at higher temperature than the ι-carrageenan step. It is a general observation that mixed carrageenans do not form mixed helices.[3]

FORMATION OF HELICAL GELS: SUGGESTED MECHANISMS

According to our current understanding of the physical gelation of polymers, reinforced by evidence from techniques such as electron microscopy, gelation occurs when the polymers associate into an infinite three-dimensional network. Network formation requires that the polymers associate, but, also, that the structures formed by association are branched. To give rise to a frequency-independent elastic modulus at low deformation frequencies (the gel caracteristics), the associations must be sufficiently long-lived, which implies that they involve a certain minimum length of the polymer; a "junction zone". At the same time, the requirement of branching means that the association must also, in some sense, be incomplete.

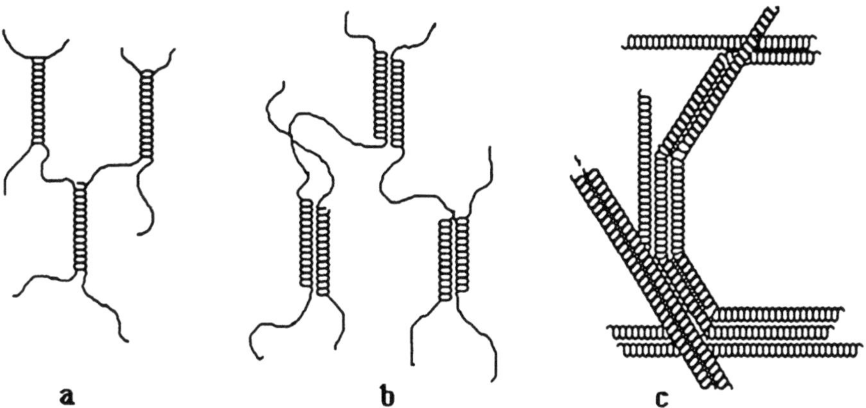

Fig. 2. Schematic drawings of possible mechanisms of network formation in helical gels.

Helix formation provides various possible mechanisms for association and branching (Fig. 2). Multiple helix formation does in itself give rise to chain association and, if the helix formation is incomplete, also to branching (Fig. 2a). We will refer to this mechanism as network formation *at the helical level*. At present, the most well-established case of network formation at the helical level is gelatin, where gelation seems to be caused by incomplete formation of triple helices.[9] Also for agarose and the carrageenans, a network formation at the helical level (by double helix formation), was originally proposed by Rees and coworkers.[10] They, furthermore, proposed that the presence of helix-incompatible "kink" units in the chains provided the molecular mechanism for branching; a kink should terminate a helical region and force the two chains to branch off in different directions.

As discussed by Higgs and Ball,[11] the formation of a network at the double-helical level is possible also for a homopolymer, provided that each chain contains, on average, two or more helical regions (cf. Fig. 2a). Higgs and Ball also showed, however, that as the fraction of helical units increases, the number of helical regions per chain passes through a maximum, so that, at a sufficiently high degree of helix conversion (the value of which depends on the chain length and on the intrinsic cooperativity of the transition), the branching is eventually destroyed. An equilibrium network formed by double-helical association of homopolymers can thus only exist over a limited range of helix conversion.

A network may also be formed *at the superhelical level*, by helix association. Rodlike molecules, such as stiff helices, have intrinsically a larger tendency to aggregate than do flexible chains, all other factors being constant, since a rod has no internal degrees of freedom, due to chain flexibility, to lose on aggregation. Therefore, helix association may be expected, and is, indeed, commonly found. Various mechanisms of branching may be envisaged at the superhelical level. If association of partially helical molecules occurs, a chain may take part in more than one aggregate (given, again, that it contains two or more helical regions), and be trapped there as the aggregation proceeds (Fig. 2b). (It may be noted, however, that in such a case it is difficult to see how the helix formation of the trapped chains may proceed towards unity.) At the other extreme, a superhelical network may be formed by the association of perfect helices into a locally ordered, but macroscopically isotropic and continuous structure (Fig. 2c). Such a mechanism is implied for the gelation of initially helical molecules, involving only the second step of scheme (I). It has been suggested that the continuous structure of such gels may be a separate phase, formed as a result of spontaneous concentration fluctuations in a spinodal phase transition.[1]

MECHANISMS OF BRANCHING IN CARRAGEENAN GELS

It thus seems important to establish, for any given helical gel, whether branching occurs at the helical level or not. For many systems, this may be difficult, since helix formation and gel formation occur simultaneously. This is normally the case also for carrageenans. There exist, however, conditions where helix formation of carrageenans may be achieved without gel formation. Such conditions may be utilized in the design of experiments where the system is taken stepwise through the transitions of scheme (I). Accordingly, we prepared a non-gelling solution of all-helical κ-carrageenan in 0.3 M LiI, which was subsequently dialyzed, at room temperature, against 0.3 M KCl. After dialysis, a gel was obtained. The gel was reversible, since it could be dissolved again by dialysis against 0.3 M LiI. During the experiment, the carrageenans should not have passed the coil state, since the helix is stable in all proportions of mixed salts of gelforming cation and gelforming anion at these conditions (cf. Fig. 3). Thus, the experiment strongly supports the notion that a carrageenan gel network is formed at the superhelical level.

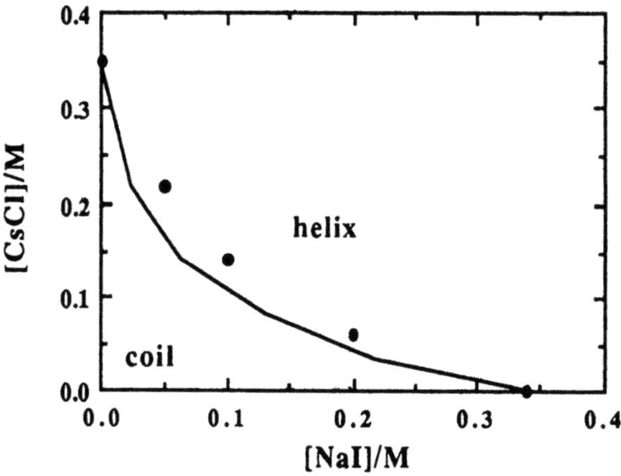

Fig. 3. Conformational stability diagram of κ-carrageenan in mixed salts of CsCl and NaI. The data indicate salt compositions which correspond to a helix onset temperature of 70°C. The pure NaI system is a helical solution, and the pure CsCl system a gel. The solid line indicates theoretical predictions by the Poisson-Boltzmann cell model, with the assumption of binding of Cs^+ and I^- ions, with intrinsic binding constants of 1.14 and 100 M^{-1}, respectively, to independent sites on the κ-carrageenan helix. Data from ref. 12.

The evidence of reversible branching at the superhelical level, which is in good agreement with pictures of carrageenan gels provided by electron microscopy,[13] does not, of course, exclude the possibility of network formation also at the helical level. One could argue that the very fact, that helical solutions exist, implies that network formation at the helical level does not occur. On the other hand, a helical solution could have passed through a network state during the transition from all coils to all helices (cf. the previous section). To check for the latter possibility, we measured the viscosity in non-gelling solutions of both κ-carrageenan (in LiI) and ι-carrageenan (0.1 - 0.5% in formamide, with 0.1 - 1M NaSCN) as a function of increasing helical content (decreasing temperature). In the event that a substantial network formation would occur, one would expect a non-monotonic variation of the viscosity through the transition. Such an effect was, however, not seen for any of the investigated systems; rather, the viscosity increased smoothly as a function of the helical content, as may be seen for κ-carrageenan in LiI in Fig. 4.

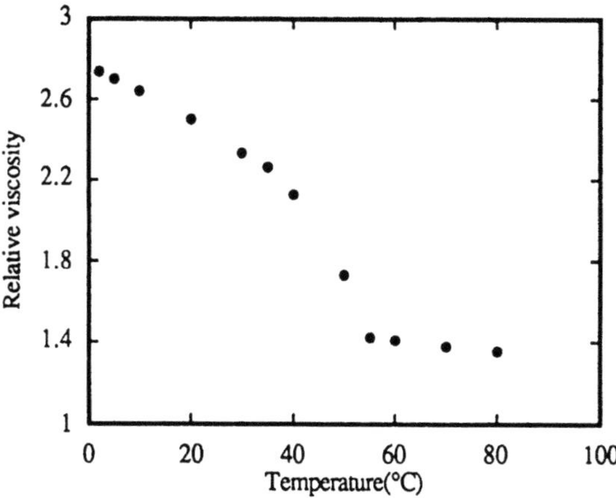

Fig. 4. Relative viscosity of 0.1 % κ-carrageenan in 0.2 M LiI as a function of temperature.

COIL-HELIX TRANSITIONS OF CARRAGEENANS

The various mechanisms of helical gelation discussed above depend on such factors as the distribution of helical units in partially helical samples. Clearly, a good understanding of the coil-helix transition is desirable for any helical gel. For carrageenans, rather limited information is provided by previous studies, which have been based on assumptions of all-or-none transitions (corresponding to a mixture of all-coil and all-helix molecules),[14] or on deductions of apparent cooperative lengths,[15] assumed not to vary with the degree of helical conversion.

Using modern coil-helix transition theory,[16] we have analyzed data obtained by Rochas et al.[7] on the coil-helix transitions of κ-carrageenan fractions of varying molecular weights. (A detailed account of this analysis will be presented elsewhere.) Fig. 5 shows that the theory, which assumes a bimolecular double-helix formation, and treats the various carrageenan samples as monodisperse homopolymers, gives a quite satisfactory description of the thermal transitions for a range of molecular weights. The remaining discrepancies should, to a large extent, be due to the polydispersity of the real samples, which invariably must broaden the transition. The theoretical fits yield the helix initiation and propagation parameters, and the probability of loop formation, which, in the model, provide a complete description of the equilibrium distribution of helical segments at any degree of conversion for any chain length.

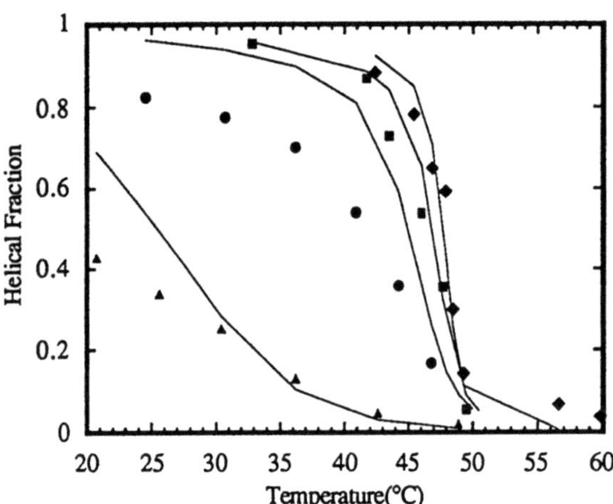

Fig. 5. Experimental (points; from ref. 7) and theoretical (lines) coil-helix transitions of κ-carrageenan fractions with average numbers of repeating disaccharides of (from left to right) 16, 78, 142, and 247.

As an example, Fig. 6 shows how the average length of a helical fraction evolves as a function of the helical content for three carrageenan fractions differing in molecular weight. Note that the helices formed initially are quite short, and that the average length increases smoothly with the helical content. Also, already very early in the transitions, the helical lengths differ for different chain lenghts, being longer for longer chains. Certainly, an all-or-none transition seems to be a very poor model for κ-carrageenan.

The question then arises why these short helical sections do not lead to the formation of a branched network at the helical level. The answer is that the probability of there being more than one helical section per molecule is always low. At low helical contents, the helical sections are few, and with a growing helical content, the growth of the existing helices greatly dominates over the formation of new helical regions, so that the number of helical regions grows only slowly.

Fig. 7 shows that the model, although based on a bimolecular helix formation, predicts a very weak dependence of the conformational transition on the carrageenan concentration, in agreement with experimental observations. Significant effects are seen only for very short chains, and then for large concentration differences. It may be noted that the absence of a concentration dependence has been invoked as an argument against a bimolecular transition.[17] Fig. 7 shows, however, that the concentration invariance is expected.

On the other hand, recent experimental studies indicate that carrageenans may, in fact, undergo an intramolecular coil-helix transition. In a detailed analysis by light scattering, Slootmaekers *et al.*[18] failed to observe the expected increase in molecular weight on the helix formation of κ-carrageenan - a result which seems to contradict the double helical structure.

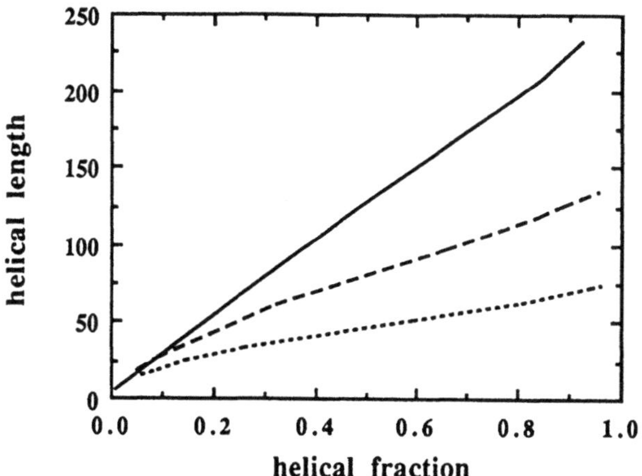

Fig. 6. Model predictions of the average number of disaccharide units in helical regions, as a function of the overall degree of helix conversion, for carrageenan fractions with 78, 142, and 247 repeating units.

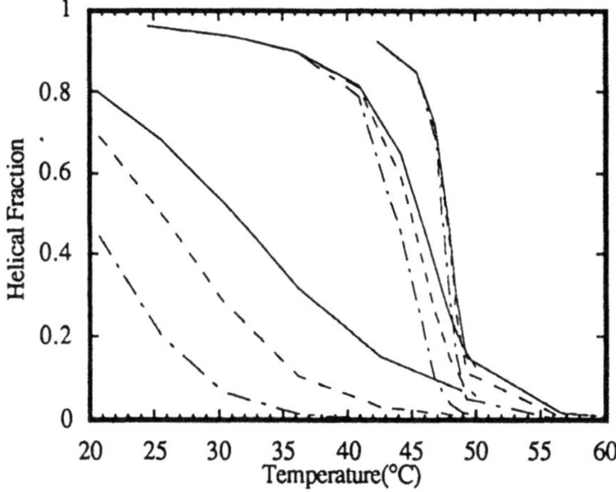

Fig. 7. Theoretical coil-helix transitions of κ-carrageenan fractions with (from left to right) 16, 78 and 247 repeating disaccharide units, at total concentrations of 13 (solid), 1.3 (dashed) and 0.13 (dotted) mM.

A clue to this dilemma is provided, however, by very recent electron micrographs by Stokke et al., demonstrating the formation of intramolecular rings of ι-carrageenan molecules.[19] Clearly, such a structure implies an intramolecular helix conformation which is still compatible with the conventional picture of a double helical structure involving parallel carrageenan chains.

THE MOLECULAR ROLE OF SMALL IONS

In a series of papers, we have used the Poisson-Boltzmann cell model to analyze both the interactions of various ions with the carrageenan molecules, and the effects of the same ions, and electrostatic interactions in general, on the coil-helix transition, the first step of scheme (I), of carrageenans.[12,20-25] The results of the analyses, which include ι-carrageenan, κ-carrageenan and furcellaran, and specific as well as non-specific cations and anions, may be summarized as follows.

All effects of mono- and divalent ions on the coil-helix transition of ι-carrageenan, and by certain monovalent ions (Na^+, Li^+, $(CH_3)_4N^+$) and most divalent ions for κ-carrageenan and furcellaran, may be explained by non-specific electrostatic interactions. The strong efficiency of divalent cations (much larger than predicted by simple ionic-strength considerations) is not due to ion binding, but is caused by the high charge density of the carrageenans.

The helix-promoting effects of specific cat- or anions on the transitions of κ-carrageenan and furcellaran (cf. Fig. 3) may be explanined by a binding of these ions to the *helical* conformation, as has also been demonstrated by spectroscopic methods. The binding sites are very similar - if not identical - for κ-carrageenan and furcellaran. Studies of systems with mixtures of specific anions and cations indicate that these ions bind to different sites on the helix, rather than compete for the same sites. This is demonstrated in Fig. 8, which shows that the binding of cesium ions initially increases on addition of iodide ions, owing to

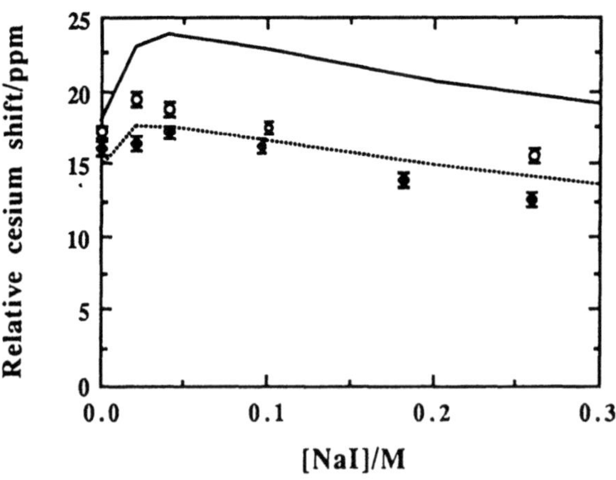

Fig. 8. Experimental (points) and calculated (lines) variation of the NMR chemical shift of cesium with the concentration of NaI in 2% Cs furcellaran at 25°C (filled symbols, dotted line) and at 5°C (open symbols, solid line). Data from ref. 12.

the increased negative charge density on the helix caused by iodide binding. Although, according to the analysis, the intrinsic affinity for the sites is much larger for the anions (binding constants of the order of $10^2 - 10^3$ M^{-1}) than for the cations (binding constants of the order of 1 - 10 M^{-1}), specific cations are, at low salt, much more effective than specific anions in inducing helix formation. This is a result of electrostatic interactions; the local concentration of ions in the vicinity of the helix, which determines the degree of binding, is much lower for negative ions than for positive ions.

It is notable that the analyses indicate that the specific ions bind to the helix itself; no helix association is needed in order to create the binding sites. This is at variance with the notion that the specific cations tie different helices together in a manner suggestive of an ion bridge. It seems, rather, that the effects of specific ions on helix aggregation are simply due to a modification of the electrostatic helix-helix repulsion. Thus, specific cations, by binding to the helix, reduce its charge density considerably, and enhance the tendency to aggregate. Conversely, the specific anions increase the charge density and the stability of the helix in solution. In principle, a quantitative treatment of these effects within electrostatic models would be possible. At present, the main obstacle to such an analysis seems to be the theoretical description of the (possibly non-equilibrium) aggregates of carrageenan helices - we still know too little about the nature of such aggregates.

CONCLUDING REMARKS

The results reviewed here indicate that the network formation in carrageenan gels occurs exclusively at the superhelical level; helix association is a prerequisite for gelation. It is noteworthy that superhelical association seems required also for ι-carrageenan, where the gel-solution transition occurs without a detectable thermal hysteresis. Thus, a lack of hysteresis does not necessarily imply an absence of superhelical association. In order to proceed further in the understanding carrageenan gelation, a better knowledge on the molecular nature of the aggregation of carrageenan helices is vital.

In general, more efforts, along the lines suggested here, to separate the two steps [scheme (1)] in the gelation, and to thoroughly analyse the distribution of helical segments during the coil-helix transition, should give valuable insights on the gel formation of carrageenans and other helical gels.

Lastly, we conclude that the similarities of helical gels implied by scheme (I) are at least partly superficial, since gelatin networks are formed at the helical level. The parentheses in the title of the present contribution are thus warranted. Still, the great similarities in the gelation of, e. g., κ-carrageenan and gellan[26] suggest that conclusions on the mechanism of network formation for carrageenans may be very relevant also for other systems.

ACKNOWLEDGEMENTS

We are grateful to Alan Parker and Bjørn Stokke for sharing their results with us prior to publication. L. P. is indebted to Madeleine Djabourov for helpful discussions and suggestions on the subject of gelation mechanisms. This work was supported by grants from the Swedish National Board for Technical Development and from the Swedish Natural Science Research Council.

REFERENCES

1. W. G. Miller, L. Kou, K. Tohyama, V. Voltaggio, Kinetic aspects of the formation of the ordered phase in stiff-chain helical polyamino acids, *J. Polym. Sci.: Pol. Symp.* 65:91 (1978).

2. P. A. Williams and G. O. Phillips, Interactions in mixed polysaccharide systems, *in* "Food Polysaccharides", A. Stephen, ed., Marcel Dekker, New York (in press).

3. L. Piculell, Gelling carrageenans, *in* "Food Polysaccharides", A. Stephen, ed., Marcel Dekker, New York (in press).

4. E. E. Braudo, Mechanism of galactan gelation, *Food Hydrocoll.* 6:25 (1992).

5. L. Piculell, S. Nilsson, and P. Muhrbeck, Effects of small amounts of kappa-carrageenan on the rheology of aqueous iota-carrageenan, *Carbohydr. Polym.* 18:199 (1992).

6. C. Rochas and S. Landry, Rheological characterisation of kappa carrageenan gels, *in* "Gums and Stabilisers for the Food Industri 4", G. O. Phillips, P. A. Williams, and D. J. Wedlock, eds., IRL Press, Oxford (1988).

7. C. Rochas, M. Rinaudo, and S. Landry, Role of the molecular weight on the mechanical properties of kappa carrageenan gels, *Carbohydr. Polym.* 12:255 (1990).

8. A. Parker, G. Brigand, C. Miniou, A. Trespoey and P. Vallée, Rheology and fracture of mixed iota/kappa gels: Two step gelation, *Carbohydr. Polym.* (in press).

9. M. Djabourov, J. Leblond, and P. Papon, Gelation of aqueous gelatin solutions, *J. Phys. France* 49:319,333 (1988).

10. D. A. Rees and E. J. Welsh, Secondary and tertiary structure of polysaccharides in solutions and gels, *Angew. Chem. Int. Ed. Engl.* 16:214 (1977).

11. P. G. Higgs and R. C. Ball, Formation of gels and complexes by pairwise interacting polymers, *J. Phys. France* 50:3285 (1989).

12. W. Zhang, L. Piculell, and S. Nilsson, Effects of specific anion binding on the helix-coil transition of lower charged carrageenans, *Macromolecules* 25:6165 (1992).

13. A.-M. Hermansson, Rheological and microstructural evidence for transient states during gelation of kappa-carrageenan in the presence of potassium, *Carbohydr. Polym.* 10:163 (1989).

14. D. S. Reid, T. A. Bryce, A. H. Clark, and D. A. Rees, Helix-coil transition in gelling polysaccharides, *Faraday Disc. Chem. Soc.* 57:230 (1974).

15. I. T. Norton, D. M. Goodall, E. R. Morris, and D. A. Rees, Dynamics of cation-induced conformational ordering in solutions of segmented iota carrageenan, *J. Chem. Soc., Faraday Trans. 1* 79:2501 (1983).

16. D. Poland and H. A. Scheraga. "Theory of Helix-Coil Transitions in Biopolymers", Academic Press, New York (1970).

17. O. Smidsrød and H. Grasdalen, Conformations of κ-carrageenan in solution, *Hydrobiologia* 116/117:178 (1984)

18. D. Slootmaekers, C. De Jonghe, H. Reynaers, F. A. Varkevisser, C. J. Bloys van Treslong, Static light scattering from κ-carrageenan solutions, *Int. J. Biol. Macromol.* 10:60 (1988).

19. B. T. Stokke, A. Elgsaeter, and S. Kitamura, Macro-cyclization of associating polysaccharides visualized by electron microscopy, *Int. J. Biol. Macromol.* (in press).

20. L. Piculell and R. Rymdén, The helix-coil transition of an ionic polysaccharide probed by counterion self-diffusion measurements, *Macromolecules* 22:2376 (1989).

21. S. Nilsson, L. Piculell, and B. Jönsson, Helix-coil transitions of ionic polysaccharides analyzed within the Poisson-Boltzmann cell model. 1. Effects of polyion concentration and counterion valency, *Macromolecules* 22:2367 (1989).

22. S. Nilsson and L. Piculell, Helix-coil transitions of ionic polysaccharides analyzed within the Poisson-Boltzmann cell model. 2. Effects salt concentration on the thermal transition, *Macromolecules* 22:3011 (1989).

23. S. Nilsson and L. Piculell, Helix-coil transitions of ionic polysaccharides analyzed within the Poisson-Boltzmann cell model. 3. Solvent effects, *Macromolecules* 23:2776 (1990).

24. S. Nilsson and L. Piculell, Helix-coil transitions of ionic polysaccharides analyzed within the Poisson-Boltzmann cell model. 4. Effects of site-specific ion binding, *Macromolecules* 24:3804 (1991).

25. W. Zhang, L. Piculell and S. Nilsson, Salt dependence and ion specificity of the coil-helix transition of furcellaran, *Biopolymers* 31:1727 (1991)

26. C. E. Manning. "Formation and Melting of Gellan Polysaccharide Gels", thesis, Silsoe (1992).

THE SPECIFICATION OF THE GUM ARABIC OF COMMERCE

Glyn O. Phillips[1] and Peter A.Williams[2]

[1]Newtech Innovation Centre
[2]Polymer & Colloid Chemistry Group
  The North East Wales Institute
  Wales, UK

INTRODUCTION

Various regulatory bodies have sought a definition of the Gum Arabic of commerce, but these vary somewhat. The Journal Officiel of France, the Food Chemicals Codex (US Academy of Sciences) and the US Pharmacopoeia, all define it as the dried gummy exudate from stems and branches of **Acacia Senegal (L.) Willdenow or other related species"** (fam. Leguminosea)[1]. The European Pharmacopoeia lists **"Acacia Senegal (L.) Willdenow and other species of Acacia of African origin"**; and the US Food and Drug Administration defines it as **"various species of the genus Acacia family Leguminosae**[2]. The Joint Expert Committee for Food Additives (JECFA) of FAO refer to a dried exudation of **Acacia senegal (L.Willdenow) or related species of Acacia Fam.Leguminosae**[3]. Proposed new specifications were announced in mid-1990 in which the word **closely** (related species) might be inserted, together with suggestions that specific **optical rotation** -26 to -34⁰ and **N content** limits (0.27 - 0.39%) might also be introduced anew.

The quite understandable objective was to characterise more specifically the Gum Arabic of commerce. The Codex Working Group on specifications met in The Hague on 1 March 1991, when numerous representations were made on the proposed changes. As a result the Working Group concluded as follows:

The Working Group devoted considerable time to discussion of comments on Gum Arabic. Along with technical comments on specifications, there were comments on Gum Arabic produced from the various Acacia species used in commerce and used in toxicological studies. Based on substantive technical comments on the specifications in question, the Working Group decided to place the specifications in Category III so that JECFA may evaluate these specifications in the context of new information on Gum Arabic produced from different Acacia.spp. The Codex Advisory Specifications for Gum Arabic remain those published in FAO Food and Nutrition Paper No.4, 1978.

The Working Group further recommended that when Gum Arabic is re-evaluated by JECFA, the test article will be totally identified both toxicologically and chemically.

Food Hydrocolloids: Structures, Properties, and Functions
Edited by K. Nishinari and E. Doi, Plenum Press, New York, 1994

Thus, the Codex Working Group in its conclusions recognised that Gum Arabic was currently produced from different Acacia species.

Most of the definitions imply that the main component of Gum Arabic is from *A.senegal*, which undoubtedly is the most available gum in the commercial producer countries of Africa, namely, **Sudan, Nigeria, Chad, Ethiopia, Senegal,** with lesser amounts available from **Ghana and Zimbabwe**. The next most available gum is *A.seyal*. Other Acacia gums might be present in smaller quantities, derived from subgenus Acacia and Aculeiferum, which correspond to the Bentham Series *Gummiferae* and *Vulgares*. Practical experience indicates that it is inevitable that more than one subgenus of the Acacia species will be present in commercial Gum Arabic. No doubt it is in recognition of this fact that the definitions have consistently been worded to include either **"various species of the genus Acacia"**, **"other species of Acacia of African origin"** or **"related species of Acacia"**, related, of course, to the major component *A.senegal*. It is our objective to identify such "related species".

THE GENUS ACACIA

Taxonomic Assessment

The genus is one of the most widespread in Africa. Until the work of Ross[5] and Vassal[6], no comprehensive taxonomic study of the genus as a whole, or of all the African species had been undertaken since that of Bentham[7]. Whereas Bentham dealt with 434 species, there are today 1100 known species. Nevertheless, Bentham's subdivisions of the genus has stood the test of time (Table 1).

Classification: Bentham recognised six series, the series being delineated primarily on foliage, on whether the plants were armed, and if armed upon whether or not the stipules were spinescent (Table 1). The inflorescence played a far less important role in his division of the genus than the vegetative characters. All the African species fell into two of the Bentham Series, namely *Gummiferae* and *Vulgares*. The former accommodating all the species with spinescent stipules, and the latter those with non-spinescent stipules.

Since Bentham's work two methods have been employed to subdivide the African species. Some authors have given preference to whether or not the stipules are spinescent as the primary character in separating the two groups, while others have employed inflorescence. Both methods have their advantages. Vassal proposed a classification based chiefly on characters of seeds and seedlings, and on the occurrence of stipular spines and pollen characters. It is fortunate that the three subgenus recognised by Vassal broadly correspond to groupings of Bentham's six series. Vassal's classifications also accords generally with that of Ross[8], but they do not fully agree with respect to the classification below the rank of subgenus.

Although the division of the African species into the two subgenus is readily accomplished, there remains the difficulty of satisfactorily subdividing each subgenus into smaller groupings of species.

Hybridization: Although hybridization has inevitably occurred, its full extent in the development of the various Acacias is unknown. Some aspects of this evolutionary pathway have been studied. El Amin's studies indicated that *A.laeta* is morphologically intermediate between *A.mellifera* and *A.senegal*, although it appears to be more closely related to *A.mellifera* than to *A.senegal*[9]. It was suggested that the closer affinity of *A.laeta* with *A.mellifera* is probably the outcome of back-crossing with it in the past. The following hybrids have also been reported *A.Kirkii x A.seyal, A.nubica x A.paolii* and *A.seyal* var.fistula x *A.xanthophloea*.

TABLE 1. Botanical Classification of Gum Acacia Trees

According to Bentham's & Vassal's main division of the genus

Australian species - subgen.Heterophyllum Vas

Series **PHYLLODINEAE BENTH.**

A. tetragonophylla, A. calamifolia, A. uncinata, A. montana, A. pruinocarpa, A. victoriae, A. aestivalis, A. bancroftii, A. mabellae, A. microbotrya, A. murrayana, A. penninervis, A. podalyriifolia, A. prainii, A. pycnantha, A. retinodes, A. rostellifera, A. rubida, A. salicina, A. saligna (syn. A.cyanophylla), A. goorginae, A. harpophylla, A. cyclops, A. latescens.

Series **JULIFLORAE Benth.**

A. pubifolia, A. acradenia, A. aneura, A. kempeana, A. coolgardiensis, A. microneura, A. auriculiformis, A. leptostachya, A. stereophylla, A. torulosa, A. holosericea, A. mangium, A. beauverdiana.

Series **BOTRYCEPHALAE Benth.**

A. dealbata, A. deanei, A. decurrens, A. alata, (syn. A. terminalis), A. filicifolia, A. leucoclada, A. mearnsii, (syn. A. mollissima), A. parramattensis, A. parvipinnula, A. silvestris, A. trachyphloia.

Indian, African & American species

Series **GUMMIFERAE Benth. = Subgen. ACACIA Vas.** also called **"Acacia seyal complex"**

A. abyssinica subsp.calophylla, A. nilotica (syn. A. adansonii and A. arabica), A. drepanolobium, A. farnesiana, A. gerrardii, A. giraffae, A. hebeclada, A. karroo, A. kirkii, A. leucophloea, A. nebrownii, A. nubica, A. reficiens, A. rigidula, A. seyal, A. sieberana (var.villosa, var. woodii), A. seyal (var. fistula, var. seyal), A. tortilis (subsp. heteracantha).

Series **VULGARES Benth. = Subgen. ACULEIFERUM Vas.** also called **"Acacia senegal complex"**

A. berlandieri, A. polyacantha subsp. campylacantha, A. catechu, A. erubescens, A. fleckii, A. goetzii subsp. goetzii, A. laeta, A. mellifera, subsp detinens, A. senegal, A. sundra.

Vassal recognised three subgenus within Acacia, namely *Acacia*, *Aculeiferum* and *Heterophyllum*. All the African species were placed in the subgenus *Acacia* and *Aculeiferum*, accommodating all the African species referred to in Bentham's series *Gummiferae* and *Vulgares* respectively.

Vassal's classification of the African species down to the rank of section is as follows:

Subgenus *Acacia* - Vassal[6] (= Series *Gummiferae* Benth.) Lectotype: *A.nilotica* (L.) Willd. ex Del. Section *Acacia.*

Subgenus *Aculeiferum* - Vassal[6] (= Series *Vulgares* Benth. and Series *Filicinae* Benth.). Type species: *A.senegal* (L.) Willd.

Section *Aculeiferum* - Vassal[6] (= *Triacanthae* Benth. and *Diacanthae* Benth. of subseries *Gerontogeae Spiciflorae* Benth.).

Section *Monacanthea* - Vassal[6] (= *Ataxacanthae* Benth. of subseries *Gerontogeae Spiciflorae* Benth.). Type species: *A.ataxacanthae* DC.

Origin: Vassal[6] considered that the ancestral form of Acacia evolved in the area presently occupied by Central America (from Mexico to Bolivia), and that the subgenera *Acacia* and *Aculeiferum* differentiated in the same area. Atchison[10] considered the *Gummiferae* (subgenus Acacia) to be the ancestral form as its members contain the morphological characters of the genus that they considered to be primitive, namely bipinnate leaves and persistent spinescent stipules. *Gummiferae* is the only section of the genus with cosmopolitan distribution. For a variety of reasons the *Gummiferae* (subgenus *Acacia)* is considered to be more advanced than the *Vulgares* (subgenus *Aculeiferum),* which nevertheless has retained primitive features. It is unlikely that any modern species would display all the characters considered to be primitive in the genus, and it is far more likely that a combination of primitive and advanced characters would exist side by side.

Regulatory Significance

On a taxonomic basis, is there any justification or need for including the additional word **closely** related species? It is our contention that on a taxonomic basis, gums obtained from the Bentham series Gummiferae and Vulgares certainly fall within the current regulatory definitions and are related to the major component *A.senegal*. They are derived from the same geographical climatic conditions, along the same evolutionary pathway and derived from the same "parent", with the Acacia subgenus giving rise to *Aculeiferum*, of which *A.senegal* is a member. In this sense they must be regarded as "closely" related. Subgenus of the same species would of necessity be **closely** and **related** species, in any event.

However, might **closely** be intended to identify only those gums within an individual subgenus section, such as *Acacia* or *Aculeiferum*? Might this call even for identification below the rank of subgenus? This is a position which might be almost impossible to achieve in practice. Also these are botanical categories, not chemical, and do not of necessity imply fundamental chemical differences. To implement such a restrictive definition, then accurate chemical or physical tests would be required to ensure that only members of the subgenus are present in the commercial product Gum Arabic. It has been suggested that setting limits of specific optical rotation $[\alpha]_D$ and N analyses is adequate for the monitoring of the newly proposed word **closely**. Both of these parameters are aggregate measurements and are undoubtedly useful indicators, but are best scientifically utilised when dealing with a single well defined molecular state. We will produce scientific evidence to show that dextro or laevo-rotatory are not specific indicators of structure or composition, in such complex variants within the Acacia species subgenus. N analyses too only indicates the overall amount of protein present, and does not take into account the distribution of individual amino acids. In our view it is more reasonable to consider both the practical and scientific considerations associated with the term Gum Arabic.

Conclusions

On a taxonomic and practical basis, it is reasonable to consider that the Gum Arabic of commerce will contain gums derived from the *A.senegal* species, with additional amounts of gums present from the related (or closely related) categories. Sub-genus Acacia (= Series *Gummiferae.* Benth.) and Sub-genus *Aculeiferum* (= Series Vulgares.Benth.).

If insertion of the word **"closely"** is meant to discriminate between these two sub-genus groups then the limits set for optical rotation $[\alpha]_D$ and N analyses should unequivocally distinguish between them. We do not consider that these parameters can so distinguish, due to the inherent variability within an individual species and to the close chemical identity between and across these two groups.

A COMPARISON OF THE PHYSICAL & CHEMICAL PROPERTIES OF GUMS DERIVED FROM THE RELATED *GUMMIFERAE* & *VULGARES* SPECIES

Table 2 shows a comparison of Acacia gums from each of the subgenus *Aculeiferum* (=Series Bentham *Vulgares*, Benth. and Series *Filicinae*, Benth.) and the subgenus *Acacia* (= Series *Gummiferae*, Benth.).

Anderson[11] has distinguished these groupings as follows:

Gummiferae:
- positive optical rotation
- low viscosity
- arabinose/galactose ratio> unity
- low rhamnose

Vulgares:
- negative optical rotation
- more acidic than Gummiferae
- more viscous
- arabinose/galactose ratio < unity
- higher proportion of rhamnose

These generalisations can be seen, at best, to be approximate by a scrutiny of Tables 2a and 2b. There are obvious anomalies. The highest recorded viscosity is *A.xanthophie* from the *Gummiferae* series. Overall viscosities range from a low value of 8 to a maximum of 20+ mlg^{-1} in **both** series. Neither is the Vulgares series consistently more acidic. Similarly, widespread arabinose contents, arabinose/galactose ratios and rhamnose contents across the two series do not support an unequivocal categorisation based on chemical composition.

Nitrogen content: Table 3 shows a comparison of the nitrogen content and amino acid composition of two authenticated gum samples from each of the two subgenera found in Africa, namely subgenus *Aculeiferum* (= Series Vulgares, Benth. and Series Filicinae, Benth.) and subgenus *Acacia* (= Series Gummiferae, Benth.). These are compared with commercial Gum Arabic samples from various geographical locations.

General observation would indicate that although there is a greater amount of protein (N 0.365 compared with 0.147) between *A.senegal* and *A.seyal*, there remains a general similarity in distribution of amino acids, despite the fact that they both derive from a different subgenus. The variation inside a subgenus is of the same order as between two different groups. The overall distribution is fundamentally comparable.

For commercial Gum Arabic samples the values of the various samples from the Sudan are broadly comparable to the authenticated specimen (also from the Sudan). Differences are evident in the samples from different geographical locations.

A.gerrardii from the Gummiferae series is way outside the proposed specification limit of 0.39% (1.8%) as is a "closely related", and, therefore, acceptable gum A.goetzii (0.89%) of the Vulgares series.

Indeed, taken as a group, the overall amino acid composition within certain limits are comparable within and between subgenus groupings.

Table 2a. A comparison of Acacia Subgenus *Acacia* (= Series Gummiferae, Benth.) and Subgenus *Aculeiferum*(= Series Vulgares, Benth. and Series Filicinae, Benth.) according to Physical Properties and Carbohydrate Composition.

Subgenus Aculeiferum (Vulgares)

	Ref.	$[\alpha]_D$	Viscosity ml/g	N%	Equiv. weight	Glucur-onic acid %	4-Me %	Gal %	Ara %	Rh. %
A.campylacantha	12	-3	16	0.28	2020	8.7		54	27	7
A.catechii	12	-32			1025	18		50	17	14
A.erubescens	13	-31	8	1.08	874	11.7	8.4	39	17	12
A.fleckii	13	-32	13	0.58	918	16.4	2.8	39	25	14
A.goetzii	14	-36	21	0.89	755	17	6	46	22	9
A.laeta	15	-35		0.65	1250	13.6		44	29	13
A.laeta	16	-35		0.73	1320	13.4		41	34	12
A.mellifera deti	13	-45	21	1.3	822	16.5	4.9	44	25	9
A.mellifera deti	13	-56	23.5	1.45	843	10.7	10.2	43	27	9
A.senegal	17	-24		0.33	1177					
A.senegal	18	-26			1430	12		44	35	9
A.senegal	19	-30		0.34		19.5		36	31	13
A.senegal	20	-30	16	0.34	1050	16	1.5	44	25	14
A.senegal		-20	15	0.34	980	18		48	25	9
A.senegal	15	-30	13.4	0.29						
A.senegal	21	-31	17	0.33	1040	15.8	1.4	46	24	13
A.senegal	22	-27	20.6	0.35	1047	16.5		46	30	16
A.senegal	23	-31	20	0.33	1290	16.7		34	25	12
A.senegal	19	-32				16		44	28	12
A.senegal	22	-31	19.9	0.33	1085	16	1.5	40	28	16
A.senegal	24	-32				28.3		30	34	14
A.senegal	25	-29	20.6	0.29	1125					
A.senegal	25	-27	21.0	0.30	1173					
A.senegal	25	-32	22.8	0.30	1183					
A.senegal	26	-28	23.1	0.34	1110					
A.senegal	27	-32		0.28		20		31	23	10
A.senegal	26	-30	23.5	0.30	1380					
A.sundra	12	-30			980	18		40	28	14

Variations Found in the Physical Properties of Acacia Senegal

Viscosity: Fresh gum was collected in Senegal[+]. The results are shown in Fig.1. The conclusions are dramatic.

[+]This investigation was conducted by J.Vassal (CNRS, Universite Paul Sabatier in partnership with P.Sall and Mm. Diorie (Institut Senegalais de Recherches Agricoles. Dakar, J.C.Fenyo, M.C.Vandevelde & S.S.Duvallet (University of Rouen).

Variations Recorded in the Literature

Anderson[37,38] has reported authenticated Acacia Senegal gum which does not confirm to the proposed JEFCA optical rotation (-26 to -34°) and N (0.27 - 0.39%) limits:

N   0.58[35], 0.46[35], 0.21[35], 0.14[35], 0.17[35], 0.40[35], 0.27, 0.70

$[\alpha]_D$ -51[34], -27[34], -13[34], -32[35], -31[35], -27[35], -40[35], -29[35], -31[35], -31, -32

Table 2b. A comparison of Acacia Subgenus *Acacia* (= Series Gummiferae, Benth.) and Subgenus *Aculeiferum* (= Series Vulgares, Benth. and Series Filicinae, Benth.) according to Physical Properties and Carbohydrate Composition

Subgenus Acacia (Gummiferae)

	Ref.	$[\alpha]_D$	Viscosity m l/g	N%	Equiv. weight	Glucur-onic acid %	4-Me %	Gal %	Ara %	Rh. %
A.arabica	12	+100	12.5	0.07	2370	7.4		36	54	0.4
A. calcigera	28	+97	15	0.15	1430	7.5	4.5	34	54	tr
A.drepanolobium	12	+75	16.6	1.12	2060	8.6		26	56	1
A.drepanolobium	15	+78		1.11	1940	7	2	38	52	1
A.ehrenbergiana	28	-9	8	0.12	810	18.5	3.5	55	13	10
A.ehrenbergiana	28	-7	7	0.09	1060	12	5	56	17	10
A.ehrenbergiana	28	-3	8	0.11	820	19	3	51	16	11
A.erioloba	29	-43		9		0	20	37	36	0
A.fischeri	30	+68	12	0.46	1210	4	11	36	44	5
A.fistula	12	+61	19.4	0.06	1530	11.5		37	50	0.4
A.gerrardii	14	+80	10	1.86	1065	7	9	40	43	1
A.giraffae	31	+28			730		41			
A.hebaciada	13	+28	13	9.4	521	18.8	15	44	14	8
A.hebeciada	32	+28		9.4		13.4	1	44	14	8
A.hebeciada	32	+70		1.9		16		26	56	1
A.hockii	28	+91	13	0.23	1460	8.5	3.5	50	30	8
A.kamerunsis	30	+29	15	0.13	670	22	4	26	36	12
A.karoo	28	+53	17	0.15	1250	11.5	2.5	50	26	7
A.karoo	19	+54				14		37	44	5
A.kirkii	21	+54	8	0.09	1817	5.6	4.1	36	46	8
A.newbrownii	13	+43	13	0.14	777	196	3.0	45	27	7
A.nilotica	15	+108	9.5	0.02	1890	3	6	44	46	0.4
A.nubica	15	+98	9.8	0.2	3030	6.5	0.5	33	59	1
A.reficiens	13	+89	12	0.65	1117	5.7	10.1	41	35	8
A.robusta	32	+36		2.8		9		40	50	1
A.seyal	15	+51	12.1	0.14	1470	6.5	5.5	38	46	4
A.sieberana	33	+108	10	0.34	2070	3.9	4.6	30	57	5
A.sieberana vsie	28	+106	12	0.35	2300	3.5	4.5	26	60	4.5
A.sieberana v.vil	28	+103	12	0.19	1230	10	4	35	47	4
A.spirocarpa	30	+65	10	1.27	920	16	3	26	47	8
A.stenocarpa	30	+14	9	0.38	1045	12	5	41	32	10
A.vortilis	32	+75		1.9		8		23	66	
A.xanthophloea	28	+35	15	0.14	1050	2.5	14.5	54	23	6
	28	+44	24	0.39	1120	4	12	61	16	7

A recent publication[38] seeks to use NMR spectroscopy to characterise Acacia Senegal from various geographical locations. This study also shows predictable variations. The samples from Uganda, for example, differ spectroscopically. Anderson

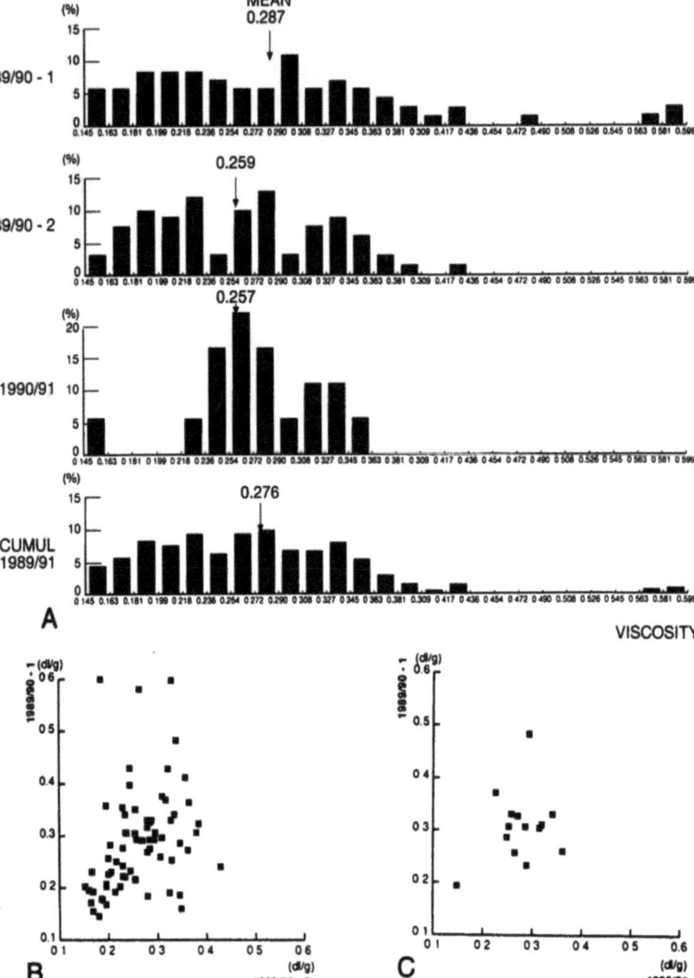

Figure 1 Intrinsic viscosity of freshly harvested gum arabic from Senegal

A.   Variations between individual samples and on storing
B.   Variations for one particular tree during ageing
C.   Variations for a second tree according to harvest year

concludes: *"Acacia Senegal is acknowledged to be a very variable species with recognised sub-species and varieties. The gum identified as Acacia Senegal in various parts of the Sahel may be either closely similar to (e.g. from Niger) or distinctly different from (e.g. Kenya) the gum from the Sudan."* The author suggests that the variations might arise from climatological or soil differences or from more extensive hybridisations in the Acacia Senegal trees themselves.

1A: Note that the intrinsic viscosity of individual samples varies considerably (from 0.14 to 0.60 dl/g). The mean intrinsic viscosity also changes somewhat with the age of the crop.

[1989/90 - 1 0.267] [1989/90 - 2 0.259] [1990/91 0.257] [Av.1989/91 0.279]

A typical value for crushed lump of *A.senegal* would be 0.16 dl/g. It is evident that the Acacia Senegal species gum from Senegal has a greater viscosity than commercial "Kordofan" Sudan gum.

1B & 1C: Notice here the variation in intrinsic viscosity of the gum obtained from the same tree. There is a very wide scatter. Evidently the treatment after picking (thermal, mechanical, chemical) tends to lower the viscosity. The values also change with the season.

Table 3. Amino Acid Compositions of Acacia Subgenera and Commercial Gum Arabic (residues/1000 residues gum)

	Subgenus Acacia (Gummiferae) A.seyal[34]	A.gerrardii[35]	Commercial Gum Arabic				(Vulgares) A.senegal[34]	A.goetzii[35]
Asp	65	77	60[36]	58[34]	39[34]	22[34]	91	91
Hydroxypro	240	320	274	252	298	351	256	215
Thr	62	55	77	78	109	109	72	62
Ser	170	80	137	162	174	178	144	121
Glu	38	49	36	40	23	15	36	56
Pro	73	58	77	71	67	64	64	64
Gly	51	59	49	54	49	42	53	41
Ala	38	42	31	28	17	11	28	56
Cys	-	-	1	-	-	-	3	-
Val	42	60	45	36	21	15	35	88
Met	-	3	-	-	-	-	2	0
Ile	16	36	14	12	6	5	11	15
Leu	85	52	75	78	105	99	70	59
Tyr	13	30	11	10	4	5	13	26
Phe	24	18	29	30	23	14	30	35
His	51	25	51	52	47	54	52	33
Lys	18	23	26	26	14	11	27	31
Arg	11	12	7	9	4	4	15	5
% N	0.147	1.86	0.35	0.327	1.06	0.384	0.365	0.89
Nitrogen Conversion Factor	6.60	7.0	6.43	6.67	6.43	6.70	6.77	6.85

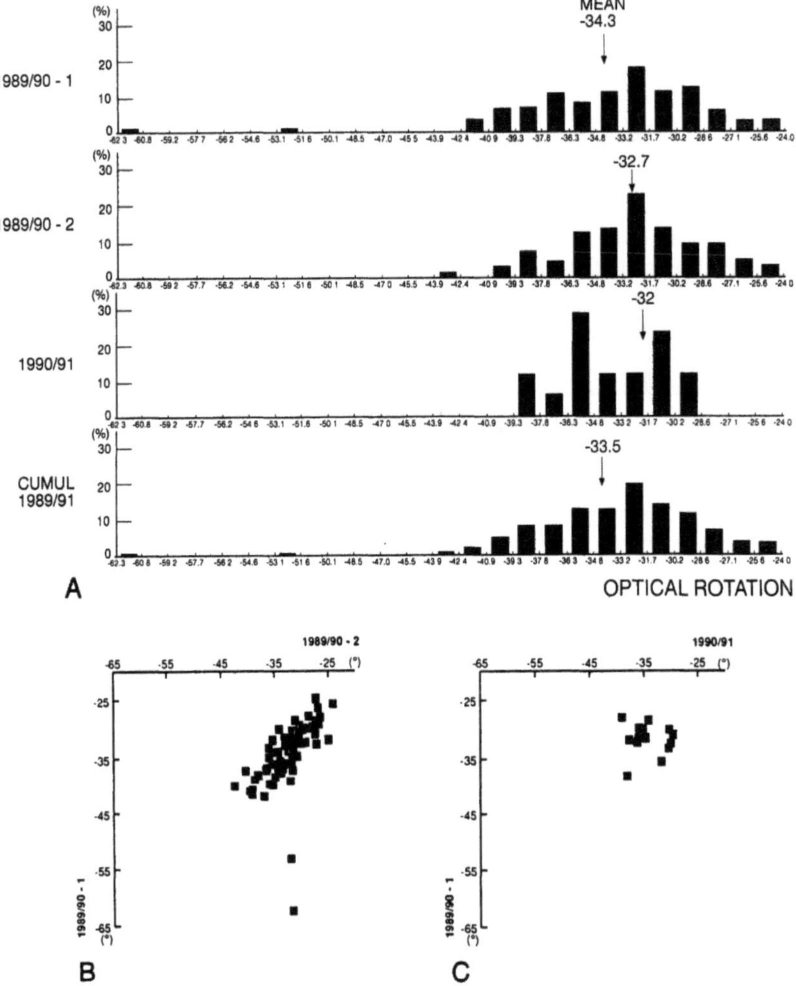

Figure 2 Specific optical rotation of gum arabic from Senegal

A. Variations between individual samples and on storing
B. Variations found for a particular tree with ageing
C. Variations found for a second tree within the harvest year

Optical rotation: Here the values range from -24 to -62°, whereas commercial samples are normally more in the region -28 to -32°. The results are shown in Fig. 2. The mean value of the fresh samples is -34.3°, with quite wide variations between individual samples (Fig. 2A). At one year old, the samples had changed in their mean value to -32.7°. Under the more draconian industrial storage conditions, even greater variation would be anticipated.

Again variations can be found for gum taken from various parts of the same tree (Figs. 2B and C). There is also variation depending on crop location and the ageing of the crop.

The average of all the values is -35.5°. However, despite the fact that this falls close to the JEFCA proposed range of values, the individual variation is so large that this single parameter cannot be regarded as diagnostic of the species.

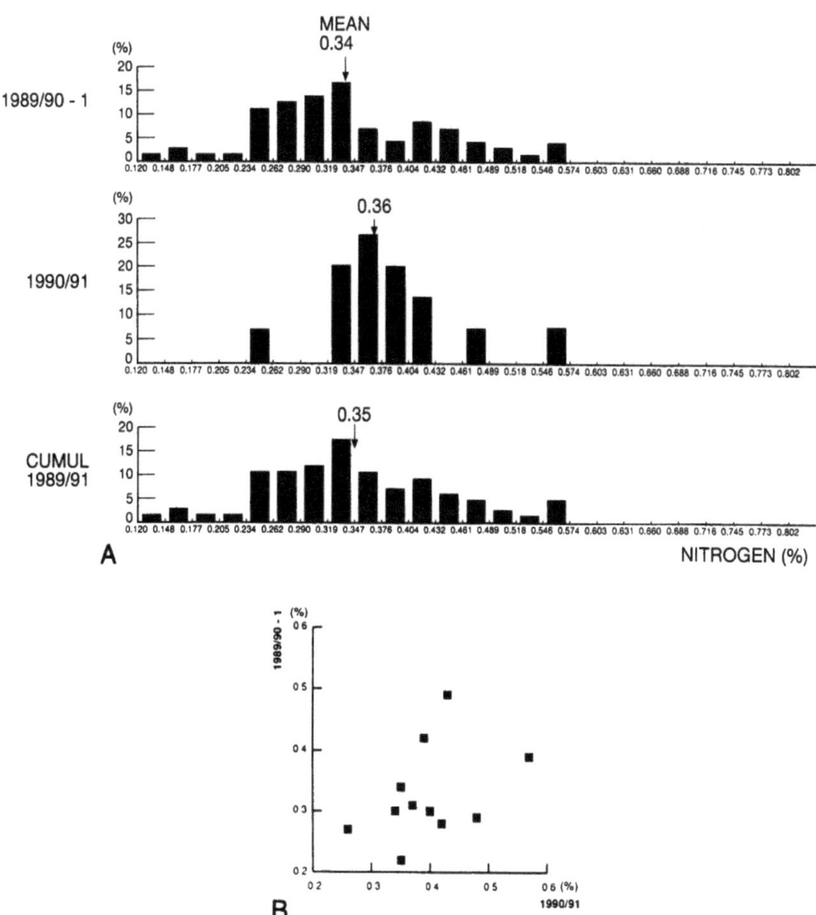

Figure 3 The nitrogen content of gum arabic from Senegal

A.   Variation between individual samples and on ageing
B.   Variations from the same tree with harvest year

Nitrogen (%): Here there is a very wide variation (Fig. 3) (from 0.12% to 0.83%), which indicates a variation in protein content of between 0.75% and 5.19% (mean 2.06%). The average N% is 0.35. There is some change with time of storage from one year to the next (from 0.34 to 0.36% in mean value).

Conclusions

Any objective assessment must conclude that on either physical or chemical grounds there are no readily identifiable parameters which unequivocally distinguish between the Vulgares and Gummiferae series. Certainly the limits of $[\alpha]_D$ and %N are inadequate, and if introduced would render inadmissable gums which fulfil the description '**A.senegal and related (or even closely related) species**'. Thus, even with the most expensive and sophisticated analytical techniques (^{13}C solid state nuclear magnetic resonance) there can still be found authenticated Acacia Senegal gum which shows the expected natural product variability. It would be unfortunate if a producer country or the gum industry generally were to be penalised for this perfectly understood natural variability. The proposed N and $[\alpha]_D$ limits as a means of characterisation are clearly inadequate.

THE CHEMICAL IDENTITY OF GUM ARABIC

A Member of a Large Family

All Acacia gums in chemical terms are Arabinogalactan - proteins (AGP)[39].

The Acacia AGP

The structural features are illustrated in Figure 4. Moving from the component sugars to the macro AGP structure, to fully characterise a particular Acacia gum, the following detail must be elucidated:

The component sugars[40-42]   These are: D-Galactose, L-Arabinose, L-Rhamnose, D-Glucuronic acid, 4-0-Methyl glucuronic acid.

The proportions differ from species to species as shown in Table 2 for the most probable gums in commercial use[43].

Structural linkages in the AG framework: The complex structure has order within it[44]. In the polysaccharide chain there are uniform blocks of (1-->3)-linked D-galactopyranosyl residues; these blocks are comparable in size to those postulated for many AG's of simpler structure, so confirming the relationship of the Acacia gums within the wider AGP family. Moreover, the unity within the various Bentham botanical classifications of Acacias can also be demonstrated on structural grounds. In Acacia Senegal and other Acacia species gums there is strong evidence for the occurrence of uniform sub-units having a molecular weight of ~8000.

Composition of the polypeptide backbone: Table 3 gives the average amino acid composition of commercial Gum Arabic authenticated A.Senegal and representatives from related Acacia species[34-36].

Point of attachment of the AG to the polypeptide: Rigorous examination of the point of attachment of the AG to the protein moiety had hitherto only been made in one instance, namely the AG from wheat endosperm. Here the hydroxyproline is linked by an alkali- stable glycosidic linkage to the β-D-galactopyranosyl residue of the reducing terminus of the AG.

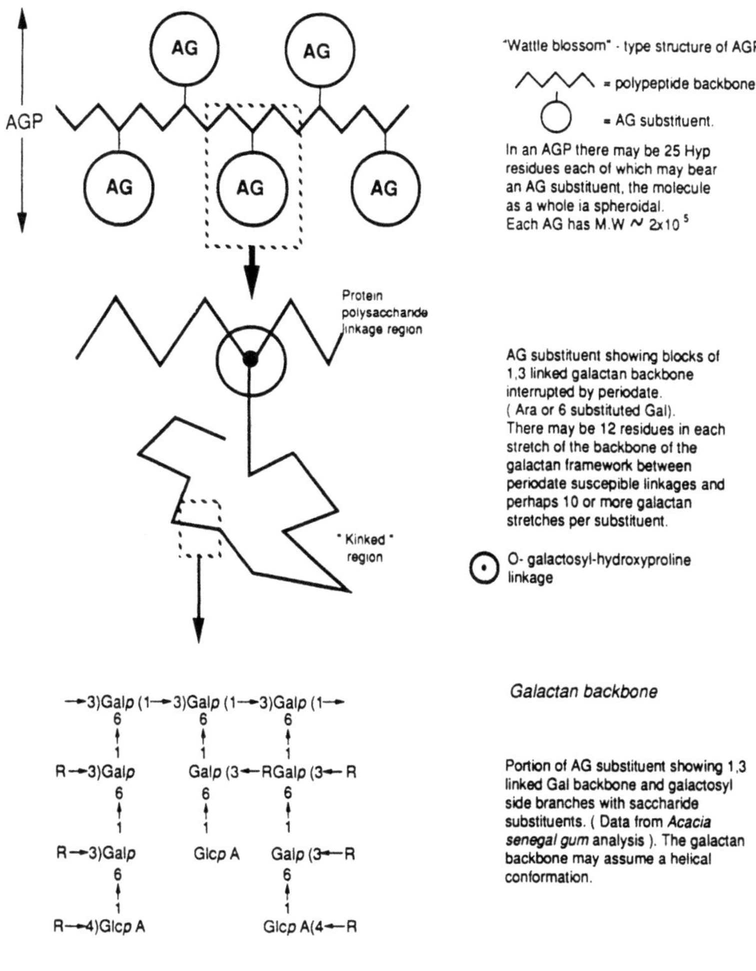

Figure 4 A schematic structure of an arabinogalactan - protein

Recently[45] it has been demonstrated that the same type of linkage, as in other AGP's (hydroxyproline linked via an alkali-stable glycosidic linkage to the β-D-galactopyranosyl residue) exists in Gum Arabic. O-Galactosylhydroxyproline was identified as the glycopeptide linkage. This new finding further demonstrates a structural identity with AGP's of different origin[46].

Molecular dimensions of the AG: Measurements of the molecular mass of Gum Arabic by various methods give widely different results. This anomaly was resolved by Fenyo[47] who showed that the commercial gum is polydisperse, containing extremely high molecular mass protein containing components ($>10^6$). After incubation with a proteolytic enzyme, pronase, to hydrolyse the protein, the AG thereafter had a uniform molecular weight of ca 2 x 10^5. Thus, the crude Acacia gum is a composite of AG molecules containing one or more polysaccharide chains linked to the protein core. These observations which support the "wattle blossom" model[48] are shown in Fig. 4.

The molecular components of A.Senegal gum: The Gum Arabic AGP is not a uniformly molecularly discrete species. Hydrophobic affinity chromatography has been used to fractionate the polydisperse aggregate[40]. The results are shown in Fig. 5.

Fractionation reveals that gum arabic consists of three distinct components, namely **AG** (fraction 1), **AGP** (fraction 2) and one or possibly two glycoproteins (fractions 3A and 3B), which we represent as **GP**. The presence of glycoproteins (**GP**) had not hitherto been shown to be present in A.Senegal, although they had been isolated from other Acacia species.

Fraction 1 represents ca 88.4% of the total, and has only small amounts (0.35%) of protein associated. (It can thus be represented as **AG**). Fraction 2, which represents 10.4% of the total, is an **APG** with more protein (11.8%). Fraction 3 (shown as the glycoprotein **GP**) represents only 1.24% of the total gum, but contains 47% protein, which is 25% of all the protein present in the whole gum. Despite its low concentration, the presence of the **APG** fraction is vital for the emulsification functionality of the gum. Without its presence, or if it is destroyed by processing methods or removed, the gum will not emulsify[40]. The presence of the **AGP** component can, therefore, be regarded as a "marker" for Gum Arabic.

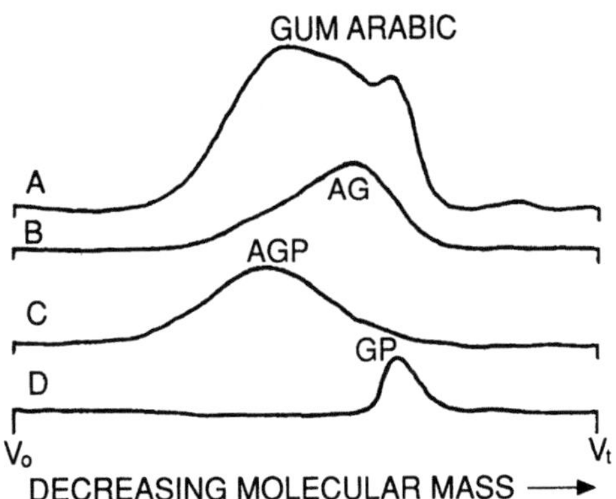

Figure 5 Gel permeation chromatograms of gum arabic and its fractions: (A) whole gum arabic: (B) fraction 1: (C) fraction 2: (D) fraction 3A

REGULATORY SIGNIFICANCE

A study has been undertaken of the presence and variation in concentration of these characteristic moieties AGP, AG and GP in samples of commercial Gum Arabic (samples 2-7), and the findings related to those for two authenticated samples of A.Senegal (sample 1) and A.Seyal (sample 8). The findings are summarised in Figure 6. The full findings have been reported[34].

From a regulatory point of view the significance is that gum which meets the new proposed specifications of $[\alpha]_D$ and N% and those which do not, all have a proportion of **AGP, AG** and **GP** present. However, the proportions vary as seen from Figure 6. Here Figs. a and b show the weight distribution of gum as identified by refractive index, whereas Figs. c and d are more indicative of the protein distribution, as identified by the ultraviolet spectra. Even of greater significance is the fact that A.Seyal also contains these same building blocks, albeit in different proportions, and also contains one other small constituent. Using an immunoassay technique, which is specific for the **AGP** component, these findings can further be confirmed[34]. Moreover, this offers a quick and practical way of evaluating this diagnostic component **AGP** of commercial Gum Arabic.

It is our submission that identifying this AGP component is a more accurate diagnostic technique for identifying commercial Gum Arabic than the aggregate properties $[\alpha]_D$ and % N, which provide frequent anomalies. Additionally, the study confirms that on a chemical basis A.Seyal can be regarded as a "closely related" gum to A.Senegal.

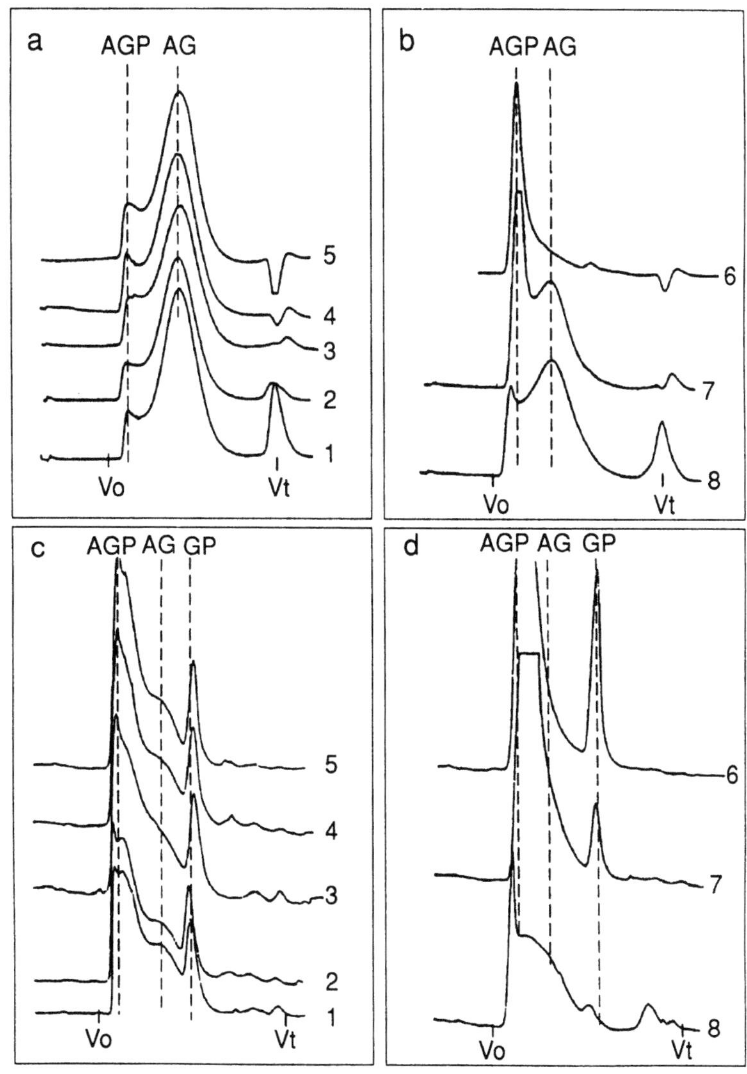

Figure 6 Gel permeation chromatography of commercial gums

Diagnostic tests have been presented to identify those gums which can be classified as *Acacia senegal and (closely) related species*. To illustrate the effectiveness of this procedure Australian Acacia gums from Series Phyllodinea were studied using the same techniques. The samples were:

	A.pycnantha	*A.microbotyra*	*A.spectabilis*
N(%)	0.23	0.06	0.11
Moisture (%)	9.2	10.4	11.5
$[\alpha]_D$	-12	+2	+5

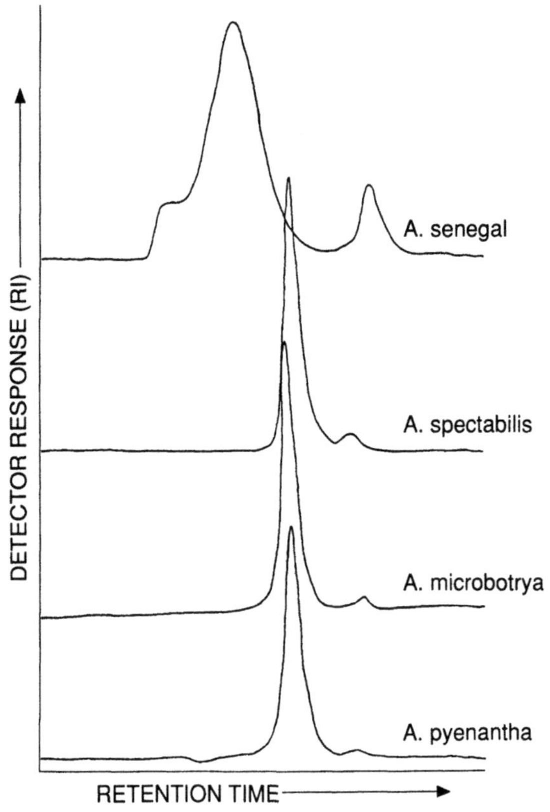

Figure 7 Size exclusion chromatography of Australian Acacia gum exudates compared with *A.senegal* gum (Kordofan, Sudan). All samples at 1% w/v in 0.5 mol.dm^{-3} NaCl

The gel permeation chromatograms in comparison with Acacia senegal are shown in Fig.7. The single peak is equivalent to a component of molecular mass of 70,000. None of the Australian Acacia gums cross-react with the Acacia senegal antibody in the immunoassay (ELISA) test. Thus, it is evident that our procedures can readily distinguish between a gum which is "related" to Acacia senegal and one from a sub-genus which is removed and "not related".

A Chemometric Method of Distinguishing Between Various Species of Arabino-galactan gums

Our work is now directed to examine whether there is a chemotaxonomic link between the various Acacia gums from the two "closely related" sub-genus Vulgares and Gummiferae, and to establish whether these can be readily distinguished from the two series of arabino-galactans which are the most likely adulterants: *Albizia* and *Combretum* gums.

Our study subjects the complex set of data available in the literature for Acacias (*Vulgares* and *Gummiferae*), *Combretum* and *Albizia* to exploratory chemometric data analysis (PCA and DCA). This computer approach to assess the multivariate data allows a separation of the classes. The data points scattered in the multidimensional space were projected on to a plane which was spanned by the two principal components with the highest amount of information, thus visualising groupings of similar categories. The results are shown in Figure 8.

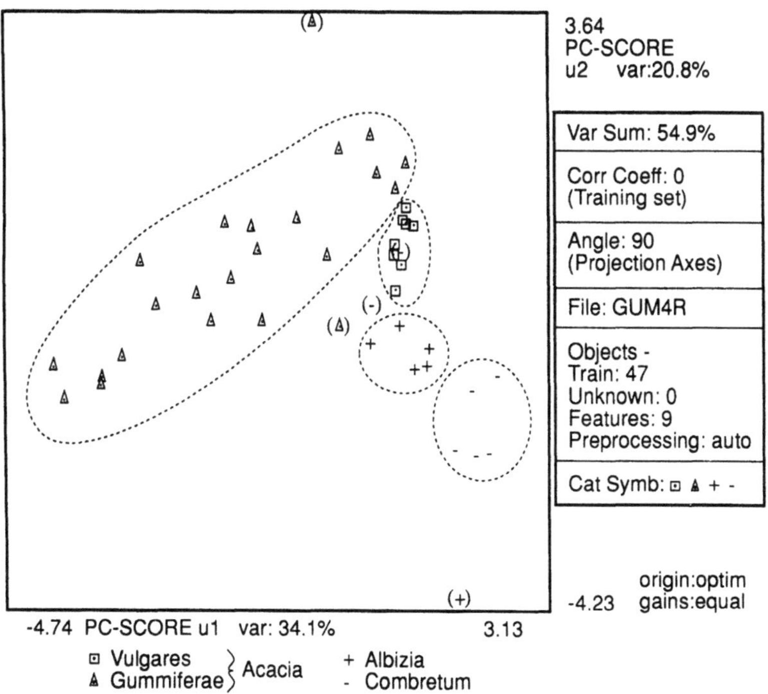

Figure 8 Principal component analysis of the physical and chemical data relating to the following arabinogalactan gums

There is clear distinction between *Albizia*, *Combretum* and the Acacia gums. At the boundaries there appears to be merging of the *Vulgares* and *Gummiferae* sub-genus gum samples, further confirming the close relationship between certain members, and illustrative possibly of a common chemotaxonomic origin and relationship. In particular, *A.seyal* and *A.senegal* fall within the same close projection, indicating the danger of giving undue emphasis on the positive and negative specific rotations of these species. When all ten parameters (specific rotation, viscosity, equivalent weight, % nitrogen, contents of the individual sugars: galactose, arabinose rhamnose, glucuronic acid, O-methyl glucuronic acid and amino acid contents) are evaluated chemometrically then this distortion is removed, and the results show the "closely related" character of *A.seyal* compared with *A.senegal*. Moreover, our chemometric study confirms also that %N is the least species characteristic feature from all the parameters described. Thus the principle of selecting physical and chemical parameters for closer specification of commercial Gum Arabic by JEFCA can be commended. However, it would be advisable to undertake a more thorough evaluation than they have provisionally proposed using all the parameters available for these "closely related" species. The full results of the chemometric study will be published elsewhere[49].

Acknowledgement: Our thanks are due to Professor J.C.Fenyo and J.Vassal for making their results on fresh gum from Senegal available to us pre-publication in the open literature, and also references to physical and chemical data on Acacia species gums.

REFERENCES

1.   The United States Pharmacopoeia, 21st revision, NF 16th ed. Official Monograph for USP XX1, 1528, (1985).

2.   USFDA GRAS, 83 (1977) and Chapter 184, 1330, (1987).

3.   JECFA - FAO Food and Nutrition Paper, Rome, No.34, (1986).

4.   JECFA - FAO Food and Nutrition Paper, Rome, No.49, (1990).

5.   J.H.Ross, A Conspectus of the African Acacia Species in Memoirs of the Botanical Survey of South Africa, D.J.B. Killick ed., Republic of South Africa, 1-151, (1979).

6.   J.Vassal, *Trav.Lab.Forest*, Toulouse, Tome 1, **8**, Art 17, (1972).

7.   G.Bentham, *Trans.Lin.Soc.* (London), **30**, 444, (1875).

8.   J.H.Ross, *Bothalia*, **11**, 107, (1973).

9.   El Amin, Sudan Silva, 3, No.21, 14, (1976).

10.   J.Atchison, *Amer.J.Bot*, **35**, 651, (1948).

11.   D.M.W. Anderson, **Food Hydrocolloids**, **4**, 327, (1987).

12.   D.M.W.Anderson & K.A.Karamalla, *J.Chem.Soc.*, (C), 762-764, (1966).

13.   D.M.W.Anderson & J.G.K. Farquhar, *Phytochemistry*, **18**, 609-610, (1979).

14.   D.M.W.Anderson & F.J.McDougall, *Food Hydrocolloids*, **1**(4), 327-331, (1987).

15.   D.M.W.Anderson, *Kew Bulletin*, **32**(3), 529-536, (1978).

16.   D.M.W.Anderson & R.N.Smith, *Carbohydr.Res.*, **4**, 55-62, (1967).

17.   A.W.Thomas & H.A.Murray Jr., *J.Phys.Chem.*, **32**, 676-697, (1928).

18.   S.C.Churms, E.H.Merrifield & A.M.Stephen, *Carbohydr.Res.*, **123**, 267-279, (1983).

19.   A.M.Stephen, *S.Afr.J.Chem.*, **40**(2), 89-99, (1987).

20.   D.M.W.Anderson, D.M.Douglas Brown, N.A.Morrison & Weiping Wang, *Food Addit. Contam.*, **7**(3), 303-321, ((1990).

21.   E.Dickinson., B.S.Murray, G.Stainsby & D.M.W.Anderson, *Food Hydrocolloids*, **2**(6), 477-490, (1988).

22.   D.M.W.Anderson, J.C.M.Dea, K.A.Karamalla & J.F.Smith, *Carbohydr.Res.*, **6**, 97-103, (1968).

23.   D.M.W.Anderson & J.F.Stoddart, *Carbohydr.Res.*, **2**, 104-114, (1966).

24.   C.L.Butler & L.H.Cretcher, *J.Am.Chem.Soc.*, **51**, 1519-1525, (1929).

25.   M.C.Vandevelde & J.C.Fenyo, *Carbohydrate Polymers*, **5**, 251-273, (1985).

26. M.J.Snowden, G.O.Phillips & P.A.Williams, **Gums and Stabilisers for the Food Industry 4,** IRL Press Oxford, 489-496, (1988).

27. R.C.Randall, G.O.Phillips & P.A.Williams, *Food Hydrocolloids,* 2(2), 131-140, (1988).

28. D.M.W.Anderson, M.M.E.Bridgeman & G.De Pinto, *Phytochemistry,* 23(3), 575-577, (1984).

29. D.W.Gammon, A.M.Stephen & S.C.Churms, *Carbohydr.Res.,* 158, 157-171, (1986).

30. D.M.W.Anderson & N.A.Morrison, *Food Hydrocolloids,* 3(1), 57-63, (1989).

31. M.Kaplan & A.M.Stephen, *Tetrahedron,* 23, 193-198, (1967).

32. S.C.Churms, A.M.Stephen & C.B.Steyn, *Phytochemistry,* 25(12), 2807-2809, (1986).

33. D.M.W. Anderson, P.C. Bell, G.H.Conant & C.G. McNab, *Carbohydr.Res.,* 26, 99-104, (1973).

34. M.E.Osman, P.A.Williams, A.R.Menzies & G.O.Phillips, *J.Agric.Fd.Chem.,* - in press.

35. D.M.W. Anderson & F.J. McDougall, *Food Hydrocolloids,* 4, 327-331, (1987).

36. D.M.W.Anderson, D.M.Brown, D.N.A.Morrison & W.Wieping, *Food Add.Contam.,* 7, 303-321, (1990).

37. D.M.W.Anderson, *Carbohydrate Polymers,* 14, 373, (1991).

38. D.M.W.Anderson, *Food Additives & Contaminants,* 2, No.3, 159-162, (1985); **Biochemical Systematics & Ecology,** 19, No.6 447-452, (1991).

39. G.B.Fincher, A.E.Clarke & B.A.Stone *Ann.Rev.Plant.Physiol.,* 34, 47-70, (1983).

40. R.C.Randall, G.O.Phillips & P.A.Williams, *Food Hydrocolloids,* 3, No.1, 65-75 (1989); *ibid.,* 2, 131-140, (1988).

41. S.C.Churms, E.H. Merrified & A.M.Stephen, *Carbohydrate Res.,* 123, 267-279, (1983).

42. D.M.W.Anderson, I.C.M.Dea, K.A.Karamalla & J.F.Smith, *Carbohydrate Res.,* 6, 97-103 (1968).

43. G.O.Phillips, *S.Afr.Suppl.Food Review* 1988, 64 & Proc.Solingen Conf. on Gum Arabic, June, 1988.

44. S.C.Churms, E.H.Merrified & A.M.Stephen, *Carbohydrate Res.,*123, 267-279, (1983), A.M.Stephen, *S.Afr.J.Chem.,* 40(2), 889-99, (1987).

45. Wu Qi, C.Fong & D.T.A.Lamport, *Plant Physiol.,* 96, 848, (1991).

46. Y.Akiyama, S.Eda & K.Kato, *Agri.Biol.Chem.,* 48(1), 235-237, (1984).

47. S.Connolly, J.-C.Fenyo & M.-C.Vandevelde, *Carbohydrate Polymers,* 8, 23-32 (1988), *Food Hydrocolloids,* 1, No.5/6, (477-4), (1987), (S.Duvallet, J.-C.Fenyo & M.-C. Vandevelde, *Polymer Bulletin,* 21, 517-521, (1989).

48. G.B.Fincher, A.E.Clarke & B.A.Stone, *Ann. Rev. Plant. Physiol.,* 34, 47-70, (1983).

49. P.Jurasek, A.Kosik & G.O.Phillips, **Food Hydrocolloids** (in the press).

HYDROGELS OF CHITIN AND CHITOSAN

Min Zhang, Eugene Kohr, and Shigehiro Hirano

Department of Agricultural Biochemistry and Biotechnology
Tottori University, Tottori 680, Japan

ABSTRACT

N-Acylchitosan gels, N-arylidene- and N-alkylidene-chitosan gels, chitosan oxalate gels, and chitosan gel were prepared from chitosan by reactions of a series of carboxylic anhydrides, aryl- and alkyl-aldehydes, and oxalic acid. A chitin gel was prepared from chitin and alkaline chitin. An intramolecular conversion of chitosan oxalate gel to N-acetylchitosan (chitin) gel via chitosan gel was demonstrated. The paper also presented a possible mechanism for the gelation of chitin and chitosan, the stability and some physical properties of these gels, the digestibility by lysozyme and chitinase, the stimulation of lysozyme activity in the blood of rabbits fed these compounds and in the culture of chicken embryo fibroblast cells in the presence of these compounds, and some possible applications in food industry.

INTRODUCTION

Hydrogels are generally composed of fibrous phase and water phase. Porous frameworks are produced by cross-linkings among polymer chains with covalent, hydrogen and ionic bonds, helixes, van der Waal's interactions, interactions among non-polar chains. Pores present in the solid frameworks are filled with water droplets to give the hydrogels.[1] As common food hydrogels,[2] neutral (e.g., curdlan, starch, and agarose) and acidic (e.g., carrageenan, alginate, and pectinate) polysaccharide gels are known, but basic (amino) polysaccharide gels are little known. Chitosan is a linear (1->4)-linked 2-amino-2-deoxy- β -D-glucan, and chitin is N-acetylchitosan (Fig. 1). As traditional foods, we have been eating mushrooms, soft shrimps and baker yeasts, all of which contain chitin and chitosan as constituents.[3] Chitin and chitosan have been officially approved as food additives in Japan. Chitin and chitosan are synthesized biologically in an estimated amount of one billion tons per year on the earth, and degraded biologically at the same amount. This is called as "Chitin cycle on the earth", which plays an important role in keeping our global ecology and environment in healthy conditions.[4]

Natural chitin has strong hydrogen bonds in the molecule, and has rigid structures.[5] In 1975, we prepared N-acetylchitosan (a regenerated chitin) from chitosan as a gel, which is considered to have week hydrogen bonds in the molecule.[6] Since then, a series of novel gels

Food Hydrocolloids: Structures, Properties, and Functions
Edited by K. Nishinari and E. Doi, Plenum Press, New York, 1994

have been prepared from chitin and chitosan. We now report chemical structures and some properties of chitin and chitosan gels, a possible mechanism for the gelation, and several applications in the field of biotechnology including food industry.

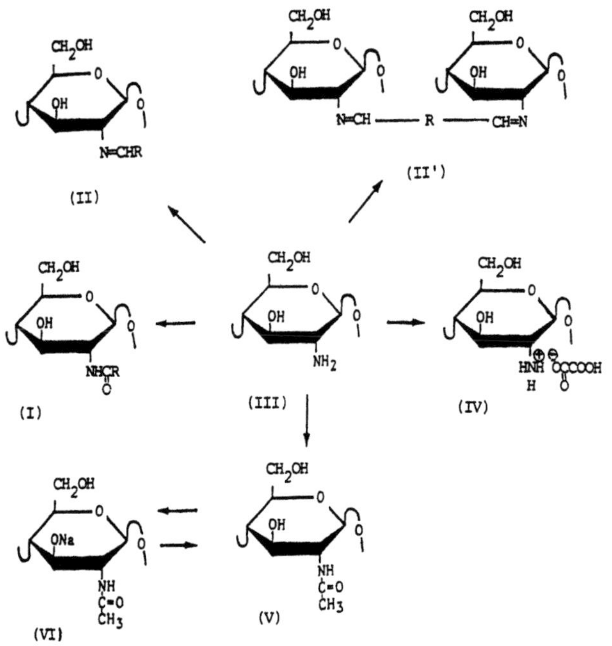

Fig. 1 The chemical structure of a repeating unit for each of N-acylchitosan (I), N-arylidene and N-alkylidene derivatives of chitosan (II) with cross-linking (II'), chitosan (III), chitosan oxalate (IV), chitin (V), and alkaline chitin (VI).

MATERIALS AND METHODS

Chitosan (d.s. 0.1 for NAc) was prepared from crab shell chitin. ^{13}C CP/MAS-NMR spectra were recorded on a Chemagenetics CMX 360 NMR spectrometer (Chemagenetics), and IR spectra (KBr) on a Jasco FT-IR 5300 spectrometer (Jasco, Tokyo). A Hitachi S-500 scanning electron microscope (Hitachi, Tokyo) was used by operating at an accelerating voltage of 20 kV. Elastic hardness and breaking points (dyne/cm^2) were measured on an Iio M-302 curdmeter (Iio, Tokyo). Lysozyme from hen egg-white, and chitinase from _Streptomyces griseus_ were commercial products (Sigma). The chemical structure of all the compounds described in the text was confirmed by IR spectra (KBr), ^{13}C CP/MAS-NMR spectra, and the elemental analysis data.

RESULTS AND DISCUSSION

N-Acylchitosan Gels

Chitosan (0.16 g) was dissolved in aqueous 2% acetic acid (10 ml), and the solution was diluted with methanol (10-20 ml). The corresponding carboxylic anhydride (3-5 mol/GlcN) was added, and the mixture solution was kept at room temperature to afford a gel. The gel was dialyzed against running water and then distilled water to afford a transparent hydrogel in over 80% yields by xerogel. The following N-acylchitosan gels were prepared: formyl, acetyl, n-propyl, n-butyryl, isobutyryl, trimethylacetyl, glycyl, alanyl, chloroacetyl, pentanoyl, hexanoyl, octanoyl, decanoyl, 9-eicosenoyl, 9-octadecenoyl, 9,10-octadecadienoyl and benzoyl derivatives. Several N-carboxyalkylacyl and N-carboxyarylacyl derivatives of chitosan were also produced in gel forms in reactions of chitosan with some intramolecular carboxylic anhydrides.[7,8]

N-Arylidene- and N-Alkylidene-chitosan Gels

To a solution of chitosan (0.16 g) dissolved in aqueous 2% acetic acid (20 ml), the corresponding aryl or alkyl aldehyde (3-5 mol/ GlcN) was added at room temperature. Produced gels were dialyzed against running water and then distilled water to afford gels in over 80% yield by dry weight. The following N-arylidene and alkylidene chitosan gels were prepared: benzylidene, 2-hydroxybenzylidene, 3-hydroxybenzylidene, 4-hydroxybenzylidene, 2-nitrobenzylidene, 3-nitrobenzylidene, 4-nitrobenzylidene, 2-methylbenzylidene, 3-methylbenzylidene, 4-methylbenzylidene, 2-chlorobenzylidene, 3-chlorobenzylidene, 4-chlorobenzylidene, 4-fluorobenzylidene, 3,4-dimethoxybenzylidene,[9, 10] N-methylene, N-ethylene, N-propylene, N-allylidene, N-1,3-propanedimethylene, N-1,3-propanalmethylene, and N-formylmethylene derivatives.[11] Some gels having intramolecular cross-linkings were produced in reactions of chitosan with glutaraldehyde and with o-, m-, and p-phthalaldehydes.[12]

Chitin Gel

1) Chitin (0.2 g) was dissolved in N,N-dimethylacetamide (40 ml) in the presence of 5% LiCl, and it was treated with a mixture of acetic anhydride (1.5 ml) and pyridine (50 ml) at 70-100°C for 6 h to give a partially O-acetylated chitin gel (d.s. 0.85 for O-Ac). The gel was treated with aqueous NaOH, followed by dialyzing against running water and distilled water to afford a gel, which was indistinguishable from N-acetylchitosan (chitin) gel.[6] This indicates that the O-deacetylation little affects the gel properties.[13]

2) A solution of 2% alkaline chitin in aqueous 14% NaOH was kept in a refrigerator for a few weeks to afford a gel. The gel was dialyzed against running water and then distilled water to afford a chitin gel, which was indistinguishable from N-acetylchitosan gel.[6]

Chitosan Oxalate Gel

Chitosan (0.97g) was dissolved in aqueous 3% oxalic acid (30ml) on a boiling water bath, and the solution cooled at room temperature to afford a gel. Under these conditions, the gel melted on heating at 80-90°C, and the gel reformed on cooling at room temperature. The heating and cooling cycle could be repeated with little affecting the gel properties. The xerogel was composed of one mol. equiv. of oxalic acid per D-glucosaminyl residue. The structure was confirmed by elemental analysis data, IR absorptions at 1700 (COOH), and 260, 1620 and 1400 (ionic salt) cm^{-1}, and ^{13}C-resonances at δ 167 and 164 (C=O of oxalic acid and salt) in ^{13}C CP/MAS-NMR spectra.[14]

N-Acetylchitosan Gel from Chitosan Oxalate Gel via Chitosan Gel

The chitosan oxalate gel (20 g by wet weight) was treated with aqueous 1N NaOH at room temperature for a few days, and then dialyzed against running water and distilled water to give an opaque chitosan gel in over 18 g yield by wet weight. The xerogel was identified as chitosan on the basis of IR and ^{13}C CP/MAS-NMR spectra, and elemental analysis data. The opaque chitosan gel was treated with acetic anhydride in an aqueous methanol solution at room temperature overnight to afford a transparent gel, which was indistinguishable from N-acetylchitosan (chitin) gel.[6]

A Possible Mechanism for the Gelation of Chitin and Chitosan

1) Chitosan chains are probably present in stretched forms in aqueous 2% acetic acid, because of a mutual repulsion among cationized amino groups in the chain. The stretched chitosan chains form frameworks during N-substitution reactions, and the gels are produced. Similarly, alkoxides in the alkaline chitin are gradually hydrolyzed in an aqueous 14% NaOH solution to afford a chitin gel.

2) The gels produced under the present conditions were composed of about 95% water and about 5% chitosan or its derivative.

3) D.s. for N-substituents in chitosan chain significantly affected the gel properties. Hard gels formed in d.s. higher than 0.7/GlcN for N-acyl, N-arylidene and N-alkylidene (monoaldehyde) groups, and soft gels or no gels formed in d.s. lower than 0.7/GlcN for these N-substituents. However, the minimum d.s. required for gelation by covalent cross-linking was as low as 0.002/GlcN in the reaction of chitosan with glutaraldehyde.

Table 1 Some properties of gels derived from chitin and chitosan.

	Gels			
	Chitosan oxalate	Chitosan	Chitin & N-acyl-chitosan	N-Arylidene- & N-alkylidene-chitosan
Color	colorless	white	colorless	colorless or brown
N-Substitution	$-NH_3^{+-}O(C=O)-$	$-NH_2$	$-NH(C=O)-$	$-N=CH-$
Transparency	transparent	opaque	transparent	transparent
Stabilities in				
heat	reversible	irreversible	irreversible	irreversible
aq. acids	stable	unstable	stable	unstable
aq. alkalis	unstable	stable	stable	stable
Elastic hardness ($\times 10^6$ dyne/cm^2)	2.8	3.0	1.5	n.d.[a]
Breaking point ($\times 10^6$ dyne/cm^2)	7.5	14.0	6.4	n.d.

[a] n.d., not determined.

4) The chemical structure of R groups in N-acylchitosan gels also affected the gel formation (Fig. 1). N-Stearoyl and N-palmitoyl derivatives of chitosan did not form in gels, and were amorphous precipitates, probably because of strong non-polar interactions between long non-polar alkyl chains. When chitosan was dissolved at elevated temperature in an aqueous solution of each of malonic acid and succinic acid, the corresponding salt gel did not form on cooling at room temperature, probably because of the steric inhibition of these dicarboxylic acids.

5) The formation of the chitosan oxalate gel was inhibited by presence of N-acetyl group (d.s. higher than 0.2/GlcN) in chitosan. In the chitosan oxalate gel, one carboxylic acid group forms a salt with the

amino group of chitosan just like a pendant, and other one is present in a free form. This was confirmed by IR and ^{13}C CP/MAS-NMR spectra, and the elemental analysis data. The structure is essentially similar to the common structure of N-acylchitosan and N-arylidene- and N-alkylidene-chitosan gels.

6) The intramolecular conversion of chitosan oxalate gel to N-acetylchitosan gel via chitosan gel strongly indicates the presence of a common architecture for the framework in these gels. This was confirmed by an electron-microscopic observation of these frameworks prepared by freeze-drying. The framework may be built up by hydrogen bonds and interactions of non-polar groups among chitosan chains.

Some Properties and Biological Functions of the Hydrogels

1) Chitosan formed ionic salt in aqueous acetic acid solutions to give a viscous solution. Higher molecular-weight chitosan showed higher viscosity, and the viscosity decreased almost linearly with decrease of the concentration.

2) N-Acylchitosan (chitin) gels underwent syneresis, but N-arylidene- and N-alkylidene-chitosan gels and chitosan oxalate gel did not. Table 1 shows the stability of these gels in water, aqueous NaOH, and aqueous acetic acid at room temperature. The elastic hardness of the gels increased almost linearly with chitosan concentration: the maximum values for its elastic hardness and breaking point ($\times 10^6$ dyne/cm^2) were 2.8 and 7.5, respectively at 3% chitosan concentration in 1.0-1.5 mol. equiv. of oxalic acid to GlcN. N-Fatty acyl, N-benzoyl, N-arylidene, and N-alkylidene derivatives of chitosan were insoluble in water and in aqueous acidic and alkaline solutions. However, N-carboxyalkylacyl and N-carboxyarylacyl derivatives of chitosan were soluble in aqueous 0.5% NaOH solution. The Schiff's base structure in N-arylidene and N-alkylidene derivatives was degraded in aqueous acids even at room temperature.

3) N-Acetylchitosan (chitin) gels and its xerogel were digested more rapidly than natural chitin by enzymes. The relative hydrolysis rates of N-acetylchitosan (chitin) gel, its xerogel, and natural chitin as control was 11:8:1 by lysozyme from hen egg-white, and 13:8:1 by chitinase from _Streptomyces_ _griseus_. This indicates that the intra- and inter-molecular interactions of the regenerated chitin gels are weaker than those of natural chitin, resulting in relatively high rates by these enzymes.

4) In the culture of chicken embryo fibroblast cells, extracellular lysozyme activity was measured by the decrease of turbidity with the use of cell walls of _M._ _lysodeikticus_ (M-3770, Sigma) as substrate. A gradual increase in extracellular lysozyme activity up to 65 U/mg of protein for 72 h was detected in the presence of chitin (10 mg/dish), but almost no intra-cellular lysozyme activity was detected even in the presence and absence of chitin. The enzyme activity was stimulated by addition of chitin oligosaccharides (d.p. higher than 7) to the medium, but little by addition of chitin oligosaccharides (d.p. lower than 6).[15] Lysozyme activity in rabbit serum was also stimulated by intravenous injection of a mixture of chitosan oligosaccharides (d.p. 2-9). These data strongly indicate that orally administrated chitin and chitosan are digested by chitinase and chitosanase, which are secreted from intestinal microorganisms. The partially digested chitin and chitosan are adsorbed in the intestinal organs, and transported to bloods, resulting in induction of blood lysozyme and in inhibiting microbial infection on tissues.[16]

Typical Applications in Food Technology

1) Waste water-treatment with chitosan for collecting acidic proteins as polyelectrolyte complexes in food industries.

2) The removal of acidic proteins from sakes, wines, vinegars, and soy sauces for improving their qualities and for controlling their viscosity and taste.

3) Chitin and chitosan for use as food-additives in cookies, noodles, breads, as growth-inhibitors of molds at low levels of NaCl in pickled vegetables and in soy sauces, and as freshness-keeping agents for fruits and vegetable.

4) Oral administration of chitosan for stimulating lysozyme activity in bloods and tissues, for improving flora of intestinal bacteria, for decreasing cholesterol levels in bloods,[17], and for enhancing lactose metabolism in animals.[18]

5) New functional food materials produced by blending chitin and chitosan with each of proteins (e.g., egg-white), polysaccharides (e. g., starch), and lipids.

REFERENCES

1. K. Ogino, Y. Osada, T. Fushimi, and A. Yamauchi. "Gels", Sangyotosho, Tokyo (1991).
2. D. A. Rees, Chem. & Ind., (London), 630 (1972).
3. S. Hirano, Ulmann's Encyclopedia of Industrial Chemistry, 6A:231 (1986).
4. S. Hirano, H. Inui, N. Hutadilok, H. Kosaki, Y. Uno, and T. Toda, Polym. Mater. Sci. Eng., 66:348 (1962).
5. R. Mink and J. Blackwell, J. Mol. Biol., 120:1167 (1978).
6. S. Hirano, S. Kondo, and Y. Ohe, Polymer, 16:622 (1975).
7. R. Yamaguchi, Y. Arai, T. Itoh, and S. Hirano, Carbohydr. Res., 88:172 (1981).
8. S. Hirano and T. Moriyasu, Carbohydr. Res., 92:323 (1981).
9. L. A. Nud'ga, E. A. Plisko, and S. N. Danilov, Zh. Obshch. Khim., 43:2752 (1973).
10. S. Hirano, N. Matsuda, O. Miura, and H. Iwaki, Carbohydr. Res., 71:339 (1979).
11. S. Hirano, N. Matsuda, O. Miura, and T. Tanaka, Carbohydr. Res., 71-344 (1979).
12. S. Hirano and M. Takeuji, Int. J. Biol. Macromol., 5:373 (1983).
13. S. Hirano and K. Horiuchi, Int. J. Biol. Macromol., 11:253 (1989).
14. S. Hirano, R. Yamaguchi, N. Fukui, and M. Iwata, Carbohydr. Res., 201:145 (1990).
15. S. Hirano, H. Inui, M. Iwata, K. Yamanaka, H. Tanaka, and T. Toda, Enhancement of serum lysozyme activity by intravenously injecting a chitosan oligosaccharide mixture in rabbits, in: "Progress in Clinical Biochemistry," K. Miyai, T. Kanno, and E. Ishikawa, eds., Excerpta Medica, Amsterdam, 1992, pp. 1009-10115.
16. A. Tokoro, M. Kobayahi, N. Takewaki, K. Suzuki, Y. Okawa, T. Mikami, and M. Suzuki, Microbiol. Immunol., 33:357 (1989).
17. S. Hirano, C. Itakura, H. Seino, N. Kanbara, Y. Akiyama, I. Nonaka, and T. Kawakami, J. Agr. Food Chem., 38:1214 (1990).
18. P. R. Austin, C. J. Brine, J. E. Castle, and J. P. Zikakis, Science, 212:749 (1981).

MOLECULAR WEIGHT DEPENDENCY OF ANTIMICROBIAL ACTIVITY BY CHITOSAN OLIGOMERS

S. Sekiguchi[1], Y. Miura[1], H. Kaneko[1], S-I. Nishimura[1], N. Nishi[2],
M. Iwase[3] and S. Tokura [2]

[1] Division of Biological Science, Graduate School of Science
[2] Division of Ecological Science, Graduate School of
   Environmental Earth Science, Hokkaido University, Sapporo 060, Japan
[3] Central Research Institute, Tamatsukuri Co. Ltd., Sapporo 004, Japan

INTRODUCTION

Chitin, a natural abundant mucopolysaccharide becomes to draw many attentions as a multifunctional polymer owing to the variety of biological activities in spite of its poor solubility. Chitosan, N-deacetylated form of chitin, is also natural mucopolysaccharide as a supporting polymer of fungi or yeast. The chemical structure of chitosan is ß-1,4 linked linear polymer of 2-acetoamide-2-deoxy-ß-D-glucose and 2-amino-2-deoxy-ß-D-glucose, respectively as shown in Scheme. As seen in Scheme, chemical structures of cellulose and chitosan are very close expect C-2 position of glucose units. Chitosan becomes amorphous with the progress of N-deacetylation and to more susceptible for chitosanase produced by a limited number of microbials. Although chitosan itself is water insoluble and be solubilized by the formation of salt with organic acids such as formic acid, acetic acid and etc., chitosan oligomers become water soluble even without organic acid. Chitosan and chitosan oligomers are reported to show several biological activities such as anticancer activity[1], antifungal activity[2] and phytoalexin elicitor activity for a plant[3]. The toxicity of chitosan was investigated through oral dose for 19 days by 18 g/Kg of mouse body weight days and confirmed to be low immunogenicity together with low acute, semi-acute toxicities and faint mutational activity, although mouse intraperitoneal injection of chitosan-acetate salt induced peritoneal macrophage activation and immunoadjuvant activity. As mentioned before, chitosan and its hydrolysates are reported to show a specific antibacterial and antifungal activities, in which a slightly hydrolyzed chitosan showed the highest antimicrobial activity followed by chitosan of high molecular weight[4]. The minimum growth inhibitions were shown by chitosan oligomers of less than 6 glucosamine residues. However, the mechanism for the antimicrobial activity is not clear enough, because regulation of molecular weight (Mw) distribution has not been achieved on

Food Hydrocolloids: Structures, Properties, and Functions
Edited by K. Nishinari and E. Doi, Plenum Press, New York, 1994

chitosan hydrolysates yet. In the present study, the molecular weight dependency of antimicrobial activity by various chitosan oligomers was mainly investigated together with regulation of Mw distribution and chemical analysis of the degree of N-deacetylation.

EXPERIMENTAL

Preparation of chitin and chitosan

Chitin was prepared from Queen Crab shell according to Hackman[5] and powdered under 60 mesh before use. N-deacetylation was achieved in 40% (w/w) NaOH aqueous solution under N_2 stream at refluxing temperature for 10 hrs. The product was rinsed with 1N-EDTA 4Na aqueous solution and water, respectively to remove heavy metals. The degree of N-deacetylation was estimated by IR spectrum according to Roberts et al[6] and elemental analysis to be 40% deacetylation (DAC-40). DAC-40 was treated with 40% (w/w) NaOH aqueous solution under similar conditions as above except time of reflux. The degree of deacetylation was also estimated to be 73% (DAC-73) by IR method.

Preparation of chitosan oligomers

Nitrous acid degradation was applied to prepare chitosan oligomers among various chemical degradations, because quantitative reaction was achieved by the amount of nitrous acid against glucosamine residue under the mild condition. Sodium borohydride was also applied immediately after the nitrous acid depolymerization to reduce the reducing end group (2,5-anhydromannose) into 2,5-anhydromannitol as shown in Scheme. Briefly, 200 g of DAC-40 was suspended in 100 ml of 10 % aqueous acetic acid. 3.7 ml of 10 % (w/v) aqueous nitrous acid was added stepwisely at ice cold temperature under the vigorous stirring for 4 hr and the pH of reaction mixture was adjusted to

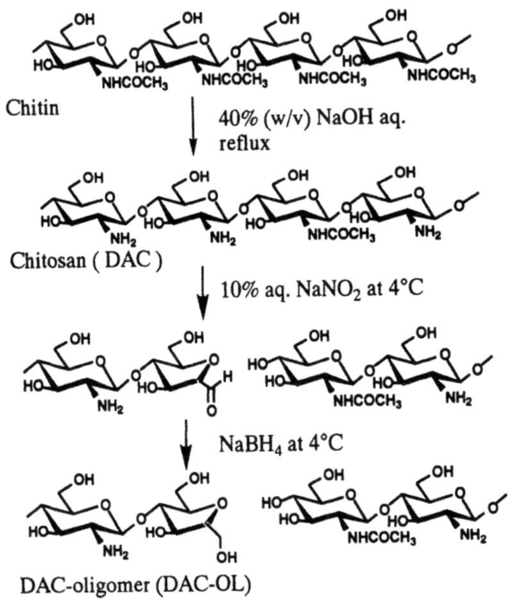

Scheme   Preparation of DAC-oligomers by depolymerization of DAC with nitrous acid.

6.0 by concentrated aqueous ammonia. 0.41 g of sodium borohydride powder (equivalent moles of nitrous acid) was added to the reaction mixture at ice cold temperature and stirred overnight at room temperature. In the case of DAC-73, reaction system is homogeneous because of good water solubility of salt form and products were fractionated by the combination of water, methanol and acetone (20 g of DAC-73 was dissolved in 200 ml of 10 % aqueous acetic acid and degradation was started by the addition of quantitative amount of 19 % nitrous acid under the similar condition for DAC-40).

<u>Quantitative analysis of nitrous acid</u>

The nitrous acid remaining in chitosan oligomers was estimated according to the direction from Food Hygienic Association[7] using sulfanilamide and naphthylethylenediamine solutions. The amount of nitrous acid was calculated by the optical density at 540 nm.

<u>Molecular weight measurements (GPC method)</u>

Molecular weight of oligomers was estimated by HPLC using GPC column at 50 °C and 0.5 M acetate buffer (pH 4.0) as a mobile phase at the rate of 0.3 ml/min to 0.5 ml/min. Pullulan (Shodex pullulan; from Showa Denko Co. Ltd.) and N-acetylchitooligosaccharides (mixture of 1 to 6 residues; purchased from Seikagaku Kogyo Co. Ltd.) were used as standard. The degree of N-deacetylation (DA) was estimated from elemental analysis, absorption ratio between 1655 and 3450 cm^{-1} of IR spectrum of chitosan oligomer and reducing end group titration (HCl-indole method)[8]. MBTH (3-methyl-2-benzothiazolone hydrazone) was also applied for the chemical analysis of the degree of N-deacetylation[9] using glucosamine HCl as a standard.

<u>Cultures for the fermentation of microbials</u>

Two kinds of culture mediums were applied to investigate the antimicrobial activities of chitosan oligomer as follows:

Agar culture :				
Casamino Acids, Bacto	16.5 g	L-Cystine	0.05 g	
Heart Infusion Agar	3.0 g	Biotin	5×10^{-6} g	
Soluble starch	1.5 g	Agar Agar	15.0 g	
Glucose	2.0 g	L-Tryptophan	0.05 g	
Liquid Culture:	NaCl	8.5 g		
	KH_2PO_4	0.3 g		
	Na_2HPO_4	0.5 g		
	Gelatin	0.1 g		
	Distilled water	1,000 mL		

<u>Measurements of growth inhibition by chitosan oligomer</u>

Microbials were fermented in liquid culture for 3 days in the presence of various concentrations (0.05-0.5 %) of chitosan oligomers. In the case of agar culture, bacteria was fermented at 37 °C for 5 days and then the diameter of colony was measured. 100 % growth was set for the fermentation without chitosan oligomer as a standard.

RESULTS AND DISCUSSION

Nitrous acid degradation of chitosan

Chitosan oligomers attaching 2,5-anhydromannose on reducing end was produced by the nitrous acid degradation at glucosamine residue. As 2,5-anhydromannose is unstable and easily forms Schiff base, reducing end group of oligomers was reduced into 2,5-anhydromannitol by sodium borohydride immediately after the degradation. The absorption at 1650 cm-1 (amide I) was

Table 1. List of Depolymerized Products of DAC-73

Samples	Yield (g)	Yield (%)	Mw	NaNO2 (10^{-6} gN/g of OL[†])
DAC-OL-0.3-12				
#1	98.6	32.9	16,100	1.50
#2	31.6	10.2	9,800	1.25
#3	60.1	20.0	4,800	0.67
#4	52.8	17.6	2,180	7.80
#1-DA	15.7	31.4	17,000	0.0
#2-DA	12.7	63.5	9,200	0.0
#3-DA	8.7	43.5	4,800	0.0
#4-DA	8.6	41.7	2,100	0.0
DAC-OL-0.5-2				
#1	46.9	23.5	14,500	16.10
#2	38.9	19.5	11,000	8.22
#3	28.0	14.0	7,400	7.32
#4	9.7	4.9	2,350	2.67
#1-DA	8.9	52.3	11,200	0.0
#2-DA	10.2	51.0	9,400	0.0
#3-DA	8.9	29.7	3,100	0.0
#4-DA	4.0	13.3	2,830	0.0
DAC-OL-0.7-2				
#1	14.8	18.5	2,430	123.4
#2	24.8	31.0	1,600	68.6

† ; Weight of nitrogen per weight of oligomers

enhanced by the amount of nitrous acid applied, and increment of absorption at 1050 cm^{-1} due to the hydroxyl group was also observed probably by the destruction of crystalline structure. Since there is little absorption at 1720 cm^{-1} due to aldehyde group produced, sodium borohydride reduction seems to be completed under present reaction condition. The fact was also comfirmed by the reducing end group analysis. The yields, molecular weights and nitrous acid contents of chitosan oligosaccharides, are shown in Table 1. As seen in the Table 1, the yields was 70-80 % when molar ratio between nitrous acid and amino group was 0.3, 62 % for the molar ratio of 0.5, and 49 % for 0.7. The nitrous acid is able to be removed from oligomers by extensive rinse with methanol and acetone mixed solvent system.

Molecular weight distribution of chitosan oligomers

The molecular weight distribution of chitosan oligomer was investigated by HPLC. Molecular weight distribution of the fractionated oligomer became narrow by applying water-methanol-acetone mixture to precipitate oligomers from reaction mixture. But the oligomers from DAC-40 was rather high molecular weight than those from DAC-73.

<u>Estimation of the degree of deacetylation</u>

DA from IR method tended to show lower values than those from other procedures, when DA was over 60 %. Since lower value was given for standard chitosan on IR analysis, KBr pellet method doesn't fit for DA analysis when DA is over 60 %. HCl-indole method seems to be too sensitive having tendency to give serious error and is able to apply only for specific DA. Thus MBTH method was applied for DA analysis in this experiment.

Table 2. Antimicrobial Activity of DAC-oligomers.

Mw & DA of oligomers	*M. flavas*	*M. lutca*	*E. coli*	*B. subtilus*	*B. cereus*
2,350 - 0.44	97	97	87	89	77
2,380 - 0.61	100	97	98	95	90
2,830 - 0.60	100	97	83	90	88
2,950 - 0.63	97	100	100	99	83
3,100 - 0.52	98	95	78	91	58
7,400 - 0.45	100	100	98	100	61
9,800 - 0.63	100	97	98	97	73
10,500 - 0.60	100	82	93	98	63
11,000 - 0.41	94	100	87	100	51
14,500 - 0.39	100	97	93	100	88
16,100 - 0.52	100	82	93	98	63
21,600 - 0.52	100	89	94	98	76

Mw: Molecular weight of oligomer esytimated by HPLC using standard pullulan
and chitooligosaccharides. DA: Degree of deacetylation.

<u>Antimicrobial activities by chitosan oligomers</u>

The antimicrobial activities by chitosan oligomers were investigated applying liquid culture for various bacteria such as *Bacillus subtilis, Staphylococcus epidermidis, Staphylococcus aureus, Pseudomonus aeruginosa, Echelichia coli* and *Candida albicans*. Antimicrobial activities on agar culture are shown in Table 2. As seen in the table, the growth of *B. cereus* was depressed by 50 % in the presence of chitosan oligomer (Mw=11,000) even at low degree of N-deacetylation. The time course of antimicrobial activities against *B. cereus* are shown in Fig.1. The activities became effective around 3 days after starting the fermentation and maintained growth inhibition for another 10 days. As the oligomer (DAC-41, Mw=11,000) shows the highest activity against *B. cereus*, the dependency of oligomer concentration on the microbial activity was investigated and 0.2-0.3 % of oligomer concentration was declared to be enough to suppress the growth of *B. cereus* as shown in Fig. 2. As the higher degree of growth inhibition was shown by the chitosan oligomer of higher molecular weight, the stacking of chitosan oligomer onto the surface of microbials would be the main factor for the antimicrobial behavior.

In the case of liquid medium, the oligomer (DAC-52, Mw=21,600) showed antimicrobial activity for *B. subtilis* and *S. epidemidis* at minimum concentration of 2 %. The oligomer (DAC-46, Mw=3,500) showed wide antimicrobial spectrum when oligomer concentration was more than 7.4

%. Other oligomers didn't show any influence on the growth of microbial even at high oligomer concentration.

Since the growth of *B. cereus* was inhibited specifically by the chitosan oligomer (DAC-41, Mw=11,000 and DAC-60, Mw=10,500; in the case of plate medium), the molecular weight seems to be the main factor of the antimicrobial acidity and DA is less effective. The prominent future of chitosan oligomer would be indicated on the application for food preservative in addition to application for immunoadjuvant.

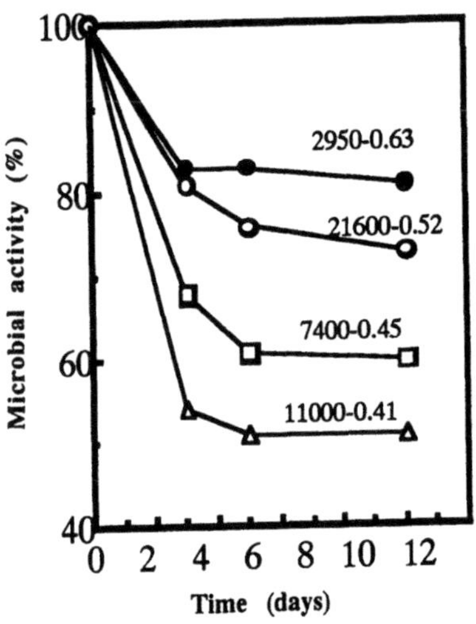

Figure 1. Time course of microbial activity against *B. cereus* by DAC-oligomers. (0.5%)

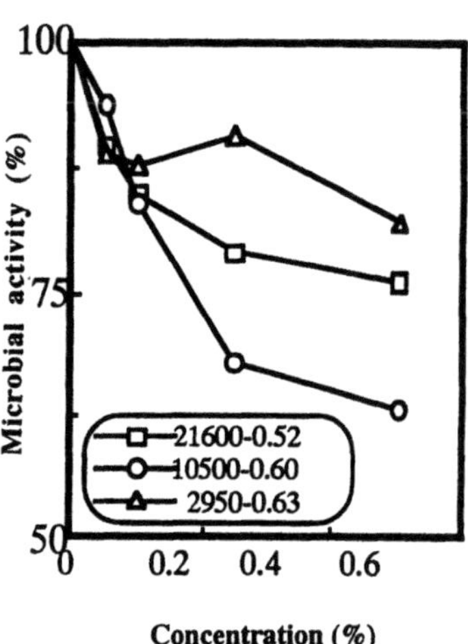

Figure 2. The concentration dependency of microbial activity against *B. cereus* by DAC-oligomers. (3 days fermentation)

REFERENCES

1. K. Suzuki , Y. Okawa, K. Hashimoto, S. Suzuki, and M. Suzuki, Microbiol. *Immunol.*, **28**, 903 (1984)

2. K. Suzuki, A. Tokoro, Y. Okawa, S. Suzuki, and M. Suzuki, *Chem. Pharm. Bull.*, **33**, 886 (1985)

3. D. Roby, A. Gadelle and A. Toppan, *Biochem. Biophys. Res. Commun.*, **143**, 885 (1987)

4. Y. Uchida, M. Izume and A. Ohtakara, *"Chitin and Chitosan"* Ed. by G. Skjak-break, T. Anthonsen and P. Sandford, Elsevier Applied Science 1988, p-373.

5. R. H. Hackman, *Austr. J. Biol. Sci.*, **146**, 251 (1958)

6. G. A. F. Roberts, *Makromol. Chem.*, **186**, 1671 (1985)

7. *"Direction for the Food Hygienic Analysis"* Japanese Food Hygienic Association, Tokyo 1978, p-706.

8. N.Ohno and T.Yatamae, *Carbohydr. Res. ,* **137**, 239 (1985)

9. A. Tsuji, *Chem. Pharm. Bull.*, **18**, 1505 (1985)

CHARACTERISTICS AND APPLICATIONS OF A POLYFRUCTAN SYNTHESIZED FROM SUCROSE BY ASPERGILLUS SYDOWI CONIDIA

Tsutomu Harada[1], Soji Suzuki[1],
Hajime Taniguchi[2] and Takashi Sasaki[2]

[1]Food Research & Development Laboratories, Ajinomoto
Co. Inc., 1-1 Suzuki-cho, Kawasaki-ku, Kawasaki 210
[2] National Food Research Institute, 2-1-2
Kannondai, Tsukuba, Ibaraki 305

A polyfructan synthesized from sucrose by the action of
Aspergillus sydowi conidia was composed of 2,1-linked beta
fructofuranoside residues with a molecular weight of over 10
million.   It was a water soluble white powder and had no taste
and flavor. Its free flowing powder characteristic seemed to be
explained from its high amount of B.E.T. monomolecular layer of
adsorbed water compared with other saccharides. The polyfructan
solution had a much lower viscosity than those of other polysaccharide solutions.   The solution was stable under moderate
heating and shearing.   The polyfructan might be used as a low
calorie bulking agent for sweeteners such as Aspartame, or as a
fat substitute in various processed foods like ice cream and
baked cheese cake.

INTRODUCTION

In recent years, an increasing number of people are suf-
fering from obesity and hyperlipemia, which often cause serious
diseases such as cancer and heart diseases. These unhealthy
conditions often arise from the over consumption of fats and
sugars. Reducing the content of fats and sugars in the processed foods have, therefore, become to be national concerns.
For this purpose, many fat and sugar substitutes have been
developed. Substitutes and bulking agents hitherto developed
are usually accompanied by some undesirable properties that
adversely affect the texture, taste and/or flavor of foods.
Recently we found that a polyfructan produced from sucrose
by the action of Aspergillus sydowi IAM 2544 conidia has quite
excellent properties as a fat substitute and/or a bulking
agent[1]. Lowenberg and Reese[2] and Kawai et al[3] reported the
production and structural study of this saccharide. The
present paper describes the physical characteristics and applications of this saccharide, together with its structural con-
firmation with NMR analysis.

Food Hydrocolloids: Structures, Properties, and Functions
Edited by K. Nishinari and E. Doi, Plenum Press, New York, 1994

MATERIALS AND METHODS

Preparation of polyfructan: Preparation of polyfructan was similar to that described by Kawai et al.[3] except for the use of solid culture for preparation of conidia. The fungus was cultured in a 5 liter Erlenmeyer flask on a solid culture medium containing wheat bran (100g), glucose (0.5g), yeast extract (0.05g) and water ('60g) at $30^{O}C$. One hundred and twenty five g of conidia were incubated with 25 l of 20% sucrose in phosphate buffer (pH 5.5) at $30^{O}C$. The reaction mixture was quantitatively analyzed with HPLC using Asahipak GS-710 for polyfructan or MCIGEL CK08EC for other saccharides.

Structure: Carbon 13 NMR spectra were observed on an AMX 600 spectrometer. NMR samples of polysaccharides were dissolved in about 0.5 ml of DMSO-d_6 with a drop of D_2O.

Molecular weight and intrinsic viscosity were determined by HPLC equipped with on line differential refractometer / viscometer Model 200 designed as a detector for gel permeation chromatography[4,5,6]. Asahipak GS-710 column was used at $25^{O}C$.

Intrinsic viscosity of the effluent was calculated from the pressure drop using Viscotek Unical 3.11 software. Pullulan samples (Shodex standard P-82) were used for the construction of the universal calibration curve.

Powder characteristics: Powder characteristics were observed by the method of B. Makower et al.[7]. Approximately 2g of dried samples were accurately weighed into 50 x 50 mm weighing bottles and placed in desiccators containing saturated solutions such as ammonium phosphate. Equilibrium moisture content at $25^{O}C$ under 93% relative humidity (RH) and adsorption isotherm at $25^{O}C$ under various RH were followed for 2 weeks.

A monomolecular layer of adsorbed water of polyfructan was computed by the Brunauer-Emmett-Teller equations[8,9].

Viscosity characteristics: Viscosity characteristics of polyfructan and other polysaccharide solutions were examined by using the Rheo-Tech Elastic Rheorometer with double concentric cylinder measuring system. All the measurements were carried out at $25^{O}C$.

Stability: Effects of pH on the stability of 5% polyfructan solutions were examined in the range of pH 4 to 6 at $80^{O}C$. Thermal stability of 5% polyfructan solution was also examined at pH 7 in the range of 90 to $120^{O}C$. Polyfructan samples were sealed in test tubes _in vacuo_ in order to prevent oxidation. The concentration of polyfructan was monitored by HPLC using Asahipak GS-710 with a refractive index detector.

Applications: As an example for using the polyfructan as a water soluble bulking agent for sweetener and fat substitute, an ice cream formulation was tested, where half of the sugar was replaced with Aspartame and polyfructan and half of the butter with 30% polyfructan paste. Furthermore, replacements of half of the cream cheese or fresh cream in baked cheese cake formulation were examined. Baking conditions were at $180^{O}C$ for 50 minutes and then at $150^{O}C$ for 20 minutes. Appearances, taste and texture of test sample were compared with those of the control sample prepared in a conventional formulation. Usually 6-8 panelists participated in the sensory analysis.

RESULTS AND DISCUSSION

Preparation of polyfructan: The conidia production was observed after 3-5 days of mycelial growth. Usually at least 8-10 days cultivation was needed for a maximum yield of conidia. In a typical cultivation, 10-14 g of conidia were produced in a 5 liter Erlenmeyer flask.

More than 80% of sucrose was consumed by the incubation for 3 days. Glucose and fructose were formed up to 40% and 10%, respectively. Polyfructan and fructo-oligosaccharides were synthesized in the reaction mixture concomitantly, as reported by Kawai et al.[3]. The yield of polyfructan based on the added sucrose was usually about 12-18%. In a typical experiment, roughly 100 g of the purified polyfructan was obtained from 1000 g of sucrose.

Structure: The 6 resonances observed with the polyfructan coincide exactly with those of inulin and are quite different from those of levan reported by Han et al[10]. Thus the polyfructan proved to be comprised of chains of beta-2,1-linked D-fructofuranoside residues as inulin.

The average molecular weight of the polyfructan was determined to be more than 10 million, which is far higher than that (about 5000) of inulin.

The intrinsic viscosity of the polyfructan was determined to be 0.16 dl/g in 7 mM NaCl solution, which is close to that of levan from Streptococcus salivarius[11]. But it is much lower than those of other polysaccharides with molecular weights of 10 million (1.59 for amylose, 2.09 for pullulan and 24.0 for xanthan)[12]. Microbial levan is known to be a beta-2,6-linked fructose polymer with some branching through beta-2,1-linkage. This branching in levan makes its hydrodynamic volume very small and is responsible for its small intrinsic viscosity in spite of its high molecular weight. Polyfructan has the same small intrinsic viscosity in spite of its high molecular weight and linear structure, so it seems to exist as a compact globular shape rather than a random coil polymer.

Powder characteristics: Polyfructan is white powder without any taste and flavor. Moisture absorption of polyfructan and other 3 sugars under RH 93% at 25°C during 2 weeks was observed and results are shown in Fig. 1. The final moisture content of polyfructan powder was slightly over 20%. Its free flowing characteristic was maintained up to 150 hours, after that slight lumping was observed. On the contrary, fructo-oligo-saccharides, maltitol and polydextrose were converted to lumping state within 2 days, and to syrupy state within one week at 25°C. Fructo-oligosaccharides, which have the same beta-2, 1-linked fructofuranoside structure but much smaller molecular weight, are very hygroscopic, sharply different from the polyfructan.

Results of adsorption isotherms of saccharides at various RH for 2 weeks showed that moisture content of all saccharides tested increased with increase in RH of circumstances. That of polyfructan varied from 5% at RH 10.5% to about 25% at RH 93.0%. On the other hand, those of maltitol and polydextrose showed negligible values at RH 10.5%, but reached 18 and 38% at RH 93%, respectively. Powder characteristics of these saccharides during above experiments were observed. Maltitol began to lump at RH 44.0% (moisture content almost 0%) and formed cake at RH 57.5% (moisture content 1%). Likewise, polydextrose formed rigid cake at RH 33.0% (moisture content 5%). Contrary to these, polyfructan did not lump at RH up to 85% (moisture content 25%). Polyfructan is, therefore, unique in keeping its free flowing character at relatively high moisture content.

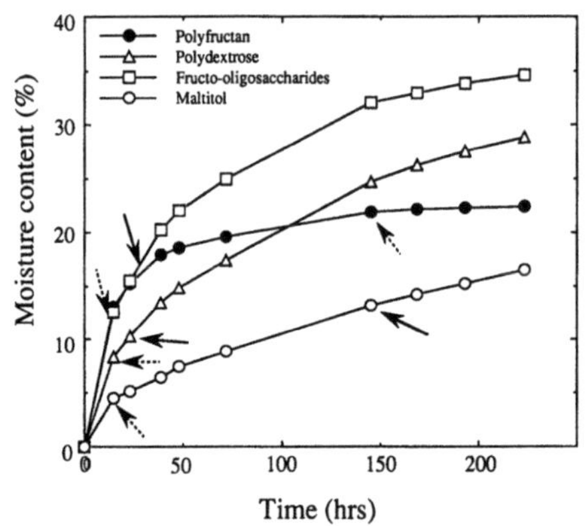

Fig.1. Moisture absorption under 93% RH as a function of storage time. Arrows indicate the time when the samples become lumping state(dotted) and syrupy state (solid).

A monomolecular layer of adsorbed water on polyfructan was calculated from above data to be 6.3%, whereas those of maltitol and polydextrose were calculated to be 0.3 and 0%, respectively. Probably this high value of polyfructan is responsible for its free flowing characteristic at high moisture content. The fact that polyfructan has such a free flowing characteristic suggests that it could be suitable as a drying aid for many dehydrated foods.

Viscosity characteristics: Polyfructan solution exhibited almost no viscosity at 10%. On the other hand, 5% levan solution had a viscosity of 190 cP. It is very interesting that polyfructan and levan had almost the same intrinsic viscosity as mentioned above, but had quite different apparent viscosities. In the case of polyfructan solution, the shear stress is directly proportional to the shear rate as observed with typical Newtonian liquids, as shown in Fig.2, and the viscosity remained constant at shear rate range of about 30 to 120^{-1}. On the other hand, solutions of glucomannan and microbial levan behaved as pseudoplastic liquids and their viscosity changed greatly depending on the shear rate.

Polyfructan solution might be suitable for ingredients of watery beverages such as tea and coffee where typical Newtonian fluid nature is required. The viscosity of polyfructan solution remains virtually at a constant level when 10% salt and/or sucrose were added. This property of polyfructan seems to be suitable for providing processed foods with body texture. Thus, using it as a bulking agent, many foods having a rich texture or volume as obtained with sucrose can be easily prepared.

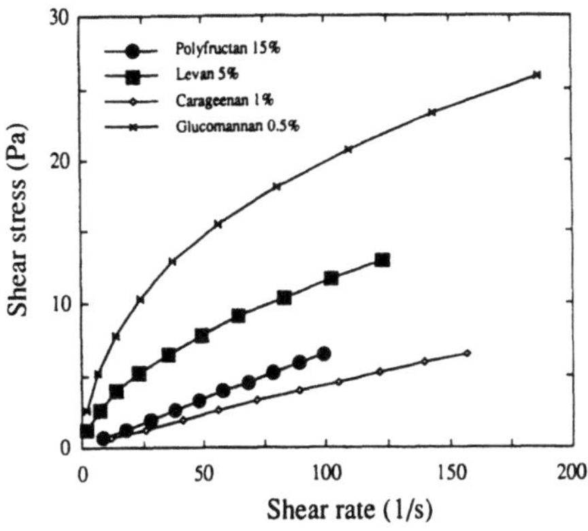

Fig.2. Shear stress as a function of shear rate.

Stability: In the pH range 5 - 7, the polyfructan solution is quite stable up to 2 hours at 80°C, but about 80 % of the polyfructan was degraded within one hour at pH 4 and formation of fructo-oligosaccharides and fructose was observed.

Stability of 5% polyfructan solution (pH 7) at various temperature is shown in Fig. 3. It was quite stable at 100°C for 120 minutes. Less than 20% of the polyfructan was degraded even at 120°C for one hour. Processed foods are usually pasteurized at 120°C up to 10 min, so the degradation of polyfructan in these foods might be minimal as long as their pH are kept at neutral.

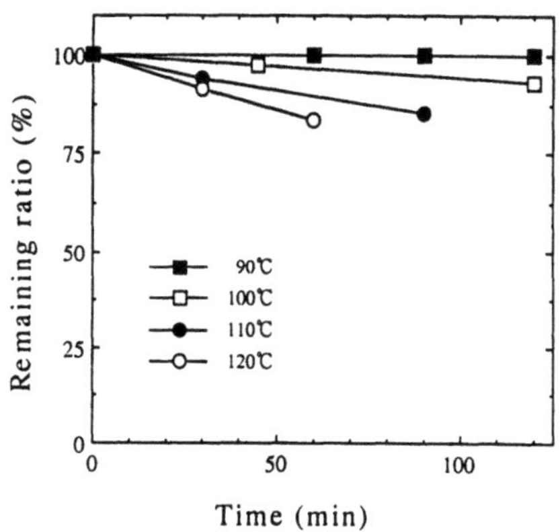

Fig.3. Thermal stability of 5% polyfructan solution.

Nutritional aspect: Polyfructan is not hydrolyzed by either of human salivary alpha amylase, gastric acid or pig pancreas alpha amylase, and negligibly hydrolyzed by intestinal acetone powders from rats[13]. These findings suggest that polyfructan remains intact in the intestine until it encounters microbial attacks in the terminal ileum and large intestine as reprted with inulin[14].

Among end products of such microbial metabolism in these
segments, short chain fatty acids are the only possible con-
tributors for host energy metabolism. Thus, production rates of
short chain fatty acids were determined by a mini scale batch
culture using pig cecal microbes[15]. The results indicated that
the caloric contribution of polyfructan is approximately 70%
(less than 1.5 Cal/g) of that for maltitol.

Applications: Ice cream prepared using polyfructan as a
fat and sugar substitute exhibited smooth mouthfeel, good
meltability on the tongue and good gel properties without any
adverse effects. Replacement of cream cheese or fresh cream
with polyfructan resulted in baked cheese cakes that had good
agreeability to the palate, meltability on the tongue, smooth
mouthfeel and smooth swallowing without any adverse effect.

Thus, using polyfructan we can overcome the defects which
are associated with conventional oil and fat substitutes, low
calorie sugars, gelation materials and thicking agents.

ACKNOWLEDGEMENTS

The authors wish to express their sincere gratitude to
professor Shuichi Kimura, Miss Seiko Hoshi of Tohoku University
and associate professor Takashi Sakata of Ishinomaki Senshu
University for their digestibility studies of polyfructan.

REFERENCES

1. T.Harada, S.Suzuki, H.Taniguchi and T.Sasaki, Characteristics and
     applications of a polyfructan synthesized from sucrose by Aspergil-
     lus sydowi conidia, Food Hydrocolloids (1992)(in press).
2. J.R.Lowenberg and R.T.Reese, Observations on microbial fructosans and
     fructosanases, Can.J.Microbiol. 3:643(1957).
3. G.Kawai, H.Taniguchi and M.Nakamura, Polyfructan and oligofructans
     synthesized from sucrose by conidia of Aspergillus sydowi IAM 2544,
     Agr.Biol.Chem. 37:2111(1973).
4. A.C.Ouano, Gel-permeation chromatography, J.Polym.Sci. 10:2169(1972).
5. D.Lecacheux, J.Lesec and C.Quivoron, High-temperature coupling of high-
     speed GPC with continuous viscosity, Appl.Polym.Sci. 27:4867(1982).
6. M.A.Haney, The differential viscometer.II.On-line viscosity detector
     for size-exclusion chromatography, J.Appl.Polym.Sci. 30:3037(1985).
7. B.Makower and W.B.Dye, Equilibrium moisture content and crystallization
     of amorphous sucrose and glucose, J.Agr.Food Chem. 4:72(1956).
8. S.Brunauer, P.H.Emmett and E.Teller, Adsorption of gases in multi-
     molecular layers, J.Am.Chem.Soc. 60:309(1938).
9. H.Salwin, Defining minimum moisture contents for dehydrated foods,
     Food Technol. 13:594(1959).
10. Y.M.Han and M.A.Clarke, Production and characterization of microbial
     levan, J.Agric Food Chem 38:393(1990).
11. E.Newbrun and S.Baker, Physico-chemical characteristics of the levan
     produced by Streptococcus salivarius,Carbohyd.Res. 6:165(1968).
12. B.Launay, J.L.Doublier and G.Cuvelier, Flow properties of aqueous
     solutions and dispersions of polysaccharides, in:"Functional Pro-
     perties of Food Macromolecules," J.R.Mitchell and D.A.Ledward,ed.,
     Elsevier Applied Science Publishers, London and New York (1986).
13. S.Hoshi and S.Kimura, personal communication.
14. K.Yazawa and Z.Tamura, Search for sugar sources for selective increase
     of Bifidobacteria,Bifidobacteria Microflora 1:39(1982).
15. T.Sakata, personal communication.

ALGINATES AND GELLAN GUM: COMPLEMENTARY GELLING AGENTS

George R. Sanderson and David Ortega

Kelco Division of Merck and Co., Inc., San Diego, California, USA

ABSTRACT

Alginates and gellan gum both have a high affinity for calcium ions. Their ability to form gels with these ions is not only well known but is also exploited in many applications. Alginate gels are usually formed in the cold, without recourse to heating, while gellan gum gels are normally prepared by heating and cooling. These two polymers can thus be considered to be complementary rather than competitive gelling agents.

The similarities and differences between the monovalent alginate salts, notably sodium alginate, and KELCOGEL gellan gum are discussed in terms of their properties and applications. This information is useful to the end user when trying to decide which of the two hydrocolloids should be used in a particular application. It has also enabled alginate and KELCOGEL to be used effectively together in some products. Specific examples are described.

INTRODUCTION

Alginates, derived from a variety of brown seaweeds, have been known for over a century. They are linear polymers composed of the salts of $\underline{D}$-mannuronic and $\underline{L}$-guluronic acid. The properties of a particular alginate depend on the relative proportions of these two monomers in the molecule and, more specifically, the size and distribution of the so-called block regions. These block regions are segments of the polymer chain composed solely of either mannuronic or guluronic acid.

Hydrocolloids are frequently classified as thickeners or gelling agents. Alginate is somewhat of an exception in that it functions both as a thickener and a gelling agent. Although all polyvalent cations with the exception of magnesium are capable of forming gels with alginate[1], the only ion of relevance for foods is calcium. It has been said that the reactivity displayed by alginate towards calcium is its strength and weakness[2]. Proper control of calcium ions can enable a wide variety of products to be formulated[3]; improper control invariably leads to singular lack of success. Fig. 1 shows the effects obtained by mixing different concentrations of calcium ions with different concentrations of an alginate composed of a high ratio of mannuronic to guluronic acid. The term "conversion" refers to the amount of calcium ion added relative to the amount required to theoretically convert sodium alginate to its calcium form. Thus, a calcium conversion of 1.0, for example, means sufficient calcium to convert sodium alginate on a stoichiometric basis to its calcium form. In practical terms, this amount is approximately 0.1 g of calcium ion per 1.0 g of sodium alginate. As implied in the figure, increasing the calcium ion concentration progressively

Food Hydrocolloids: Structures, Properties, and Functions
Edited by K. Nishinari and E. Doi, Plenum Press, New York, 1994

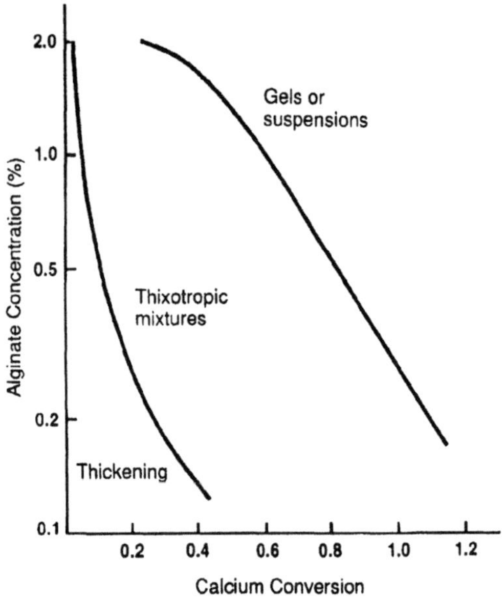

Fig. 1   Influence of calcium ion level on alginate rheology

increases intermolecular association resulting in a progressive increase in viscosity, subsequently the formation of thixotropic solutions, and ultimately gel formation, in which interchain association is permanent.

In the preparation of alginate gels, calcium ion control is essential in order to avoid premature gelation and the formation of undesirable, broken gels. There are three general methods of making gels, namely diffusion setting, internal setting and setting by cooling. The first procedure is the simplest and, as the name implies, gels are formed by allowing calcium ions to diffuse into a solution of alginate. Diffusion setting is used in the production of fabricated onion rings and structured pimiento strip for stuffed olives, and to encapsulate various core materials in a gelled alginate skin. A simplified illustration of the formation of blackcurrants consisting of blackcurrant puree encapsulated by an outer skin of alginate gel is shown is Fig. 2 [4]. The rate determining step in the diffusion process is the time taken for the calcium ions to diffuse through the alginate solution. The effectiveness of

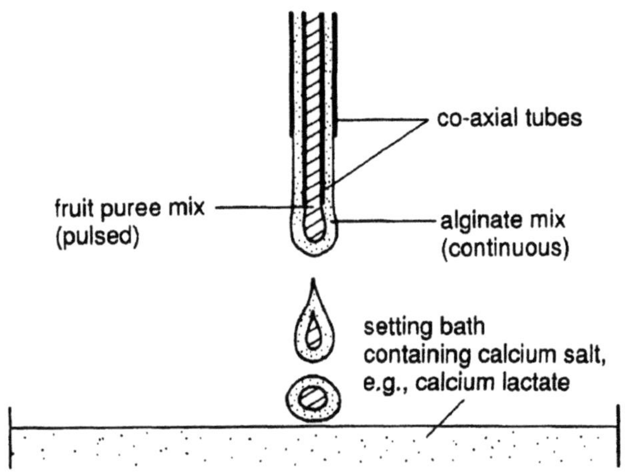

Fig. 2   Co-extrusion system for preparing blackcurrants

the technique is therefore limited to the setting of thin films, through which diffusion time is short. In situations in which diffusion is not practical, internal setting can be used. This is done by controlled release of calcium ions from particles of a calcium salt dispersed throughout the system. Rate of calcium release depends on the inherent solubility of the calcium salt used, the amount present, particle size and pH. Salts commonly used include dicalcium phosphate anhydrous and calcium sulfate dihydrate. Calcium release can be further controlled by inclusion of calcium sequestrants and the use of slowly dissolving acids and short mixing times. Internal or bulk setting can be successfully used to prepare a wide variety of food products. These include products which are prepared by reconstitution of a dry mix using water or milk. A prerequisite to success is designing the dry mix to ensure that the alginate has time to hydrate before sufficient calcium to cause gelation is released.

Table 1. Alginate dessert gel formulation.

Ingredients	%
Water	84.75
Sugar	13.92
Sodium alginate (KELTONE HV)	0.49
Adipic acid	0.41
Sodium tripolyphosphate	0.31
Calcium carbonate	0.08
Color	0.01
Flavor	0.03
	100.00

A third but less common method of forming alginate gels is setting by cooling. Table 1 shows a simple formulation for an alginate dessert gel, prepared by adding a mixture of the dry ingredients to boiling water and cooling. Although useful, the resulting gel is weak, and a limitation to preparing alginate gels by cooling is the fact that it is not possible to achieve the high degree of conversion from sodium to calcium alginate needed to give strong gels since the required levels of calcium cause the alginate to gel, even at elevated temperatures. These gels, however, tend to be thermodynamically more stable than internal or diffusion set gels since the calcium required for gelation is present throughout the system in soluble form and is not fed progressively to the alginate from an external source as in the case of diffusion setting or by diffusion from the surface of a dissolving particle of calcium salt as in internal setting.

Unlike alginates, gellan gum has only been known for over a decade. Despite this short existence, it has already been extensively researched and is approved for foods in Japan. Approval in non-standardized jams and jellies and icings and frostings has also been granted in the U.S. and approval for general food use is anticipated in the near future.

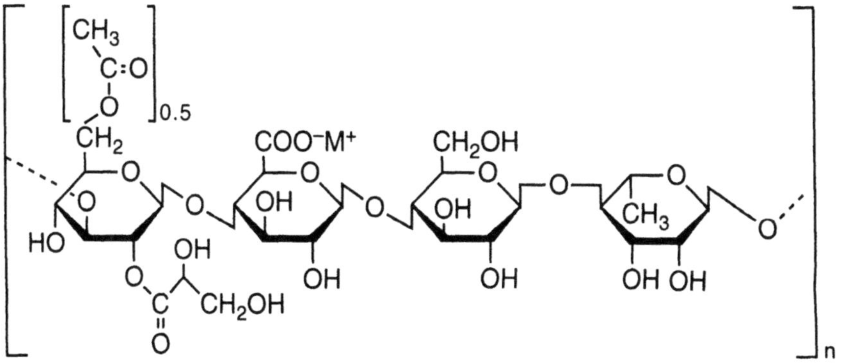

Fig. 3   Native gellan gum primary structure

Gellan gum, a microbial polysaccharide derived from the organism *Pseudomonas elodea*, is a linear polymer with a tetrasaccharide repeat unit consisting of glucose, glucuronic acid, glucose and rhamnose[5,6]. As shown in Fig. 3, there is also one glycerate substituent per repeat and one acetate approximately every second repeat. The form of the gum in which the acyl substituents have been removed is marketed by Kelco as KELCOGEL gellan gum.

KELCOGEL reacts with mono- and polyvalent cations to form a gel. These gels are normally prepared by adding the appropriate ion to a hot solution of the gum and cooling. Divalent ions are much more efficient than monovalent ions and, by way of example, gels of optimal gel strength require around 0.4% sodium but only in the region of 0.016% calcium. Since low levels of dissolved calcium promote gel formation and hence inter-molecular association, they also prevent chain dissociation or hydration. This potential problem is easily overcome by the inclusion of a calcium sequestrant such as sodium citrate,[7,8] and, by manipulation of the relative concentrations of dissolved calcium and sequestrant, it is possible to hydrate KELCOGEL at any desired temperature. In other words, if desired, KELCOGEL can be hydrated at room temperature. These solutions, like their alginate counterparts, can be converted to a gel by diffusion setting. However, in the case of the KELCOGEL solutions, the gelling ion can be not only calcium but other ions such as sodium, potassium or magnesium.

Preparation of gels by internal setting, although not impossible[9, 10], is more difficult to achieve with KELCOGEL than alginates. This is because inclusion of a sequestrant to control the release of the divalent ions in the system, so effective in preparing internal set alginate gels, frequently in itself prevents hydration since the amount of sequestrant required gives rise to too much sodium.

In summarizing a comparison and contrast of alginates and KELCOGEL in the context of food systems, the following key issues emerge. They both form gels with calcium ions. Unlike alginates, KELCOGEL also forms gels with other ions, notably sodium, potassium and magnesium. Excess of these ions can cause precipitation which can also be induced by excess acid, i.e., hydrogen ions. (It has been suggested that precipitation with calcium and/or acid, as practiced in alginate production, could be used in the manufacture of gellan gum.) Setting by cooling, diffusion setting and internal setting can all be used to produce gels using either alginates or KELCOGEL. However, setting by cooling is usually the method of choice with KELCOGEL while setting in the cold is frequently preferred with alginates.

In view of the similarities between the two polymers, it seemed desirable to investigate whether or not the two in combination could be used to advantage. In this respect, an obvious area to study was mixed gels formed by cooling hot solutions.

MATERIALS AND METHODS

Gel Preparation

Gellan gum (KELCOGEL, Lot No. 86-0082) and sodium alginate (KELTONE HV, Lot No. 85120A), alone and in combination, were dispersed in cold deionized (Arrowhead) water and the dispersions/solutions heated to 90°C under agitation in a Helmco-Lacy Hot Cup. Calcium or magnesium ions were added using the appropriate quantities of 0.5M stock solution. The hot polysaccharide solutions were then poured into ring molds (13mm height, 28mm inside diam.), covered with plastic cover slips and allowed to stand overnight at room temperature to cool.

Specific gum concentrations evaluated were in % by weight: 0.2% KELCOGEL, 0.6% KELTONE HV, and 0.2% KELCOGEL + 0.6% KELTONE HV.

The specific calcium ion concentrations used were: 0, 2, 4, 6, 8 and 10mM; these same concentrations were also used for the magnesium ion.

Gel Texture Measurement

After overnight storage at room temperature, the gel discs, in the cases where gels had formed, were carefully removed from the ring molds and evaluated using an Instron Universal Testing Machine, Model No. 4201, as previously described[11].

RESULTS AND DISCUSSION

Using calcium ions, gels were formed with KELCOGEL. The textural parameters for these gels are shown in Figures 4 – 6. The results are totally as expected[11, 12]. Hardness (Fig. 4) rapidly increases to a maximum and then gradually declines with increasing, calcium ion concentration. Modulus (Fig. 5) shows a more symmetrical increase and decrease while gel brittleness (Fig. 6) increases (lower % brittleness = more brittle) fairly

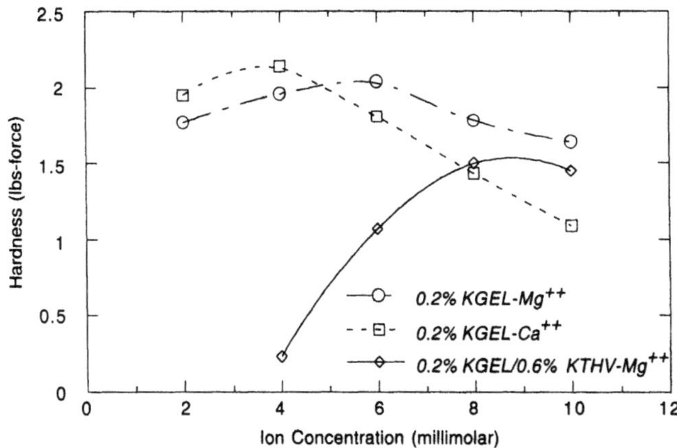

Fig. 4   Influence of ion concentration (Ca^{++} or Mg^{++}) on gel hardness

rapidly up to a calcium ion concentration of around 6mM and then starts to level off at higher calcium. With solutions of 0.6% KELTONE HV, attempts to prepare gels with between 2 and 10mM calcium were unsuccessful since as little as 2mM calcium resulted in immediate precipitation of the alginate from the hot solution. Additional tests showed that, even by reducing the alginate concentration to 0.2% and the added calcium ion concentration to 0.5mM, gelation could not be prevented. These results demonstrate that, unlike KELCOGEL, unbuffered sodium alginate has minimal to zero tolerance for calcium ions at elevated temperatures. In other words, gelation cannot be prevented by use of

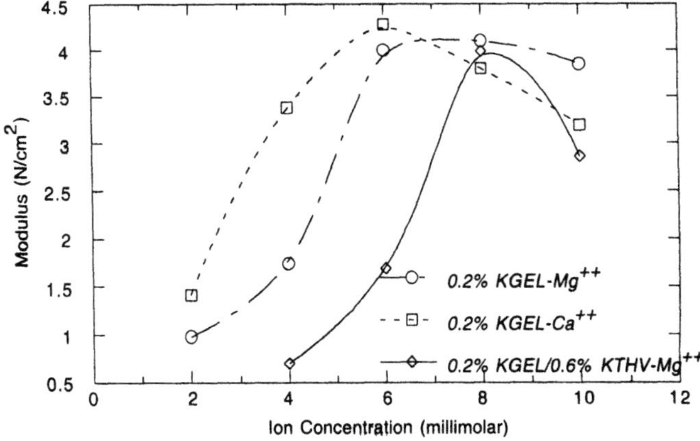

Fig. 5   Influence of ion concentration (Ca^{++} or Mg^{++}) on gel modulus

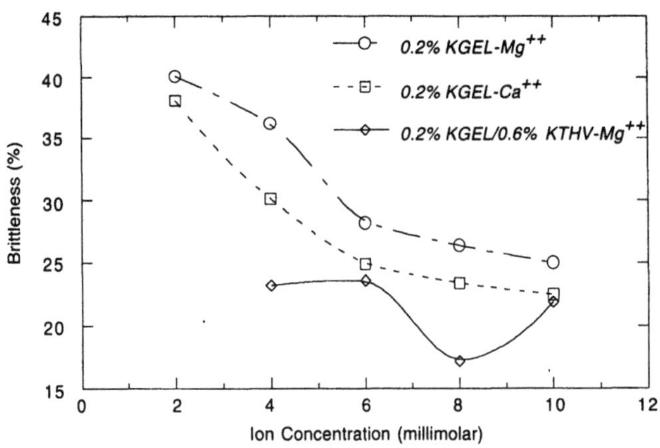

Fig. 6   Influence of ion concentration (Ca^{++} or Mg^{++}) on gel brittleness

elevated temperatures. Precipitation was also observed when 2 – 10mM calcium was added to hot solutions of 0.6% KELTONE HV/0.2% KELCOGEL. At the higher levels of added calcium, precipitation was accompanied by gelation on cooling. These results strongly suggested that the alginate was again precipitating from solution while, at higher levels of added calcium, sufficient remained unbound by the alginate and available to cause the KELCOGEL to gel on cooling. Thus, even in the presence of KELCOGEL, alginate still displays minimal tolerance to added calcium at elevated temperature.

Interpreted superficially, these findings would argue against the practicality of preparing useful alginate and mixed alginate/gellan gels by cooling in the presence of calcium ions. However, in the case of alginates, commercial products prepared in this manner, such as the example in Table 1, are well known. Useful combinations of alginate and gellan gum that produce dessert gels on cooling have also been recently developed[13]. A basic difference between these practical systems and the model gels described in this paper is that the former are buffered with a sequestrant such as citrate or phosphate which controls the calcium availability to the gelling polysaccharides. Another difference is that, in the commercial gels, the calcium is frequently introduced through controlled dissolution of calcium carbonate dispersed throughout the system while, in the model gels, the required calcium is added in concentrated, predissolved form from a stock solution. Although, as indicated, dessert gels containing both alginate and KELCOGEL have been formulated, the nature of the resulting gels and how the calcium ions partition between the alginate, KELCOGEL and sequestrant have still to be determined. This work is in progress.

With magnesium as the gelling ion, results are quite different. As anticipated, no gels were formed with these ions and alginate. However, gels were obtained with both 0.2% KELCOGEL and 0.2% KELCOGEL/0.6% KELTONE HV. Figures 4 – 6 show that the gels produced with magnesium and KELCOGEL are similar, although not identical, in texture to those obtained with calcium, confirming that magnesium and calcium ions interact similarly with KELCOGEL[14]. When magnesium ions were added to the gellan/alginate combination, the gel strength peaks, for both hardness and modulus (Fig. 4 and 5), shifted to higher ion concentrations relative to the respective peaks for magnesium and KELCOGEL. Similar behavior is observed for gels made with KELCOGEL in combination with a sequestrant such as sodium citrate. It therefore appears that the magnesium ions, although not able to promote interchain association and gelation of the alginate, are nevertheless binding with the alginate to form a soluble complex. In other words, the alginate is functioning as a sequestrant for magnesium ions. Inspection of the viscosities of alginate solutions containing 0 – 10mM magnesium (Table 2) shows that viscosity decreases with increasing magnesium ion concentration. This suggests increased binding as the ion concentration is increased, with an accompanying increase in charge screening and reduction in viscosity. The viscosity measurements also show that the magnesium/alginate interaction is not time dependent, in keeping with magnesium's inability to promote interchain association.

Table 2. Effect of magnesium ion concentration on the viscosity of a 0.6% KELTONE HV solution.

| | Brookfield LVT Viscosities (6 rpm, spindle no. 1) | | |
Added Mg^{++} (mM)	0 hrs.	2 hrs.	24 hrs.
0	101	98	105
2	75	71	74
4	65	58	65
6	54	51	56
8	51	50	53
10	53	49	53

These studies highlight the difficulty associated with forming alginate gels by cooling. High temperature does not prevent the strong association between alginate and calcium ions and, unless a buffer is included to reduce the free calcium to a low level, precipitation of the alginate results. It has also been shown that, when magnesium is used instead of calcium, gels can be produced by cooling solutions containing KELCOGEL in combination with alginate. In these gels, the alginate functions as a magnesium ion sequestrant and texture modifier.

In summary, although KELCOGEL rather than alginate is the more logical choice for gels formed by cooling, it is possible, from an understanding of the principles of gelation, to make useful gels by cooling solutions containing both KELCOGEL and alginate. For completeness, it should also be mentioned that diffusion setting in the cold, normally the province of alginate, can be used to form films from solutions of KELCOGEL or KELCOGEL and alginate[15]. A particular advantage of KELCOGEL is that these films can be formed using sodium ions, from a source such as common salt, to bring about setting.

REFERENCES

1. R.H. McDowell. "Properties of Alginates," Kelco International, London (1986).
2. W. Bryden, personal communication.
3. W. Bryden and G.R. Sanderson. "Structured Foods with the Algin/Calcium Reaction," Kelco Division of Merck and Co., Inc., San Diego (1982).
4. M.E. Sneath, Artificial fruit berries, *British Patent* 1,484,562 (1977).
5. M.A. O'Neill, R.R. Selvendran, and V.J. Morris, Structure of the acidic extracellular gelling polysaccharide produced by *Pseudomonas elodea, Carbohydrate Res.* 124:123 (1983).
6. P.E. Jansson, B.Lindberg, and P.A.Sandford, Structural studies of gellan gum, an extracellular polysaccharide elaborated by *Pseudomonas elodea, Carbohydrate Res.* 124 :135 (1983) .
7. V.L. Bell, D. Ortega, and G.R. Sanderson. "The Preparation of KELCOGEL Gellan Gum Gels," Kelco Division of Merck and Co., Inc., San Diego (1989).
8. V.L. Bell, D. Ortega, and G.R. Sanderson, A comparison of gellan gum, agar, κ-carrageenan, and algin, *Cereal Foods World* 34 :991 (1989).
9. J.K. Baird and J.L. Shim, Non-heated gellan gum gels, *U.S. Patent* 4,503,084 (1985).
10. J.K. Baird and J.L. Shim, Non-heated gellan gum gels, *U.S. Patent* 4,563,366 (1986).
11. G.R.Sanderson, V.L. Bell, R.C. Clark, and D. Ortega, The texture of gellan gum gels, *in*: "Gums and Stabilisers for the Food Industry 4," G.O. Phillips et al., eds., IRL Press, Oxford (1988).
12. G.R. Sanderson, Gellan gum, *in*: "Food Gels," P.Harris, ed., Elsevier Applied Science, London (1990).
13. Anon., Use of gellan/algin gum combinations in water dessert gels, *Research Disclosure* 333:69 (1992).
14. G.R. Sanderson and R.C. Clark, Gellan gum, *Food Technology* 37:63 (1983).
15. W.F. Chalupa and G.R. Sanderson, patent pending (1992).

KELCOGEL and KELTONE are trademarks of Merck and Co., Inc. (Rahway, New Jersey), Kelco Division, U.S.A., and are registered in the U.S. Patent and Trademark Office.

GELLAN GUM: PRODUCTION AND PROPERTIES

Claire T. Vendrusculo, José L. Pereira and Adilma R. P. Scamparini

Food Science Department. Food Engineering Faculty, UNICAMP
13081 Campinas, SP., Brazil

ABSTRACT

Gellan gum, the extracelullar polysaccharide obtained by the aerobic fermentation of the microorganism, Pseudomonas elodea, which forms higly viscous aqueous solutions and under appropriate conditions forms thermoreversible gels. Gellan gum has potential industrial applications as a high viscosity biogum, a suspending agent and a gelling agent. This report is part of survey undertake to assess the commercial potential of extracellular polysaccharides for made by bacteria isolated from soil, aquatic plants and lakes, in So Paulo State, Brazil. In the screening programme over 48 places were investigated. The organism isolated were rods, gram negative, aerobic with motility and submitted for the examination on the basis of yield and physical properties of the extracellular polysaccharide produced.

The cultures isolated were incubated for 3 days at 32°C in media containing sucrose as carbon source at pH 6.8-7.3 in a shaker incubator during a period of 96 hours. The recovery of extracellular polysaccharide was obtained by precipitation of the fermentation liquor at 25°C with 3 volumes of 99.5% ethanol. The fibrous precipitate was recovered with sieves and dried in forced-air tray dryer at 55°C until constant weight.

INTRODUCTION

Several strains of Pseudomonas sp produce extracellular polysaccharides under certain culture conditions, resulting in extremely viscous broth during submerged cultivations. Fermentation studies using batch culture indicate that exopolysaccharides production by Pseudomonas sp in a chemically defined medium increase under high C/N ratios at the pH range 6.7-8 and at 25°C[1]. Polysaccharides are formed from a variety of carbon sources and their composition are not affected by the nature of the carbohydrate source[2] and their production depends on the phosphate content of the medium[3].

Food Hydrocolloids: Structures, Properties, and Functions
Edited by K. Nishinari and E. Doi, Plenum Press, New York, 1994

Some species of Pseudomonas produce copious amounts of an exocellular heteropolysaccharide (PS-60) which is composed of approximately 46% glucose and 30% rhamnose, in addition, contains 21% uronic acid and 3% O-acetyl. Upon deacetylation by a mild alkaline treatment, PS-60 produces a brittle, firm, and optically thermoreversible clear gel[4]. Gellan gum can be also used as an agar substitute on bacterial growth[5].

The work reported were is part of an investigation about isolation of Pseudomonas which produces extracellular polysaccharides. The strain studied was isolated from the soil in the agricultural region of Ribeiro Preto, So Paulo State, Brazil. In this study we report on the exopolysaccaride production and physical characteristics of the polymer.

MATERIALS AND METHODS

Bacterial strain. The strain RPB430 was isolated from soil surface of Ribeiro Preto, So Paulo State. It was selected for ability to produce mucoid growth on solid media. The organisms was tentatively identified as Pseudomonas sp. The bacterial strain was maintained on YM agar medium and some were liofilized.

Exopolysaccharide production. Two media were used in this experiment and their composition were as follows:

1- YM tryptose medium: Yeast extract 0.02%, sucrose 2%, tryptose 0.2% and KH_2PO_4 0.2%.
2- Fermentation broth: Yeast extract 0.02%, sucrose 2%, tryptose 0.2%, KH_2PO_4 0.2%, $MgSO_4$ 0.3%, $FeSO_4$ 0.001%, $MnSO_4$ 0.001% and $CoCl_2$ 0.001%.

Two procedures for fermentation were used in the first process, cells growth in nutrient broth at $30^{\circ}C$ for 24 hours, were innoculated in YM tryptose medium without salt and maintained for 24 hours at $30^{\circ}C$ in a shaker incubator at 150 rpm. The cells were transfered for YM containing salts and incubated for 24 hours in shaker incubator at $30^{\circ}C$ for a period of 24 hours.

The fermentation broth were innoculated with this suspension and incubated for a period of 48 hours at $30^{\circ}C$ and 240 rpm.

In the second procedure, the two initial steps were supressed. We got two kinds of polysaccharides, A produced in the first process and B in the second.

Exopolysaccharide isolation and purification. The cells in the broth fermentation were killed by thermal treatment (15 min, $75^{\circ}C$) and, centrifugated at 23000 x g for 45 min for removing the cells. The supernatant were treated with 3 vol. of ethanol 99.5% to precipitate the polysaccharide. The precipitate was recovery by filtration and dissolved in distilled water. The solution was subjected at an extensive dialysis against distilled water at $25^{\circ}C$ and the polysaccharide was reprecipitated by the same way as before. The material was dried in forced air tray drier at $55^{\circ}C$ until constant weight. The flow behavior of exopolysaccharides A and B solutions were measured with a Haake rheometer CV20C/ZA15-Haake rot 2.3 at 25, 45 and $65^{\circ}C$ and at pH 7.0 and 10.0.

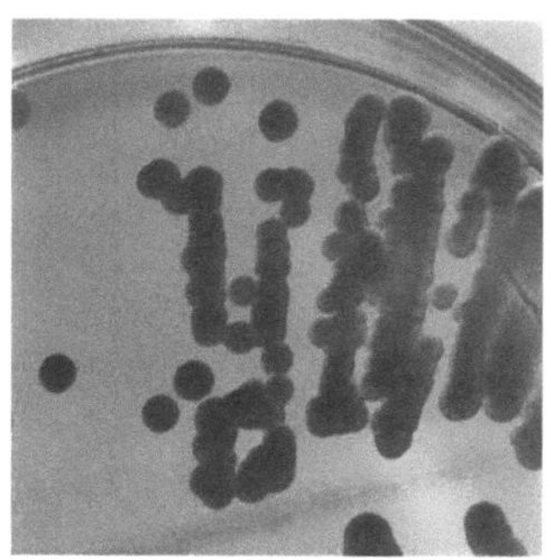

Fig. 1. Colony morphology of strain RPB430, after 24 h of growth on YM agar at 30°C

RESULTS AND DISCUSSION

The strains isolated was alocated in the family <u>Pseudomonadaceae</u>, according its morphological and physiological characteritics, it was Gram negative strictly aerobic rods, did not produce gas or acid from carbohydrates, grew in poor media, showed motility and did not grow over 30°C[6]. The typical colony morphology of strains RPB430 is shown in Fig. 1.

Exopolysaccharide production was dependent of the fermentation process[7]. The second process showed more yield in polysaccharide production in comparison with the first process probably because of its salt composition.

Apparent viscosities of 1% exopolysaccharides A and B aqueous solutions decreased reversible when the temperature rose at pH 7.0 and 10.0 (Table 1). These results suggest that the molecules of both polysaccharides may adopt a randon coil conformation at high temperatures[8].

The rheograms of 1% exopolysaccharides aqueous solutions at 25, 45 and 65°C are showed in Fig. 2 (pH 7.0) and Fig. 3 (pH 10.0). Aqueous solutions of polysaccharides A and B showed Newtonian behavior when heated at 65°C, at pH 7.0 and 10.0.

Table 1 - Apparent viscosities (cp) of exopolysaccharides solutions at differents pH values and temperatures.

Exopolysaccharide	Apparent viscosity (cp)					
	Temperature (oC)					
	25		45		65	
	pH					
	7.0	10.0	7.0	10.0	7.0	10.0
A	172.8	102.1	103.9	97.4	81.7	24.7
B	254.4	144.5	108.7	68.1	30.6	33.2

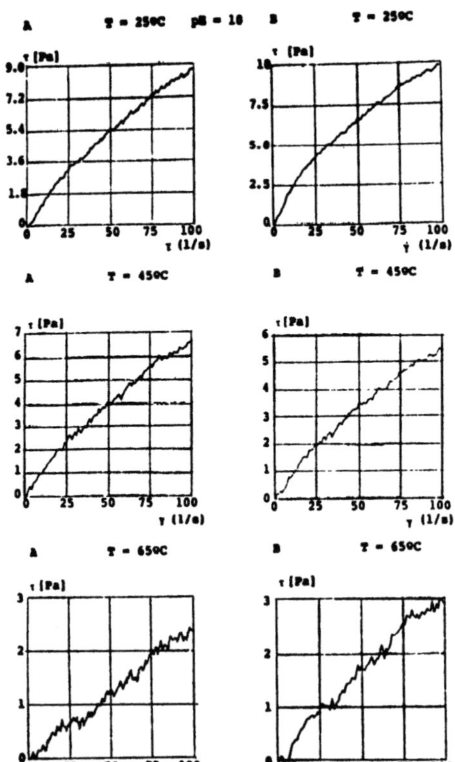

Fig.2. Rheograms A and B exopolysaccharides solutions (1% in deionized water) at 25, 45 and 65°C, at pH 7.0.

Fig.3. Rheograms of A and B exopolysaccharides solutions (1% in deionized water) at 25, 45 and 65°C, at pH 10.0.

REFERENCES

1. F. Congregado, I. Estañol, M.J. Espuny, M.C. Fusté, M.A. Manresa, A.M. Marqués, J. Guinea and M.D. Simon-Pujol, Preliminary studies on the production and composition of the extracellular polysaccharides synthesized by *Pseudomonas* sp EPS5028, Biotechnology Letters 7: 883 (1985).

2. A.G. Williams and J.W.P. Wimpenny, Exopolysaccharide production by *Pseudomonas* NCIB11264 grown in batch culture, J. of General Microbiology 102: 13 (1977).

3. A.M. Marqués, I. Estañol, J.M. Alsina, C. Fusté, D. Simon-Pujol, J. Guinea and F. Congregado, Production and rheological properties of the extracelullar polysaccharide synthesized by *Pseudomonas* sp strain EPS-5028, Applied and Environmental Microbiology 52: 1221 (1986).

4. K.S. Kang, G.T. Veeder, P.J. Mirrasoul, T. Kaneko and I.W. Cottrell, Agar-like polysaccharide produced by a *Pseudomonas* species: Production and basic properties, Applied and Environmental Microbiology 43: 1086 (1982).

5. D. Shungu, M. Valiant, V. Tutlane, E. Weinberg, B. Weissberger, L. Koupal, H. Gadebusch and E. Stapley, Gelrite as an agar substitute in bacteriological media, Applied and Environmental Microbiology 46: 840 (1983).

6. N.J. Palleroni, Pseudomonadaceae, in: "Bergey's Manual of Systematic Bacteriology", J.G. Holt, ed., William and Wilkins, Baltimore (1986).

7. S. Goto, T. Murakawa and S. Kuwahara, Slime production by *Pseudomonas aeruginosa* II. A new synthetic medium and cultural conditions suitable for slime production by *Pseudomonas aeruginosa*, Japan, J. Microbiol. 17: 45 (1973).

8. M. Tako, A. Sakae and S. Nakamura, Rheological properties of gellan gum in aqueous media, Agric. Biol. Chem. 53: 771 (1989).

CONFORMATION AND PHYSICAL PROPERTIES OF
TWO UNUSUAL MICROBIAL POLYSACCHARIDES:
RHIZOBIUM TRIFOLII CPS AND LEVAN

Stefan Kasapis and Edwin R. Morris

Cranfield Institute of Technology
Silsoe College, Silsoe
Bedford MK45 4DT, UK

ABSTRACT

The neutral capsular polysaccharide (CPS) from *Rhizobium trifolii* converts from a low-viscosity solution to a stiff gel on cooling through a narrow temperature range. Coil overlap in solution occurs at ~2.0% w/v, about 60 times higher than the minimum critical gelling concentration ($c_0 \approx 0.035\%$ w/v). CPS gels are crosslinked by aggregation of intra-molecular helices, causing thermal hysteresis (of ~7.5°C), which decreases sharply below c_0. Bacterial levan has a densely-branched structure of high molecular weight but very low hydrodynamic volume, allowing solutions to be prepared to high concentrations (>25% w/v). The concentration-dependence of viscosity is unusually steep ($\propto c^{9.3}$ above 20% w/v). The exclusion properties of levan (demonstrated by its ability to promote ordering of CPS) may be of potential practical value in mixed systems with other hydrocolloids.

INTRODUCTION

The uncharged capsular polysaccharide (CPS) from many strains of Rhizobia (nitrogen fixing bacteria from soil) forms thermally-reversible gels at very low concentrations.[1] CPS has the repeating sequence →4)-α-D-Glc*p*-(1→3)-α-D-Man*p*-(1→3)-β-D-Gal*p*-(1→ with sidechains of α-D-Gal*p* and β-D-Gal*p*-(1→4)-β-D-Gal*p* attached at positions 2 and 6, respectively, of each glucose residue, to give an overall hexasaccharide repeat.[1,2]

Levan is a non-gelling polysaccharide produced by a wide variety of plant-pathogenic bacteria, and may have a role in creating a favourable environment for the pathogen within the intercellular spaces of the plant.[3] Bacterial levan is highly branched, consisting of β-D-fructofuranose residues linked predominantly (2→6), with (2→1) linkages at branch points.

We report here a brief investigation of the rheological properties of these two novel biopolymers, in relation to their primary structure and conformation, to assess the potential for commercial exploitation as industrial hydrocolloids.

Food Hydrocolloids: Structures, Properties, and Functions
Edited by K. Nishinari, Plenum Press, New York, 1994

"

MATERIALS AND METHODS

CPS from *Rhizobium trifolii* (strain TA-1) and levan from *Pseudomonas syringae* pv. *phaseolicola* were kindly donated by, respectively, Professor L.P.T.M. Zevenhuizen (University of Wageningen, The Netherlands) and Dr. M. Gross (University of Göttingen, Germany). Solutions of both materials were prepared in deionised water by mechanical stirring (typically for 20 minutes at 70 $^{\circ}$C).

Rheological measurements under low-amplitude oscillation and steady shear were made using cone and plate geometry on a Sangamo Viscoelastic Analyser. Dilute solution viscosity was measured using cup and bob geometry on a Contraves Low Shear 30 rotational viscometer. Conformational transitions were monitored by optical rotation (Perkin-Elmer 241 polarimeter) and DSC (Setaram microcalorimeter).

RESULTS

Conformation and Gelation

Gelation of CPS on cooling from the solution state at high temperature is accompanied by a sharp change in optical rotation (Fig. 1), reflecting the formation of ordered intermolecular junctions.[4] The temperature-course of conformational ordering can also be followed by a sharp exotherm in DSC. The reverse transition on heating occurs at appreciably higher temperature and can again be monitored both by optical rotation and by the accompanying DSC endotherm. Levan shows no evidence of conformational change by either technique.

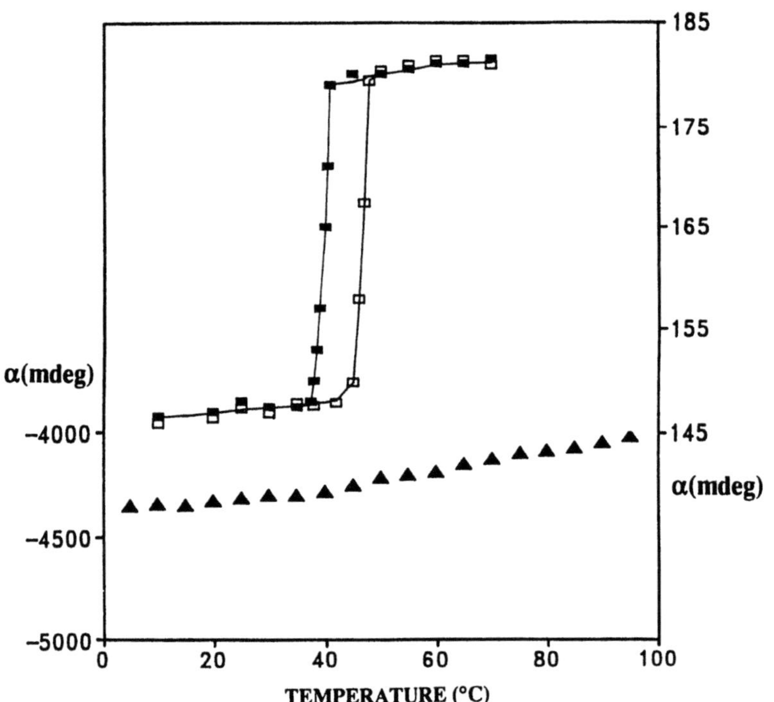

Fig. 1. Temperature-dependence of optical rotation (365 nm; 1 cm pathlength) for levan (left-hand axis) at 25% w/v (▲) and for CPS (right-hand axis) at 0.55% w/v on heating (□) and cooling (■).

The variation of rigidity modulus (G') with concentration (c) for CPS gels (Fig. 2) fits well to the general form anticipated from cascade theory,[5] with the best-fitting model yielding a "functionality" of three equivalent binding sites per chain and a minimum critical gelling concentration of $c_0 = 0.035\%$ w/v. Gelation at even lower concentrations has been reported previously[2] for a different preparation of CPS. At concentrations above c_0 (i.e. sufficient for gel formation) the degree of thermal hysteresis between the disorder-order and order-disorder transitions (Fig. 1) remains constant at about 7.5°C, but at sub-gelling concentrations it decreases steeply (Fig. 2).

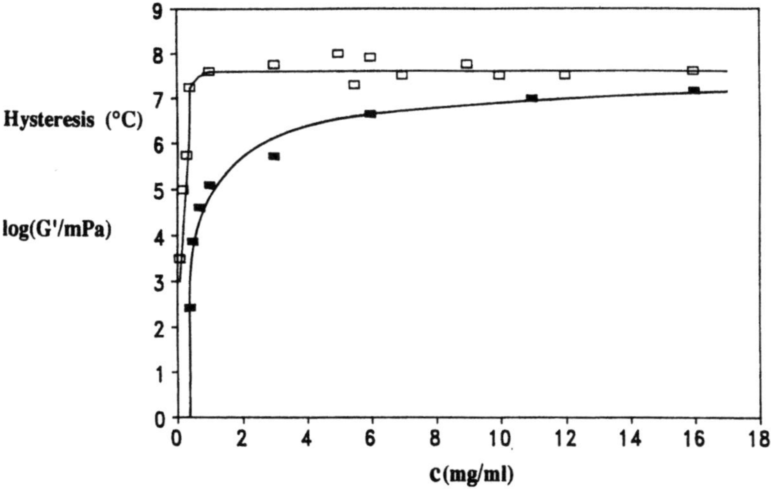

Fig. 2. Concentration-dependence of thermal hyteresis (□) and G' at 10°C (■) for CPS.

Addition of levan to gelling concentrations of CPS has no effect on the transition enthalpy ($\Delta H \approx 18$ J/g) or on the magnitude of thermal hysteresis, but shifts both transitions to higher temperature as the concentration of levan is increased (Fig. 3). Such behaviour is an expected consequence of non-specific thermodynamic incompatibility[6] between the two polymers driving CPS into its more compact ordered form.

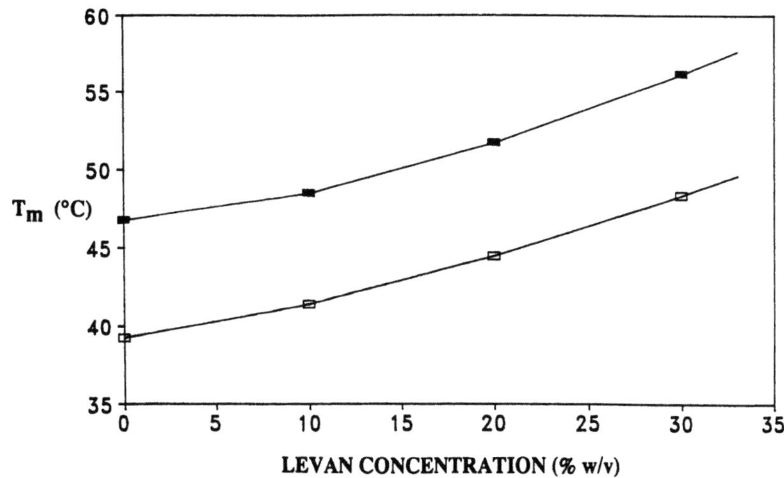

Fig. 3. Effect of levan on transition temperature (T_m) of CPS (0.5% w/v) on heating (■) and cooling (□).

Solution Properties

Polymer viscosity can be usefully expressed as specific viscosity ($\eta_{sp} = (\eta - \eta_s)/\eta_s$, where η is solution viscosity and η_s is solvent viscosity), and the volume occupied by the individual chains may be characterised by the intrinsic viscosity (η_{sp}/c, extrapolated to zero concentration), with the dimensionless product $c[\eta]$ providing an index of the total volume occupied by the polymer. The relationship between space-occupancy and specific viscosity (at low shear rate, before the onset of shear thinning) is the same[7] for most disordered polysaccharides: double-logarithmic plots of η_{sp} *vs.* $c[\eta]$ show two linear regions, with an abrupt change of slope from ~1.4 to ~3.3 at $c[\eta] \approx 4$ and $\eta_{sp} \approx 10$, corresponding to the transition from a dilute solution, in which the chains are free to move independently, to an entangled network of overlapping coils.[8]

As shown in Fig. 4, this generality of behaviour is followed closely by CPS in the disordered form (at 55°C). The intrinsic viscosity, however, is unusually low ($[\eta] = 1.8$ dl/g, in comparison with 5-25 dl/g for typical commercial polysaccharides). For levan (at 20°C) the intrinsic viscosity is far lower ($[\eta] = 0.17$ dl/g), and the concentration-dependence of viscosity is different in form. The onset of very steep concentration-dependence occurs at about the same degree of space-occupancy as for CPS ($c[\eta] \approx 3.6$), but the first divergence from normal dilute-solution behaviour is at $c[\eta] \approx 0.75$, with log η_{sp} *vs.* log $c[\eta]$ having a slope of about 2.2 between the two "break points". The most striking difference, however, is that the high-concentration slope for levan is ~9.3, rather than the usual value of ~3.3, so that doubling concentration would increase viscosity by about a factor of 600, in comparison with the 10-fold increase observed for CPS and for most other disordered polymers.

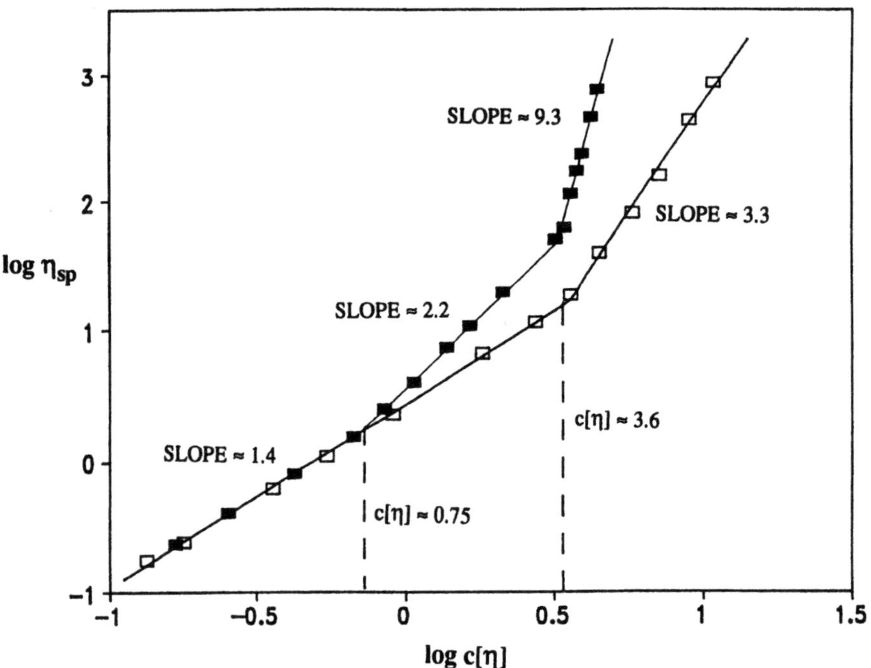

Fig. 4. Variation of "zero-shear" specific viscosity (η_{sp}) with degree of space-occupancy ($c[\eta]$) for CPS in the disordered form at 55°C ($\square$) and for levan at 20°C ($\blacksquare$).

One possible interpretation of such extreme concentration-dependence would be formation of intermolecular junctions[4] as well as, or instead of, physical entanglement. However, concentrated solutions of levan (and of disordered CPS), in common with those of entangled polysaccharide coils,[7] show close superposition of steady-shear viscosity (η) from large-deformation (rotational) measurements and complex dynamic viscosity (η^*) from small-deformation oscillatory measurements at equivalent values of shear rate ($\dot{\gamma}$ /s^{-1}) and frequency (ω /rad s^{-1}). Enthalpic interactions which survive small-deformation but are broken down at higher strain, by contrast, normally give $\eta^* > \eta$ (as observed, for example, in xanthan "weak gels").[9]

Another characteristic feature of entangled polysaccharide networks is that they show a general form of shear-thinning behaviour[10] that gives linear plots of η vs. $\eta\dot{\gamma}^{0.76}$, whereas enthalpic interactions (i.e. intermolecular junctions) produce a different form of shear-thinning, with gross departures from linearity when plotted in the same way. As shown in Fig. 5, concentrated solutions of both levan and disordered CPS give convincing linearity, indicating that molecular interactions are limited to physical entanglement.

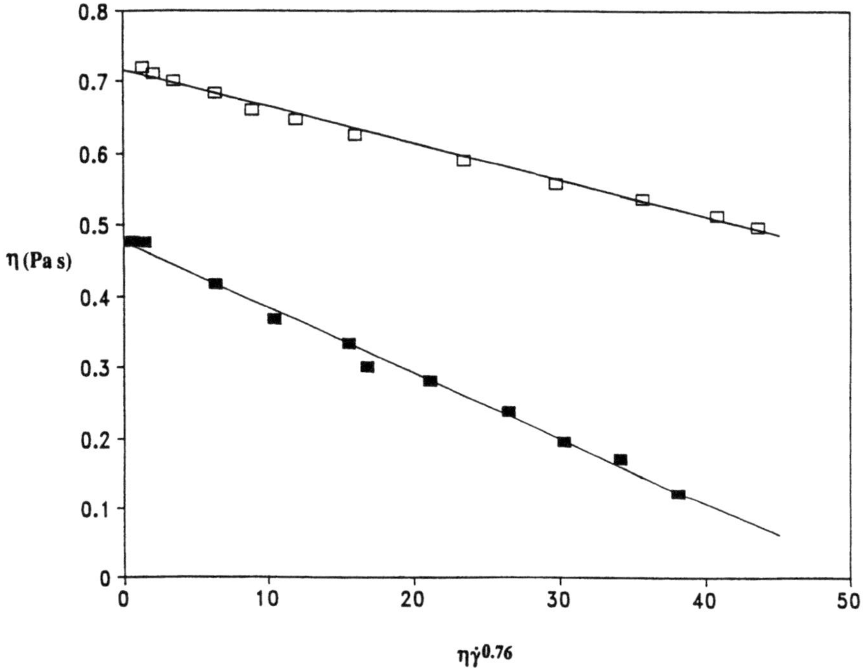

Fig. 5. Characteristic shear-thinning of entangled coils,[10] illustrated for 6% w/v CPS at 55°C (■) and for 26.5% w/v levan at 20°C (□).

DISCUSSION

A common feature of the two materials studied here is their low intrinsic viscosity. This is not due to low molecular weight, since light scattering studies (U. ter Meer and M.J. Gidley, unpublished) gave a molecular weight of ~1.2 x 10^6 for the CPS sample used in the present work, and bacterial levans with intrinsic viscosities close to that of the present sample have been found[11] (by ultracentrifugation) to have molecular weights of ~2 x 10^7.

The very low hydrodynamic volume of levan at such extremely high molecular weights can be readily explained by extensive branching giving rise to a densely packed spheriodal structure. For CPS, two residues in the trisaccharide repeating sequence of the polymer backbone (mannose and galactose) have linkages at positions 1 and 3, and in the third residue (glucose) the linkages at positions 1 and 4 are, respectively, axial and equatorial. Such non-parallel bonding arrangements introduce a systematic change in chain direction,[4] promoting compact coil geometry, in contrast to the 1,4-diaxial or 1,4-diequatorial linkage patterns that give extended coil dimensions in, for example, alginates or galactomannans.

The extremely steep increase in viscosity of levan at high concentrations can be explained qualitatively by the reptation theory of de Gennes,[12] which envisages flow occurring by polymer chains moving through a hypothetical tube created by the entangled network of neighbouring molecules. The additional constraints on movement of a densely branched species such as levan are obvious. Similar behaviour has been observed for "star branched" polymers[13] with four or six long, flexible chains joined at a single polyfunctional residue, but in these systems slopes comparable to the value of 9.3 observed for levan are attained only at vastly higher viscosities (5-7 orders of magnitude greater).

The onset of coil overlap in solutions of disordered linear polymers can be detected by sensitive techniques such as light scattering[14] at volume-occupancy of $c[\eta] \approx 1$ or less, but, as demonstrated here for CPS, sufficient entanglement for appreciable departure from dilute-solution behaviour does not normally occur[7] until $c[\eta] \approx 4$. The earlier break for levan (at $c[\eta] \approx 0.75$) indicates that, as a consequence of their densely-branched "hedgehog-like" structure, significant restriction of flow occurs as soon as the individual levan molecules touch.

The absolute concentrations of levan at the two break-points shown in Fig. 4 are about 4% and 20% w/v, and solutions can easily be prepared at concentrations of 25% or more, suggesting possible practical applications as a bulking agent. High concentrations of materials of low hydrodynamic volume (e.g. maltodextrins[15]) can also be usefully exploited in creation of biphasic systems with other hydrocolloids. In such systems, the effective concentration of the individual polymers within their respective phases can be raised substantially above their overall nominal concentration. The effectiveness of levan in polymer exclusion (Fig. 3), coupled with its extreme sensitivity to changes in concentration (Fig. 4), might therefore generate mixed systems with versatile rheology and unusual scope for manipulation of end-use characteristics.

Evidence from x-ray fibre diffraction[16] indicates that the ordered structure of CPS is an intramolecular "pseudo double helix" stabilised by packing of sidechains along the polymer backbone. The thermal stability of the ordered structure, however, is raised by partial debranching,[17] suggesting additional stabilisation by aggregation of the individual pseuso double helices. Helix-helix aggregation would also provide a mechanism for interchain association in gel formation. This interpretation is consistent[4] with the thermal hysteresis observed (Fig. 1) between gelation and melting (with aggregation stabilising the helices to temperatures above those at which they will form initially), and with the coincidence of loss of hysteresis and loss of network structure (Fig. 2).

Because of its low hydrodynamic volume, the onset of entanglement in solutions of CPS (Fig. 4) does not occur until comparatively high concentration (~2.0% w/v). This is about 60 times higher than the minimum gelling concentration ($c_0 \approx 0.035\%$), showing conclusively that coil overlap in solution is not a necessary precursor of gel formation. The important consequence for potential practical applications is that dilute solutions of CPS (e.g. at about 0.2 - 0.3% w/v), which visually are virtually indistinguishable from water, convert to rigid gels on cooling through the very narrow temperature-range (Fig. 1) of the disorder-order transition.

ACKNOWLEDGEMENTS

We thank Professor S.B. Ross-Murphy for advice and discussions, and for generously providing a computer program to implement the cascade fit of modulus/concentration data. We are also grateful to Professor L.P.T.M. Zevenhuizen and Dr. M. Gross for provision of samples, and to Dr. M.J. Gidley for communication of research results prior to publication.

REFERENCES

1. L.P.T.M. Zevenhuisen and A.R.W. van Neerven, Gel-forming capsular polysaccharide of *Rhizobium leguminosarum* and *Rhizobium trifolii, Carbohydr. Res.* 124:166 (1983).
2. M.J. Gidley, I.C.M. Dea, G. Eggleston, and E.R. Morris, Structure and gelation of *Rhizobium* capsular polysaccharide, *Carbohydr. Res.* 160:381 (1987).
3. M. Gross and K. Rudolph, Demonstration of levan and alginate in bean plants (*Phaseolus vulgaris*) infected by *Pseudomonas syringae* pv. *phaseolicola, J. Phytopathology,* 120:9 (1987).
4. D.A. Rees, E.R. Morris, D. Thom, and J.K. Madden, Shapes and interactions of carbohydrate chains, *in:* "The Polysaccharides", Vol. 1, G.O. Aspinall, ed., Academic Press, New York (1982).
5. A.H. Clark, The application of network theory to food systems, *in:* "Food Structure and Behaviour", J.M.V. Blanshard and P. Lillford, eds., Academic Press, London (1987).
6. V.B. Tolstoguzov, Functional properties of protein-polysaccharide mixtures, *in:* "Functional Properties of Food Macromolecules", J.R. Mitchell and D.A. Ledward, eds., Elsevier, London (1986).
7. E.R. Morris, A.N. Cutler, S.B. Ross-Murphy, D.A. Rees, and J. Price, Concentration and shear rate dependence of viscosity in random coil polysaccharide solutions, *Carbohydr. Polym.* 1:5 (1981).
8. W.W. Graessley, "The Entanglement Concept in Polymer Rheology", Springer-Verlag, Berlin (1974).
9. S.B. Ross-Murphy, V.J. Morris, and E.R. Morris, Molecular viscoelasticity of xanthan polysaccharide, *Faraday Symp. Chem. Soc.* 18: 115 (1983).
10. E.R. Morris, Shear-thinning of "random coil" polysaccharides: characterisation by two parameters from a simple linear plot, *Carbohydr. Polym.* 13: 85 (1990).
11. E. Newbrun and S. Baker, Physico-chemical characteristics of the levan produced by *Streptococcus salivarius, Carbohydr. Res.* 6:165 (1989).
12. P.J. de Gennes, Brownian motions of flexible polymer chains, *Nature,* 282:367 (1979).
13. W.W. Graessley, Effect of long branches on the flow properties of polymers, *Acc. Chem. Res.* 10:332 (1977)
14. J.G. Southwick, M.E. McDonnell, A.M. Jamieson, and J. Blackwell, Solution studies of xanthan gum employing quasi-elastic light scattering, *Macromolecules,* 12:305 (1979).
15. S. Kasapis, E.R. Morris, and I.T. Norton, Physical properties of maltodextrin/gelatin systems, *in:* "Gums and Stabilisers for the Food Industry 6", G.O. Phillips, P.A. Williams, and D.J. Wedlock, eds., IRL Press, Oxford (1992).
16. E.J. Lee and R. Chandrasekaran, The "pseudo-double-helical" structure of the gel-forming capsular polysaccharide from *Rhizobium trifolii, Carbohyd. Res.* 231:171 (1992).
17. M.J. Gidley, G. Eggleston, and E.R. Morris, Selective removal of α-D-galactose sidechains from *Rhizobium* capsular polysaccharide by guar α-D-galactosidase: effect on conformational stability and gelation, *Carbohyd. Res* 231:185 (1992).

RHEOLOGICAL AND THERMAL PROPERTIES
OF GELLAN GUM GELS

Hatsue Moritaka[1], Katsuyoshi Nishinari[2],
Nobuko Nakahama[3] and Hiroyasu Fukuba[1]

[1]Dept.of Home Economics,Showa Women's University,
Taishido,Setagaya-ku,Tokyo 154,Japan
[2]National Food Research Institute,Tsukuba,
Ibaraki 305,Japan
[3]Dept.of Home Economics,Nihon Women's University,
Mejirodai,Bunkyou-ku,Tokyo 112,Japan

ABSTRACT

Complex Young's modulus (E'), compressive break-force (Ph)
and curves of differencial scanning calorimetry (DSC) of gellan
gum solution, with and without salt, were observed gellan gum
gels as a function of salt and polymer concentration, Cs and Cg.
E' and Ph increased up to a certain Cs and then decreased with
increasing Cs. The gelling and melting temperatures shifted to
higher temperatures with increasing Cs and Cg. The sol-gel
transition of samples was examined numerically by using zipper
model approach. The introduction of cations increases the number
of junction zones or zipper N and decreases the rotational
freedom G of parallel links.

INTRODUCTION

Gellan gum cotains D-glucose, D-glucuronic acid and L-
rhamnose in the molar ratios 1:2:1 and it has a carboxyl side
group[1] . Gellan gum is an extremly good gelling agent. Gelation
is sensitive to the type, valency, and concentration of cations
present in salts added to the dispersion before gelation[2,3] In the
present work, the rheological and thermal properties of gellan
gum gels were observed in order to clarify the mechanism which
interferes with the strengthening of gels by addition of salt.

MATERIALS AND METHODS

Gellan was supplied by San-Ei Chemical Industries
Co.,Ltd(Osaka,Japan). The details of experimental procedure were
described previously[2,3].

Food Hydrocolloids: Structures, Properties, and Functions
Edited by K. Nishinari and E. Doi, Plenum Press, New York, 1994

RESULTS AND DISCUSSION

Dynamic Elastic Constants

The tempetature dependence of the dynamic Young's modulus (E') for gellan gels with and without salt from 50°C to 10°C is shown in Fig.1. E' increased with increasing polymer concentration (Cg) due to the increased formation of the gellan network structure for gellan gels without salt. E' increased up to a certain Cs and then decreased with increasing Cs. The molecular association in gellan solutions was promoted by cations due to the electrostatic shield of repulsion between carboxyl groups in gellan molecules. Divalent cations are far more effective than monovalent cations, even by a very small amount. E' decreased monotonically with increasing temperature. The temperature dependence of E' of gellan gels could be understood by a reel chain model[4]. According to this treatment, the monotonical decrease of the elastic modulus may be attributed to the small value of the bond energy of gellan gels. E' of 0.4% (w/w) gellan gels containing KCl or NaCl was less temperature-dependent than that of more concentrated gellan gels without salt. For 0.2% (w/w) containing divalent cations, E' was less temperature-dependent than that of more concentrated gellan gels without salt.

Gel Strength

Fig.2 shows the gel strength Ph of gellan gels at various times after the solution was poured into a mould. Concentrated gellan solution formed gels faster than dilute solutions. In the presence of monovalent cations, the gelation speeded up more than in the case of gellan without salt. As in the case of divalent cations, gelation was accelerated with increasing salt concentration at low Cs, but Ph decreased for high Cs examined. The gels with larger E' at 20°C showed greater Ph when left staning for 120 min for all samples. The ratio of E' and Ph of gellan gels was found to be about ten to one.

Cooling DSC

Cooling DSC curves for gellan gels of various Cg were analyzed. Only one exothermic peak was observed for gellan gels of low concentrations, but the peak splite into multiple peaks for gellan gels of higher concentrations than 1.3% (w/w). The mid point transistion temperature (Ts) shifted to higher temperatures with increasing Cs for gels of the same Cg (Fig.3). The gellan solution containing KCl showed a higher Ts than that containing NaCl of the same concentration. Uedaira[5] reported that a potassium ion is a structure-breaking ion while a sodium ion is a structure-making ion. Therefore, potassium ions may promote the helix formation and association of helices, i.e. strengthen the structure of junction zones more effectively than do sodium ions.

Heating DSC

The lowest endothermic peak temperature (Tm) was around 30°C (Fig.4). The Tm shifted to higher temperatures and the endothermic enthalpy increased with increasing Cg. Only one endothermic peak was observed for a 0.3% (w/w) gel, but many peaks were observed for gels of higher concentrations. The number of endothermic peaks increased and shape of these endotherms

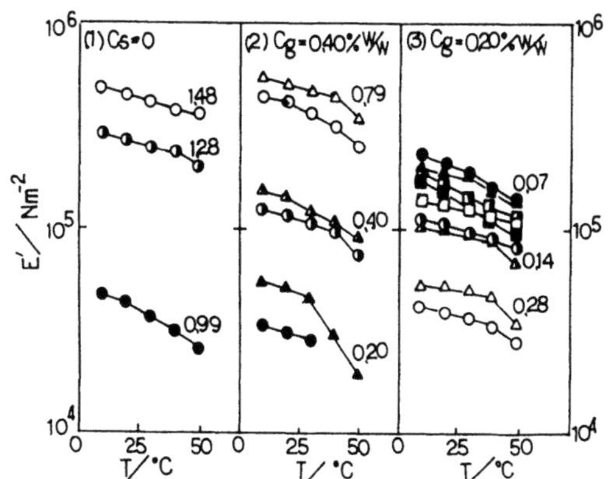

Fig.1.The temperature dependence of storage Young's modulus of gellan gels.(1)Gellan gels without salt;(2)NaCl($\bullet$),KCl($\blacktriangle$),(3)MgCl$_2$ ($\blacktriangle$), CaCl$_2$ ($\bullet$),Ca(CH$_3$CHCOO)$_2$ ($\blacksquare$).

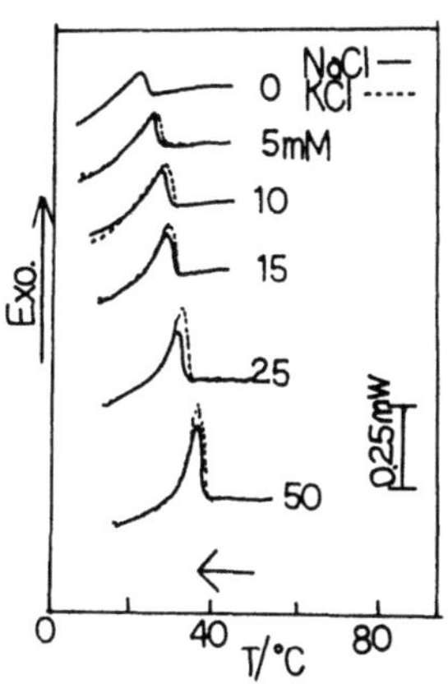

Fig.3.Cooling DSC curves of 0.5%(w/w) gellan solutions of containing KCl or NaCl. cooling rate 1.0°C/min.

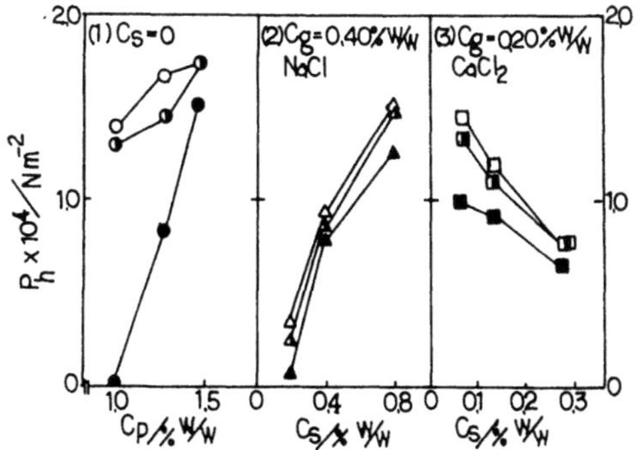

Fig.2.Gel strength Ph of gellan gels at various times at 25°C. $\bullet$,5min;$\circleddash$,15min;O,120min.

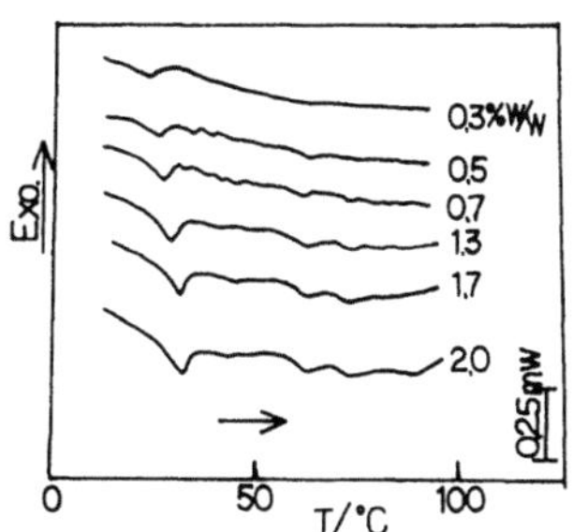

Fig.4.Heating DSC curves of gellan gels. heating rate 1.0°C/min.

became complicated with increasing Cs. Potassium is far more effective than sodium in shifting the Tm to higher temperaures.

Heat of Reaction of Cross-Links

Fig.5 shows the Eldrige-Ferry plot[6] of gellan gels with NaCl. The slope of the line is proportinal to the heat released on forming 1 mol of junction zones. The slope increased with increasing Cs. The gellan gel containing KCl had a much larger ΔH than that containing NaCl of the same concentration – as in the case of the exothermic enthalpy.

Application of a Zipper Model

DSC curves of 0.5% (w/w) gellan gels containing KCl of various concentrations (Fig.6). Solid curves represent the observed cooling DSC curves and dotted curves are obtained by curve fitting[7]. When ε is 2000 K and N is 100, the number of zippers (N) increased and the rotational freedom of the constituting links (G) decreased with increasing Cs. With gellan solutions without salt, N increased and G decreased with increasing Cg. Compared to sodium ions, potassium ions expedited gelation. The increase in N by the addition of salt is consistent with rheological results. Thus, the introduction of cations increases the number of junction zones or zippers N and decreases the rotational freedom G of parallel links. This makes the structure of junction zones more heat resistant, and increases the elastic modulus of the gel.

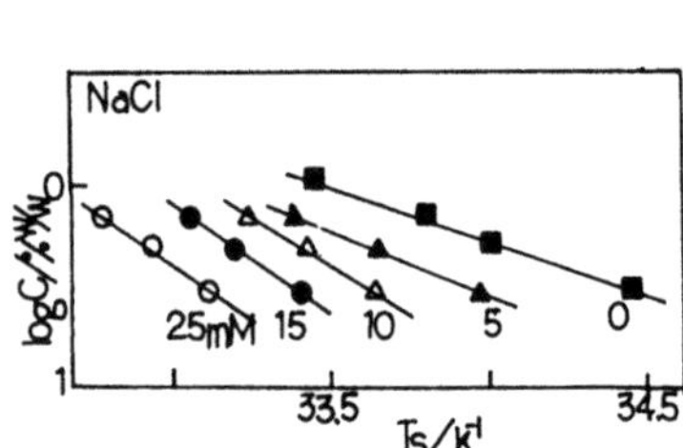

Fig.5.Plot of logarithm of gellan concentration vs. the reciprocal of the setting temperature.

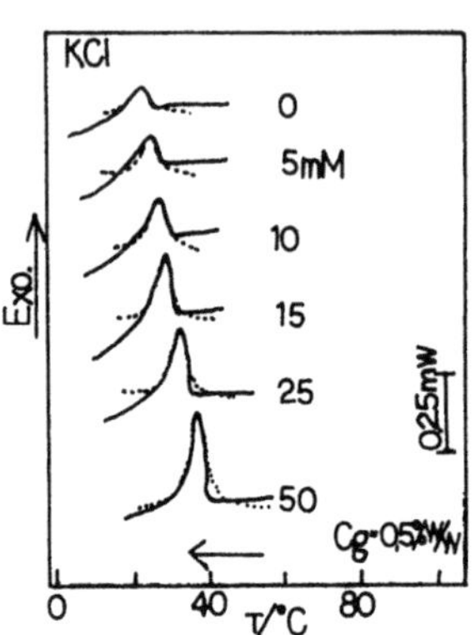

Fig.6.The solid curves represent the observed cooling DSC curves. The dotted curves were obtained by curve fitting[7]. (1) with N=100, ε=2000k. (1)G=870,N=1900; (2)G=820,N=2700; (3)G=785,N=3500; (4)G=760,N=4700; (5)G=620,N=7500.

REFERENCE

1.M.A.O'Neill,R.R.Selvendran,and V.J.Morris,Structure of the Acidic Extracellular Gelling Polysaccharide Produced by *Pseudomonas Elodea, Carbohydr.Res.*,124:123(1983).
2.H.Moritaka,H.Fukuba,K.Kumeno,N.Nakahama,and K.Nishinari,Effect of Monovalent and Divalent Cations on the Rheological Properties of Gellan Gels,*Food Hydrocolloids*,4(1991).
3.H.Moritaka,K.Nishinari,N.Nakahama,and H.Fukuba,Effects of Potassium Chloride and Sodium Chloride on the Thermal Properties of Gellan Gum Gels,*Biosci.Biotech.Biochem*,56:4 (1992).
4.K.Nishinari,S.Koide,and K.Ogino,Synergistic Interaction of Xanthan Gum with Glucomannans and Galactomannans,*J.Phys. France*,46:793(1985).
5.H.Uedaira,and A.Ohsaka,"Water in Biological Systems,"Publisher, Kodansha,(1990)(in Japan)
6.J.E.Eldrige,and J.D.Ferry,Studies of the Cross-Linking Process in Gelatin Gels.I.Dependence of Melting Point on Concentration and Molecular Weight,*J.Phys.Chem.*,58:(1958).
7.K.Nishinari,S.Koide,P.A.Williams,and G.O.Phillips,A Zipper Model Approach to the Thermoreversible Gel-Sol Transition, *J.Phys.France*,51:1759(1990).

SELECTION OF GUM PRODUCING <u>ALCALIGENES</u> SP STRAINS

Daniela M. Mariuzzo, Francisco J.S. Santucci, Doraci H.D. Machado,
José L. Pereira and Adilma R.P. Scamparini

Science Department - Food Engineering Faculty - UNICAMP
13081 Campinas, SP., Brazil

ABSTRACT

The bacteria <u>Alcaligenes</u> excrets an extracelullar anionic polysaccharide (welan gum) produced under submerged fermentation procedures.

Aqueous dispersions of welan gum have high viscosity at low-shear rates and exhibit shear-thinning behavior.

The high viscosity is maintained at elevated temperatures and the polysaccharide has high potential in industrial applications as thickening agent and for supporting particles in suspension.

In this work, we isolated and characterized nine hundred strains selected from different habitats like water (sea, lakes, river), soil and aquatic plants from several regions of Brazil. The isolation medium used was YM.

About 40% of the strains isolated produced gum in media of pH 7.0, containing sucrose or glucose as well as potassium phosphate and magnesium sulphate, when incubated at 28°C, at 150 rpm during 96 hours.

INTRODUCTION

The genus <u>Alcaligenes</u> is widely distributed in nature. It is found in soil, fresh water, marine sources as well as clinical specimens[1]. Members of the genus <u>Alcaligenes</u> are characterized as aerobic gram negative motile cocci with a G + C in the range 56 to 70%. Some strains can perform anaerobic respiration in the presence of NO^-_3 or NO^-_2[2]. Some strains produce a non-gel-forming extracelullar polysac-charide (welan gum). This polysaccharide shows shear-thinning behavior at 0.1% concentration and plastic behavior at concentrations above 0.3% at 25°C. The dynamic viscoelasticity increases directly with the polysaccharide concentration but is not temperature-dependant[3].

Food Hydrocolloids: Structures, Properties, and Functions
Edited by K. Nishinari and E. Doi, Plenum Press, New York, 1994

The main chain branching prevents the formation of Ca^{2+} bridges with different molecules[4].

In the present work nine hundred isolates from several environment such as aquatic plants, soil and water from lakes, rivers and the sea, were found to produce extracelullar polysaccharydes under specific culture conditions.

MATERIALS AND METHODS

Collection of Samples

The samples were collected at locations where gum producing bacterias were expected to occur.

Each sample of soil of about 1 kg was obtained by the mixing of ten separate sub samples obtained from an area of 5 m^2 of homogeneous soil. The samples of water were picked: at different points of the river beds; at residue treatment plants; at different points of the lake surface; at different points of the sea surface and at fish-farm. When aquatic plants were present, they were collected at the same places where the water was collected.

Isolation of Strains

Isolation from samples of soil. 1 g of soil was suspended in 50 mL of YM medium in tubes and the tubes incubated at 32°C for 4 days. After the incubation period the content of each flask was streaked in plates containing the same enrichment medium plus 1.4% agar. The plates were incubated at 32°C during 2 days.

Isolation from water obtained at lakes and rivers. Flasks containing 10 mL of nutrient broth were innoculated with 1 mL of water and incubated at 30°C for 3 days. After the incubation period the content of each flask was streaked on plates containing YM medium and the plates incubated at 30°C for 2 days. The same procedure was followed with the plants from the aquatic environment using 1 g of plant for each enrichment nutrient broth medium.

Samples from sea water. Flasks containing 10 mL of nutrient broth plus 3% NaCl were innoculated at 22°C for 3 days. After the incubation period the content of each flask was streaked on plates containing YM medium plus 3% NaCl and the plates incubated at 22°C for 2 days.

Gum Production

Each bacterial colony was placed in 100 mL of liquid Yeast Malt medium containing 2% of glucose and keeped at 30°C, at 150 rpm during 2 days. After enrichment, the broth was

Table 1. Isolated and theirs sources.

isolated numbers	source	
001-169	(FTA)	fish-farm - Atibaia - SP.
170-329	(TMC)	tanks mud - Campinas - SP.
330-509	(WLC)	water lakes - Campinas - SP.
510-569	(ARC)	mud and water from Atibaia River - Campinas-SP
570-649	(SCU)	UNICAMP soil - Campinas - SP.
650-789	(SWG)	sea water - Garujá - SP.
790-809	(SWS)	sea water - Santos - SP.
810-869	(SWR)	sea water - Rio de Janeiro - RJ.

transfered to Fernback flasks containing 150 mL of fermentation medium consisting of 2% sucrose or glucose, 0.01% $MgSO_4$ and 0.5% K_2HPO_4 at pH 7.0. The flasks were incubated at $28^\circ C$ and skaked at 150 rpm for 4 days. At the end of the fermentation period, the cells were separated by centrifugation at 15.000 rpm and the gum precipited with ethanol $96^\circ GL$. The precipited was dried at $48^\circ C$.

Determination of the Viscosity

The viscosity was determined using 0.1% gum solutions in a Brookfield microviscometer RVT (spindle 18) at 25, 50 and $75^\circ C$.

RESULTS AND DISCUSSION

Strains FTA001, WLC330 and WLC500 produced the highest concentration of biopolymer. Solutions of the gum produced by these strains showed higher viscosities when compared with other strains under study.

The three higher gum production isolates were biochemically characterized. These strains were gram negative, cocci, oxidase positive and grew in a medium containing citrate as the sole carbone source. These strains did not hydrolyse gelatin and did not produced H_2S in the presence of the amino acids cistine or cisteine.

The colony morphology of strains FTA001, WLC330 and WLC500 are showed in Fig. 1. The three strains showed mucoid colonies on YM sucrose or glucose medium, but the biopolymer was produced only in YM glucose broth. An increased of the bipolymer field was obtained by the addition of K_2HPO_4 and $MgSO_4$. Good biopolymer production was obtained in YM sucrose broth containing K_2HPO_4 and $MgSO_4$.

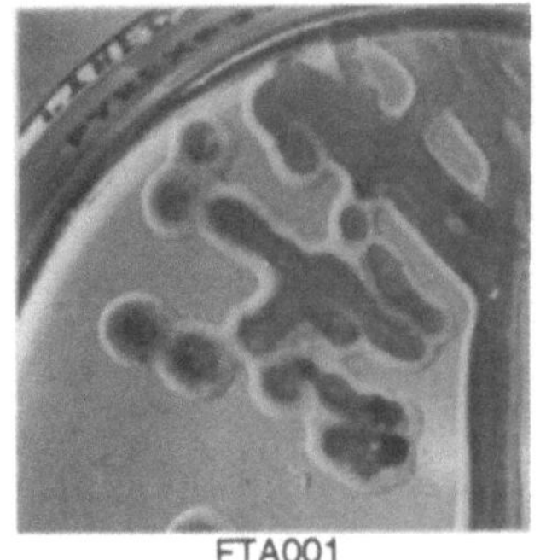

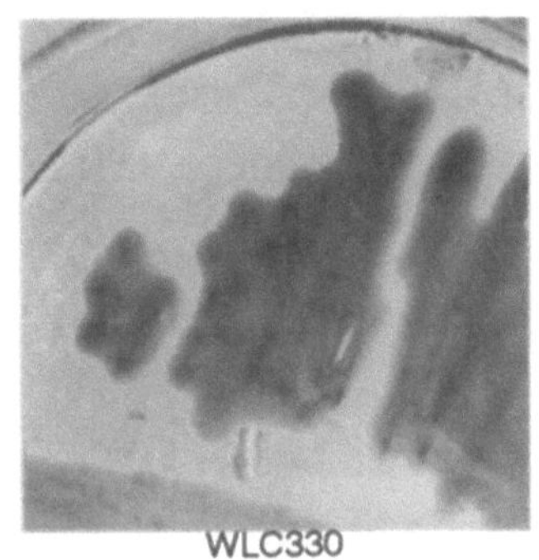

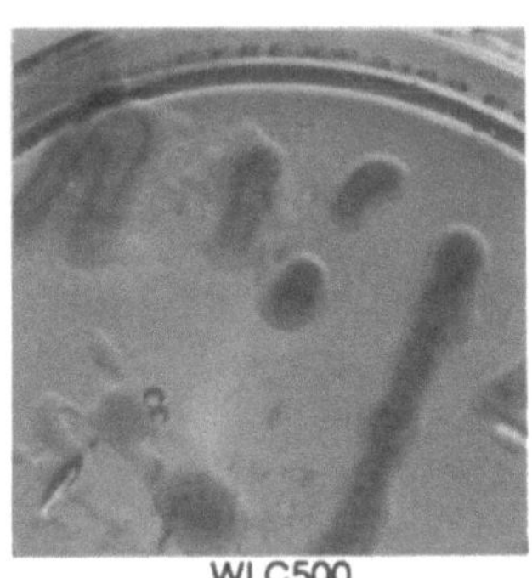

Fig. 1. Colony morphology of strains FTA001, WLC330 and WLC500 after 24 h of WLC500 growth on YM at 25°C.

Viscosity solutions containing 0.1% of biopolymer produced by the strains FTA001, WLC330 and WLC500 were not affected by temperature or pH changes.

REFERENCES

1. K. Yamasato, M. Akagawa, N. Oishi and H. Kuraishi, Carbon   substrate assimilation profiles and other taxonomic   features of Alcaligenes faecalis, Alcaligenes ruhlandii   and Achromobacter xylosoxidane, J. Gen. Appl. Microbiol.   28: 195 (1982).

2. K. Kerters and J.D. Ley, Genus Alcaligenes, In: "Bergey's   Manual of Systematic Bacteriology", J.G. Holt, ed.,   William and Wilkins, Baltimore, (1986).

3. K.S. Kang, G.T. Veeder and I.W. Cottrell, Some novel   bacterial polysaccharides of recent development, Prog.   Ind. Microbiol. 19: 231 (1983).

4. T. Masakuni and K. Masahiro, Rheological Properties of   Welan Gum in Aqueous Media, Agric. Biol. Chem. 54: 3079   (1990).

MOLECULAR ASSOCIATION AND DISSOCIATION IN FORMATION OF CURDLAN GELS

Akira Konno,[1] Kenji Okuyama,[2] Atuo Koreeda,[3] Akira Harada,[4] Yoshimi Kanzawa,[5] and Tokuya Harada[5]

[1]Kinran Junior College, Suita, Osaka, 565, [2]Faculty of Technol. Tokyo Univ. of Agr. and Technol. Koganei, Tokyo, 184, [3]Inst. Sci. Ind. Res. Osaka Univ., Ibaraki, Osaka, 565, [4]Faculty of Science, Osaka Univ., Toyonaka, 560, [5]Kobe Women's Univ., Suma, Kobe, 654 Japan

ABSTRACT

We examined the formation of curdlan gels by differential scanning calorimetry and electron microscopy. When curdlan heated in water to various temperatures was cooled, an exo-thermic peak appeared at about 40 °C, with formation of hydrogen bonds involving water to give gel. When the cooled preparations were heated again, endo-thermic peaks re-appeared at about 60 °C and exo-thermic peaks also appeared at higher temperatures. However, no exo-thermic peaks were observed on cooling and then heating above 145 °C. Thus, the gel formation caused by heating to 60-120 °C and cooling is reversible, but that caused by heating to above 145 °C and cooling is irreversible as shown by the heat analysis. In neutralized gel and the gel formed by heating at 55 or 60 °C and cooling, microfibrils of endless length consisting of fibrils of about 100 nm length and 10-25 nm width appeared. In gels obtained by heating at 70, 80, and 100 °C and cooling, microfibrils with released fine stubs appeared. Gels cooled after heating at 120 and 145 °C contained spindle shaped microfibrils of about 100 nm length and 30 nm width. Preparations obtained by heating to 170 °C and cooling gave microfibrils like those formed by unheated preparations.

INTRODUCTION

Harada *et al.* (1-3) found that curdlan, a bacterial polysaccharide composed entirely of $(1\rightarrow3)$-β-D-glucosidic linkages, forms two types of gel. One type is formed by neutralizing an alkaline solution or heating an aqueous suspension to about 60 °C and then cooling (4). This type of gel has much lower gel strength and syneresis than the other type obtained by heating an aqueous suspension to above 90 °C.

Kasai's (5) and Okuyama's (6) groups demonstrated by X-ray diffraction analysis that single stranded helices in curdlan are converted to triple stranded helices by heating at a higher temperature and cooling. Recently, Konno and Harada (7) studied the thermal properties of curdlan gel by differential scanning calorimetry (DSC). DSC curves of curdlan in aqueous suspension showed a sharp endo-thermic peak at 50-64 °C, a broad

Food Hydrocolloids: Structures, Properties, and Functions
Edited by K. Nishinari and E. Doi, Plenum Press, New York, 1994

endo-thermic peak between 70-120 °C and an exo-thermic peak in the range of 140 to 170 °C. Harada and his colleagues (3,8) studied the ultra-structure of curdlan by transmission electron microscopy(TEM) and found that micrographs of curdlan in neutralized gel and gel formed by heating at higher temperatures are different. This paper deals with the association of molecules of curdlan on heating in water at various temperatures and cooling studied by DSC and TEM.

MATERIALS AND METHODS

Spray-dried curdlan powder was provided by Takeda Chemical Ind. The average degree of polymerization(DPn) of the polymer determined according to Manner's method was 450.

Differential Scanning Calorimetry (7). DSC was carried out in a Super High Sensitive DSC instrument (Seiko Instruments Inc.). A homogenate of curdlan (4%) prepared with a Waring blender was sealed in a silver pan of 70 µl. Distilled water was used as a reference. The amount of water used corresponded to that in the curdlan suspension to obtain a flat base line. Calibration for temperature of the specimen was raised from 20 or 40 °C to 170 °C at a rate of 1 °C/min.

Preparation for Electron Microscopy (3,8). A carbon film (~15 nm thick) formed by vacuum evaporation on a freshly cleaved mica sheet was transferred to a TEM grid, dried in air and annealed by heating at 100 °C for 4 h. The carbon film was then subjected to ion-cleaning by glow discharge at low pressure to remove oily material that might cause abnormal assembly of the test material. The sample (10 µl of 0.1% gel) deposited on the carbon film was dried in air and a droplet (10 µl) of 2% uranyl acetate solution at pH 4.4 was placed on its surface. After a few minutes, residual solution was removed by absorption with filter paper, and the negatively stained sample was dried in air in a dust-free fume hood. The preparation was then examined in a Hitachi H-600FE electron microscope equipped with a field emission gun and operated at an accelerating voltage of 100 kV. Micrographs were taken at an original magnification of 50,000x.

Measurements of gel strength. Gel strength was determined according to our method previously reported (4). For heating at temperature above 100 °C, a stainless reactor tube (Taiatsu, Scientific Glass Co.) was used.

Determination of the Rate of Syneresis. Syneresis was calculated as a percentage of the weight of the original suspension of curdlan as previously reported by us (4).

RESULTS AND DISCUSSION

Previously, we examined curdlan by DSC (7). The DSC curve (Fig. 1) showed an endo-thermic peak at 50-64 °C, caused by swelling of curdlan due to the breakage of some hydrogen bonds, and a broad endo-thermic peak between 70-120 °C. A large exo-thermic peak appeared at 120 to 153 °C, resulting from the formation of spindle-shaped microfibrils with a pseudo-crystalline form. The appearance of a small endo-thermic peak at about 160 °C may be due to breakage of the firm gel structure.

In the present study, we first examined the cooling curves of curdlan gel obtained by heating it to (A) 65 °C, the completely swelling stage, (B) 100 °C the boiling stage, (C) 120 °C, the stage when spindle-shaped microfibrils started to form, (D) 145 °C, the stage of the maximum exo-thermic peak, and (E) 170 °C, the final stage of the exo-thermic peak (Fig. 2). During cooling, an exo-thermic peak appeared in the range of about 40 °C in all preparations except those cooled down from 170 °C. At the stage of appearance of the exo-thermic peak, swelling or gelatinizing curdlan forms hydrogen bonds involving water resulting in gelation. The peak area decreased with an increase in the heating temperature, because the proportion of molecules capable of forming hydrogen bonds decreased as a result of hydrophobic interactions on heating above 70 °C. The preparation cooled from 170 °C gave peaks at about 82 and 133 °C. The reason for the appearance of these peaks and a shoulder at about 125 °C in the preparation cooled from 145 °C are unknown.

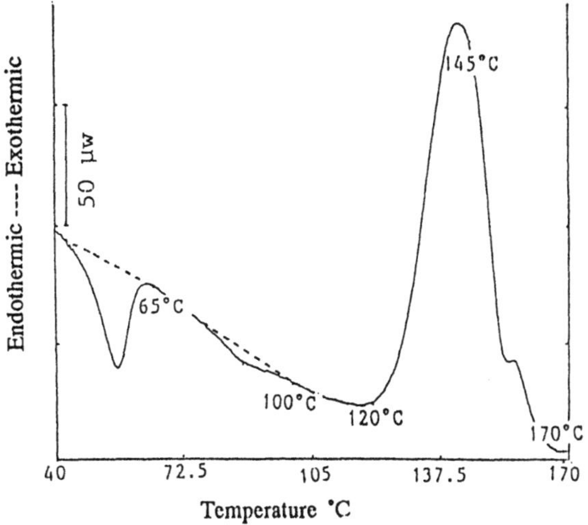

Fig. 1. DSC curve of curdlan in water.

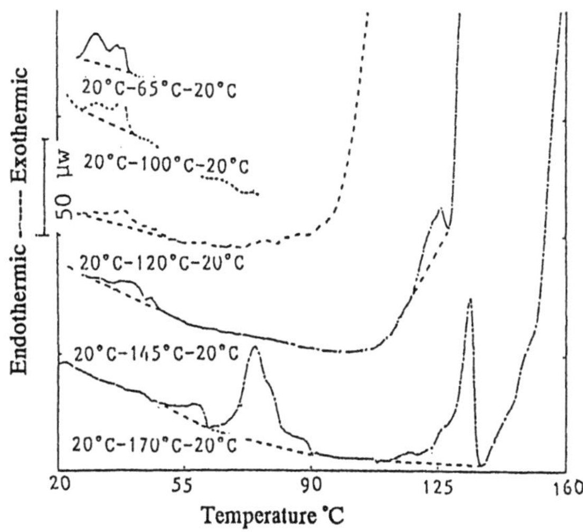

Fig. 2. Cooling curves of curdlan gels after heating to various temperatures.

Next, these kinds of preparations whose cooling curves had been examined were heated to 170 °C at a rate of 1 °C/min. Figure 3 shows their DSC curves. When the gel preparation (C), which had been cooled to 20 °C after heating to 120 °C was re-heated to 170 °C, an endo-thermic peak appeared at about 60 °C, the temperature of swelling of the neutralized curdlan, and exo-thermic peaks appeared from 120 to 155 °C as in regenerated curdlan without heating. In preparation (D), obtained after heating to 145 °C, a small peak appeared at about 60 °C, but no clear exo-thermic peak appeared at 155 °C. In preparation (E) obtained after heating to 170 °C, no endo-thermic peak at about 60 °C and no exo-thermic peak from 120 to 155 °C appeared, but an endo-thermic peak appeared at about 140 °C. These results show that after heating to below 120 °C, the formation of a gel is almost completely reversible, whereas after heating to above 145 °C, gel formation is irreversible as shown by heat analysis. However, we have shown that gel formation caused by neutralization of an alkaline solution or by heating to 60 °C is reversible (7)., whereas the gel formation caused by heating to above 80 °C is irreversible (8, 9). The difference between the previous and the present findings may be due to the slight development or absorption of heat caused by formation or break-down of hydrophobic interaction. During the heating, hydrophobic bonds are formed, but the broad endo-thermic peak is small.

As shown in Fig. 3, the exo-thermic peak at temperature of 120 to 150 °C is very large. In the experiments mentioned above, the temperature was increased at a rate of 1 °C/min. Next, we examined the effect of prolonging the incubation times. Stepwise heating for 60 min at 120 °C and subsequently for 20 min at 125 °C resulted in comparatively small increase in the exo-thermic peak, but heating for 45 min at 140 °C resulted in marked increase of the peak. Thus, heating at above 130 °C seemed to be required for significant increase in the exo-thermic peak. This was confirmed by an experiment in which the preparation was heated at 135 °C for 90 min.

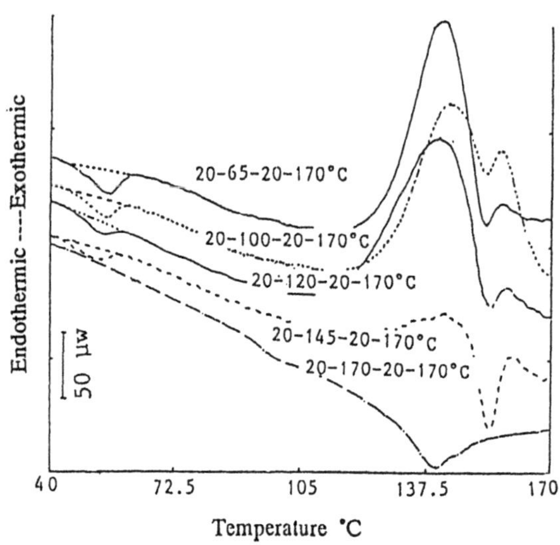

Fig. 3. Heating curves of curdlan gels cooled after heating to various temperatures.

Using TEM, we examined electron micrographs of curdlan preparations obtained after heating curdlan in water to various temperatures (Fig. 4). Clear microfibrils were seen in the unheated preparation, and similar microfibrils were seen in the preparations heated to 55 and 60 °C. This is because, as shown in DSC experiments, the gel formed on cooling after heating to 55 and 60 °C, is reversible, and so similar microfibrils were found to those without heating. In contrast, heat-treatment at 70 and 80 °C gave microfibrils with released fine stubs. Hydrophobic interactions occur at these temperatures. Heat-treatment at 120 and 145 °C resulted in formation of pseudo-crystalline forms of 100 nm length and 30 nm width. At these temperatures, the structure of curdlan began to change from single helices to triple helices (5, 6, 11). The presence of triple helices in high-set gel has been confirmed by Chuah *et al.* (12) and Fulton and Atkins (13). The appearance of a large exo-thermic peak at 120 to 155 °C may be caused by the formation of spindle-shaped microfibrils composed of triple helices. However, heating to 160 °C changed the structure to similar microfibrils to those in unheated curdlan.

The strengths and syneresis of gels formed by heating curdlan in water to various temperatures and then cooling them were examined (Table 1). As already reported (14), the gel strengths of gels formed by heating at 60 °C were similar. These gel strengths are due to the formation of microfibrils in the above preparations. During heating at these temperatures, the rate of syneresis of the gels was quite low. In these gels, single helices associate to form microfibrils, but hydrophobic interaction does not participate in the molecular association. This would explain why these gels are substantially reversible. The gels formed on heating at 70 and 80 °C showed a higher syneresis and contains microfibrils with fine released fibrils, but the gel strength was similar to that of gels heated to below 60 °C.

The gel formed by heating at 100 °C and then cooling showed higher gel strength and syneresis. This gel contained microfibrils with a net-like structure. In the 70-100 °C gels,

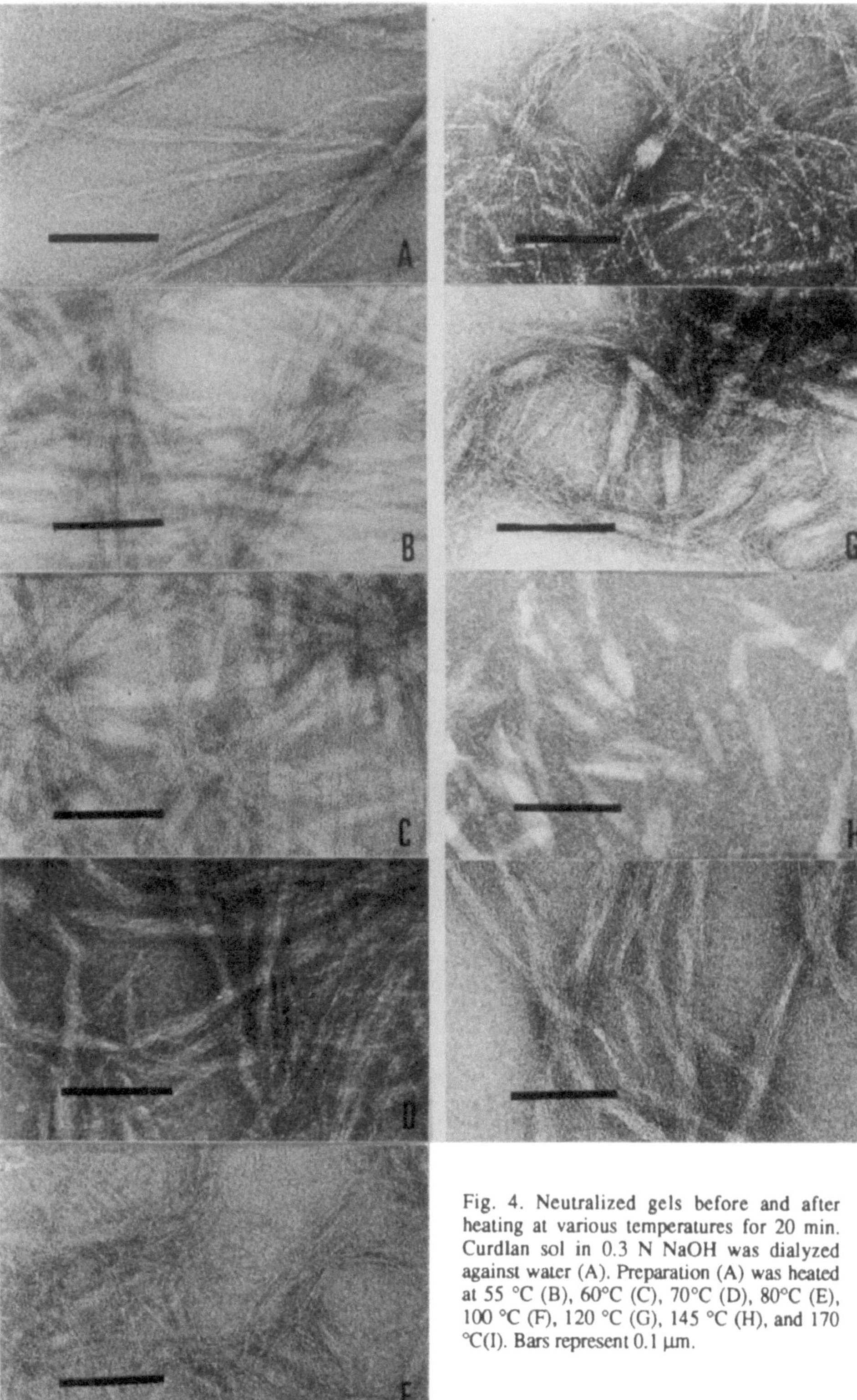

Fig. 4. Neutralized gels before and after heating at various temperatures for 20 min. Curdlan sol in 0.3 N NaOH was dialyzed against water (A). Preparation (A) was heated at 55 °C (B), 60°C (C), 70°C (D), 80°C (E), 100 °C (F), 120 °C (G), 145 °C (H), and 170 °C(I). Bars represent 0.1 μm.

molecules bind to each other by hydrophobic interactions. The preparations heated at 120 and 145 °C showed much higher gel strength and syneresis. In these gels molecules associate closely with each other as triple helices, and form pseudo-crystalline microfibrils. The preparations obtained by heating at 160 and 170 °C had the highest gel strength and syneresis. Kuge et al. (15, 16) showed by experiments in a glass-vessel autoclave that on heating at about 160 °C firm gel melts and then when cooled forms a gel again. The firm gel breaks at 160 °C and when cooled make microfibrils like those in unheated gel. However, they showed that the resultant gel was irreversible because the gel did not melt when heated again to 160 °C. We have found that powder prepared by dehydrating the gel obtained on cooling after heating to 160 °C did not form a gel when heated at 120 or 170 °C(unpublished data). Curdlan in this powder has been shown to have a triple helical structure by ^{13}C NMR (17) and X-ray studies(Okuyama *et al.* unpublished).

Table 1. Gel strength and syneresis of gels of curdlan(3%) obtained after heating at various temperatures in water.

Heating temperature(°C)	Gel strength g/cm^2	Syneresis %
None	620	2
55	310	2
60	530	4
70	540	8
80	630	10
100	940	16
120	1100	18
145	1200	20
160	1800	29
170	2200	34

The heating time was 20 min. The gel strength was measured at 30 °C

ACKNOWLEDGEMENT

The authors thank prof. N. Kasai in Kobe Women's University for valuable advice.

REFERENCES

1. T. Harada, "Trends in Glycosciences and Glycotechnology," 4:309(1992).
2. T. Harada, M. Terasaki, and A. Harada, in "Industrials Gums," 3rd ed. by R. L. Whistler, Academic Press, Inc. p427(1993).
3. T. Harada, Y. Kanzawa, K. Kanenaga, A. Koreeda, and A. Harada, *Food Structures*, 10: 1(1991).
4. Y. Kanzawa, A. Koreeda, A. Harada, and T. Harada, Agric. *Biol. Chem.*, 53: 979(1989).
5. T. Takeda, N. Yasuoka, N. Kasai, and T. Harada, *Poly. J.*, 10: 365(1978).
6. K. Okuyama, M. Otsubo, M. Ozawa, T. Harada, and N. Kasai, *Carbohydr. Chem.*, 10: 645(1991).
7. A. Konno and T, Harada, *Food Hydrocolloids*, 5: 427(1991).
8. T. Harada, A. Koreeda, S. Sato, and A. Harada, *J. Electron. Micros.* 28: 147(1979).
9. Y. Kanzawa, T. Harada, A. Koreeda, A. Harada, and K. Okuyama, *Carbohydr. Polym.*, 10: 299(1989).
10. A. Konno, H. Kimura, T. Nakagawa, and T. Harada, *Nogei Kagaku-Kaishi*, 6:247(1978).
11. H. Saito, Y.Yoshioka, M. Yokoi, and J. Yamada, *Biopolymers*, 19: 1689(1991).
12. C. T. Chuah, A. Sarko, Y. Deslandes, and R. H. Marchessault, *Macromolecules*, 16:1375(1983).
13. W. S. Fulton and A. D. Atkins, ACS Symp. Ser. 141. i n "Fiber Diffraction Methods," ed by A. D. French and K. H. Garner, p386(1980).
14. I. Maeda, H. Saito, M. Masada, A. Misaki, and T. Harada, *Agric. Biol. Chem.*, 31:1184(1967).
15. T. Kuge, N. Suetsugu, and K. Nishiyama, Scientific Reports Kyoto Prefectural University, Agriculture, 31: 170(1979), in Japanese.
16. T. Kuge, N. Suetsugu and K. Nishiyama, *Agric. Biol. Chem.*, 41: 1315(1979).
17. Y. Yoshioka, N. Uehara, and H. Saito, *Chem. Pharm. Bull.*, 40: 1221(1992).

FOOD APPLICATIONS OF CURDLAN

Masanori Miwa, Yukihiro Nakao and Kiyoshi Nara

Vitamin & Food Research Laboratories
Takeda Chemical Ind., Ltd.
17-85, 2 Cho-me, Juso-Honmachi, Yodogawa-ku, Osaka, Japan

ABSTRACT

Curdlan is a polysaccharide produced by bacteria and a homopolymer of glucose with ß-1,3-glucosidic linkage. Its unique characteristic is its ability to form a firm, resilient and thermo-irreversible gel when heated in an aqueous suspension higher than 80°C. Also, the curdlan gel does not melt even when it is heated to above 100°C, for example, during processes such as thermal sterilization. Moreover, curdlan gel, which is different from the gels made from other gelling agents, such as carrageenan, agar-agar and konjac, is stable against freezing and thawing.

Some examples of food applications for curdlan were shown as follows:
1. Food additive uses: to improve, modify and stabilize the physical properties of the products.
Texture modifier in noodles and fish paste products such as kamaboko; Water-holding agent in meat products such as sausages, hams and hamburgers, etc.
2. Food ingredient uses: to develop the new products.
Noodle-shaped tofu; Tofu for retorting, freezing or freeze-drying; Frozen noodle-shaped konjac-like gel food; Low-fat sausage (coarse-cut), substituting curdlan gel containing vegetable oil for pork fat; Non-fat whipped cream (analogue), as a milk fat substitute, etc.

INTRODUCTION

Curdlan, discovered by Harada et al. in 1966, is a polysaccharide produced by bacteria such as *Alcaligenes faecalis* var. *myxogenes*. It has unique gelling characteristics such as forming a heat resistant gel which is stable even when heated to above 100°C. The production, chemical properties, chemical structure and rheological properties of curdlan as well as some potential uses in foods have been reported and reviewed.[1,2,3]

Reagent grade curdlan powder has been distributed by Wako Pure Chemical Ind., Ltd. (Osaka, Japan). In 1989, Takeda Chemical Ind., Ltd. (Osaka, Japan) commercialized curdlan powder to supply for the food industry.

In this report, typical properties and food applications of curdlan are shown, laying emphasis on applications to retorted foods or frozen foods.

Food Hydrocolloids: Structures, Properties, and Functions
Edited by K. Nishinari and E. Doi, Plenum Press, New York, 1994

PROPERTIES OF CURDLAN

Curdlan is a linear homopolymer of D-glucose with ß-1,3-glucosidic linkage. An aqueous suspension of curdlan forms a firm, resilient and thermo-irreversible gel when heated to above 80°C. Curdlan is a white powder with high fluidity. It is very stable in its dry form as it maintains its ability to form a gel even after storage for a long period of time. Curdlan is insoluble in water, alcohol and most organic solvents, but soluble in alkaline solutions such as sodium hydroxide and trisodium phosphate. Curdlan is proved to be non-toxic from extensive safety evaluations and nutritionally inert, like agar.

PROPERTIES OF CURDLAN GELS

Two types of curdlan gels can be obtained depending on heating temperatures. The relationship between heating temperature and gel strength or viscosity is shown schematically in Fig. 1. The viscosity of the aqueous suspension of curdlan shows a rapid increase at a temperature of about 60° C. When this suspension is cooled down to below 40° C, it

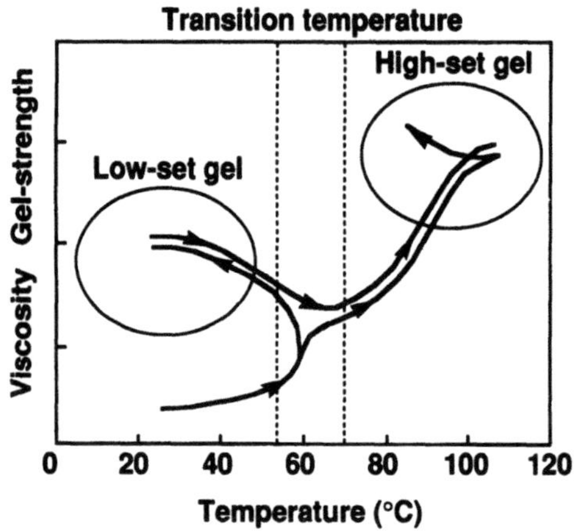

Fig. 1. Formation of high-set gel and low-set gel (schematic figure).

forms a gel. This gel is a thermo-reversible gel, similar to agar-agar and gelatin gels, and is called a low-set gel. On the other hand, the same aqueous suspension of curdlan forms another type of gel when it is heated to above 80°C. This gel is a thermo-irreversible gel and is called a high-set gel. Furthermore, it is possible to transform the low-set gel into a high-set gel by re-heating it to above 80°C. In this report, "curdlan gel" refers to the high-set gel unless otherwise specified.

As mentioned above, an aqueous suspension of curdlan forms a thermo-irreversible gel when heated to above 80°C. It has also been known that the gel strength increases as the temperature increases up to 100°C.[4] A recent study shows the effects of heating at temperatures higher than 100°C on the properties of gels.[5] Even when an aqueous suspension of curdlan is heated to 100-130°C, temperatures applied to many foods packaged in retort pouches, the gel strength increases with increasing heating temperature. Also, gel strength increases as heating time increases.

Another unique property of the curdlan gel is its freeze-thaw stability. To demonstrate the unique characteristic of the gel, the behavior of the curdlan gel in frozen state is compared with those of carrageenan, agar-agar and konjac gels.[5] The gel strength and syneresis of various gels after freezing and thawing are shown in Table 1. The gel strength of curdlan gel is affected least by the freeze-thaw process, while a considerable increase in syneresis is observed. On the other hand, the gel strength of carrageenan and agar-agar gels decreases significantly as their gels exhibit sponge like texture after freeze-thaw process. The syneresis of carrageenan and agar-agar gels without freezing is low, but their measure- ment after freezing and thawing is impossible because the gels release water when pressed lightly. In the case of the konjac gel, the gel strength increases to about three times upon freezing and thawing, although the syneresis is practically equivalent to that of the curdlan gel, resulting in hardening and loss of elasticity.

Table 1. Effects of freezing and thawing on gel strength and syneresis of curdlan, carrageenan, agar-agar and konjac gels.

Sample (wt%)	After 20 hr at 4°C		After freezing and thawing	
	Gel strength (g/cm^2)	Syneresis (%)	Gel strength (g/cm^2)	Syneresis (%)
Curdlan				
4%	2,190	10.3	2,496	20.5
Carrageenan				
3.5%	2,307	1.4	509	—a
Agar-agar				
3%	2,409	0.6	188	—a
konjac				
3.5%	2,312	10.3	6,010	21.0

a: Measurement of syneresis was impossible because water ran when gel was pressed.

As to the syneresis of curdlan gel, it has been proved that addition of food starch to curdlan gel for the purpose of decreasing syneresis is most effective, even after freezing and thawing.[6] When waxy corn starch is used, the syneresis of curdlan gel decreases with increasing starch concentration. When the addition level of the starch is up to 10%, syneresis of the curdlan gel is hardly observed after 20 hours at 4°C or freeze-thaw.[5]

FOOD APPLICATIONS OF CURDLAN

When curdlan is utilized in food applications, in general, there are two categories (Table 2). The first category is its usage as a food additive for quality improvement purposes and the second is its usage as an essential ingredient in newly developed foods that utilize the unique gelling characteristics of curdlan.

Table 2. Potential food applications of curdlan.

Food	Function	Approx. use level
<u>Usage as a food additive:</u>		
•Noodles	Texture modifier	0.2-1%
•Kamaboko (boiled fish paste)	Texture modifier	0.2-1%
•Sausages, Hams	Texture modifier, Water holding	0.2-1%
•Processed cooked foods	Binding agent, Improvement in moisture retention and product yield	0.2-2%
•Processed rice cake	Retention of shape	4-6%
•Cakes	Retention of moisture	0.1-0.3%
•Ice cream	Retention of shape	0.1-0.3%
<u>Usage as an essential ingredient:</u>		
•Jellies	Gelling agent stable against heating and freezing-thawing	1-5%
•Fabricated foods Noodle-shaped tofu Processed tofu (frozen, retorted, freeze-dried) Thin-layered gel food (frozen) Konjac-like gel food (frozen) Heat-resistant cheese food	Gelling agent stable against heating and freezing-thawing	1-10%
•Edible films	Film formation	30-100%
•Dietetic foods	Low-energy ingredient	

Usage As a Food Additive

In this case, a relatively small amount or concentration of curdlan is added to existing food products to improve, modify and stabilize the physical properties of the products.

Noodles. The addition of 0.2-1% curdlan to Chinese noodles improves the texture of noodles. As the curdlan level increases in noodles, the Young's modulus, the breaking strength and elongation at breaking increase.

Kamaboko (Boiled Fish Paste). The addition of 0.5-1% curdlan to kamaboko increases the breaking strength and the elasticity of the product, and furthermore, decreases the expressible water because of the increased water holding capacity.

Hamburgers. The addition of 0.2-1% curdlan to hamburgers gives a soft and juicy texture and low cooking loss of the product.

Hams and Sausages. The addition of 0.2-1% curdlan to hams or sausages gives a soft and juicy texture and improves water holding capacity of the product.

Usage As an Essential Ingredient

In this case, curdlan is the main ingredient and essential to develop new food products.

Noodle-shaped tofu. Noodle-shaped tofu is a new product which is soy milk processed into the shape of noodles. Tofu, an extensively consumed soybeen food in Japan with a soft characteristic texture, is prepared by coagulating soy milk with coagulants such as calcium sulfate. Since we considered that a different texture would be obtained if tofu were processed into the shape of noodles, the noodle-shaped tofu is prepared using curdlan. Soy milk is heated up about 60°C and then mixed with curdlan and coagulant. After starch and ice are added to the suspension to form a low-set gel, the mixture is homogenized and deaerated. As a result of these processes, a high viscous suspension is prepared. Then, the suspension is extruded directly into hot water through a spinneret with many holes to form a high-set gel. It is possible to make noodle-shaped tofu with various shapes and sizes by exchanging the extruding holes. This noodle-shaped tofu is served in cold dishes, and it can be served in heated, boiled or fried dishes because of resistance to heating. Moreover, this noodle-shaped tofu, which differs from traditional noodles made from flour, is free from excessive swelling.

Tofu for retorting, freezing or freeze-drying. Traditional tofu is rarely used in processed foods which are subject to retorting because of hardening due to syneresis during the retort process. Therefore, tofu for retorting which can be used in foods packaged in retort pouches has been developed by new formulations and preparation methods using curdlan. An example of processed foods with this product is retorted Mabo-dofu (a Chinese dish, soybean curd with spicy minced meat). Ordinary retorted Mabo-dofu sauce does not contain tofu because of the above reason. Accordingly, before serving, consumers have to add tofu and heat. Usage of the tofu using curdlan can overcome this problem as the texture of the tofu is scarcely changed even after retorting.

Also, ordinary tofu is rarely used in frozen processed foods because it becomes a sponge like texture after freezing. Therefore, by new ingredients and preparation methods using curdlan, tofu for freezing has been developed. An example of processed foods with this product is frozen Agedashi-dofu (deep-fried tofu in stock). Its tofu retains the soft and smooth texture even after freezing.

Likewise, curdlan can be utilized for freeze-dried tofu which is made by the same preparation methods as the tofu for freezing. The formulation is slightly different, and the product is freeze-dried instead of normal freezing. This product is easily reconstituted with hot water and has smoother and softer texture than conventional products. Electron micrographs show that freeze-dried conventional tofu has a thick, irregular and layered structure, while freeze-dried tofu using curdlan has a thin, uniform and porous structure. Consequently, the product with curdlan maintains the soft and smooth texture when it is soaked in hot water. This freeze-dried tofu can be used in instant foods.

Frozen noodle-shaped konjac-like gel food. Konjac gel can not be used in frozen processed foods, because the texture deteriorates after freeze-thaw as described previously. Curdlan can be used to create a freezable noodle-shaped gel with a konjac-like texture. This product is prepared by extruding curdlan suspension into hot water in the same preparation methods as the noodle-shaped tofu and it has an elastic texture like konjac gel even after freezing and thawing. This frozen noodle-shaped konjac-like gel food can be utilized in frozen processed foods.

Low-fat sausage. In Japan, coarse-cut sausages are very popular because they have a characteristic juicy, soft and elastic texture. However, the high fat content of these sausages is a problem. Accordingly, low-fat sausage has been developed by substituting curdlan gel containing 15% vegetable oil for pork fat. The curdlan gel containing vegetable oil is prepared by emulsifying curdlan suspension and vegetable oil, deaerating under vacuum and then heating. The manufacturing process is the same process as ordinary coarse-cut sausages except for substituting the above emulsified gel for pork fat. This product has a similar appearance and texture to its full-fat counterpart, and acheives 44% fat reduction and 30% calorie reduction. As a matter of course, cholesterol is also reduced.

Non-fat whipped cream (analogue). As an example of curdlan based fat replacers, non-fat whipped cream (analogue) has been developed by substituting curdlan gel for milk fat. In this case, it is necessary to incorporate a whipping agent in the formulation, which is similar to that of the various substitute toppings made from vegetable oil, and the curdlan gel must be thoroughly homogenized until the particle size becomes small enough to be perceived creamy in the mouth. The creamy texture and the retention of shape of the whipped cream are caused by the particles of curdlan gel as a substitute for milk fat globules. This product results in a significant reduction in fat and about one-fourth reduction in calories when compared to traditional whipped cream.

FUTURE OUTLOOK

Curdlan, a polysaccharide produced by bacteria, is a very interesting food ingredient and it has the potential to create new and innovative food products. Research on food applications of curdlan has been still carried out. We expect a lot of efforts will be made to develop new foods with curdlan.

REFERENCES

1. T. Harada, Curdlan: a gel-forming ß-1,3-glucan, in: "Polysaccharides in Food", J.M.V. Blanshard and J.R. Mitchell, ed., Butterworths, London (1979).
2. M. Glicksman, Curdlan, in: "Food Hydrocolloids Volume 1", M. Glicksman, ed., CRC Press, Florida (1982).
3. T. Harada, S. Sato and A. Harada, Curdlan, *Bull. of Kobe Women's Univ.* 20:143 (1987).
4. I. Maeda, H. Saito, M. Masada, A. Misaki and T. Harada, Properties of gels formed by heat treatment of curdlan, a bacterial ß-1,3 glucan, *Agric. Biol. Chem.* 31:1184 (1967).
5. Y. Nakao, A. Konno, T. Taguchi, T. Tawada, H. Kasai, J. Toda and M. Terasaki, Curdlan: Properties and application to foods, *J. Food Sci.* 56:769 (1991).
6. F. Takahashi and T. Harada, Resistance of curdlan gel against freeze-thawing, *J. Home Economics of Japan.* 37:251 (1986).

RHEOLOGY AND DSC OF CURDLAN - DMSO - WATER SYSTEMS

M. Watase and K. Nishinari*

Chemical Research Laboratory, Faculty of Liberal Arts,
Shizuoka University, Ohya, Shizuoka 422, Japan

*National Food Research Institute, Tsukuba 305, Japan
Present address: Department of Food and Nutrition,
Faculty of Science of Living, Osaka City University
Sumiyoshi, Osaka 558 Japan

ABSTRACT

Effects of dimethyl sulfoxide (DMSO) on the gelation of curdlan
were studied by dynamic viscoelastic measurements and differential
scanning calorimetry (DSC). Elastic modulus E' of curdlan gels prepared
by heating curdlan-DMSO-water suspension as a function of measuring
temperature decreased with increasing temperature whilst E' as a
function of DMSO content showed a maximum at around 0.277 mole fraction
DMSO at the temperature range from 15 to 85°C. An endothermic peak in
heating DSC curve around 50~60°C for curdlan-DMSO-water suspension was
attributed to the gel formation, and the peak shifted to higher
temperatures with increasing DMSO content. It was suggested that
curdlan gels are formed mainly by hydrophobic interaction but the
contribution of hydrogen bonds is not negligible.

INTRODUCTION

Curdlan is a water insoluble polysaccharide produced by Alcaligenes
faecalis var. myxogenes. The chemical structure of curdlan is β-1,3
glucan. Heating the suspension of curdlan gives rise to gels; the
heating at the temperature range from 60°C to 70°C and cooling it leads
to a so-called low-set gel which is thermo-reversible whilst heating
above 80°C leads to a so-called high-set gel which is thermo-

Food Hydrocolloids: Structures, Properties, and Functions
Edited by K. Nishinari and E. Doi, Plenum Press, New York, 1994

irreversible. Konno and Harada[1] reported differential scanning calorimetry (DSC) results on curdlan-water suspension.

DSC and rheological studies were carried out to clarify the details of gelling mechanism and the relation between structure and properties in the present work.

EXPERIMENTAL

Curdlan (Lot. 13140) was purchased from Takeda Chemical Industries, Ltd (Osaka, Japan). Storage Young's modulus E' and mechanical loss tangent $\tan\delta$ as a function of temperature at 2.5Hz were determined from the observation of longitudinal vibrations of cylindrically molded gels.[2] Test pieces of the gels were prepared as follows: curdlan-DMSO-water systems were poured into a test tube of 28mm inner diameter, and stirred gently, and heated gradually. The temperature was kept at 90℃ for 40min, and then lowered to room temperature. Cylindrical gels were taken from molds and kept at 2℃ in silicone oil overnight before measurement.

DSC was carried out by use of a sensitive DSC apparatus SSC5200. DSC 120 (Seiko Electronics. Inc., Tokyo). The silver DSC pan of 70μL was heated at 250℃ for 30min before the measurements and then cooled. Then, about 50mg of curdlan-DMSO-water suspension was sealed in the DSC pan hermetically.[2] The temperature was kept at 2℃ for 30min and then raised at 2℃/min.

RESULTS AND DISCUSSION

Fig. 1 shows the measuring temperature dependence of storage Young's modulus E' and the mechanical loss tangent $\tan\delta$ for curdlan-DMSO-water gels with various polymer concentrations and DMSO contents. E' decreased gradually with increasing measuring temperature whilst $\tan\delta$ increased gradually. It is believed that junction zones in a so-called high-set gel of curdlan are formed by hydrophobic interaction because the gel is formed by heating above 65℃.[3] Since the elastic modulus as a function of measuring temperature decreases with increasing measuring temperature, however, the contribution from hydrogen bonds may not be negligible.

E' as a function of DMSO content showed a maximum at around 0.277mole fraction (mf) DMSO, and $\tan\delta$ became minimum at this DMSO content. The decrease of E' with increasing temperature was least conspicuous at this DMSO content.

The relation between E' at 25℃ and curdlan concentration C was represented by $E'=kC^n$, and the exponent n ranged from 1.3 to 1.7. The value of n became minimum (1.3) around DMSO content 0.277mf. According to the modern theories on concentration dependence of elastic modulus of gels by Clark and Ross-Murphy[4] and by Oakenfull[5], the slope of the

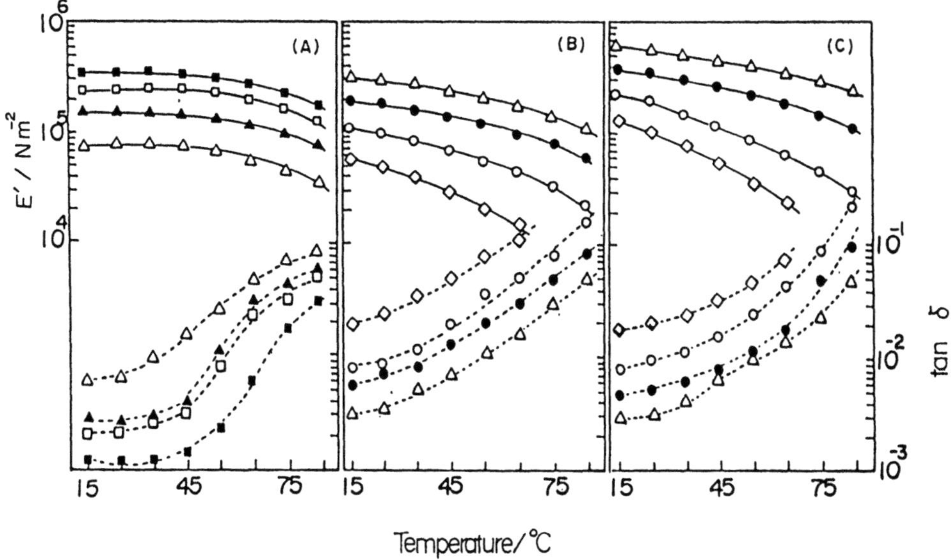

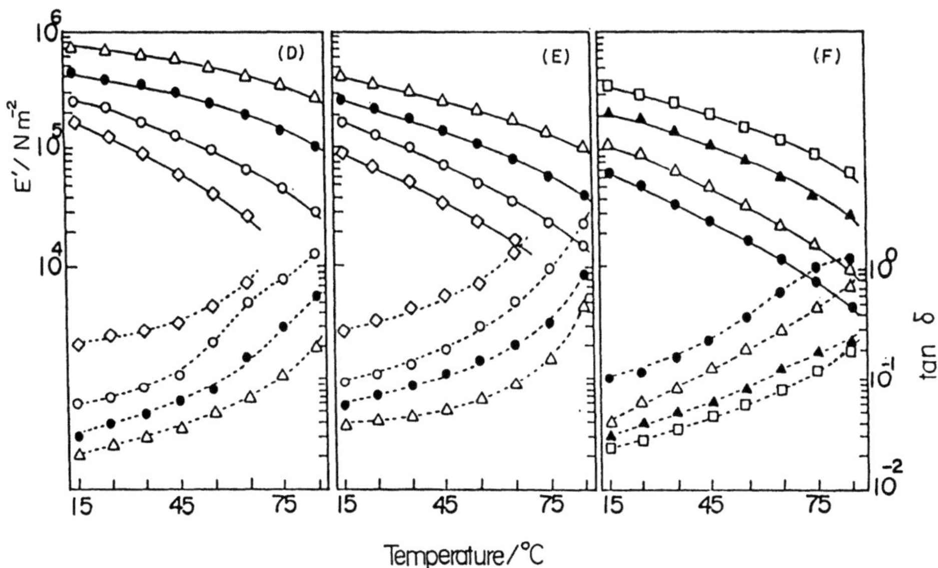

Fig. 1　Temperature dependence of storage Young's modulus E' and mechanical loss tangent tan δ　for curdlan-DMSO-water gels of various curdlan concentrations. Curdlan concentration: $\diamondsuit$, 3wt%; $\bigcirc$, 5wt%; $\bullet$, 7wt%; $\triangle$, 10wt%; $\blacktriangle$, 15wt%; $\square$, 20wt%; $\blacksquare$, 25wt%.　DMSO content: (A), 0mf; (B), 0.095mf; (C), 0.168mf; (D), 0.277mf; (E), 0.45mf; (F), 0.64mf.

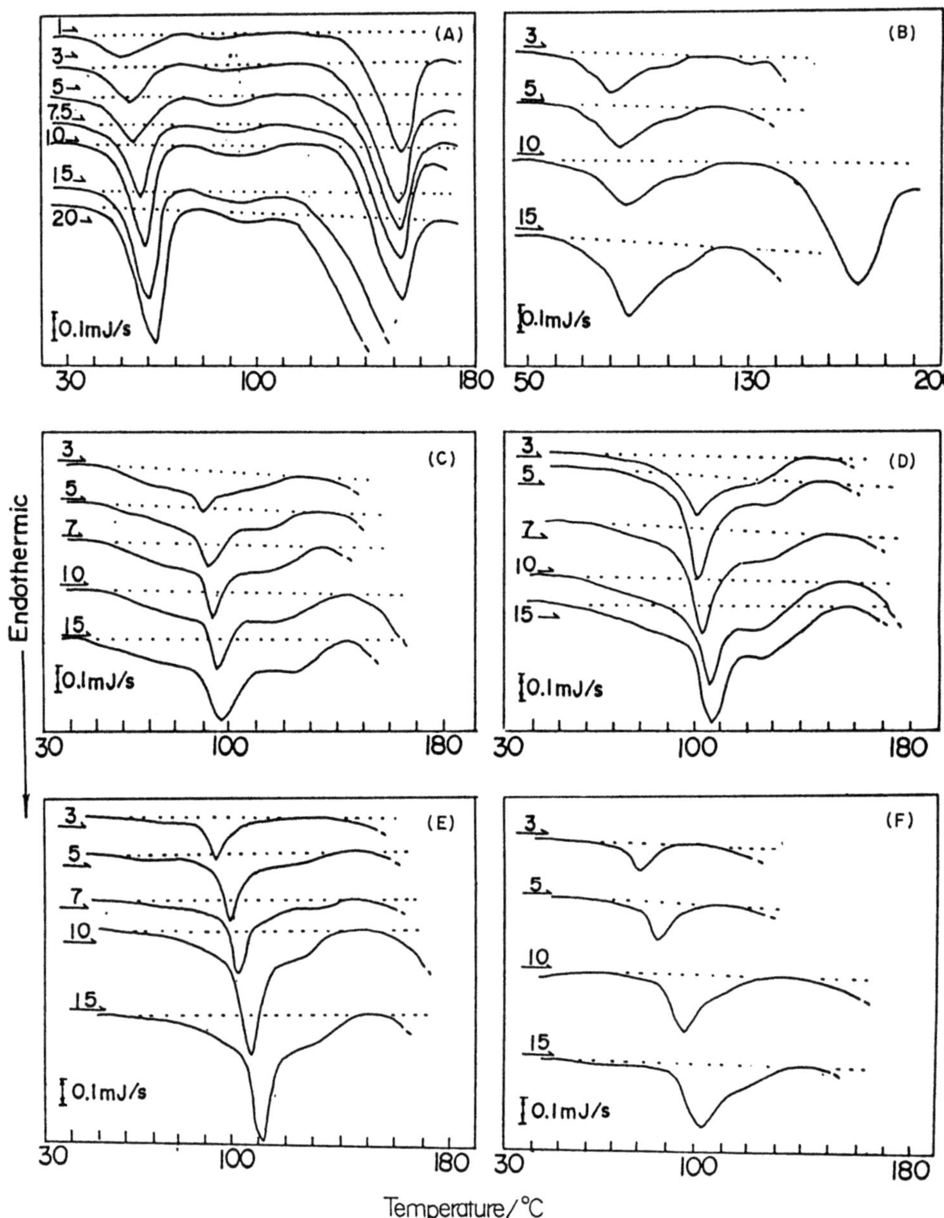

Fig. 2　Heating DSC curves of curdlan-DMSO-water gels of various curdlan concentrations. Figures beside each curve represent the curdlan concentration in wt%.　DMSO content: (A), 0mf; (B), 0.095mf; (C), 0.168mf; (D), 0.277mf; (E), 0.45mf; (F), 0.64mf.

128

double logarithmic plot of E' against C/Co, where Co is the critical
concentration, is larger in the dilute region than in concentrated
region. The experimental findings that n becomes minimum at 0.277mf
DMSO suggest that Co becomes minimum at this DMSO content.

Heating DSC curves for curdlan-DMSO-water suspension are shown in
Fig.2. An endothermic peak around 50°C~60°C shifted to higher
temperatures with increasing concentration of curdlan. The peak
temperature as a function of DMSO content shifted to higher temperatures
with increasing DMSO content and was highest for 0.277mf DMSO. The heat
absorbed on forming one mole of junction zones determined from Eldridge-
Ferry plot i.e. the plot of the logarithm of the polymer concentration
against the inverse of the melting temperature (endothermic peak
temperature) as well as the endothermic enthalpy determined from the
area enclosed by the endothermic peak and the baseline showed maximum at
0.277mf DMSO. These results suggest that the curdlan-DMSO-water forms
the most heat resistant gel at 0.277mf DMSO. It is well known that DMSO
interacts strongly with water, and forms hydrate. The structure of the
hydrate was thought as $(CH_3)_2 SO \cdot 2H_2O$ from viscosity[6] and ultrasonic
sound velocity[7] measurements, and as $(CH_3)_2 SO \cdot 3H_2O$ from the neutron
scattering.[8] The DMSO content 0.277mf is just between these two
models. Below this DMSO content, the free water content decreases with
increasing DMSO content, whilst above this DMSO content, the free DMSO
content increases with increasing DMSO content. Generally speaking, the
gel consists of the balance between the crystallinity and the
solubility. The increase in free water or DMSO leads the system to the
direction of solubility whilst the decrease in these leads to that of
crystallinity. Therefore, the curdlan-DMSO-water system forms the most
stable gel with highest elastic modulus and with highest heat resistance
at 0.277mf DMSO.

Broad endotherms appeared around 100°C and these shifted to higher
temperatures with increasing DMSO content. Another endothermic peak
appeared around 150°C and this peak did not return to the baseline. The
first peak around 60°C in curdlan-water systems was attributed to the
gelation. The third peak might be attributed to the structural change
of curdlan gels by further heating because when the DSC pans were opened
after heating beyond 150°C, the color of the gel was changed into brown.

REFERENCES

1. A.Konno and T.Harada, Food Hydrocolloids, 5: 427 (1991).
2. M.Watase and K.Nishinari, Makromol. Chem., 188: 2213 (1987).
3. A.Konno, H.Kimura, T.Nakagawa and T.Harada, Nippon Nogeikagakukaishi,
   52: 247 (1978) in Japanese.
4. A.H.Clark and S.B.Ross-Murphy, Adv. Polym. Sci., 83: 60 (1987).
5. D.Oakenfull, J. Food Sci., 49: 1103 (1984).
6. S.A.Schichman and R.L.Amey, J. Phys. Chem., 75: 98 (1971).
7. D.E.Bowen, M.A.Priesand and M.P.Eastman, J.Phys.Chem., 78: 2611 (1974).
8. D.H.Rasmussen and A.P.MacKenzie, Nature, 220: 1315 (1968).

STUDIES ON XANTHAN STRUCTURE BY SCANNING TUNNELLING MICROSCOPY

Michael Wilkins, Martyn C. Davies, David E. Jackson, Clive J. Roberts,
Saul J. B. Tendler, John R. Mitchell and Sandra E. Hill

Departments of Pharmaceutical Sciences and ABFS
University of Nottingham
Nottingham
NG7 2RD, U.K.

INTRODUCTION

Scanning tunnelling microscopy (STM) is a technique for the direct 3D imaging of surface atoms, molecules and biomolecular shapes. The use of the equipment for the imaging of polysaccharides is still in its infancy and techniques have to be established that allow critical evaluation of the images produced by STM. Images of the microbial polysaccharide xanthan using preparation techniques applicable to both the STM and electron microscopy (EM) have been produced, thus enabling comparison of the two techniques.

METHODS AND MATERIALS

Materials Xanthan samples, tradename Keltrol (supplied by Kelco International Ltd. London U.K.) were dissolved in 0.1M ammonium acetate buffer to achieve an approximate concentration of 1mg/ml. Debris was removed by centrifugation. The solutions were further diluted to a working concentration of 80µg/ml with 60% glycerol.

Deposition and Coating Aliquots of xanthan were sprayed or added as droplets (8µl) onto freshly cleaved mica discs, gold or highly oriented pyrolytic graphite (HOPG) then dried in air or under vacuum (approx. 10-6 Torr for 1 hr). These samples were investigated by STM.

Food Hydrocolloids: Structures, Properties, and Functions
Edited by K. Nishinari and E. Doi, Plenum Press, New York, 1994

Other samples prepared as above on mica discs were rotary shadowed with a 0.6nm to 0.7nm thick Pt/C (95%:5%) layer at an incident angle of 5°. An additional coating, of 6 to 7 nm thick carbon backing layer at an incident angle of 90°, was applied to the samples with the Pt/C coating as in standard EM practice. These replicates were transferred to copper grids and HOPG substrates and examined by EM and STM respectively.

Sample Analysis STM images were obtained under ambient conditions using a VG STM 2000 (VG Microtech, Uckfield, UK) employing a negative biased, mechanically cut platinum/iridium (80:20) tip. The microscope was operated in the constant current mode, with a tunnelling current and voltage being set at 30pA and 1.5V respectively.

A Philips EM 400T electron microscope was also used to image the xanthan samples. This was operated in transmission mode at 80keV accelerating voltage with a $50\mu m$ aperture and nominal electron magnification between 17000x and 46000x.

RESULTS

Attempts to image the xanthan molecules without the coating showed only the underlying substrate irrespective of the type of substrate or the concentration of xanthan solutions applied. This suggested that the absorbed molecules were being swept out of the scan area by the tip.

The Pt/C followed by carbon overcoating of the molecules ensured mechanical stability of the samples so that replica samples could be imaged by both the EM and STM. The correlation between the the images obtained from both techniques was excellent and thus validates the use of the STM to look at the polysaccharide. The STM images show the isolated molecules have the polydispersity of chain length and the entanglements reported by EM studies (Stokke and Elgsæter, 1991).

However, the overcoating does reduce the resolution that should be possible for the STM. The grain size of the carbon backing layer ranged from 16 to 20 nm in diameter. It was found that the Pt/C coating alone stabilised the xanthan sufficiently for imaging by the STM and this allows much improvement in the resolution. The grain size of the coating was 5 ± 1 nm and the molecular dimensions measured were 8 ± 1 nm.

The samples deposited on the substrate as droplets showed dense and non uniform networks of xanthan molecules, as demonstrated in Figure 1a. At higher resolution the individual xanthan molecules can be seen (Figure 1b). Again there was excellent correlation between the images obtained by STM and those seen by EM. It was thought that the network could have been caused by the advancement of the drying front during the sample preparation.

Images of the xanthan which had been sprayed onto a mica surface would seem to indicate that the molecules were isolated and laterally dispersed across the surface rather than combining in a network. It was noted that networks did occur when spraying the xanthan onto the HOPG. This agglomeration on the surface of the hydrophobic HOPG may reflect the differing adhesion forces between the xanthan and the solid interface.

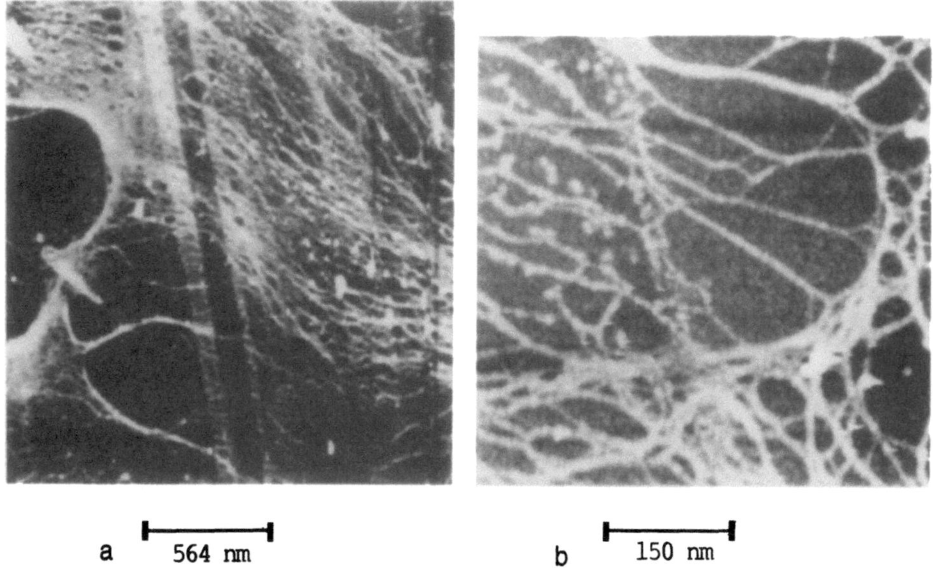

Figure 1. STM topographs showing xanthan solutions deposited as droplets onto HOPG and coated by Pt/C. Figure 1a is a 1945x1945 nm scan and Figure 1b is at a higher resolution where the scan represents 591x591nm.

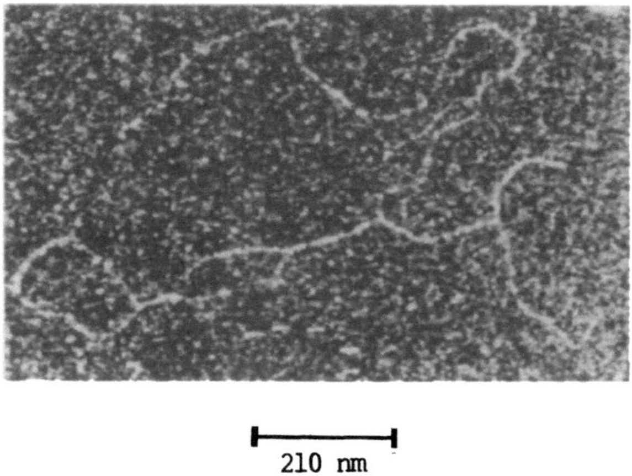

Figure 2. A 996nmx591nm scan of xanthan sprayed onto a mica disc and rotary shadowed with a Pt/C layer. The xanthan molecule was imaged directly onto the mica substrate and has a maximum diameter of 8nm. The molecule forms two loops suggesting the presence of single and double strand conformations.

Figure 2 shows an STM image of Keltrol sprayed onto mica and coated with Pt/C and shows a single strand and a double strand conformation. The double strand is wider than the single strand and the measurements, allowing for coating, are in agreement with the EM studies.

It is possible to use the background determination technique to reveal the xanthan network. Image noise can be removed by median filtering and the background removed by subtraction of the results from a minimum ranking filter. Subsequent "thresholding" to remove all data with heights below 10% of the maximum can be performed to reveal the non-substrate information.

CONCLUSION

It is possible to obtain images from STM which are similar to those produced for EM for the polysaccharide xanthan. The method of deposition of the sample has a major influence on the images produced. The molecules are seen to form networks, demonstrate differences in chain length and chain width. There is indication that double and single chains occur. Although, a coating layer is required to visualise the molecule at the present time, the coating can be thinner that that used in electron microscopy and hence the resolution is higher.

Studies are in progress to measure the persistence lengths of the molecules for a range of xanthan molecules and relate these to other characteristics of the gums. Further work is being carried out to try and image individual polysaccharides molecules and combinations of polysaccharides without coating thereby producing the resolution required to visualise structure of the molecules.

ACKNOWLEDGEMENTS

We wish to thank the MAFF/DTI hydration (HYDRA) Link programme for their support for these studies. We would also like to acknowledge the support given to us by Dr. Stokke at the Norwegian Biopolymer Laboratory, Trondheim.

REFERENCE

Stokke, B.T. and Elgsæter, A., 1991, Electron microscopy of carbohydrate polymers, Advances in Carbohydrate Analysis. 1:195-247.

Williams, P. M., Davies, M. C., Jackson, D. E., Roberts J. C., Tendler, S. J. B. and Wilkins, M. J., 1992, Biological applications of scanning tunnelling microscopy: novel software algorithms for the display, manipulation and interpretation of STM data, *Nanotechnology* 2:172-181

XANTHAN GUM: A MULTIFUNCTIONAL STABILISER FOR FOOD PRODUCTS

Ian A. Challen

Kelco International Ltd.
Westminster Tower
3 Albert Embankment
London, SE1 7RZ
United Kingdom

ABSTRACT

Xanthan gum is the exocellular biopolysaccharide produced by the bacterium _Xanthomonas_ _campestris_. During commercial production the gum is made by a pure culture viscous fermentation process[1] The commercial scale production process was developed by the Kelco Company (now a Division of Merck & Co. , Inc.). Since it was granted FDA food approval in 1969, KELTROL xanthan gum has found functional applications in a very diverse range of food products worldwide. Even today, when we might expect that all avenues for its application have been explored and exploited, new applications continue to be found for it. It is noteworthy that, now, for the majority of applications in which xanthan gum is the product of choice, it is being used almost exclusively for the functional benefits that it provides. Its functionality is a direct consequence of its unique chemical structure.

Xanthan gum binds water more strongly than the other hydrocolloids; it is more tolerant to pH, salts, temperature, alcohol and enzymes than these polymers. It also enters into strong synergistic interactions with the galactomannans.

During the lecture, examples of food products will be discussed in which one or more of the properties of xanthan gum are being used to provide the functional benefits.

STRUCTURE

The structure of the xanthan gum molecule is well documented and will not be considered in detail in this paper[2,3]. However, the uniqueness of the functionality of the polymer lies in its structure: a cellulose backbone substituted with a trisaccharide side chain on alternate glucose residues in the backbone. In solution the side chains wrap around the backbone thereby protecting the β -1, 4 linkages from hydrolysis and subsequent loss of solution viscosity.

Food Hydrocolloids: Structures, Properties, and Functions
Edited by K. Nishinari and E. Doi, Plenum Press, New York, 1994

Preparation

Xanthan gum is completely soluble in cold water where full viscosity can be achieved provided the normal precautions for making solutions of hydrocolloids are taken. The use of heat during solution preparation will have very little effect upon the time taken to hydrate the gum. At 20° C the gum will typically generate a viscosity of 1200-1600 mPa.s (1% solution, Brookfield LVT viscometer, 60rpm).

Flow Properties

'Xanthan gum solutions at low ionic strength undergo a thermal transition which was first demonstrated by Jeanes et al.[1] as a signoidal change in viscosity of 1% salt-free solutions (t_m = 55° C). Subsequently Morris et al.[4] showed that optical rotatory and circular dichroic transitions are coincident with the viscosity change, indicating a conformational transition of the molecule. These data are consistent with the unwinding of an ordered structure like a helix into a random coil, with a consequent decrease in effective hydrodynamic volume and, therefore, viscosity[5] (Fig. 1).

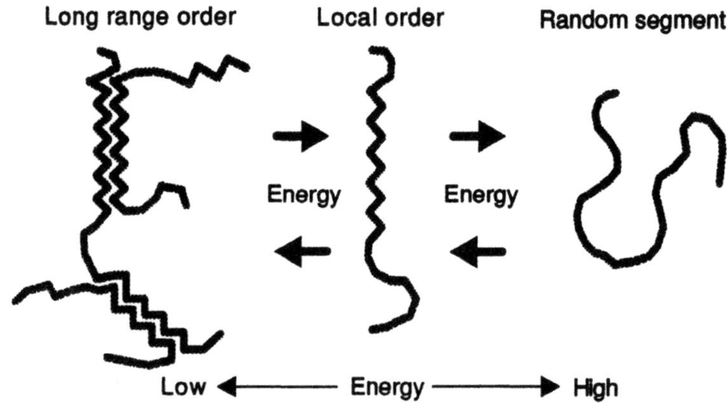

Fig. 1 Conformational ordering of xanthan gum in solution

Xanthan gum solutions exhibit pseudoplastic flow properties. At rest, or in the absence of shear, the solution has a very high apparent viscosity. As the shear rate is increased, so the measured viscosity decreases. Eventually, at very high shear rates, the measured viscosity levels out. When the shear is removed, the viscosity returns to its original value almost instantaneously. This behaviour is in marked contract to other common hydrocolloids, such as the alginates and guar gum, whose flow behaviour is more Newtonian at low rates of shear.

When a food product is eaten the shear exerted in the mouth is considered to be 50-200s[-1]. At such shear rates the xanthan gum in the product effectively exhibits a much lower viscosity than it does at lower shear rates thus; not only does the product taste less viscous but the release of flavour is thereby enhanced.

Suspension

The property of pseudoplasticity offers benefits to the product developer and to the food product alike: the high viscosity at low shear provides the ideal conditions for suspensions especially since the high viscosity is regenerated immediately after the shear is removed.

This property is made use of in many types of food products to suspend solid particles such as flour in batters oil, in emulsified dressings and air in cake mixes and other aerated products such as ice cream.

Emulsions are good examples of liquid systems that require stabilisation and xanthan gum is particularly useful in oil-in-water emulsions. Emulsions stabilised with xanthan gum exhibit very high viscosity at rest and are exceptionally stable under the low shear conditions encountered during transport, storage and subsequent use. For an oil-in-water emulsion to be stable the continuous phase must be viscous enough to prevent coalescence and separation of the disperse oil phase.

It should be noted that while xanthan gum is not an emulsifier as such it will help to stabilise an emulsion once it has been formed. The xanthan gum produces high viscosity in the aqueous phase at low shear rates, thereby ensuring that the individual oil droplets remain separate.

COMPATIBILITY

Xanthan gum produces solutions having a wide range of compatibility with many food and non-food ingredients and additives. In some cases the degree or level of compatibility is dependent upon the concentration of xanthan gum used, as we shall see later.

Salt

The presence of ionic material stabilises the three-dimensional xanthan gum network, thereby providing additional stable viscosity at salt (sodium chloride) levels greater than about 0.1%. Since many foodstuffs contain salt at >0.1% for reasons of taste and preservation, the benefit of viscosity stability is evident. It should be noted that a minimal salt concentration is needed to generate the viscosity stability.

From ca. 0.1% up to ca. 20% sodium chloride xanthan gum will provide long term viscosity stability to many food products such as soy sauce and marinades for meat and 'half-fabrics' or partial food preparations.

Some xanthan gums are available that can be hydrated in 20% sodium chloride solution -a phenomenon that is normally considered to be very difficult to achieve.

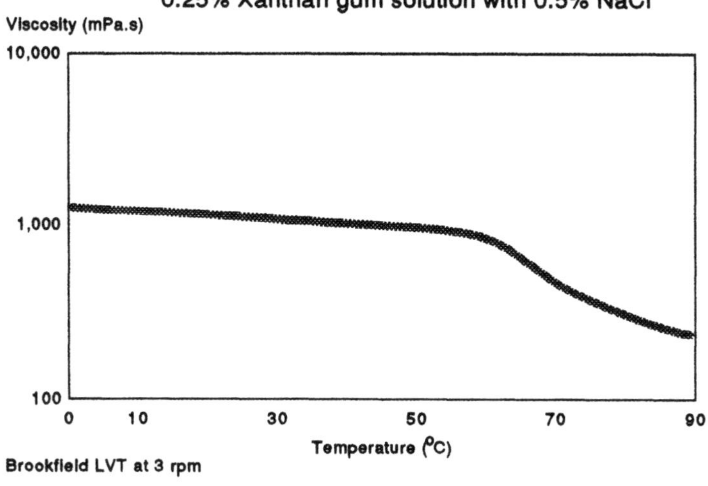

Fig. 2 Viscosity/Temperature profile of 0.25% xanthan gum solution

Temperature

The effect of temperature on solution viscosity is gum-concentration dependent. At low gum levels (eg 0.25%) there is a point of inflexion in the viscosity curve at ca. 40oC; this point moves as the temperature is increased so that, at a gum level of 1% it is about 90oC. (Fig.2)

In effect this means that typical heating regimes used during manufacturing processes (such a pasteurisation and sterilisation) are accomplished when the gum is exhibiting relatively low viscosity, hence heat penetration is faster and the process is more economical.

The other most important point concerning the effects of temperature is the recovery of the viscosity when the product or solution is cooled. Over the entire pH range of foodstuffs, full viscosity is regenerated after pasteurisation. This property is unique among the food hydrocolloids.

These effects are clearly demonstrated in products like gravy or sauces, where the xanthan gum binds the water so strongly that it helps to delay the formation of a surface skin. This keeps the gravy in a good condition for a longer time. In acidic fruit sauces, which are used with hot meats, the xanthan gum will prevent the heat from the meat causing the sauce to loose all of its viscosity.

pH

The effects of pH on solution viscosity are also dependent upon the gum concentration. At levels of 1% gum, there is virtually no change over the pH range 1.5 to 11. However, at a gum level of 0.25% maximum viscosity is achieved between pH 6 and 8; there is a slight decrease on either side of this range. This predictable viscosity stability provides great scope for the product developer who can make a product at one pH and then change the pH knowing that any viscosity change in the product will be caused by ingredients other than the xanthan gum. It should also be noted that the viscosity generated by the xanthan gum will be stable throughout the shelf life of the food product.

Typically, xanthan gum is used in fruit juices to help to suspend the fruit particles, to prevent oil ringing and to provide good viscosity without masking the natural flavour of the fruit. It is also used in acid relishes to prevent syneresis from the cut vegetable pieces; in all types of salad dressings for temporary or permanent emulsion stability, for herb suspension, ease of pouring and cling to the salad.

Alcohol

Xanthan gum can be used to increase the viscosity of food alcohols up to a level of 60% w/w alcohol. Above this alcohol level the gum becomes incompatible with the alcohol and a gel results. At levels of alcohol below 60% thickened solutions are produced whose viscosity is determined by the level of gum used.

This thickening property is made use of in industrial alcohol preparations, fillings for chocolate confectionery and for ready-to-drink and dry mix alcoholic beverages.

Enzymes

Xanthan gum solutions are not depolymerised by enzymes commonly encountered in food products and raw materials. Thus this gum is frequently the product of choice for stabilising food products that are noted for residual enzyme activity such as pineapple preparations and various spicy sauces such as hot pepper sauce where it is used to prevent solid separation and to provide additional viscosity.

Reactions with the Galactomannans

Xanthan gum will interact synergistically with the galactomannans to produce increased solution viscosity and/or gels. At all concentrations of guar gum and at low concentrations of locust bean gum (LBG) viscosity increases are observed. At higher levels of LBG, thermoreversible gels are formed.

The interaction between xanthan and guar gum will occur at low temperature since both gums are fully hydrated in cold water. However, the use of normal processing temperatures generally enhances the viscosity synergism effect. The strength of the interaction depends upon both the ratio of the two gums and on the total gum concentration. For example, at a total gum level of 0.2% maximum, increased viscosity is obtained at a ratio of 50:50. When the total gum level is raised to 0.5% the maximum viscosity increase occurs at a ratio of 70:30 guar: gum: xanthan gum.(Fig.3) While the interaction produces enhanced viscosity, the rheology of the solution is also modified, the typical pseudoplastic flow properties of the xanthan gum solution becoming more Newtonian in behaviour.

The viscosity synergism is reduced, or even eliminated, in the presence of some common materials, eg acids and salts. Hence this synergism is of limited value, especially in acidic or saline conditions.

Xanthan Gum / Guar Gum Interactions

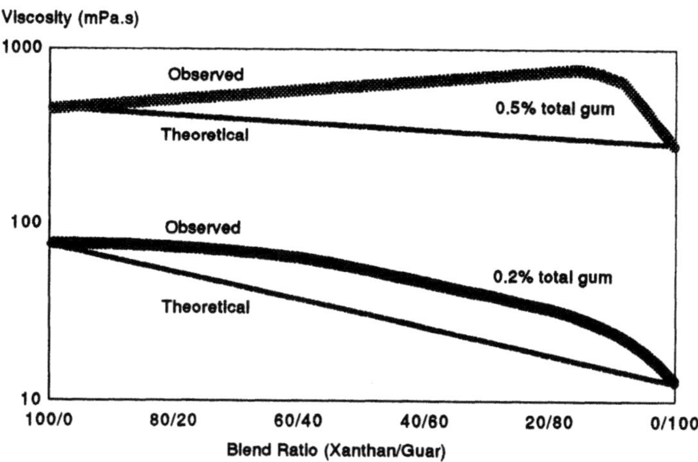

Fig. 3 Synergistic viscosity of xanthan gum/guar gum solutions

LBG is not hydrated until it is heated to >80°C for 5-10 minutes. Thus the interaction between xanthan gum and LBG can only take place in hot solution, the synergism becoming evident on cooling. At neutral pH, the maximum viscosity synergism is generated with equal parts of the two gums at a total gum level of ca. 0.1%. Again, solutions show modified pseudoplastic flow properties, as would be expected.

Above a level of >0.1% total gum concentration, thermoreversible gels are formed. The gel is formed by allowing a hot solution of the two gums to cool. Gelation begins at about 50°C and the gel strength increases down to about 25°C where some 95% of the final gel strength will be obtained. With time, the full gel strength will be achieved. These gels are very elastic and cohesive and possess good freeze-thaw stable properties. Again, the presence of ionic and/or acidic materials will reduce or even inhibit the interaction. The gel is stronger at neutral pH than at acid pH (Fig.4).

Gel Modification

In many cases the texture of the xanthan gum/LBG gel is not optimal for a given food product. This situation is not unique since many processed food gels contain a variety of materials used to generate not only the required texture, but also to build in other necessary properties such as freeze-thaw stability or bakestability.

This same approach can equally well be used for the xanthan gum/LBG gels. Firstly, the concentration of the interacting gums can be altered - this will affect the gel texture as well as other parameters. Alternatively, the ratio of the gums can be altered, thereby altering the strength of the interaction. Adjusting the pH will effect both the strength of the interaction as well as the gel strength in gelling conditions. A change of temperature will also change the gel strength - up to the temperature at which the gel will start to melt. Finally, other thickening or gelling agents can be included into the system to help to generate the desired effect.

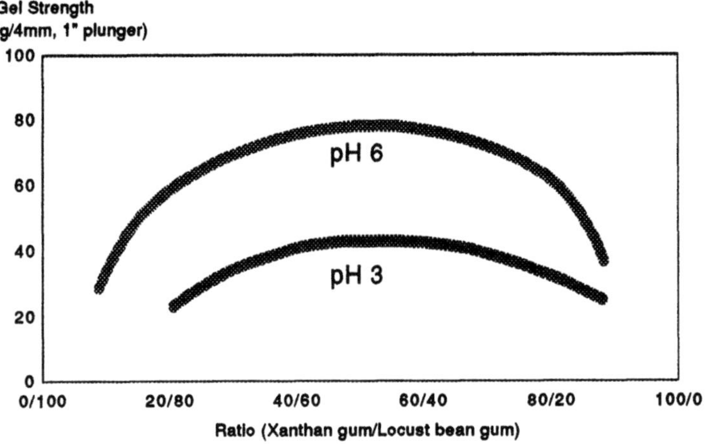

Fig.4 Xanthan Gum/Locust Bean Gum gel strength - effects of pH

REFERENCES

(1) Jeanes, A. R., Pittsley, J.E. and Senti, F.R. 1961. Polysaccharide B - 1459: A new hydrocolloid produced from glucose by bacterial fermentation. J. Appl. Sci. 5: 519-26

(2) Jansson, P. E. , Kenne, L. and Lindberg B. 1975. Structure of the extracellular polysaccharide from <u>Xanthomonas campestris</u>. Carbohydrate Research 45: 275-282

(3) Melton, L. D., Mindt, L., Rees D. A. and Sanderson G.R.. 1976. Covalent structure of the polysaccharide from <u>Xanthomonas campestris</u>: evidence from partial hydrolysis studies. Carbohydrate Research 46(2): 245-257

(4) Morris, E.R., Rees, D.A., Young, G. Walkinshaw, M. D., Darke, A. 1977. Order-disorder transition for a bacterial polysaccharide in solution: a role for polysaccharide conformation in recognition between <u>Xanthomonas</u> pathogen and its plant host. Journal of Molecular Biology 110(1):1-16

(5) Xanthan Gum - Natural Biogum for Scientific Water Control. Third Edition. Kelco Division of Merck & Co., Inc.

ERS STUDY ON SPIN-LABELED XANTHAN GUM

Shoji Takigami[1] Machiko Shimada[1], Peter A. Williams[2] and Glyn O. Phillips[2]

[1]Faculty of Engineering, Gunma University, Kiryu, Gunma 376, Japan
[2]Faculty of Science, Health and Medical Studies, North East Wales Institute
Deeside, Clwyd CH5 4BR, U.K.

ABSTRACT

The thermally induced conformational change of aqueous xanthan solutions has been studied using ESR spectroscopy.

Nitroxide spin-label was covalently attached to carboxyl groups on the side-chains of xanthan gum. Aqueous spin-labeled xanthan solutions at various NaCl concentrations gave rise to composite spectra at room temperature due to the presence of isotropic and anisotropic components. The ESR spectra gradually changed with rising temperature and motionally-narrowed three-lined isotropic spectra were observed. The disappearance of the anisotropic component was attributed to the change from an ordered to disordered structure. The transition temperature of the conformational change increased with salt concentration. It was also found that in the disordered form the spin labeled xanthan in dilute salt solution had a smaller value of the rotational correlation time than that in high concentration solution. This was attributed to the presence of some ordered regions along the xanthan chains or to intermolecular chain association.

INTRODUCTION

Xanthan gum is an exocellular polysaccharide produced by the bacterium *Xanthomonas Campestris*. It consists of a β-D glucose backbone with a charged trisaccharide side-chain consisting of two D-mannose residues and a D-glucronic acid residue.

The thermally induced conformational change has been studied by many workers using a variety of techniques[1-4]. Recently, ESR spectroscopy has been used to monitor the rotational motion of polymers by attaching a nitroxide spin label and is able to monitor molecular motion in the 10^{-7} to 10^{-11} sec timescale range. The spin-label technique has been employed to monitor conformational changes of biopolymer molecules, but few studies of xanthan gum[5] have been undertaken.

In this work, nitroxide spin-label agent was covalently attached to the side-chains of xanthan gum. The objectives of this study are to investigate the thermally induced conformational change of xanthan gum in aqueous solution and salt solutions by ESR, and to investigate the effect of salt on the conformational transition.

Food Hydrocolloids: Structures, Properties, and Functions
Edited by K. Nishinari and E. Doi, Plenum Press, New York, 1994

EXPERIMENTAL

Materials

The xanthan gum used was KELTROL T supplied by Kelco International Ltd.. The xanthan was freed of cell debris by centrifugation (at 12,000 g for 1 h) and precipitated using acetone. 4-amino 2,2,6,6-tetramethyl piperidino -1-oxy (4-amino TEMPO) and 1-ethyl-3-(3-dimethylamino propyl) carbodiimide (WSC) were purchased from Sigma Chemical Co. Ltd. and used without further purification.

Spin-labeling of Xanthan Gum

Xanthan gum (1 g) was dissolved in distilled water (450 ml) at 65 °C and the solution was cooled to 4 °C. A solution of WSC (0.85 g in 35 ml of water) was added and the pH was adjusted to 5.0 using 0.1 M HCl solution. The mixture was stirred at 4 °C for 15 min, and then a solution of 4-amino TEMPO (0.48 g in 15 ml of water) was added. The pH of the mixture rose to 9.7. The mixture was stirred for 1 h, and was allowed to stand for 93 h at 4 °C.

In order to disrupt ionic bonding between xanthan and free spin label, 4M NaCl solution was added to the reaction mixture to give a salt concentration of 2 M, then the solution was dialysed against 2 M NaCl aqueous solution for 1 day and against distilled water until the dialysate showed no chloride ions. To remove free spin label reagent completely, this process was repeated twice, and then the xanthan solution was freeze-dried. The spin label was covalently attached to the side chains of xanthan through carboxyl groups. The degree of spin labeling estimated from the double integration method was 1 spin label per several hundred sugar residues.

ESR Spectroscopy

0.3 % spin label xanthan solutions were prepared without heating. The ESR spectra of spin labeled xanthan were measured using a Bruker ESP 300 spectrometer equipped with a temperature control apparatus. The measurements were carried out in the range of 20 °C to 95 °C using a flat quartz cell suitable for aqueous solutions. The modulation frequency, the microwave power and modulation amplitude were 100 kHz, 9.9 mW and 0.56 gauss (G), respectively.

RESULTS AND DISCUSSION

Typical ESR spectra of spin-labeled xanthan in aqueous solution are shown in Fig. 1. The spectrum obtained at 25°C (spectrum *a*) contained both isotropic and anisotropic components indicating restricted mobility. Since the spin label is attached to the xanthan side-chains this is consistent with current models of the xanthan ordered structure with the side chains associating with the backbone. On increasing the temperature to 70°C (spectrum *b*) a motionally-narrowed three-lined isotropic spectrum was observed indicating increased mobility of the spin label. This was attributed to the fact that the xanthan chains now adopted a disordered structure with the side-chains no longer associated with the backbone. On reducing the temperature back to 25°C (spectrum *c*), the spectrum again showed an anisotropic component but the proportion was less than before suggesting that spectrum *a* may also signify that chain aggregation occurs. The spectra obtained in 1 x 10^{-3} and 1 x 10^{-2} M NaCl showed a similar trend but in 1 x 10^{-1} M NaCl (Fig. 2) and 5 x 10^{-1}M NaCl even at 95°C (spectrum *b*), the spectrum did not become as motionally-narrowed as in the absence of electrolyte.

In order to illustrate the effect of salt concentration on the conformational change of xanthan, the ratio of the peak height of the central line (h_0) to that of the high field line (h_{-1}) was used as an arbitrary measure of mobility. The ratio (h_0/h_{-1}) for each system changed linearly with reducing temperature in a different manner, and an inflection point, which might reflect the transition temperature (Tm) of the conformation change, was observed. Fig. 3 shows which was attributed to the relationship between the Tm and logarithm of the

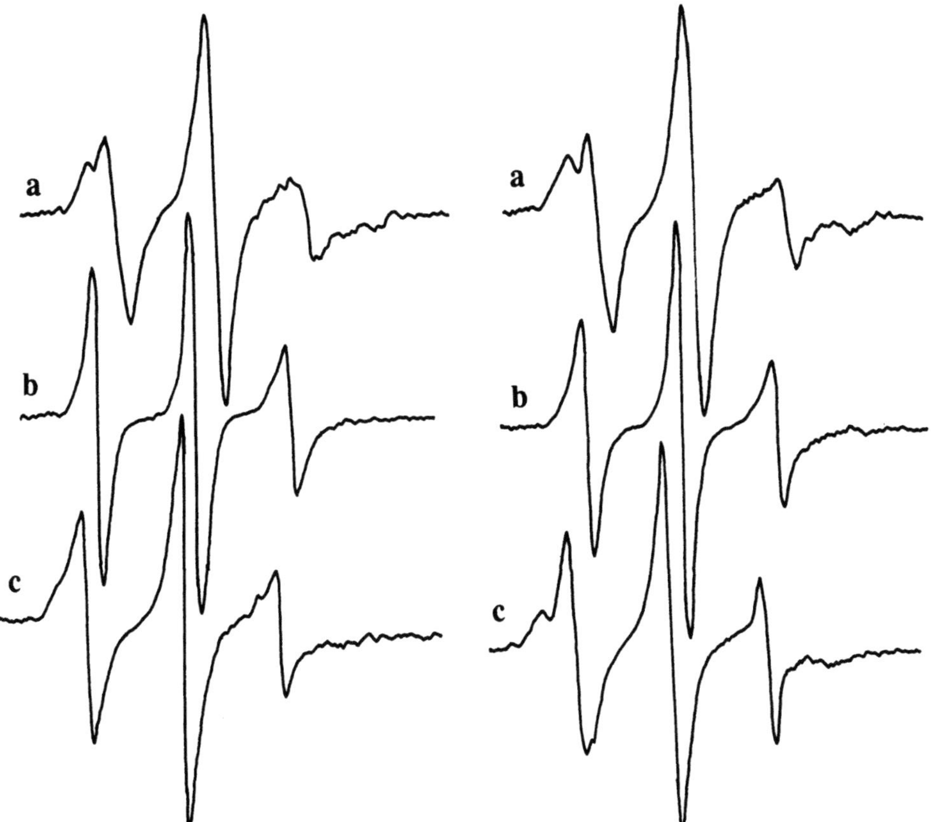

Fig.1. ESR spectra of spin-labeled xanthan in aqueous solution: (a) 25°C; (b) 70°C; (c) 25°C (after cooling).

Fig. 2. ESR spectra of spin-labeled xanthan in 1 x 10-1 M NaCl solution: (a) 30°C; (b) 95°C; (c) 20°C (after cooling).

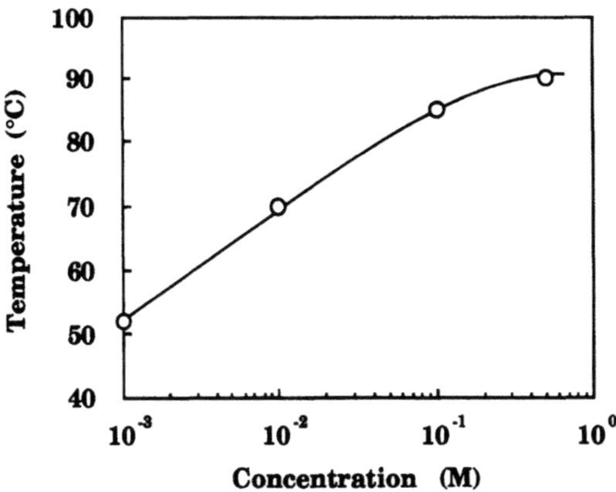

Fig. 3. Relationship between transition temperature of conformation change and logarithm of concentration of NaCl.

concentration of NaCl.   The Tm increases linearly with the logarithm of the salt concentration up to 1 x 10^{-1} M, and then increases gradually. This means that the unfolding of the helix by heating is depressed by addition of salt and the thermal stability of xanthan is enhanced by an increase of the ionic strength. A similar tendency was reported using the optical rotation[2] and DSC[6] techniques.

Since the ESR spectra were essentially isotropic at high temperatures, above Tm, the rotational correlation time (τ_c) was calculated by the following equation which can be applied only to isotropic spectra:

$$\tau_c = K\, \omega_0\, \{\ (h_0/h_{+1})^{1/2} + (h_0/h_{-1})^{1/2} - 2\}\qquad(1)$$

where ω_0 is the line width of the central line, and h_{+1}, h_0, and h_{-1} are the peak heights of the low field, central and high field lines, respectively. K is constant and the value used is 6.08 x 10^{-10} according to Zhao et al.[7]. The values of τ_c thus obtained are listed in Table 1.

Table 1. Relationship between correlation time and salt concentrations for spin-labeled xanthan in various concentration of NaCl solution.

samples	Correlation time (x 10^{-10} sec)							
Temp (°C)	40	50	60	70	80	85	90	95
Aq. Soln.	14.6	12.5	10.4	8.2	-	-	-	-
1 x 10-3 M	12.2	12.3	9.3	7.7	6.7	-	-	-
1 x 10-2 M	*	*	10.2	8.8	6.0	-	-	-
1 x 10-1 M	*	*	*	11.7	12.5	12.6	10.9	11.1
5 x 10-1 M	*	*	*	*	11.9	-	11.1	12.9

* : impossible to determine because of overlapping peaks

\- : unmeasured

τ_c for xanthan in aqueous solution decreased with rising temperature and a value of 8.2 x 10^{-10} s was obtained at 70 °C. Similar trends were observed for the spin-labeled xanthan in 1 x 10^{-3} M and 1 x 10^{-2} M aqueous NaCl solutions, and the values of the τ_c were   6.7 x 10^{-10} s and 6.0 x 10^{-10} s at 80 °C, respectively. On the other hand, in 1 x 10^{-1} M and 5 x 10^{-1} M aqueous NaCl solutions, the τ_c values measured at 95 °C were 1.1 x 10^{-9} s and 1.3 x 10^{-9} s. This means that the rotational motion of the spin label attached on the side chains is somewhat restricted in a high concentration salt solution even when the xanthan molecules have adopted a disordered structure.   This suggests that ordered sections exist along the xanthan chains or that there is intermolecular chain association. Liu et al.[3] have proposed that xanthan in 1 x 10^{-2}M aqueous NaCl solution takes the disordered conformation at 80 °C in which four wormlike chains are joined together by a very short double-helical segment. The value of τ_c of xanthan gum in dilute salt solution is consistent with their result. Accordingly, it was found that xanthan gum in aqueous solution and aqueous NaCl solution with less than 1 x 10^{-2} M takes the partially unmelted helical segments even when disordered at high temperature.

REFERENCES

1. E.R. Morris, D.A. Rees, G. Young, M.D. Walkinshaw, and A. Darke, Order-disorder transition for a bacterial polysaccharide in solution. A role for polysaccharide conformation in recognition between xanthomonas pathogen and its plant host, *J. Mol. Biol.* 110: 1(1977)
2. G. Holzwarth, Conformation of the extracellular polysaccharide of xanthomonas campestris, *Biochemistry* 15: 4333(1976)
3. W. Liu, T. Sato, T. Norisuye, and H. Fujita, Thermally induced conformational change of xanthan in 0.01M aqueous sodium chloride, *Carbohydr. Res.* 160: 267(1987)
4. P. Fuoss, B.T. Stokke, and O. Smidsrod, Thermal stability and chain conformational studies of xanthan at different ionic strengths, *Carbohydr. Polym.* 7: 421(1987)
5. M. Yalpani and L.D. Hall, Some chemical and analytical aspects of polysaccharide modifications. I. Nitroxide spin-labelling studies of alginic acid, cellulose, and xanthan gum,*Can. J. Chem.* 59: 3105(1981)
6. S. Kitamura, T. Kuge, and B.T. Stokke, A differential scanning calorimetric study of the conformational transition of xanthan in aqueous NaCl, *in:*"Gums and   Stabilisers for the Food Industry 5", G.O. Phillips, D.J. Wedlock, and P.A. Williams, ed., IRL Press, Oxford(1990)
7. F. Zhao, M.T. Rosen, and N.-L. Yang, Spin probe studies of the microviscosity gradient in the middle phase of three-phase micellar systems, *Colloids and Surfaces* 11: 97(1984)

XANTHAN GUM PRODUCTION FROM BRAZILIAN STRAINS

A.C. Mochi and Adilma R.P. Scamparini

Food Science Department. Food Engineering Faculty, UNICAMP
13081 Campinas, SP., Brazil

ABSTRACT

Xanthan gum is an extracelullar polysaccharide obtained during fermentation of bacteria Xanthomonas sp. It is a biopolymer and consists of D-glucose, D-mannose and D-glucuronic acid units in the ratio of 3:3:2. It also contains acetyl and pyruvic acid groups. Xanthan gum dissolves readily in cold or hot water, turning into very viscous solution either if the concentration is small. Temperature and pH have very little effect on viscosity of the solutions. Due to these characteristics, xanthan gum is used as a thickening agent in foods and cosmetics, as an edible film-forming agent. It is used as and emulsifier in a wide variety of industrial processes, and is specially effective in secondary oil recovery operations.

In this paper we utilized xanthan gum obtained from Xanthomonas campestris pv manihots that showed to produce high quality gums in terms of viscosity, pH and temperature stability. Two strains were selected, and their growing and fermentation parameters were determined.

To grow the culture, two stages, of 24 hours each, were carried out in a medium containing glucose, malt extract, yeast extract and peptone at 25°C and 250 rpm.

The fermentation was carried out at 250 rpm, during 96 hours in a medium containing sucrose, dipotassium phosphate and magnesium sulphate. The temperature was set to 30°C and the pH to approximately 7.

INTRODUCTION

Xanthan gum is a microbial polysaccharide produced by bacteria of the genus Xanthomonas. A problem arising in screening microorganisms for polysaccharide production is the lack of suitable techniques for selection of good potential strains. Colony diameter did not correlate with the viscosifying capacity of the polymer[1].

Food Hydrocolloids: Structures, Properties, and Functions
Edited by K. Nishinari and E. Doi, Plenum Press, New York, 1994

Solutions of xanthan gum are pseudoplastic. Aqueous solutions containing more than 0.75% xanthan gum have a yield point. That is, the solution behaves as solid until a minimum shearing force is reached. At shearing stresses above this minimum, the system behaves as a liquid[2].

Viscosity of aqueous xanthan gum dispersions declines with increasing shear rates and approaches a constant infinit shear viscosity, over a shear rate range of 1000-250000/s[3].

Xanthan gum is the best overall acid tickener for use in well stimulation. Model studies indicate that 15% HCl gelled with xanthan gum retains its viscosity during treatments at bottomhole temperatures up to 220°F[4].

Interactions between sucrose and xanthan must affect sucrose diffusivity which very probably would decrease availability of sucrose to the microorganism during fermentation development. The culture stability for xanthan formation is higher under conditions which are favorable for cell growth with limited xanthan production[6].

MATERIALS AND METHODS

Bacterial strains. Two strains (DCA450 and DCA459) of <u>Xanthomonas campestris</u> pv <u>manihots</u> were originally isolated from cassava in Campinas, So Paulo State, Brazil. Strains were lyofilized when first isolated and also maintained on YM agar slopes in screw-capped vials by monthly subculture.

Xanthan gum production. The production of xanthan gum was made in two steps:

1- Cell production. The YM medium contained 1% glucose, 0.3% yeast extract, 0.3% malt extract and 0.5% peptone. The pH was adjusted to 7.0 with sodium hydroxide before sterilization. The cells were cultivated in 1000 mL erlenmeyer flasks containing 220 mL YM medium on a New Brunswick rotary shaker at 250 rpm and 25°C, during 24 hours.

2- Xanthan gum production. Seed cultures (220 mL) were innoculated in 450 mL of the medium contaning 2% sucrose, 0.5% K_2HPO_4 and 0.01% $MgSO_4$ (pH = 7.0) and kept at 30°C and 250 rpm, during 48 hours. After dilution with 3 vol. of sterile distilled water, the fermentation broth were centrifuged (7,000 x g for 30 min). The supernatant was treated with 3 vol. ethanol 99.5% to precipitate the polysaccharide. This precipitate was resuspended and dialyzed against running tap water for 72 hours and then was dried at 55°C until constant weight.

Determination of the viscosity.

Viscosity vs. temperature. Viscosity values of 0.5% xanthan gum dispersions (pH 7.0) were measured with a Brookfield microviscometer, type RVT (spindle 18, 12 rpm) at 10, 20, 50 and 70°C. Viscosity values of 0.5% xanthan gum dispersions containing 2% NaCl were measured as previously described.

Viscosity vs. pH. Viscosity values of 0.5% dispersions at various pH's were measured with a Brookfield microviscometer, type RVT (spindle 18, 12 rpm) at 26°C.

RESULTS AND DISCUSSION

The typical colony morphology of strains DCA450 and DCA459 are shown in Fig. 1. The colonies of both strains were very mucoid in nature, to the extent that the bacterial growth would drip down onto the lid of the plate if incubated in a inverted position. This mucoid growth pattern was seen on several different solid growth media.

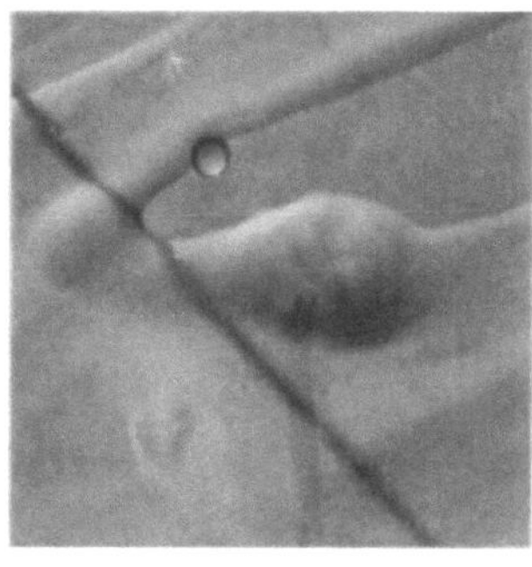

A

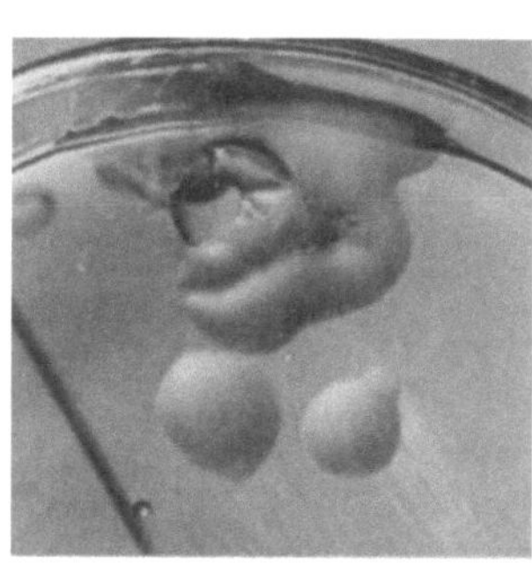

B

Fig. 1. Colony morphology of strains DCA450 (A) and DCA459 (B), after 72 hours of growth on YM agar at 30°C

The decline of viscosity of xanthan gum dispersions when the temperature increases are shown in Table 1. With DCA459 xanthan gum, a dramatic rise in viscosity was observed; whereas with DCA450 xanthan gum only a slight decrease was seen. The addition of NaCl increased the viscosity of the solutions of both xanthan gums and reduced the temperature effect on the decrease of viscosity.

Table 1 - Viscosity of 0.5% dispersions of DCA450 and DCA459 xanthan gum, with and without added NaCl, at various temperatures at pH 7.0.

Temperature (^{o}C)	Viscosity (cp)			
	DCA450	DCA459	DCA450 with NaCl 2%	DCA459 with NaCl 2%
10	208.8	217.5	235.0	245.0
20	203.8	212.2	228.0	232.5
50	186.3	135.0	205.5	210.0
70	162.5	80.0	173.0	175.0

The decline of viscosity when the temperature increase can be interpreted as an indication that a conformational changes (order-disorder transition) is taking place in solution. The rise in ionic strenght causes an increase in the transition midpoint temperature of the conformational change which occurs with xanthan gum in solution[7], consequently the effect of temperature on the viscosity decrease.

The viscosity of xanthan gum DCA450 solutions was declined when pH values were increased, while the viscosity of the xanthan gum DCA459 solutions was increased when pH values were rose to 12.02, at 26°C.

Table 2 - Viscosity of 0.5% dispersions of DCA450 and DCA459 xanthan gum at various pH's at 26°C.

| | Viscosity (cp) | |
pH	DCA450	DCA459
2.55	281.3	210.0
4.05	192.0	210.5
6.70	203.8	211.3
9.75	187.5	222.5
12.02	174.3	237.5

REFERENCES

1. B. Torrestiana, L. Fucekovsky and E. Galindo, Xanthan production by some Xanthomonas isolates, Letters in Applied Microbiology 10: 81 (1990).

2. J.K. Rocks, Xanthan gum, Food Technology 25: 476 (1971).

3. K.R. Holme, L.D. Hall, R.A. Speers and M.A. Tung, High shear rate flow behavior of xanthan gum dispersions, Food Hydrocolloids 2: 159 (1988).

4. C.W. Crowe, R.C. Martin and A.M. Michaelis, Evaluation of acid-gelling agents for use in well stimulation, Society of Petroleum Engineers Journal 8: 415 (1981).

5. B. Torrestiana and E. Galindo, Cooperative binding of sucrose in xanthan gum solutions, Biotechnology Progress 4: 14 (1988).

6. P.K. Lee, H.M. Chang and B.H. Kim, Xanthan production by Xanthomonas campestris in continuous fermentation, Biotechnology Letters 11: 573 (1989).

7. M. Rinaudo and M. Milas, Xanthan properties in aqueous solutions, Carbohydrate Polymers 2: 264 (1982).

EFFECT OF SOLUBLE SOYBEAN POLYSACCHARIDE ON
DISPERSION STABILITY OF ACIDIFIED MILK PROTEIN

Iwao Asai[1], Yumiko Watari[1], Hiroki Iida[1],
Kenji Masutake[1], Takashi Ochi[1], Shiro Ohashi[1],
Hitoshi Furuta[2] and Hirokazu Maeda[2]

[1]San-Ei Gen F.F.I., Inc. 1-11, Sanwa-cho 1-chome, Toyonaka
City, Osaka 561. Japan, [2]Fuji Oil Co., Ltd, 4-3, Kinunodai,
Yawara, Tsukuba-gun, Ibaraki 300-24. Japan

ABSTRACT

The effect of soluble soybean polysaccharide (hereinafter referred to as SSHC) obtained from so-called bean-curd refuse, a by-product of bean curd and soybean protein, on the dispersion stability of yoghurt drinks was examined by measuring the viscosity, zeta potential, median diameter, precipitation quantity after centrifugation, and absorbance at 660 nm of the supernatant fluid.

As a result, it was found that yoghurt drinks were dispersed and stabilized with a viscosity (less than 10 mPa·s) lower than that of pectin, sodium carboxymethylcellulose (hereinafter referred to as CMC) and propylene glycol alginate (hereinafter referred to as PGA) at pH 3 to 4.6 and an SSHC concentration of 0.3% or above, and at milk solid non-fat concentrations (hereinafter referred to as MSNF) of 3% and 8%.

INTRODUCTION

Acidified milk is classified into two groups : one is milk juice drinks where the juice is artificially added to the milk, and the other is yoghurt drinks where the milk is fermented by lactic acid bacteria. All of these products are prepared with a pH range lower than the isoelectric point (about pH 4.6) of milk protein. Consequently, milk protein may be denatured into hydrophobic particles, which precipitate during storage, and cause whey separation. To disperse and stabilize milk protein at a pH lower than the isoelectric point, pectin,[1, 2, 3] CMC[4,5] and PGA[6] are currently used. These hydrocolloids, on the basis of their characteristics, tend to give a high viscosity to finished products.

SSHC obtained from "Okara" (bean-curd refuse), a by-product of "tofu" (bean-curd) or soybean protein, whose main components are galactose, arabinose and galacturonic acid,

Food Hydrocolloids: Structures, Properties, and Functions
Edited by K. Nishinari and E. Doi, Plenum Press, New York, 1994

has an average molecular weight of hundreds of thousands. Its dietary fiber content by the AOAC Official Method is about 60%. Furthermore, it has an extremely low viscosity compared to other polysaccharides, which enables the preparation of a highly concentrated solution (about 70%)[7]. Using low viscosity SSHC, we examined the effect in yoghurt drinks on dispersion stabilization of protein by measuring the viscosity, zeta potential, median diameter of the milk protein particle, amount of precipitation after centrifugation, and absorbance of the supernatant fluid at 660 nm.

MATERIALS AND METHODS

Materials

SSHC was obtained from Fuji Oil Co., Ltd., Tokyo. The other materials used were commercial 50% lactic acid, food additive standards (Daiichi Seiyaku Co., Ltd., Tokyo), sodium hydroxide, first-class reagent (Kishida Chemical Co., Ltd., Osaka), pectin, Genu Pectin Type JM-150-J (The Copenhagen Pectin Factory Ltd., Copenhagen), CMC, Cellogen F-SB (Dai·Ichi Kogyo Seiyaku Co., Ltd., Kyoto) and PGA, Duck loid PF (Kibun Food Chemifa Co., Ltd., Tokyo). Powdered skim milk and starter were commercial preparations.

Methods

1. Preparation

Various hydrocolloids solution

0.5% SSHC, pectin, CMC and PGA were added to cold deionized water. After heating to 80℃ and allowing to dissolve for 10 min, the solution was cooled to 20℃ and adjusted to the pH range between 2 and 6 with 0.1N lactic acid and sodium hydroxide.

Fermented milk

20% powdered skim milk solution was pasteurized at 90℃ for 15 min., and then the solution was cooled to 45 ℃. Thereafter, 5% starter was added to the above solution, which was fermented to pH 4.3 at 42℃. After fermentation, the system was cooled to 10 ℃ and stirred until a sol was obtained.

Yoghurt drinks

7% sucrose and SSHC were added to water. After the mixture was heated to 80℃ and held for 10 min to dissolve, it was then cooled to 10℃. Thereafter, fermented milk at 10℃ in the sol state was added to the mixture of 7% sucrose and SSHC. The solutions were adjusted to the pH between 3 and 5 with 0.1 N lactic acid and sodium hydroxide. Then the solution was homogenized at 150 kg/cm^2 with a homogenizer, Model HV-OA-3-5.5S (Izumi Food Machinery Co., Ltd., Osaka).

2. Measurement

Viscosity

The viscosity was measured at 20℃ with a B type Viscometer, Model BL (Tokimec Inc., Tokyo).

Zeta potential

Each solution was diluted to measurable concentrations (1 to 500 times). Measurement was performed at 20℃ with a Lazer Zee Meter, Model 501 at a voltage of 100 V (Pen Kem Inc., USA).

Particle size

Each solution was diluted to measurable concentrations (100 to 200 times), and particle size was measured at 20℃ with a Shimadzu Laser Diffraction Particle Size Analyzer, SALD-1100 (Shimadzu Corporation, Kyoto). On the basis of the smallest particle size, a 50% particle size is considered as the median diameter.

Absorbance

After each solution had been centrifuged at 1400G for 10 min., supernatant fluid, which was diluted to measurable concentrations (1 to 100 times), was taken for measurement at 660 nm with a Shimadzu UV-Visible Recording Spectrophotometer (Shimadzu Corporation, Kyoto). Absorbance is the product of dilution rate and measured value.

Precipitation quantity

After each 50 g solution had been centrifuged at 3000G for 20 min., the supernatant was decanted and the residue allowed to stand for 20 min. Additional remaining supernatant was decanted before measurement of the precipitation quantity.

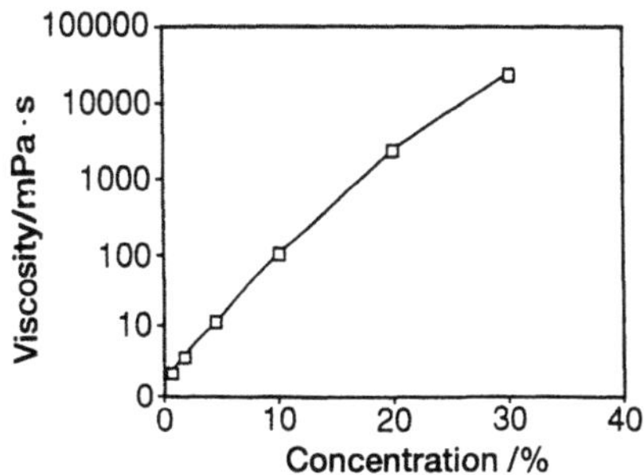

Fig.1 Effect of SSHC concentration on viscosity at 20℃

RESULTS AND DISCUSSION

Viscosity of SSHC

As shown in Figure 1, the viscosity of SSHC is about 100 mPa·s at 20℃ even for 10% aqueous solution.[7] Its viscosity is extremely low compared to other water soluble polysaccharides and increases almost linearly with increase in concentration[7]. The structure of SSHC has not been fully determined. However, as proposed by Lavabitch et al.[8] and Misaki et al.,[9] the structure of the principal element is presumed to consist of galactan and arabinan linked to the galacturonic acid main chain.[7] Also, the infrared spectrum of SSHC bears a resemblance to that of arabinogalactan (unpublished result). Gum molecules of arabinogalactan are thought to be spherical or highly branched from viscosity studies.[10] Consequently, since SSHC has a similar structure, SSHC is presumed to show a low viscosity behavior similar to that of arabinogalactan.

Zeta potential of SSHC

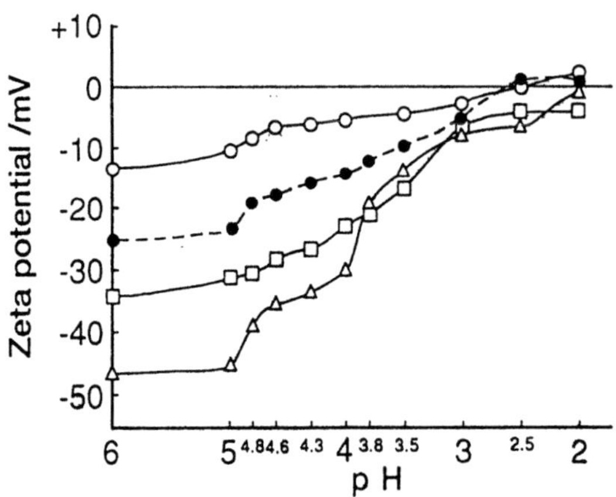

Fig.2. Zeta potential at pH 2 to 6 of 0.5% aqueous solutions of PGA, SSHC, pectin and CMC

O—O : PGA □—□ : pectin
●···● : SSHC △—△ : CMC

As shown in Figure 2, negative zeta potential increased in the order of CMC > pectin > SSHC > PGA with increase in pH. This is attributable to the fact that all of these polysaccharides have a carboxyl group and its degree of dissociation increases with increase in pH.

However, at pH 3 or below, all of these polysaccharides show a small negative zeta potential, because the dissociation of the carboxyl group is depressed.

Effect of the concentration of SSHC on dispersion stability of yoghurt drinks

As shown in Figure 3A, viscosity and median diameter decreased significantly at the SSHC concentration of 0.3% or above, while negative zeta potential increased with increase in SSHC concentration.

In Figure 3B, absorbance at 660 nm increased remarkably at the SSHC concentration of 0.6% or above, while precipitation quantity decreased considerably at the SSHC concentration of 0.4% or above. These phenomena are thought to be due to the following events : SSHC, on addition, caused adsorption of a hydrophobic protein particle on its surface and the positive zeta potential of the protein particle was neutralized by 0.1% SSHC. On further addition of SSHC, negative zeta potential increased with increase in SSHC concentration and dramatically increased on the addition of more than 0.3% SSHC.

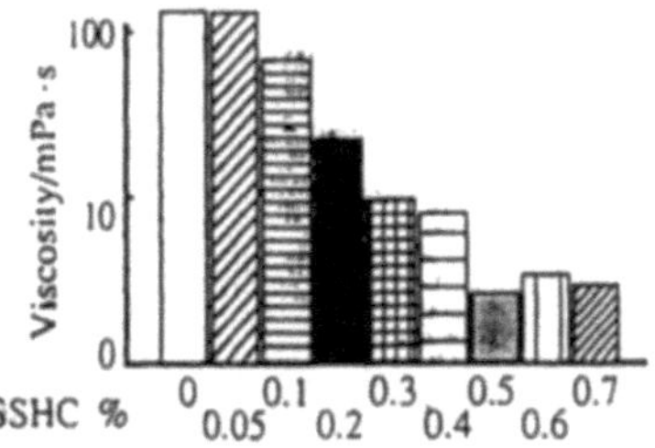
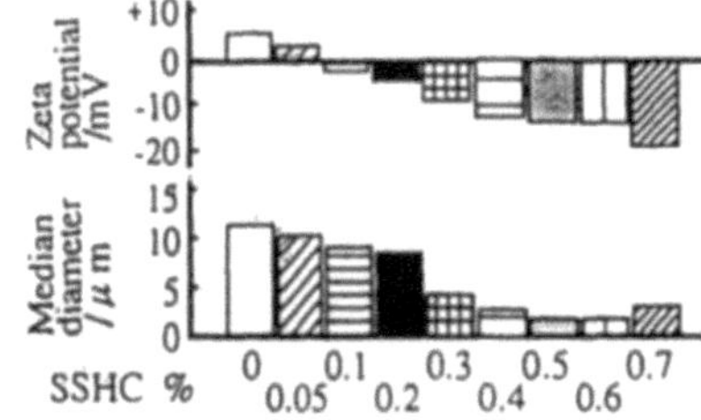

Fig.3A Effects of concentration of SSHC on viscosity, zeta potential and median diameter for yoghurt drinks with 8% MSNF at pH 4.3

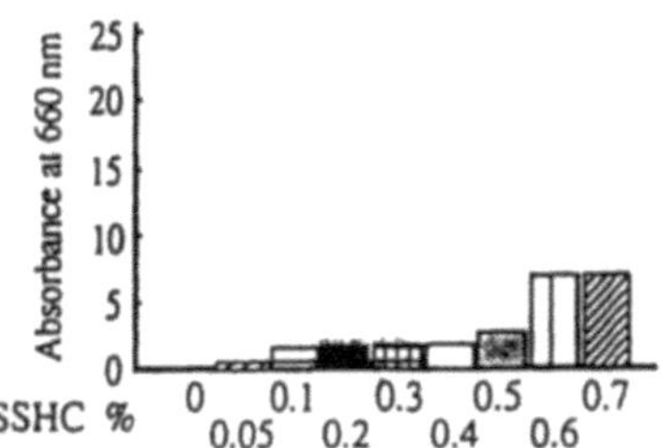
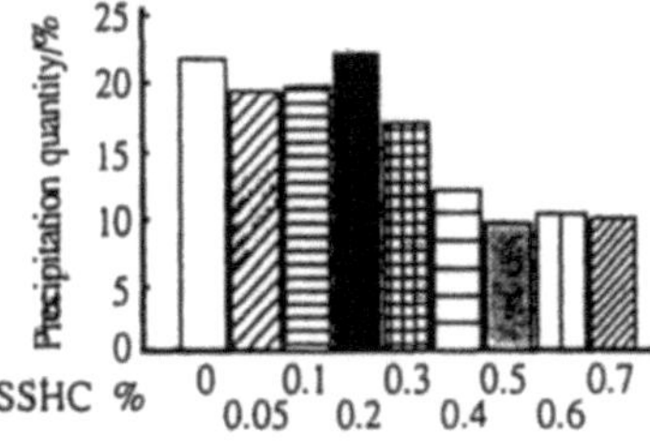

3B Effects of concentration of SSHC on the absorbance and precipitation quantity for yoghurt drinks with 8% MSNF at pH 4.3

These means to inhibits intermolecular aggregation due to increasing electric repulsion and increased hydrophilicity of the protein particle, thus resulting in stabilization. Dispersion and suspension of the protein particles is particularly pronounced when the SSHC concentration is 0.3% or above.

Effect of pH on the stability of yoghurt drinks with 3% and 8% of MSNF in the presence of 0.5% SSHC

As shown in Figure 4A, viscosity and median diameter increased significantly at pH 5.0 in 3% MSNF and at pH 4.6 and 5.0 in 8% MSNF. However, viscosity was less than 10 mPa·s at pH 3.5 to 4.3 in 3% and 8% MSNF. Negative zeta potential increased gradually with increase in pH.

In Figure 4B, absorbance decreased markedly at pH 5.0 in 3% MSNF and at pH 4.6 and 5.0 in 8% MSNF. Precipitation quantity increased remarkably at pH 5.0 in 3% MSNF and at pH 4.6 and 5.0 in 8% MSNF.

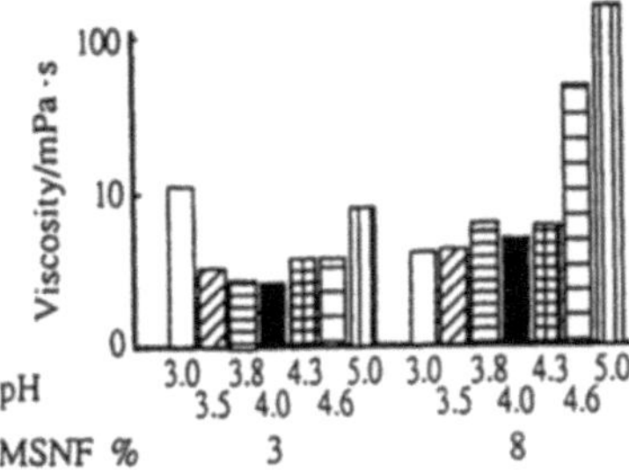
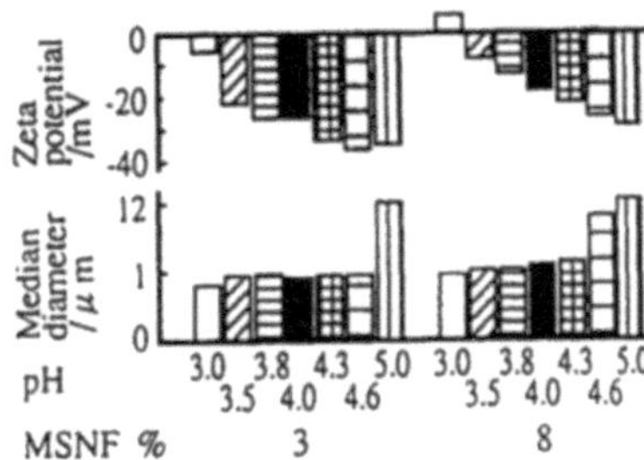

Fig.4A Effects of pH on viscosity, zeta potential and median diameter for yoghurt drinks with 3% and 8% of MSNF in the presence of 0.5% SSHC

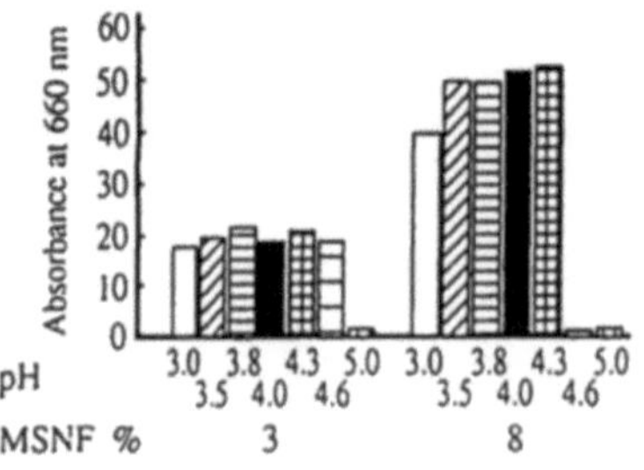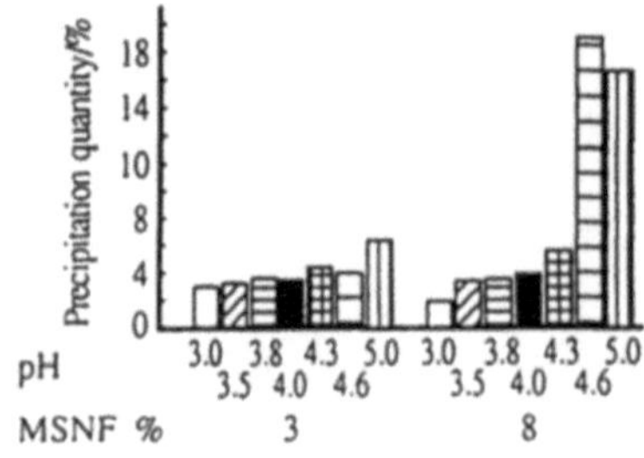

Fig.4B Effects of pH on absorbance and precipitation quantity for yoghurt drinks with 3% and 8% of MSNF in the presence of 0.5% SSHC

Since the isoelectric point of milk protein is pH 4.6 and protein is nearly in the noncharged state at pH 4.6 to 5.0, it seems to not interact with SSHC. Consequently, even if SSHC is present among protein particles in the noncharged state, no electric repulsion between the protein particles occurs. Therefore, protein particles mutually aggregate and this aggregation is attributable to the extreme increase in viscosity, median diameter and precipitation quantity and the extreme decrease in absorbance at pH 4.6 and 5.0. Since there are many protein particles in 8% MSNF, compared to 3% MSNF, the distance between particles is short. Accordingly, since the particles have a tendency to stick easily to each other, it is thought that aggregation can occur even at lower pH (pH 4.6).

REFERENCES

1. J.J. Does Burg and L, De Vos, 5 *Internationaler Fruchsaftkongress*, Wien 306 Juni 32037 (1959).
2. P.E. Glahn, Hydrocolloid stabilization of protein suspensions at low pH, in : *"Prog. Fd. Nutr. Sci. 6,"* G. O. Phillips, D. J. Wedlock, and P. A. Williams, eds., Pergamon Press, Oxford (1982).
3. H. Exler, *Germ. Pat.* 270938 (1969).
4. Anonym, *Food Eng.* April, 97 (1971).
5. A.K. Quereshi, A.S. Mahmood, Mahmoodul-Hassan, and M. Elahi, *Sci. Ind.*, 9 : 211 (1972).
6. H. Luck and J. Govtone, *S. Afr. J. Dairy Technol.*, 5 : 47(1973).
7. H. Maeda, *3rd Food Hydrocolloid Symposium* in Japan : 32 (1992).
8. J.M. Lavabitch, L.E. Freeman, and P.A. Albersheim, *J. Biol. Chem.*, 251 : 5904 (1976).
9. A. Misaki, K. Higashira, and H. Kaku, *Japan Society of Starch Science lecture outline collection* : 259 (1987).
10. M.F. Adams and B.V. Ettling, Larch arabinogalactan, in : "Industrial Gums," R.L. Whistler and J.N. Bemiller, eds., Academic Press, New York (1973).

"AI-GEOK", A UNIQUE GEL-FORMING POLYSACCHARIDE OF
Ficus awkeotsang: STRUCTURE, GELATION MECHANISM
AND UTILIZATION OF THE OLIGOSACCHARIDES

Kozo Komae[1] and Akira Misaki

Department of Food and Nutrition
Osaka City University, Sumiyoshi
Osaka 558, Japan
[1]Present address: National
Agriculture Research Center
Tsukuba, Ibaraki 305, Japan

ABSTRACT

A Water-extract of the seed of *Ficus awkeotsang* forms spontaneously an edible gel, known as "AI-GEOK". A gel-forming polysaccharide located on the surface of seed was purified from the water-extract of heat-treated seeds. Structural analysis showed that it is a highly methylesterified(DE, 63.6%), α-1,4-galacturonan(M.Wt, 3.4×10^5). The intrinsic pectinesterase(EC 3.1.1.11), responsible for gelation of AI-GEOK, located solely in the red tepals, was purified by column chromatography. The purified enzyme showed monomeric nature with M.Wt. 4.2×10^4, pI 4.4, and optimal pH 5.4, which sifted to 7-8 in the presence of K^+. When 0.5% AI-GEOK solution was incubated with the enzyme and mixture of metal ions, containing Ca^{2+}, Mg^{2+}, K^+, and Na^+, it gradually formed a gel. These results confirmed that the gel formation mechanism must be involved in, a) de-esterification of highly methylesterified galacturonan by the intrinsic esterase, and b) gradual formation of "egg-box" structure by ionic interaction between Ca^{2+} and regularly arranged α-1,4-linked galacturonan. Apart from the gelation problem, AI-GEOK appears to be a good sourse of phytoalexin-elicitor. Among oligogalacturonides prepared by enzymic degradation, 4,5-unsaturated hexa-galacturonide was found to exhibit a potent activity as well as deca-galacturonide.

INTRODUCTION

Ficus awkeotsang Makino is grown on the mountainsides of the central part of Taiwan. The seeds of its fruit, called jelly fig, are used to make home-made jelly cake in some oriental countries. The water-extract spontaneously form a pudding-like gel at room temperature. Previously, Oda and Tanaka[1] suggested that the gel-forming polysaccharide may be α-1,4-linked D-galacturonide. Moreover, the presence of some neutral sugar components, *i.e.*, arabinose, galactose and rhamnose, was also reported. With regard to the mechanism underlying such spontaneous gel-formation, Haung *et al.*[2] suggested that the

Food Hydrocolloids: Structures, Properties, and Functions
Edited by K. Nishinari and E. Doi, Plenum Press, New York, 1994

action of endogenous pectinesterase may be involved in the gel-formation in the process of seed extraction.

In the course of studies on the structure and gel-formation mechanism of AI-GEOK, we became aware that the native acidic polysaccharide can be prepared, without de-esterification or de-polymerization, by water-extraction from heat-treated awkeotsang seeds, and endogenous pectinesterase is localized in red tepal attached to the pedicel of the seed. This enable us to study on the enzymic gel-formation of AI-GEOK. In addition, this paper reports a utilization of oligosaccharides prepared from AI-GEOK for studies on phytoalexin-elicitor activity.

MATERIALS AND METHODS

Gel-forming polysaccharides were extracted sequentially with cold and hot water, and the aqueous ammonium oxalate from awkeotsang seeds(50g), pre-heated at 95°C for 90min in drying oven for inactivation of any pectinolytic enzymes. Water-extracted pectic substance was then fractionated by anion-exchange chromatography on Q-Sepharose for its physiochemical characterization[3]. Degree of methylesterification(DE) of pectic substance was determined quantitatively by an enzymic method involving as alcohol oxidase/ formaldehyde dehydrogenase system[3].

Endogenous pectinesterase was extracted from red tepals(1.2kg) with 1M NaCl(16L) and then purified by mainly ion-exchange column chromatography[45].

AI-GEOK oligosaccharides, used for studies on phytoalexin-elicitor activity in soybean cotyledon, was prepared using purified endo-pectate lyase from *Erwinia carotovora* and then fractionated by column chromatography on QAE-Sephadex A-25. The purity, structure and molecular size of the paticular oligosaccharide fraction with potent activity were confirmed by both HPLC and FAB/MS analysis[6].

RESULTS AND DISCUSSION

Characterization of AI-GEOK. Table 1 shows the results of extraction of pectic substances from seeds of *F. awkeotsang* Makino. Three pectic substances were obtained with the successive extraction procedure; cold water-extracted acidic polysaccharide (Fraction A, 2.6g, DE=63.6%) was the main product, with small amount of hot water-extracted pectin(Fraction B, 0.15g) and ammonium oxalate-extracted pectin(Fraction C, 0.75g). Fraction A, corresponding to AI-GEOK, was composed almost entirely of galacturonic acid, with trace amount of arabinose, mannose and galactose. It had a molecular weight of 3.4×10^5, estimated by gel-filtration on Sepharose CL-6B.

Table 1. Sugar composition of materials solubilized during the extraction of seeds of *Ficus awkeotsang* Makino.

Fraction	Yield (%)	Sugar composition(%)							Protein (%)
		Rham	Ara	Xyl	Man	Gal	Glc	GalUA[1]	
A	5.2	tr.[2]	0.9	tr.	0.8	1.8	tr.	87.5	5.5
B	0.3	7.6	11.0	1.4	2.8	10.9	8.4	56.5	2.4
C	1.5	9.6	16.4	2.1	1.1	19.2	0.6	50.3	0.7

[1]Uronic acid is expressed as galacturonic acid(GalUA).
[2]tr.=trace.

The major polysaccharide(Fraction A) was further fractionated by anion-exchange column chromatography on Q-Sepharose, which gave three fractions; fraction P-1(yield, 59%), fraction P-2(18%), and fraction P-3(4.8%). The acid-soluble fraction, obtained on treatment of fraction P-1 with hot 1M HCl, was subjected to methylation analysis according to the method described in previous report[3]. This methylation analysis revealed the presence of 2,3,6-Me$_3$-galactose, together with a very small amount of 2,3,4,6-Me$_4$-galactose, the molar ratio being 42.5 : 1.0. This confirmed that the purified acidic polysaccharide has the linear structure of an α-1,4-linked D-galacturonan and is devoid of L-rhamnose residues, indicating that AI-GEOK cannot be classified as a constructive pectic substance containing a rhamnogalacturonan region reported by Lau et al[7].

Localization and Characterization of Pectinesterase. Figure 1 shows the changes in the methyl ester contents of water-extracts of awkeotsang seeds with and without red tepals. The methyl ester content of polygalacturonide(DE, 53.8%) extracted from heat-untreated seeds with red tepals decreased rapidly with the incubation time, and reached to 18.1% by 60min, accompanied with spontaneous gel-formation, which was observed distinctly after 25min. On the contrary, incubation of the extract of heat-treated seeds with or without red tepals under the same conditions resulted in only a slight decrease in the methyl ester content. This result confirmed that pectinesterase involved in spontaneous gel-formation of AI-GEOK is located in red tepals attached to the seeds. Pectinesterase extracted from red tepals was highly purified by column chromatography and then chracterized as shown in Table 2. The purified enzyme showed a characteristic

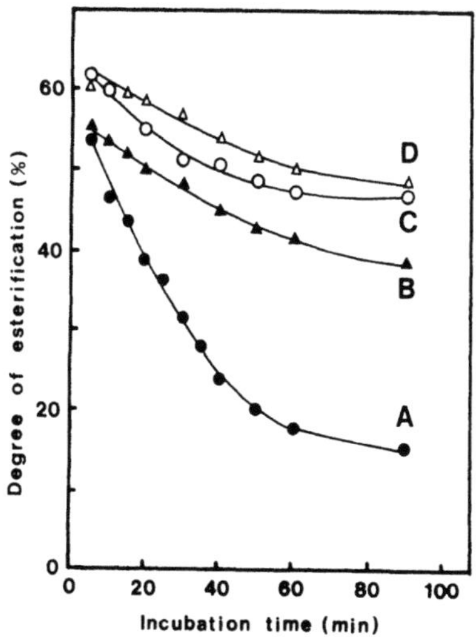

Fig. 1. Time course of Methyl ester contents of Water-extracts of awkeotsang seeds with and without red tepals.
A, heat-untreated seeds with red tepals; B, heat-treated seeds with red tepals; C, heat-untreated seeds without red tepals; D, heat-treated seeds without red tepals.

Table 2. Characterization of purified pectinesterase.

Molecular weight 42,000	Isoelectric point 4.4
Metal ions for stimulation of enzyme activity	Ca>Mg>K>Na
Optimal pH in the absence of metal ions	5.4
in the presence of metal ions	7.5
Km at pH 5.4 in the absence of metal ions	2.94 mg/ml
at pH 7.5 in the presence of 50mM KCl	2.78 mg/ml

optimal pH for the esterase activity toward the purified native AI-GEOK, which depended on the pH as well as metal ions. Interestingly, the optimal pH for the activity in the absence of metal ions was found to be 5.4, like that of fungal esterases, e.g., *Corticium rolfsii*[8] and *Clostridium multifermentans*[9]. However, when calcium or potassium ions were present, the optimal pH shifted to 7-8, similar to those for other plant pectinesterase.

The Enzymic Gel-formation Test. An aqueous solution of AI-GEOK(0.5% GalUA eq.; 30ml, pH 5.4) containing KCl(15.8mg), NaCl(3.2mg), $CaCl_2 \cdot 2H_2O$(27.9mg) and $MgSO_4$(4.6mg) was placed in a jelly glass, and then incubated with the purified enzyme at 25°C for 45min. As shown in Table 3, gel-formation in the reaction mixture proceeded gradually with the incubation time, vary similar to in the case of the spontaneous gel-formation occuring with the aqueous extract of heat-untreated seeds with red tepals. After incubation for 45min, the firmness of the gel had reached 2.52×10^4 dyn/cm^2, the breaking strength being 1.16×10^5 dyn/cm^2. These results suggested that spontaneous gel-formation of AI-GEOK is initiated by de-esterification of high methylesterified polygalacturonide solution with intrinsic pectinesterase, which is activated by potassium or other metal ions, leading to the gradual formation of a rigid gel through ionic binding to calcium, being so-called "egg-box" structure, as proposed by Grant et al.[10].

Table 3. Gel strength of AI-GEOK on treatment with the purified pectinesterase.

Incubation time (min)	Firmness ($\times 10^4$ dyn/cm^2)	Breaking strength ($\times 10^4$ dyn/cm^2)
25	1.04	4.25
30	1.82	8.19
35	2.17	8.97
40	2.26	10.80
45	2.52	11.60

Phytoalexin-elicitor activities of AI-GEOK oligosaccharides. AI-GEOK oligosaccharides prepared by enzymic degradation using endo-pectate lyase were precisely fractionated by anion-exchange column chromatography, and then used for assay of phytoalexin-elicitor activity. Among the oligogalacturonides with DP 3 to 12, it was found that 4,5-unsaturated hexa-α-1,4-galacturonide was a potent elicitor as well as 4,5-unsaturated deca-α-1,4-galacturonide previously reported by Davis et al.[11].

REFERENCES

1. Y. Oda and R. Tanaka, *Agric. Biol. Chem.*, 30:406 (1966).

2. Y.C. Huang, W.P. Chen, and Y.P. Shao, *Chaina Hort.*, 4:117 (1980).

3. K. Komae and A. Misaki, *Agric. Biol. Chem.*, 53:1237 (1989).

4. K. Komae, Y. Sone, M. Kakuta, and A. Misaki, *Agric. Biol. Chem.*, 53:1247 (1989).

5. K. Komae, Y. Sone, M. Kakuta, and A. Misaki, *Agric. Biol. Chem.*, 54:1469 (1990).

6. K. Komae, A. Komae, and A. Misaki, *Agric. Biol. Chem.*, 54:1477 (1990).

7. J.M. Lau, M. Mcneil, A.G. Darvill, and P. Albersheim, *Carbohydr. Res.*, 137:111 (1985).

8. O. Yoshihara, T. Matsuo, and A. Kaji, *Agric. Biol. Chem.*, 41:2335 (1977).

9. L. Miller and J.D. Macmillan, *J. Bacteriol.*, 102:72 (1970).

10. G.T. Grant, E.R. Morris, D.A. Rees, P.J.C. Smith, and D. Thom, *FEBS Letters*, 32:195(1973)

11. K.R. Davis, A.G. Darvill, P. Albersheim, and A. Dell, *Plant Physiol.*, 80:568 (1986).

GELLING PROPERTIES OF ACID-MODIFIED STARCHES

Karin Autio and Kaisa Poutanen

VTT, Food Research Laboratory
P.O. Box 203
02151 Espoo
Finland

ABSTRACT

The gelling mechanism of most biopolymers, including starches, is not completely understood. A wide spectrum of starches with different gelling properties can be prepared by varying the starch source, type and degree of modification. Acid modification is widely used to produce thin-boiling starches for food industry.

This report describes the effect of mild acid modification on the viscoelastic and viscous properties and microstructure of maize and barley starch dispersions at different temperatures. The gel-formation of amylose weakened with the increasing degree of hydrolysis if heating was to 90°C, but not if heating was to 98°C. Amylose gel formation in native maize starch was hindered by addition of a small amount of hydrolyzed waxy maize starch. Microstructural studies suggested that soluble amylopectin interfered with the formation of amylose gel.

When the degree of acid modification was low, the starch pastes assumed a phase-separated structure at both heating temperatures. The most hydrolyzed barley starch dispersions formed an elastic network structure with good mechanical properties if preheating was to 98 instead of 90°C.

INTRODUCTION

The application of starches in the food industry is based on their ability to thicken or gel aqueous solutions. From a rheological point of view, the most important factor is the structure of the continuous phase in the starch dispersion. When the structure of a starch granule is broken down during heating in an aqueous medium, amylose and amylopectin are redistributed. Studies carried out with native starches indicate that, depending on pasting

Food Hydrocolloids: Structures, Properties, and Functions
Edited by K. Nishinari and E. Doi, Plenum Press, New York, 1994

behaviour, the continuous phase may be composed of amylose (Miles et al. 1985; Ring, 1985; Langton and Hermansson, 1989; Autio, 1990) or of a mixture of amylose and amylopectin (Autio, 1990).

10% wheat or barley starch dispersions which have been heated to 95°C can be described as filled gels, in which the swollen starch granules and particles are dispersed in an amylose network. This type of structure is typical of commercial wheat and barley starch preparations, which contain mainly large A-type granules. Gels with swollen granules embedded in amylose network have good mechanical properties provided that amylopectin fragmentation has not occurred, since amylopectin fragments interfere with the gelation of amylose (Autio, 1990; Svegmark and Hermansson, 1991). Recent studies have indicated that small, B-type barley starch granules exhibit different heat-induced changes (Poutanen et al., 1992), similar to those of oat starch. Whereas in other cereal starches amylose is solubilized first, in oat starch pastes amylose and amylopectin are solubilized at the same time (Doublier et al., 1987). As a result a mixed gel composed of amylose and amylopectin is formed. Because of the lack of favourable interactions between these polymers, a phase-separated structure is formed (Kalichevsky and Ring, 1987). The mechanical properties of gels composed of both amylose and amylopectin are dependent on the ratio of amylose to amylopectin, since amylose is a much better gelling agent than amylopectin (Leloup et al., 1991).

Acid modification is one of the most widely used conversion methods for the production of thin-boiling starches. The fact that starch remains in the granule form even after acid treatment suggests that the heating conditions during application of these products have a significant role in determining the structure of the hydrolyzed starch gels. The aim of the present study was to monitor the viscoelastic behaviour of hydrolyzed maize and barley starch pastes during cooling and relate it to the microstructure of the starch gels. Special emphasis was placed on the effect of heat treatment.

MATERIALS AND METHODS

Commercial barley and maize starches were obtained from Raisio Company, Finland and from CPC-Cerestar, Denmark, respectively. Acid hydrolysis was performed with 1M HCl for 1 h at 40°C as described earlier (Pessa et al., 1992). Details of the rheological measurements and microstructural characterization have been described elsewhere (Autio, 1990; Pessa et al., 1992; Autio et al., 1992). Total and apparent amylose content was determined by the method of Morrison and Laignelet (1983) and lipid content from phosporus analysis (Morrison, 1964; Morrison, 1988; Tester and Morrison, 1990).

RESULTS AND DISCUSSION

Chemical Analysis

The lysophopholipid and total amylose contents were higher in the barley starch granules (Table 1). The difference between total and apparent amylose was larger in barley starch indicating that slightly more of the amylose was lipid-complexed in the barley starch granules. The lysophospholipid content, as determined from the amount of phosphorus, was considerably higher for barley starch. This difference was already observed by Morrison (1988), who also pointed out that the amount of free fatty acids in maize starch was much higher than in barley starch.

Table 1. Chemical properties of barley and maize starch.

Starch	Total amylose (%)	Apparent amylose (%)	Lipid-complexed amylose (%)	Lysophospho-lipids (%)
Barley	29.0	21.2	7.8	0.97
Maize	27.5	22.1	5.4	0.32

Microstructural Characterization

Acid modification induced considerable changes in the microstructure of barley and maize starch dispersions. In native starch pastes heated to 90 °C, starch granules were dispersed in the amylose network which stained blue with iodine (results not shown). The appearance of the continuous phase was totally different in acid-modified starch dispersions. This difference was due to the fact that acid-induced degradation of starch granules starts in the amylopectin moiety of starch (Pessa et al., 1992). The network around the granules was composed of both amylose and amylopectin. When the hydrolysis time was shorter (0.5-1.0 h instead of 4 h in this study) the heated starch pastes assumed a phase-separated structure (Pessa et al., 1992; Autio et al., 1992). Significant reduction in the molecular weight of amylopectin was observed when the hydrolysis time was 4 h and phase separation of amylose and amylopectin was not observed any more, probably since the degradation of amylopectin to smaller molecules resulted in an increase in the miscibility between amylose and amylopectin.

As a result of acid modification a great change was also observed in the granule structure. In the hydrolyzed barley starch dispersion, the paste contained violet- or brown-stained granule residues. In the hydrolyzed maize starch, especially the violet-stained granule residues were more degraded and the amylopectin droplets more separated from the granule residues. It was concluded that the violet-stained granule residues were composed of amylose or mixtures of amylose and amylopectin, since acid-modified waxy maize starch stained brown (Pessa et al., 1992).

Heating at 98°C for 10 min caused fragmentation of amylopectin in both native maize and barley starch dispersions. The granule structure of native maize starch was more degraded than that of barley starch. A very interesting structural change was observed in hydrolyzed barley starch dispersion between 90 and 98°C. The residual amylose was liberated from the granule and formed a dense network structure. Such a network was not observed in hydrolyzed maize starch dispersion.

Rheological Studies

Amylose aggregated readily in barley and maize starch dispersions preheated at 90°C (Fig. 1,curves NB and NM), the storage modulus (G') of barley starch gel being higher than that of maize starch. Hydrolysis caused a decrease in the storage modulus (Fig. 1, curves HB and HM) and an increase in phase angle (Table 2), which means that the gels became less rigid and more viscous. It was shown earlier that in native barley and maize starch pastes amylose alone formed the continuous phase, whereas in hydrolyzed starch pastes the continuous phase was composed of both amylose and amylopectin. In native maize starch the amylose gel formation was totally hindered by the addition of a small amount of hydrolyzed waxy maize starch, suggesting that amylopectin interfered with the gelation of amylose (Pessa et al., 1992).

Increasing the heating temperature to 98°C considerably increased the effect of granule degradation. In comparison with pastes heated at 90°C the storage modulus of the native starch pastes was lower and the phase angle was higher, indicating that the gels were less elastic and less rigid. By contrast heating to 98°C instead of 90°C increased the storage modulus (Fig. 2) and decreased the phase angle of the gel made of hydrolyzed barley starch. The low phase angle of this gel indicated the existence of a strong network structure (Table 1). The micrographs suggest that heating to 98°C liberated the residual amylose from the granule residue to the continuous phase. In the case of hydrolyzed maize starch a less rigid gel structure was formed. The structure of the continuous phase of hydrolyzed barley and maize starch dispersion differed greatly. It is suggested that in the case of hydrolyzed barley starch, the lipid-complexed amylose plays a more important role than in hydrolyzed maize starch.

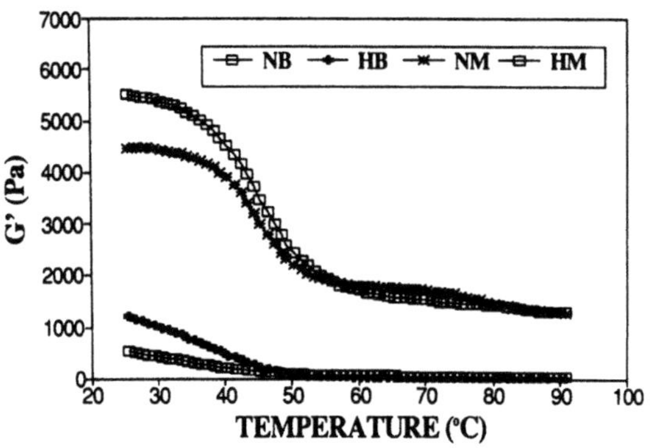

Fig.1. Storage modulus (G') during cooling of 15% starch dispersions heated to 90°C: NB) native barley starch dispersion, HB) hydrolyzed barley starch, NM) native maize starch, HM) hydrolyzed maize starch.

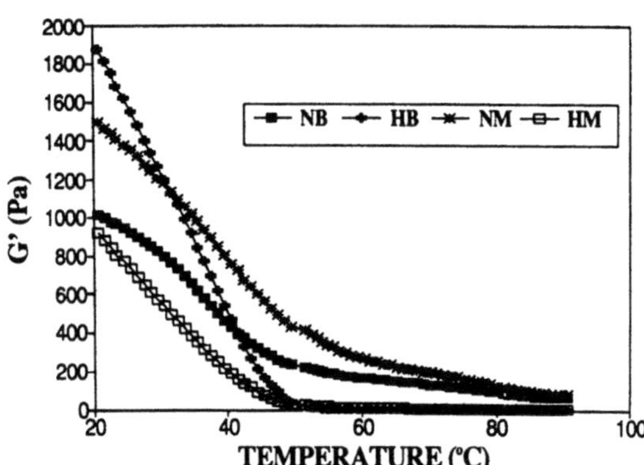

Fig.2. Storage modulus (G') during cooling of 15% starch dispersions heated at 98°C for 10 min. Legends as in Fig.1.

Table 2. Effect of acid modification on the phase angles of the gels.

Sample	Phase angle at 25°C preheating at 90°C	Phase angle at 25°C preheating at 98°C
Native barley	0.4±0.1	5.7±0.2
Acid-modified barley	5.3±0.3	1.5±0.1
Native maize	0.6±0.2	2.9±0.1
Acid-modified maize	4.7±0.1	1.7±0.1

ACKNOWLEDGEMENTS

We thank Erkki Pessa for preparing the hydrolyzed starch samples and Liisa Änäkäinen for expert technical assistance.

REFERENCES

Autio, K., 1990, Rheological and microstructural changes of oat and barley starches during heating and cooling, *Food Structure* 9:297.

Autio, K., Poutanen, K., Suortti, T. and Pessa, E., 1992, Heat-induced structural changes in acid-modified barley starch dispersions, *Food Structure* 11:315.

Doublier, J.L., Paton, D. and Llamas, G., 1987, A rheological investigation of oat starch pastes, *Cereal Chem.* 64:21.

Kalichevsky, M.T. and Ring, S., 1987, Incompatibility of amylose and amylopectin in aqueous solution, *Carbohydr. Res.* 162:323.

Langton, M. and Hermansson, A.M., 1989, Microstructural changes in wheat starch dispersions during heating and cooling, *Food Microstructure* 8:29.

Leloup, V.M., Colonna, P. and Buleon, A., 1991, Influence of amylose-amylopectin ratio on gel properties, *J. Cereal Sci.* 13:1.

Miles, M.J., Morris, V.J., Orford, P.D. and Ring, S.D., 1985, The roles of amylose and amylopectin in the gelation and retrogradation of starch. *Carbohydrate Research* 135:271.

Morrison, W.R., 1964, A fast, simple and reliable method for the microdetermination of phosphorus in biological materials, *Anal. Biochem.* 7:218.

Morrison, W.R. and Laignelet, B., 1983, An improved colorimetric procedure for determining apparent and total amylose in cereal and other starches, *J. Cereal Sci.* 1:9.

Morrison, W.R., 1988, Lipids in cereal starches: a review. *J. Cereal Sci.* 8:1.

Pessa, E., Suortti, T., Autio, K. and Poutanen, K., 1992, Molecular weight characterization and gelling properties of acid-modified maize starches. *Starch/Stärke* 44:64.

Poutanen, K., Pessa, E., Myllärinen, P. and Autio, K., Composition and heat-induced structural changes of small and large barley starch granules, *in* "Barley for Food and Malt", ICC/SCF International Symposium, September 7-10, 1992, Uppsala, Sweden.

Ring, S.G., 1985, Some studies on starch gelation, *Starch/Stärke* 37:80.

Svegmark, K. and Hermansson, A.M., 1991, Changes induced by shear and gel formation in the viscoelastic behaviour of potato, wheat and maize starch dispersions, *Carbohydr. Polymers* 15:151.

Tester, R.F. and Morrison, W.R., 1990, Swelling and gelatinization of cereal starches. I. Effects of amylopectin, amylose and lipids, *Cereal Chem.* 67:551.

THE EFFECT OF LOW LEVELS OF ANTIOXIDANTS ON THE SWELLING AND SOLUBILITY OF CASSAVA STARCH

Sandra. E. Hill, Dzulkifly B. Mat Hashim, John R. Mitchell and
John M. V. Blanshard

Department of Applied Biochemistry and Food Science
University of Nottingham
Sutton Bonington Campus
Loughborough
Leics LE12 5RD, U.K.

INTRODUCTION

When starch is heated in an excess amount of water above its gelatinization temperature, the granule swells and carbohydrate is released. The extent of swelling and the composition of the material that leaves the starch granule depend upon the botanical source of the starch as well as the heating and shearing procedure. It seems possible that a significant contribution to swelling and subsequent disintegration of the granule at high temperatures is the depolymerisation of the carbohydrate. Breakage of some linkages particularly in the amylopectin component, could facilitate the loss of this material from the swollen granule structure. Extensive work on xanthan (Wellington, 1983) and galactomannans (Mitchell et al., 1991) has demonstrated that the use of binary antioxidant systems, consisting of sodium sulphite and propyl gallate, were effective in reducing viscosity loss that normally occurred on heating the biopolymers. If an effect of heat on starch is due to oxidative reductive depolymerisation (ORD) reactions, it may be expected that this process could be controlled by the binary antioxidant system. This idea prompted us to investigate the effect of sodium sulphite and propyl gallate and their combinations on the swelling volume and carbohydrate release from cassava starch pasted at high temperature.

Food Hydrocolloids: Structures, Properties, and Functions
Edited by K. Nishinari and E. Doi, Plenum Press, New York, 1994

MATERIAL AND METHODS

Materials. Cassava starch (M4 variety) (moisture 10.4%, amylose 20.9%) was obtained from the Central Tuber Crop Research Institute (CTCRI), Trivandrum, India. Sodium sulphite (AR grade) was obtained from Fisons PLC and n-propyl gallate from Sigma Chemical Company. Solutions of different concentrations of sodium sulphite and n-propyl gallate were prepared in distilled water and the starch was then dispersed in these solutions.

Measurement of Swelling Volume, Total Solids, Carbohydrate Content and Absorbance. The swelling volume was determined by heating 15 ml of an aqueous 1% (w/v) starch suspension in screw top 40 ml Universal sample bottles. The samples were heated in a 95 °C water bath with gentle shaking to ensure the granules remained in suspension until gelatinization occurred and then left in the the water bath for a further period of one hour. After cooling, the samples were transferred into 15 ml conical centrifuge tubes and centrifuged. The swelling volume was obtained directly by reading the volume and is quoted as the volume of sediment per 100 ml of starch sample. The absorbance of the supernatant was measured at 330 nm with distilled water as the reference sample.

The total solids content in the supernatant was measured by drying about 3 g of the supernatant in an air oven at 105°C for 24 hours. Total carbohydrate content of the supernatant was determined after dilution using an autoanalyser with 70% sulphuric acid containing 1% orcinol as reagent. The samples were measured against glucose standards.

RESULTS

Slight changes in pH (mean pH 7.5) occurred as a result of addition of the antioxidants and gelatinisation. However, any change in pH was not greater than one pH unit.

Effect of Sodium Sulphite. Figure 1 displays the effect of sulphites alone on the swelling volume. There is a pronounced minimum at around 0.01 % or 100 ppm sulphite concentration and for levels higher than this there is a recovery in the swelling volume to a value approaching that of samples containing no sulphite. The reduction in swollen volume is accompanied by an increase in the amount of solids, as measured gravimetrically, and by the amount of carbohydrate found in the supernatant. If all the solid material was evenly distributed throughout the centrifuge tube then a solids content of about 9 mg/ml would be expected. This is what is observed at 0.01% sulphite addition. There are also changes in the clarity of the supernatant as evidenced by the change in absorbance.

Effect of Propyl Gallate. In contrast, the addition of gallate at these low concentrations did not influence the swelling volume levels. However, it did reduce the amount of material released into the supernatant. Since the amylose content of cassava is approximately 20%, it appears that only this component was released during the pasting process when gallate was incorporated whereas, in its absence, some amylopectin may have been leached out. In order to be certain, it is necessary to characterize the amylose-amylopectin ratio.

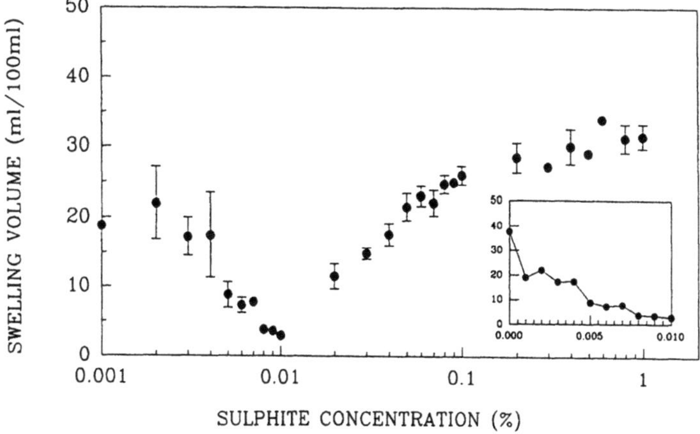

Figure 1. The effect of sodium sulphite addition on the swelling volume of cassava starch. The error bars are standard errors of three replicates. The inset graph is for values of swelling volume for sulphite concentrations between 0 and 0.001%.

Effect of Gallate and Sulphite. When gallate was added to systems containing 0.01% sulphite, the pronounced reduction in swelling volume that was observed in the case of 0.01% sulphite alone, was prevented. The levels of gallate required to do this were only around 0.002% (Figure 2). The increase in swelling volume was accompanied by a decrease in the amount of carbohydrate released into the supernatant.

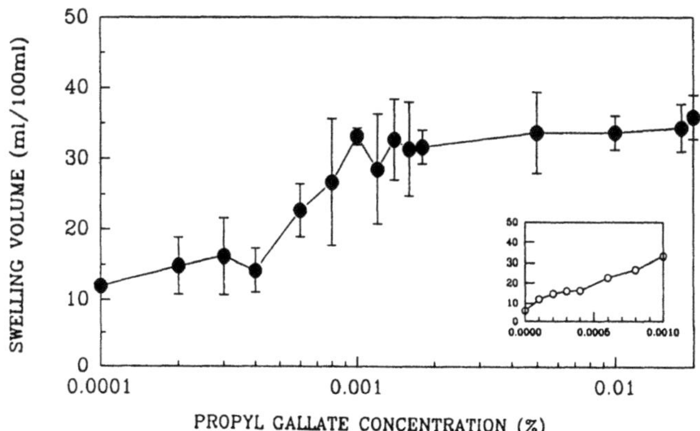

Figure 2. The effect of propyl gallate addition on the swelling volume of cassava starch containing 0.01% sodium sulphite. The inset graph is for values of swelling volume for propyl gallate concentrations between 0 and 0.001%.

DISCUSSION

It would seem that the addition of low levels (0.01% or 0.79 mmol) of sulphite to cassava starch resulted in more carbohydrate leaching out of the swollen granule. This resulted in collapse of the swollen granule and near total solubilization of the polysaccharide into the continuous phase. A possible role of sulphite could be its interaction with oxygen resulting in the removal of the oxygen but leading to the formation of sulphite free radicals and the superoxide ion ($\cdot O_2^-$).

Oxygen levels in water at 90°C are in the region of 0.2 mmol and it would therefore seem that there is an excess of sulphite in the reaction mixture to ensure the complete removal of the oxygen. However, it could be envisaged that the incomplete removal of the oxygen would lead to the production of superoxide ions and sulphite free radicals. The sulphite radicals are also capable of many interactions independent on the presence of oxygen; reactions with double bonds, with the disulphide bonds in protein and nucleophilic attack of carbonyl groups are all known reactions of the sulphur (IV) oxoanions (Wedzicha, 1984).

At higher levels of sulphite the oxygen could be totally scavenged from the system and although both the sulphite and the superoxide radical are able to initiate chain propagating reactions, their maintenance depends on the formation of the superoxide and if this species is removed then the chain reaction will cease. This could explain the recovery in the swollen volume at higher sulphite levels.

It was demonstrated by Dev and Jain (1961) that propyl gallate inhibits the oxidation of sulphite. Propyl gallate is the most polar of the food allowed phenolic antioxidants and therefore will be effective in aqueous solutions as a free radical terminator inhibiting ORD reactions.

It would appear that under the pasting conditions employed the granule is close to complete disintegration since in the absence of the antioxidant about 40% of the total carbohydrate in the starch granule is found in the supernatant. This partially explains why very low levels of antioxidants have a dramatic effect on the state if the granule. It would also be expected that the results will be temperature dependent.Although we have interpreted these results in terms of limited degradation of carbohydrate polymers through ORD reactions, the sulphite can promote lipid oxidation and this may inhibit the formation of amylose-lipid complexes which in the absence of sulphite have a stabilising role on the thermal degradation of starch.

In our view this work has some interesting implications both in terms of industrial applications of starch and development of our understanding of the changes taking place when the starch granule is heated.

REFERENCES

Dev, B. and Jain, B.D., 1961, Inhibitors of the autoxidation of sodium sulphite solution, *J. Sci. Ind. Res.* 20:461-462.

Mitchell, J.R., Reed, J., Hill, S.E. and Rogers, E., 1991, Systems to prevent loss of functionality on heat treatment of galactomannans, *Food Hydrocolloids.* 5:141-143.

Wedzicha, B.L., 1984, "Chemistry of Sulphur Dioxide in Foods," Elsevier, London.

Wellington, S.L., 1983, Biopolymer solution viscosity stabilization - polymer degradation and antioxidant use, *Soc. Pet. Eng. J.* 9296:901-912.

PHYSICAL PROPERTIES OF STARCH PASTES

S. Akuzawa, S. Sawayama, and A. Kawabata

Department of Nutrition, Faculty of Agriculture
Tokyo University of Agriculture, 1-1-1 Sakuragaoka
Setagaya-ku, Tokyo 156, Japan

INTRODUCTION

Starch paste is used in cooking and processing as a viscosity enhancer and viscosity stabilizer.[1,2] But few studies have been conducted on its spinnability, which is a visco-elastic phenomenon. We have jointly developed a Rheoner RE-33005 with Yamaden Co. [3] to assess spinnability. The area of the tip of the rod in this equipment is a very small cross-sectional area to minimize the effects of surface tension. The spinnability of various starch pastes was compared with this apparatus, and the results suggest that the characteristics of spinnability are important indicators of viscoelastic behavior. The relationship between spinnability and dynamic viscoelasticity was developed, and the stress relaxation was investigated.

MATERIALS AND METHODS

Materials

The starch materials used in this study were corn, sago, *katakuri* (Dogtooth violet), sweet potato, *kuzu*, edible canna, cassava, indian lotus root, arrowroot and potato. All these starches were purified in our laboratory.

Preparation of the starch pastes

Starch paste samples of 150g each were prepared at 4% concentrations (w/w on the basis of the anhydride). The proper amount of starch was measured into a beaker, and demineralized water was added to bring the total to 150g. The starch suspension was allowed to swell for 15min, before being stirred by hand with a glass rod at about 120rpm while being heated in a thermostatic tank held at 85° C, after which it was heated for 20min under saturated steam conditions in a steam oven (RO-3700, Mitsubishi Denki Co. Tokyo) The concentration (w/w) was corrected with demineralized water and each sample degassed, before being kept in a thermostatic enclosure for 1.5 hr at 20° C until needed.

Food Hydrocolloids: Structures, Properties, and Functions
Edited by K. Nishinari and E. Doi, Plenum Press, New York, 1994

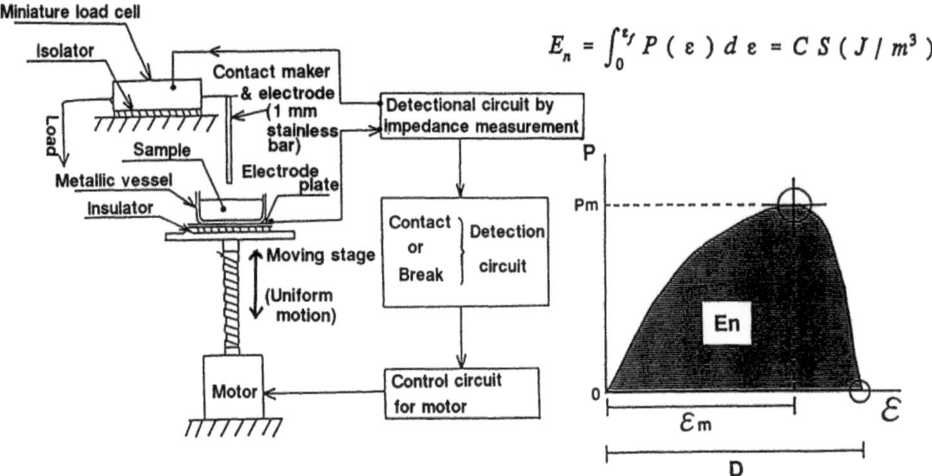

$$E_n = \int_0^{t_f} P(\varepsilon)\, d\varepsilon = C\,S\ (\,J\,/\,m^3\,)$$

Figure 1. Schematic diagram for spinnability and the method for calculating spinning energy En; Spinning energy (J/m³), P; Stress (N/m²), ε; Strain (m/m²), $^\varepsilon_f$; End point of spinnability, D; Apparent spinning distance, C; Conversion coefficient, S, Area, εm; Strain at the point of maximum strength.

Measurement of spinnability

The proper amount of each starch paste sample at a 4% concentration was transferred to a stainless steel petri dish 50mm in diameter and 14mm in depth. The spinning distance and the stress from spinning at 20° C at tensile velocities of 0.1, 1.0, 3.0 and 5.0mm/sec were measured by a Rheoner RE–33005 (Yamaden Co.), which having been modified for measuring spinnability.

Fig. 1 shows a diagram of the measuring apparatus and a stress–strain curve plotted from the measurements.

Measurement of dynamic viscoelasticity and stress relaxation

Dynamic viscoelasticity was measured by a RFS–II Fluid spectrometer (Rheometrics Far East Tokyo) under the following experimental conditions: conical plate of 0.04 rad; 2ml of starch paste, 5% strain, 20° C temperature, and 0.16–100rad./sec frequency.

We obtained stress–time curves to observe the stress relaxation phenomenon in each starch paste for 100 sec.

RESULTS

Spinnability of the starch paste

All the samples had a greater spinning distance with higher tensile velocity, this being particularly evident with the cassava and potato starches. As the tensile velocity for cassava was increased from 2.21mm to 6.35mm, while potato at 4% concentration showed a large increase in the spinning distance with tensile velocity, the spinning distance increased from 2.69mm to 7.56mm. The spinning energy also increased with increasing tensile velocity. Potato showed very high dependence on the tensile velocity, but due to its relatively high spinning energy, katakuri showed low dependence on the tensile velocity.

Dynamic viscoelasticity of the starch paste

The storage modulus (G') and loss modulus (G") both increased with increasing frequency, but were characteristic for each starch paste. Potato and cassava starch pastes were similar in the rate of change of G' and G" with frequency. With corn starch paste, G' changed only slightly but G" increased greatly, while *katakuri* and edible canna showed similar characteristics. Changes in the loss tangent, tanδ (=G"/G') for both potato and cassava starches were similar with only slight change with increasing frequency; potato and cassava starch pastes gave tanδ values of around 0.5 and 0.6, respectively, which were larger than the values for the other starch pastes. *Kuzu* and sweet potato had similar constant values of about 0.3, while *katakuri* and edible canna produced similar curves of frequency dependency.

Stress relaxation of the starch paste

The stress relaxation of potato and cassava were analyzed corresponding to a Maxwell model of four elements. The other starch pastes were analyzed corresponding to a five–element model, on the assumption that they had an apparent equilibrium stress. The viscoelastic constants of each starch paste are shown table 1.

Table 1. Viscoelastic constants of each starch paste

	Modulus of elasticity (N/m²)				Viscosity (Pa·s)			Relaxation time (s)		
	$E\infty$ ($\times 10^{-1}$)	E_{M1} ($\times 10$)	E_{M2} ($\times 10$)	Σ_{EM} ($\times 10$)	η_{M1} ($\times 10^2$)	η_{M2} ($\times 10^2$)	$\Sigma_{\eta M}$ ($\times 10^2$)	τ_{M1}	τ_{M2}	$\Sigma_{\tau M}$
Potato		0.44	0.78	1.22	4.86	1.91	6.77	109	24	133
Cassava		0.96	0.22	1.18	12.62	0.38	13.00	131	17	148
Sago	2.10	0.86	0.12	1.00	0.27	0.06	0.33	32	5	37
Arrowroot	2.90	0.10	0.34	0.47	0.40	0.52	0.92	40	15	55
Indian lotus root	2.80	0.16	0.54	0.73	0.67	1.04	1.71	42	19	61
Corn	5.00	0.07	0.10	0.22	0.25	0.12	0.37	36	13	49
Kuzu	2.80	0.10	0.46	0.59	0.40	0.95	1.15	40	21	61
Sweet potato	4.00	0.56	0.47	1.07	0.20	0.42	0.62	36	9	45
Edible canna	9.70	0.16	0.80	1.06	0.60	1.32	1.92	37	16	53
Katakuri	8.20	0.18	0.76	1.02	0.67	1.03	1.70	37	14	51

DISCUSSION

It is well known that gelatinized starch paste can form threads when a rod in contact with the surface of the starch paste is pulled away, this property being known as "spinnability". Nakagawa [4] has described how spinnability was produced by a well-balanced combination of viscosity and elasticity, i.e., by the degree of viscoelastic balance. In this study, with the objective of classifying starch pastes of different origin according to their spinnability, the characteristics of each starch paste were calculated from the characteristic spinnability value at the prominent 4% concentration; i.e., the spinning distance, maximum stress, strain at the point of maximum stress, and spinning energy. Fig. 2 shows the dendrogram obtained from a cluster analysis by the maximum distance method, and the spinning distance for each of these parameters (stress–strain curve). Four clusters were obtained, with the spinning pattern of each cluster being characteristic. When measuring the spinnability, the stress dropped sharply after the point of maximum stress, while the spinning distance was observed to increase. We consider this to have been due to a reduction in viscosity once the point of maximum stress had been reached. Spinnability is a viscoelastic phenomenon, and the appearance of spinnability is a type of web-like structure forming in the liquid, [5] so that the

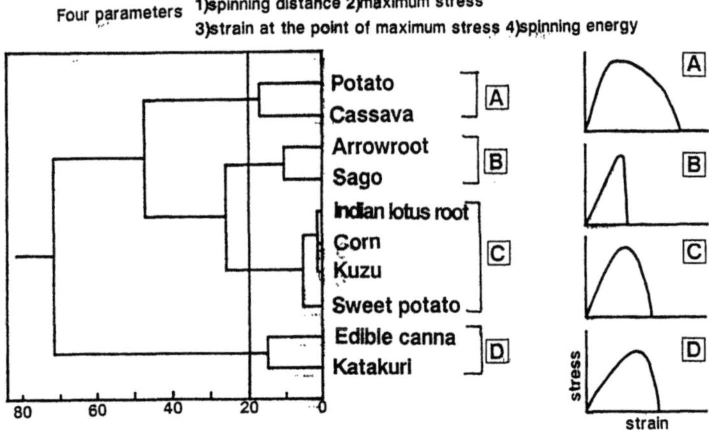

Figure 2. Dendrogram of spinnability for the starch pastes by a cluster analysis and some curves of spinnability for 4% concentration at 20°C.

that different spinnability of each starch would be due to the different molecular structure of each starch paste. The results of dynamic viscoelasticity and stress-relaxation measurements enabled the starch pastes to be classified: (A) Potato and cassava showed similar changes in G' and G", with balance maintained in G' and G" values, in spite of the stress applied. G" represents the effect on elasticity by internal friction, while the mechanism for η is represented by G". (C) G" for corn decreased immediately, and we consider that this is related to the greater spinning distance at higher tensile velocity of cassava and potato. The tanδ values for, potato and cassava suggest that these starch pastes were more fluid-like than others. The stress relaxation results for potato and cassava suggested that they were in a different state from others, so that they were analyzed corresponding to a Maxwell model of four elements. When the spinning distance was plotted against the total relaxation time, we found a linear relationship, with a regression line $Y=0.73X+2.19$ and a correlation coefficient of 0.80. This result indicates that the spinning distance was highest with a comparatively long relaxation time. This is similar to the results of Takahashi,[6] Kando[7] and Murayama [8]. However, the magnitude of time difference is large: spinnability is from 1 to 2 sec, and relaxation time is from 37 to 148 sec. Consequently, these relaxation times need to be studied to discover the effect of viscosity for various kinds of starch paste.

REFERENCES

1. R. Takahashi, Physicochemical properties and utilization of the starch, J. Jpn. Soc. Starch Sci., 21: 51 (1974).
2. M. Okada, The utilization of starches for seafoods, Food Ind., 11: 25 (1968).
3. S. Akuzawa, S. Sawayama, and A. Kawabata, Spinnability of starch pastes, Biosci. Biotech. Biochem, 56: 932 (1992).
4. T. Nakagawa, Spinnability of Liqiud, Bull. Chem. Soc. Japan, 25: 88 (1952).
5. T. Nakagawa, "Rheology", 2nd Ed, Iwanami Shoten, Tokyo (1977)
6. R. Takahashi, Spinnability of starch pastes, Die Starke, 12: 315 (1969).
7. S. Kando and Y. Itou, Spinnability of Liqiod, Jpn. Chem Soc., 78: 686 (1957).
8. Y. Murayama, M. Kobayashi, H. Akabane, and N. Nakahama, Spinnability of starch pastes, J. Jpn. Soc. Food Sci. Technol., 33: 274 (1986).

EFFECTS OF SACCHARIDES ON STABILITIES OF RICE STARCH GELS – CORRELATION BETWEEN STRUCTURE OF SACCHARIDES AND STABILIZING EFFECTS ON GEL STRUCTURE

Keiko Katsuta,[1] Akio Nishimura[2] and Makoto Miura[2]

[1]Faculty of Education
Niigata University
8050 Ikarashi Ni-No-Cho
Niigata 950-21
[2]Research Center
Mitsubishi Kasei Corporation
1000 Kamoshida-Cho
Midori-Ku, Yokohama 227
Japan

Effects of hydrogenated saccharides (sugar alcohols) on the stabilities of rice starch gels were investigated using the retrogradation kinetics on the basis of creep data of starch gels. The retrogradation process of rice starch could be expressed as the first-order reaction, in the initial stage of retrogradation during storage at 0°C .

In the hydrogenated saccharides, sorbitol (G_1OH) through maltotriitol (G_3OH), the most effective saccharides to stabilize the gel structure was sorbitol. The stabilizing abilities of saccharides on starch gel structure may be complicated and do not depend only on the mean number of equatorial hydroxyl group or the dynamic hydration number but also are influenced by conformational structure of saccharides molecules.

INTRODUCTION

In the series of our studies on retrogradation of rice starch,[1-3] we proposed that saccharides gave the stabilizing effect to water structure in the gels, hence the gel matrix was strengthened and the flexibility of starch chains was reduced. Thus the retrogradation of starch is interfered by saccharides. We examined the effects of conformation and degree of polymerization of saccharides on retrogradation of rice starch in previous papers.[2,3] In this

Food Hydrocolloids: Structures, Properties, and Functions
Edited by K. Nishinari and E. Doi, Plenum Press, New York, 1994

study, we investigated the effects of various hydrogenated saccharides on stabilities of starch gels, and the stabilizing abilities of hydrogenated saccharides were compared to corresponding saccharides.

MATERIALS AND METHODS

Materials

Non-glutinous rice starch was obtained from Matsutani Chemical Co. Ltd. (Tokyo, Japan). Reagent grade saccharides and hydrogenated saccharides were used. Maltotriitol was provided from Mitsubishi Kasei Corp. (Tokyo, Japan) and its purity was 99%.

Preparation of starch gels

Thirty percents (W/W) of rice starch gels were prepared by the procedure which had been reported previously.[1-3] The concentration of hydrogenated saccharides added to starch-water system was 6% against starch weight.

The samples stored at 0℃ (up to 50 hrs) were carefully removed from the tubes, cut into cylindrical blocks (11ϕ × 10 mm) using a Ultrasonic Sample-Cutter (USC-3305, Yamaden Co., Ltd., Tokyo, Japan) to obtain the smooth surface and submitted to creep measurement.

Static viscoelastic measurements

A Creep Meter (Rheoner RE-3305, Yamaden Co., Ltd.) was used for measurements of static viscoelastic properties (creep behavior).

Gel samples were set in a thermostatted chamber connected to a water bath to maintain the measuring temperature (25℃). Silicone oil was used to fill the chamber to prevent the evaporation of water in the samples. Measurements were carried out under uniaxial compression for 300 sec, while ensuring that subsequent deformation was maintained at <10%, as the linearity between stress and strain was determined.[1-3]

Kinetic treatments

We proposed previously[1-3] that the retrogradation process of rice starch gels in the initial stage could be expressed by the first-order kinetic equation as follows:

$$\log J = -k \cdot t + \log J_0 \tag{1}$$

where k is the rate constant, i.e. retrogradation rate and t the storage time at 0℃ . J_0 is initial creep compliance, i.e., J when storage time was zero (t=0). In this study creep compliance, J, was used, defined as follows:

$$J = J(300) - t_m / \eta_N \tag{2}$$

where J(300) is the value of creep compliance at 300 sec, t_m
the measuring time and η_N the Newtonian viscosity.

The rate constants k of 30% starch gel (starch-water
system: control) and starch-water saccharides systems
containing various hydrogenated saccharides were calculated.

RESULTS AND DISCUSSION

Effects of hydrogenated saccharides on the rate constants

The values of rate constants k for gels prepared with
addition of hydrogenated saccharides decreased less than
gels without sugar alcohols (Table 1). The results show that
all the hydrogenated saccharides impede the retrogradation
of starch, i.e., hydrogenated saccharides make the more
stable gel matrix as the same as saccharides[1-3] rather than
the gel prepared by starch alone. The most effective
hydrogenated saccharide, in this study, was sorbitol, while
the worst one was maltotriitol.

Table 1. The rate constants, k, of starch-water
system and starch-water-hydrogenated saccharides
systems.

	Rate constants $(\times\ 10^{-4} \cdot min^{-1})$
Control	4.20
Xylitol	2.45
Sorbitol	2.02
Galactitol	2.38
Maltitol	3.14
Maltotriitol	3.61

Comparison of hydrogenated saccharides and saccharides

In the hydrogenated monosaccharides, the values of k
for starch-water-sorbitol system was smaller than those for
starch-water-xylitol and -galactitol systems. The rate
constants k for starch-water-hydrogenated monosaccharides
systems increased in the order:

$$sorbitol < galactitol \leqq xylitol \qquad (3)$$

The results were coincident with our previous study on
monosaccharides[2] in which k for gels containing the
monosaccharides increased in the order:

$$glucose\ (fructose) < galactose \leqq xylose \qquad (4)$$

On the other hand, the values of k and J_0 for gels
containing hydrogenated oligosaccharides increased with

increasing chain length of hydrogenated saccharides in the
order:

$$sorbitol < maltitol < maltotriitol \qquad (5)$$

In our previous work[3] on the effects of degree of
polymerization of saccharides, the longer the saccharide
molecule was (up to trisaccharide), the smaller the k value
possessed in the order:

$$glucose > maltose > maltotriose \qquad (6)$$

In this study, however, the ability for stabilizing gel
structure decreased with increasing chain length of
hydrogenated saccharides as seen in eq. (5). The results
suggest that there are different mechanisms between
monosaccharides and their hydrogenated ones, and between
oligosaccharides and the corresponding hydrogenated one to
stabilize the water structure and/or gel matrix.

The stabilizing properties of starch gel matrix are
also influenced by hydrogenated saccharides as the same as
saccharides. The stabilizing effects of hydrogenated
monosaccharides were larger than those of corresponding
monosaccharides. The ability of saccharides to stabilize the
gel matrix increased with increasing the degree of
polymerization up to maltotriose (DP=3),[3] but converted
results were obtained from the hydrogenated
oligosaccharides. Hydrogenated oligosaccharides were not so
effective to stabilize the gel matrix and to interfere the
retrogradation of rice starch.

Acknowledgments

We thank Miss Y. Yoshida for her technical help.

REFERENCES

1. K. Katsuta, M. Miura, and A. Nishimura, Kinetic
   treatment for rheological properties and effects of
   saccharides on retrogradation of rice starch gels.
   Food Hydrocolloids, 6:187(1992).
2. K. Katsuta, A. Nishimura, and M. Miura, Effects of
   saccharides on stabilities of rice starch gels. 1.
   Mono- and di-saccharides. Food Hydrocolloids, 6:in
   press (1992).
3. K. Katsuta, A. Nishimura, and M. Miura, Effects of
   saccharides on stabilities of rice starch gels. 2.
   Oligosaccharides. Food Hydrocolloids, 6:in press
   (1992).

EFFECTS OF MOLECULAR WEIGHT ON THE RETROGRADATION OF AMYLOSE

Shinichi Kitamura, Kanako Hakozaki, and Takashi Kuge

Department of Agricultural Chemistry, Kyoto Prefectural
University, Shimogamo, Kyoto 606, Japan

ABSTRACT

We have examined the effects of molecular weight on the
morphological features of amylose precipitates, especially the
particle size and size distribution, during retrogradation of
amylose using scanning electron microscopy. The amylose precipitates consisted of particles with a surprising uniformity in
size, and the volume of each particle increased proportionally
with the degree of retrogradation. These results suggest that
the retrogradation of amylose proceeds via a nucleation process, which is completed within a short time after the onset of
retrogradation; this is then followed by growth of the nuclei.
At a given retrogradation percentage, the particle size is much
larger for the particle made up of low-molecular-weight amy-
lose(DP=172) than that of high-molecular-weight amylose
(DP=2600), indicating that the number of nuclei formed in the
initial stage is much smaller for the particle of low-molecular-weight amylose. In conjunction with the result on the rate
of retrogradation, it was found that the growth rate of nuclei
is much higher for the low-molecular-weight amylose, implying
the shorter amylose chain is easily incorporated into the
particle.

INTRODUCTION

Amylose is metastable in neutral aqueous solution. Solutions of low concentration become increasingly cloudy due to
the progressive association of the amylose molecules into
larger particles. This phenomenon is often referred to as
retrogradation. Many factors such as the molecular weight and
branching structure of amylose, amylose concentration, temperature, pH, and additives affect the rate of retrogradation.
In earlier work,[1] we examined the effects of KCl concentra-

Food Hydrocolloids: Structures, Properties, and Functions
Edited by K. Nishinari and E. Doi, Plenum Press, New York, 1994

"

tion on the retrogradation. The results obtained by the study showed that the retrogradation of amylose proceeds through a two step process; the first step is the formation of stable nuclei in the solution, and the second is the further growth of the nuclei by the addition of amylose molecules to their surfaces. KCl reduces the number of nuclei formed in the initial stage, and this effect results in a decrease in the rate of retrogradation.

Here we report preliminary results on the effects of molecular weight on the rate of retrogradation and, on the morphological features of amylose precipitates, especially particle size and size distribution, during retrogradation as observed by scanning electron microscopy.

MATERIALS AND METHODS

Two amylose samples were synthesized enzymatically using potato phosphorylase from maltopentaose as a primer and glucose-1-phosphate as a substrate. The detailed procedure for the preparation of synthetic amyloses was given previously.[2,3] The number-average degree of polymerization, DPn, of a synthetic amylose(L-amylose) was estimated to be 172 from the amount of phosphate liberated during the phosphorylase reaction and the concentration of the primer used. The weight-average degree of polymerization, DPw, of another sample(H-amylose) was determined to be 2600 by light-scattering measurements.

Aqueous amylose solutions were prepared by dissolving the amyloses in 1 N KOH under N_2 gas overnight at 5 °C, followed by dilution with distilled water and neutralization with 1 N HCl. The pH of the solutions was adjusted to 4.5 by adding 1 M K-phosphate buffer solution. The final concentrations of KCl and buffer were 0.4 and 0.01 M, respectively. The final concentrations of amylose were controlled to be 0.2 and 0.4 %(w/v) by changing the amount of amylose dissolved in aqueous KOH. The aqueous amylose solutions were placed in a water bath thermostatically controlled at 10 ± 0.1 °C. After an appropriate period, the solution was centrifuged for 5 min at 3000 rpm and the retrograded amylose collected. The precipitate was washed with cold water, dehydrated with ethanol, washed with ether and dried *in vacuo* at room temperature. The samples thus prepared were submitted to study by scanning electron microscopy. The concentration of amylose remaining in the supernatant was determined by the phenol-sulfuric acid method. The percentage of the amylose retrograded was calculated from this and initial concentration of amylose in the solution.

Scanning electron microscopic observations of retrograded amyloses were performed according to the method described previously.[1] Several photomicrographs were taken for each sample. The diameters of retrograded amylose particles in photomicrographs were measured manually with a rule, and classified into subgroups with size differences of 0.125 μm interval to obtain the histograms. The number average volume is defined as follows:

$$Vn = \frac{\Sigma N_i D_i^3}{\Sigma N_i}$$

where N_i and D_i are the number of particles in each sub-group, i, and the specified diameter of sub-group, i, of the histogram, respectively.

RESULTS AND DISCUSSION

Fig. 1 represents the time courses of retrogradation of amyloses. The rate of retrogradation for L-amylose is much higher than that for H-amylose.

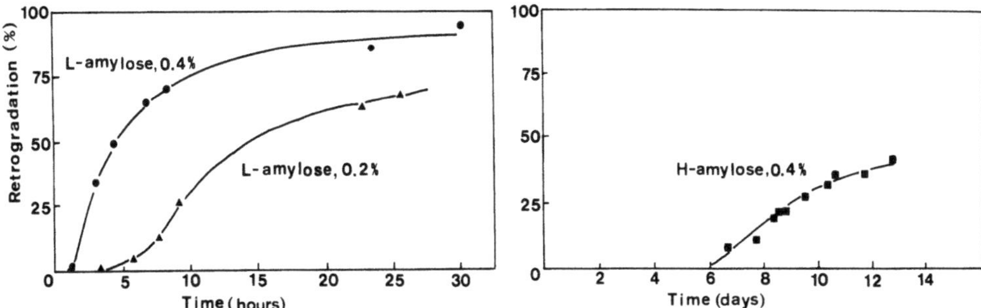

Fig. 1. Time courses of the retrogradation of amylose.

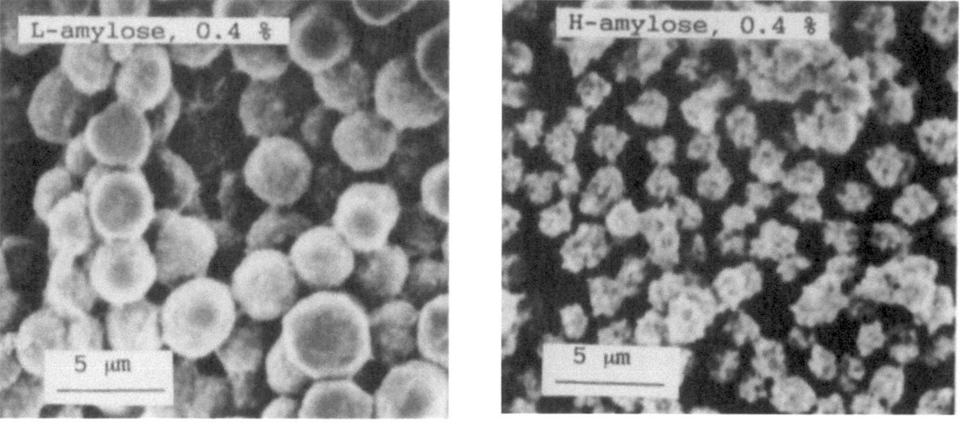

Fig. 2. Amylose particles retrograded from aqueous solution.

Fig. 2 displays typical retrograded amylose particles as seen using scanning electron microscopy. The surfaces of the particles are not smooth, but the shapes as a whole may be approximately spherical. These photomicrographs represent at the stage of retrogradation of 49 % for L-amylose, and 56 % for

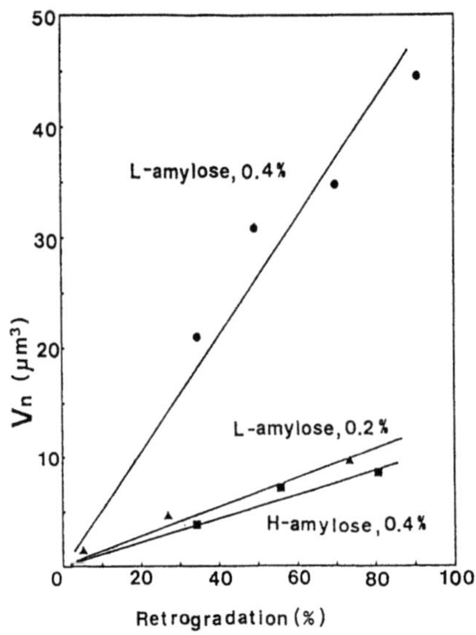

Fig. 3. Plots of of V_n versus percentage of retrogradation.

H-amylose, respectively. The changes in the particle size and particle-size distribution during the period of retrogradation was investigated. The histograms of size distribution for the amylose particles demonstrated that every amylose sample has a similar monomodal distribution with relatively little spread and the number average particle diameter increases with the period of retrogradation. Fig. 3 shows the plots of V_n versus percentage of retrogradation. It is seen that all the case examined give linear relations and have a common intercept on the ordinate close to zero. Thus, the particle volume increases proportionally with increasing percentage of retrogradation. Fig. 3 also shows that the particle volume is dependent on the DP. Comparison of each V_n at the same percentage of retrogradation indicates that the particle volume is much larger for the particle made up of L-amylose(0.4 %) than that of H-amylose(0.4%). This fact suggests that the number of nuclei formed in the initial stage is much smaller for the particle of the shorter amylose. In conjunction with the result on the rate of retrogradation (Fig. 1), it is considered that the growth rate of nuclei is much higher for L-amylose, i.e., the shorter amylose chain is easily incorporated into particle. We are extending this work to cover the wide range of DP.[3]

REFERENCES

1. S. Kitamura, S. Yomeda, and T. Kuge, Study on the retrogradation of starch. I. Particle size and its distribution of amylose retrograded from aqueous solutions, Carbohydr. Polym. 4:127(1984).
2. S. Kitamura, H. Yunokawa, S. Mitsuie, and T. Kuge, Study of polysaccharide by the fluorescence method. II.Micro-Brownian motion and conformational change of amylose in aqueous solution, Polym. J. 14:93(1982).
3. B. Pfannemuller, H. Mayerhofer, and R.C. Schulz, Conformation of amylose in aqueous solution:Optical rotatory dispersion and circular dichroism of amylose-iodine complexes and dependence on chain length of retrogradation of amylose,Biopolymers 10:243(1971).

EFFECT OF DMSO ON THE GELATION OF AMYLOSE

Hidenobu Takeyama,[1] Masatoshi Kobayashi,[1] Hirofumi Yajima,[1] Ryuichi Endo,[1] Kaoru Kohyama,[2] and Katsuyoshi Nishinari[2]

[1]Department of Applied Chemistry, Faculty of Science, Science University of Tokyo, Tokyo, 162 Japan
[2]National Food Research Institute, Tsukuba, Ibaraki, 305 Japan

ABSTRACT

Turbidity measurement is one of the useful methods to observe the phase separation. Cloud time (Ct) at which the turbidity begins to develop was measured at various aging temperatures. The relationship between the value of $Ct^{-0.5}$ and the aging temperature seems to be linear at various amylose concentrations. Critical cloud temperature (Tc*) above which the aggregation does not occur was determined by extrapolation of the $Ct^{-0.5}$ to zero. The heat (ΔH°) absorbed on forming one mole of junction zones was calculated by Eldridge-Ferry plot (1/Tc* vs Log C). If the functionality constant f in the Eldridge-Ferry's equation is assumed to be 4, then ΔH° can be obtained by the curve fitting. Effects of DMSO on ΔH° were examined. It was found that ΔH° became minimum at 10% (v/v) DMSO content.

INTRODUCTION

Starch consists of two main polysaccharides, amylose and amylopectin, both of which are based on α-(1,4)-linked α-D-glucose unit. Amylose is essentially a linear macromolecule, whereas amylopectin is a highly branched chain with α-(1,6)-linkage. The gelatinization occurs when aqueous suspension of starch is heated at a certain temperature. The retrogradation occurs after the gelatinized solution is cooled to a certain temperature. There have been many investigations on these phenomena.[1,2] It is generally believed that the gelation of amylose in aqueous solution occurs at the first stage of the retrogradation of starch.[3-5] Recently, it has been suggested[3,6,7] that the gelation process of amylose occurs through the phase separation caused by aggregation of amylose. Therefore the aggregation of polymers is an important process in gelation. There has been much discussion concerning the gelation process of amylose.[2-6] We have studied factors affecting the gelation process of amylose-DMSO-water system, such as DMSO content, amylose concentration, and cooling rate. Then it was found that gelation process of amylose was strongly dependent on these factors. But the behavior of the critical state is not

wellunderstood in detail about amylose-DMSO-water system. The purpose of our research
is to clarify this point.

EXPERIMENTAL

Materials

The amylose sample prepared from corn starch was purchased from Wako Pure
Chemical Industry Ltd (Osaka).   The intrinsic viscosity in DMSO at 25°C was 80mL/mg,
yielding a viscosity-average molecular weight (Mv) of 220,000 according to the equation
proposed for amylose in pure DMSO solution.[8]

Preparation of Sample Solutions

Solutions were obtained by heating appropriate mixtures of amylose and DMSO, sealed
in the test tubes at 40 °C for 15 min, and then heated at 100 °C for 30 min. Distilled water
was added in the DMSO solution after heating. The temperature of the solution was kept at
100 °C for 90 min.

Turbidity Measurement

The development of turbidity was monitored by measuring absorbance variations as a
function of time at 625 nm with a PC2101-UV Spectrophotometer of Shimadzu Seisakusho
(Kyoto). DMSO-water mixed solvent was used for the reference. Temperature was
controlled by the same way as the rheological measurement.

RESULTS AND DISCUSSION

Figure 1 shows the turbidity for 4.7% (w/w) amylose containing 10% (v/v) DMSO as
a function of time at various aging temperatures. Cloud time (Ct) in Fig.1 was determined
as a time at which the plateau line and the maximum tangent line of turbidity curve crossed.

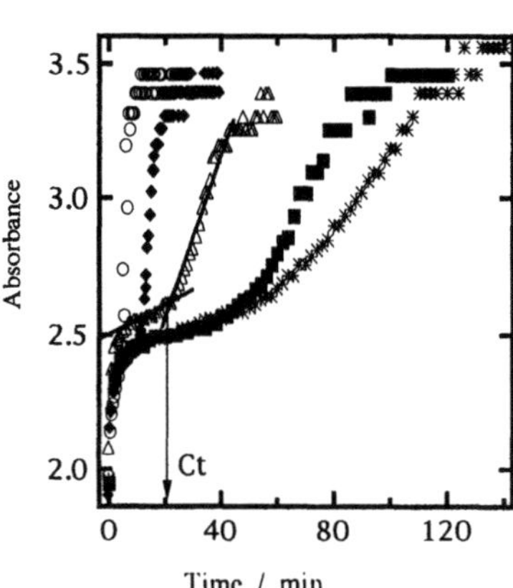

Fig.1   Turbidity curve for 4.7% (w/w) amylose
10% (v/v) DMSO water system at variousaging
temperature ○ 20°C ◆ 25°C △ 30°C
■ 33°C ✳ 35°C

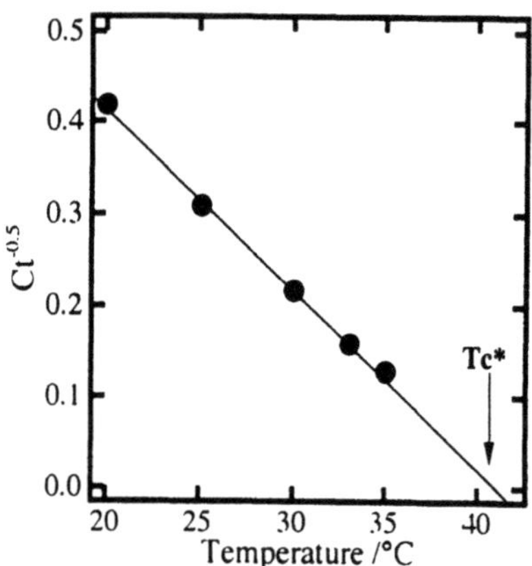

Fig.2 Relationship between Ct$^{-0.5}$ and
temperature for 4.7% (w/w) amylose
10% (v/v) DMSO water system.

Ct was prolonged with increasing aging temperature. It was suggested that gel time (Gt) required before the commencement of gelation of gelatin correlated with the aging temperature by Nijenhuis.[9,10] Figure 2 shows the relationship between $Ct^{-0.5}$ and aging temperature. It was found to be linear, and the critical cloud temperature (Tc*) above which the aggregation does not occur was obtained by the extrapolation of $Ct^{-0.5}$ to zero as illustrated in Fig.2. The critical cloud temperatures obtained for concentration range of amylose from 1.9% (w/w) to 4.7% (w/w) are given in Table 1.

Table.1   The critical cloud temperature Tc* for 10% (v/v) DMSO water amylose solutions as a function of the amylose concentration

C%(w/w)	4.72	3.81	2.89	2.51	1.94
Tc*(°C)	41.3	38.6	34.4	33.1	29.9

The relationship between the melting temperature (Tm) of gels and the concentration give by Eldridge and Ferry[11] can be rewritten as

$$\frac{\partial \log C}{\partial(1/Tc^*)} = \frac{-\Delta H^\circ}{2.303R(1-f/2)} \quad (1)$$

where Tc* is the critical cloud temperature in stead of Tm, C is polymer concentration, f is functionality, R is the gas constant, and ΔH° is the heat of reaction. It can be applied to obtain the value of ΔH°, which is the heat required for the aggregation of a mole of polymer chain in the amylose-DMSO-water system. $f/2$ potential junction zones make f cross-links with getting and losing the ΔH°. If we determine the f value, ΔH° is estimated from $\partial \log C / \partial(1/Tc^*)$. Nijenhuis[10] suggested that the functionality f of gelatin gel is assumed to be 6, in view of the fact that one cross-link in gelatin gels is formed by three polymer chains involved in the formation of triple helix. In the case of amylose, it has been suggested[12] that aggregation of amylose makes double-helix structure, followed by helix-helix aggregation. If the aggregation of amylose occurs by this mechanism, functionality f can be assumed to be equal to 4. On the assumption that the f is equal to 4, then ΔH° can be obtained.

In Fig.3, log C is plotted against the reciprocal critical cloud temperature at 10% (v/v) DMSO content From the slope of the straight line, the value of ΔH° was obtained to be - 60 kJ / mol. The ΔH° values for other DMSO contents were obtained in the same way.

Figure 4 shows the ΔH° as a function of DMSO content for amylose-DMSO-water system. It was found that ΔH° took a minimum around 10% (v/v)

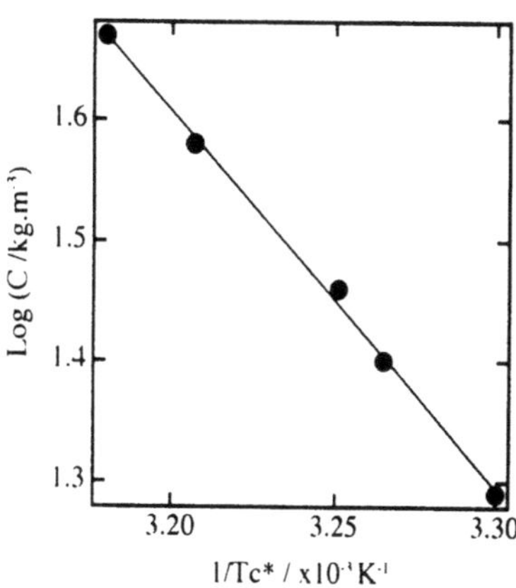

Fig.3 Logarithm concentration of amylose plotted against reciprocal critical cloud temperature.

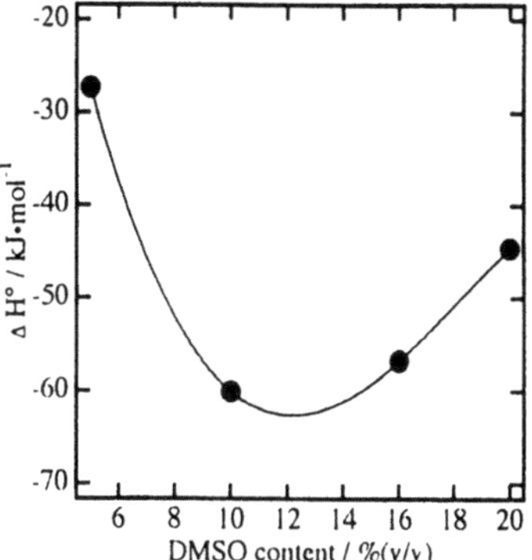

Fig.4 Relationships between ΔH° and DMSO content.

DMSO content. A similar feature for $\Delta H°$ in the heating process appears for agarose-DMSO-water system[13] and PVA-DMSO-water system,[14] where $\Delta H°$ shows a maximum at the DMSO mol fraction of 0.277 in the solvent corresponding to ca. 60 % (v/v) DMSO content. The appearance of maximum in those system was interpreted in terms of the concept that DMSO forms a hydrated compound such as $(CH_3)_2SO \cdot 2H_2O$ or $(CH_3)_2SO \cdot 3H_2O$.[13,14] However, this concept will not be applied for our system because of the fact that the region of DMSO content around 10% does not correspond to the region where hydrated compounds form effectively.

This discrepancy results presumably from the difference in the solubility between amylose and agarose in water. The increase in $\Delta H°$ with increasing DMSO content above 10% (v/v) is suggested to be involved in the junction formation with disordered structure due to high solubility of amylose in DMSO. However, a clear-cut interpretation on the increase in $\Delta H°$ below 10% (v/v) has not yet been established. Probably, this behavior is attributable to the difference of amylose conformation in water rich solution.

At present, the dynamic light scattering study is in progress to clarify the gel structure.

REFERENCES

1. S.G.Ring, Staerke, 37: 80 (1985).
2. M.J.Miles, V.J.Morris, P.D. Orford, S.G.Ring, Carbohydr.Res.,135: 271 (1985).
3. M.Guenet, B.Lots, J.C.Wittmann, Macromolecules, 18: 420 (1985).
4. L.Doublier, G.Llamas, L.Choplin, Macromol. Chem., Macromol. Symp, 39: 171 (1990).
5. J.L.Doublier, L.Choplin, Carbohydr. Res. 193: 215 (1989).
6. M.J.Miles, V.J.Morris, S.G.Ring, Carbohydr. Res. , 135: 257 (1985).
7. H.S.Ellis, S.G.Ring, Carbohydr. Polym., 5: 201 (1985).
8. W.W.Everett, J.F.Foster, J.Am.Chem.Soc., 18: (1959).
9. K.te.Nijenhuis, Colloid and Polym. Sci, 259: 522 (1981).
10. K.te.Nijenhuis, Colloid and Polym. Sci, 259: 1017 (1981).
11. J.E.Eldridge, F.D.Ferry, J. Phys.Chem., 58: 992 (1954).
12. M.J.Gidley, Macromolecules, 22: 351 (1989).
13. M.Watase, K.Nishinari, Polym.J, 21, 8: 597, (1989).
14. M.Watase, K.Nishinari, Polym.J, 20, 12: 1125 (1988).

PHYSICOCHEMICAL CHARACTERISATION OF ALGINATE
FROM MALAYSIAN BROWN SEAWEEDS

Fasihuddin B. Ahmad[1], Siraj Omar[1], and Peter A. Williams[2]

[1]Faculty of Science and Natural Resources
Universiti Kebangsaan Malaysia, Sabah Campus
L.B. No. 62, 88996 Kota Kinabalu, Sabah, Malaysia

[2]Faculty of Science and Innovation
The North East Wales Institute
Connah's Quay, Deeside, CLWYD
CH5 4BR, United Kingdom

ABSTRACT

Nine species of brown seaweeds have been studied to obtain information on the percentage yield of alginate and their physicochemical properties. The percentage yield varied between 18 to 35% when extracted in 0.1 M Na_2CO_3 for 24h. Higher yields were observed when the seaweed samples were extracted in EDTA, CDTA, sodium phosphate and sodium citrate solution. Intrinsic viscosities and molecular weights varied between 2.15 to 14.40 dl/g and 1.05 to 9.49 x 10^5 dalton, respectively. Pre-treating the seaweeds with formaldehyde and ethanol prior to extraction with sodium carbonate, or extraction under neutral conditions with selected calcium ion sequestrants, gave alginate with higher intrinsic viscosity and molecular weight. The M/G ratio for all the samples studied were in the range of 0.42 to 2.10. Based on the physicochemical properties and percentage yield, *Sargassum* spp. are found to be good sources for alginate production.

INTRODUCTION

Alginate is a type of polysaccharide thickener used in the food industry to impart certain textural characteristics to food products. Alginates are found in almost all brown seaweed and the major sources are *Macrocystis pyrifera*, *Laminaria hyperborea* and *Ascophyllum nodosum*. Although certain species of bacteria such as *Azotobacter vinelandii* and *Pseudomonas aeruginosa* can produce alginate[1], the alginates produced are not effective gelling agents due to variable degrees of acetylation and very low polyguluronate content.

Food Hydrocolloids: Structures, Properties, and Functions
Edited by K. Nishinari and E. Doi, Plenum Press, New York, 1994

Alginate is a (1—>4) linked block-like co-polymer of α-L-guluronate and β-D-mannuronate, arranged in regions of constant uronic acid composition with some regions of random uronic acid composition. The structure consists of homopolymeric buckled ribbons of poly-α-L-guluronate which plays a major role in the gelation process[2], and poly β-D-mannuronate with sequences that approximate to alternating guluronate mannuronate repeating units which are not involved in gelation[3].

Gel properties depend on these block structures and the ratio of poly β-D-mannuronate to poly α-L-guluronate referred as M/G ratio. The lower the M/G ratio, the stronger and more brittle is the gel that is formed. Gelation occurs through the formation of an 'egg-box' like structure by the poly-α-L-guluronate blocks which strongly chelate Ca^{2+} ions[2].

In view of the fact that different species of brown seaweed produce alginate of different composition and structure, it is important to characterize alginates from each source of algae separately, in order to evaluate their commercial potential as stabilizers and gelling agents. In this study, nine species of brown seaweeds from Malaysian waters were studied to characterise their physico-chemical properties especially the percentage yield, intrinsic viscosity, molecular weight and M/G ratio.

MATERIALS AND METHODS

Alginates were extracted from various species of brown algae collected from coastal water of Kudat, Sabah, Malaysia using SCUBA and washed thoroughly with tap water and finally with distilled water and dried in the air. All seaweed samples were ground to fine powder prior to extraction.

Extraction of Alginate with Various Solvents

The seaweed samples were extracted for 6 hours and 24 hours using 0.1 M Na_2CO_3, 0.1 M EDTA (disodium ethylenediamine tetraacetic acid), 0.1 M CDTA (1,2-diaminocyclohexane N,N,N,N-tetraacetic acid), 0.1 M sodium phosphate or 0.1 M sodium citrate. The extraction was performed at room temperature with constant stirring. The viscous solution obtained was centrifuged at 2800 x g for 15 minutes. The alginate was finally precipitated from the viscous supernatant using 3 volume equivalent of ethanol. The precipitate was extensively washed with ethanol and dried in a vacuum oven at 60°C to constant weight and the percentage yield was determined.

Pretreatment with Ethanol and Formaldehyde Prior to Extraction with Na_2CO_3.

The ground seaweed samples were soaked in ethanol (95%) or HCHO (40%) for 24 hours at 28°C. The amount of solvent used was just sufficient to wet the seaweeds and filtered. The residue was air dried and extracted with 0.1 M Na_2CO_3 as described above.

Instrinsic Viscosity

Intrinsic viscosities for all samples were determined in a Canon-Ubbelohde, semi-micro dilution viscometer size 75. The solvent used was 0.2 M NaCl and the alginate samples were dialysed to equilibrium against 0.2 M NaCl for 48 h with constant stirring. Final concentration of alginates was determined using the phenol-sulphuric acid assay[4]. The intrinsic viscosities were calculated in the usual manner[5].

Laser Light Scattering

Molecular weight of alginate samples was determined by multiangle laser light scattering which was performed on a Photon Correlation Spectrometer (K7027 Malvern Instruments, Worchester, UK) in total intensity counting mode using a 632.8 nm, 10 mN, He-Ne laser. The angular dependence of scattering was determined at 10° intervals between 20° and 150°. 0.2M NaCl was used as a solvent. All samples were dialysed against 0.2 M NaCl, and dialysate was used as diluent.

Mannuronate/Guluronate Ratios

Mannuronate/guluronate ratios were determined as described previously[5].

RESULTS

Percentage yield for samples studied are given in Table 1 which shows that extracting alginates from brown seaweeds in EDTA, CDTA, phosphate and citrate gives higher yield compared to Na_2CO_3 when extracted for 6 h, while for samples extracted for 24 h no significant differences were observed.

Table 1. Percentage yield of alginate extracted in different extracting medium

Sources of alginate	Percentage yield				
	Na_2CO_3 0.1 M	Citrate 0.1 M	EDTA 0.1 M	CDTA 0.1 M	Phosphate 0.1 M
Colpomenia sinuosa	13(25)	25(24)	25(28)	28(27)	26(25)
Dictyota dentata	10(24)	26(25)	28(27)	27(26)	25(26)
Hydroclathrus clathratus	12(25)	25(27)	23(29)	24(29)	26(28)
Hormophysa triquetra	10(19)	25(26)	26(26)	23(24)	24(27)
Padina minor	10(18)	23(22)	22(25)	21(26)	20(24)
Sargassum kuetzingii	12(30)	30(34)	30(34)	29(35)	31(34)
Sargassum duplicatum	13(32)	30(35)	32(38)	32(37)	32(35)
Sargassum bicorne	15(35)	28(37)	30(37)	32(40)	30(39)
Turbinaria conoides	10(25)	23(26)	24(28)	22(29)	25(27)

Values in parentheses correspond to samples extracted for 24 h and other values correspond to samples extracted for 6 h.

M/G ratio, molecular weight and intrinsic viscosity for all the alginate samples extracted in 0.1 M Na_2CO_3 are given in Table 2. The results clearly show that samples extracted for shorter period (6 h) give higher intrinsic viscosity and higher molecular weight due to less degradation in alkaline medium compared to samples extracted for longer period (24 h). *Sargassum kuetzingii* gives highest intrinsic viscosity, that is 14.40 dl/g and 11.05 dl/g when extracted for 6 h and 24 h respectively. M/G ratios for all the samples studied are below 1.00 except for *Dictyota dentata* with M/G ratio 1.28 and *Turbinaria conoides* with M/G ratio of 2.10. *Sargassum kuetzingii* gives the lowest M/G ratio, that is 0.42.

Table 2. M/G ratio, intrinsic viscocity and molecular weight for alginate samples extracted in 0.1 M Na_2CO_3.

Sources of alginate	M/G ratio	Intrinsic viscosity, dl/g	Molecular weight x 10^{-5} dalton
Colpomenia sinuosa	0.70	5.75 (3.05)	5.75 (1.67)
Dictyota dentata	1.28	4.54 (2.52)	2.89 (1.31)
Hydroclathrus clathratus	0.72	5.24 (2.99)	5.35 (1.60)
Hormophysa triquetra	0.80	5.04 (2.75)	5.10 (1.52)
Padina minor	0.65	4.32 (2.15)	2.59 (1.05)
Sargassum kuetzingii	0.42	14.40 (11.05)	9.49 (6.24)
Sargassum duplicatum	0.88	12.25 (9.55)	8.25 (6.17)
Sargassum bicorne	0.66	11.95 (9.00)	8.01 (5.76)
Turbinaria conoides	2.10	6.90 (4.38)	5.07 (2.49)

Values in parentheses correspond to samples extracted for 24 h and other values correspond to samples extracted for 6 h.

Table 3 gives results for the effect of HCHO and ethanol pre-treatment on viscosity of alginate solutions. The results clearly show that pre-treatment of the seaweeds with HCHO or ethanol resulted in significantly higher intrinsic viscosities compared to samples which were not pre-treated.

Table 3. Effect of HCHO and ethanol pretreatment on intrinsic viscosity for alginate samples (All samples were extracted for 24 h).

Sources of alginate	Intrinsic viscosity dl/g		
	extracted with 0.1 M Na_2CO_3	pretreated with HCHO followed by extraction with 0.1 M Na_2CO_3	pretreated with ethanol followed by extraction with 0.1 M Na_2CO_3
Colpomenia sinuosa	3.05	5.99	6.10
Dictyota dentata	2.52	5.10	5.25
Padina minor	2.15	4.45	4.50
Sargassum kuetzingii	11.05	22.05	23.45
Sargassum duplicatum	9.55	19.55	20.41

DISCUSSION

Nine species of brown seaweeds have been studied to evaluate their potential for alginate production. Percentage yield, M/G ratios, intrinsic viscosity, molecular weight, effect of pretreatment on intrinsic viscosity and effect of different extracting medium on yield have been determined. The percentage yield varies between 10 to 32% when extracted for 6 h and 18 to 40% when extracted for 24 h in different extracting medium (Table 1). Higher yields were observed when the samples were extracted for 6 h in citrate, EDTA, CDTA and phosphate solution compared to Na_2CO_3 solution, while for samples extracted for 24 h no significant differences were observed in the percentage yield. Higher yields when extracted in CDTA,

EDTA, citrate and phosphate are due to more efficient extraction because CDTA, EDTA, citrate and phosphate are better sequestering agents for polyvalent cations such as Ca^{2+} and Sr^{2+} that bind to alginate in the cell wall. The percentage yield especially for *Sargassum* spp. is high and comparable to other seaweeds used commercially[6], such as *Ascophyllum nodosum* (20-30%), *Macrocystis pyrifera* (13-24%) and *Laminaria hyperborea* (10-35%).

Table 2 gives results on M/G ratio, intrinsic viscosity and molecular weight. The M/G ratio for all the samples studied are below 1.00 except for *Turbinaria conoides* and *Dictyota dentata*. *Sargassum kuetzingii* gives the lowest M/G ratio that is 0.42. Compared to commercial[5] samples such as *A. nodosum* with M/G ratio 1.52 and *L. hyperborea* with M/G ratio 0.60 the alginate extracted from *Sargassum* spp. can be used as a good gelling agent because it is rich in guluronate residues.

Intrinsic viscosity for all the samples studied varies between 4.32 to 14.40 dl/g and 2.15 to 11.05 dl/g when extracted for 6 h and 24 h respectively; while molecular weights are in the range of 2.59×10^5 to 9.49×10^5 and 1.05×10^5 to 6.24×10^5 when extracted for 6 h and 24 h respectively. *Sargassum kuetzingii* gives the highest intrinsic viscosity. The lower intrinsic viscosity and molecular weight obtained when samples were extracted for longer periods are probably due to degradation in the alkaline medium. All samples that were extracted in EDTA, CDTA phosphate and citrate gave higher molecular weight and intrinsic viscosity compared to samples extracted in Na_2CO_3 (unpublished data). The samples can also be extracted under neutral conditions to give high intrinisic viscosity but this will involve unacceptably long extraction times[7].

Results on the effect of pre-treatment of the seaweeds with HCHO and ethanol are given in Table 3. The results clearly show that pre-treating the seaweed with HCHO or ethanol prior to normal extraction procedure resulted in much higher intrinsic viscosity. From previous observations[7] it was found that the intrinsic viscosity also depends on the concentration of HCHO and 40% HCHO was chosen in this study because it gives highest value. The increase of intrinsic viscosity is probably the result of the removal of the phenolic compounds during pre-treatment. Phenolic compounds present in most brown seaweeds can contribute to significant degradation of alginate during extraction by an oxidative - reductive depolymerization. Phenolic compounds in brown algae can combine with HCHO to gives insoluble polymeric products. The seaweeds that are free of phenolic compounds can be extracted as usual to give high intrinsic viscosity samples.

Results in Table 1 and 2 clearly show that *Sargassum* spp are good sources for alginate production based on the high yield and low M/G ratio and high intrinsic viscosity. Pre-treating the seaweed samples with HCHO or ethanol prior to normal extraction procedures (using 0.1M Na_2CO_3) can produce alginate with very high intrinsic viscocity and molecular weight.

REFERENCES

1. A. Haug, B. Larsen and O. Smidsrod, Uronic acid sequences in alginate from different sources, *Carbohydr. Res.* 32:217 (1974).
2. E.R. Morris, D.A. Rees and D.D. Thom, Chiroptical and stoichiometric evidence of a specific primary dimerization process in alginate gelation, *Carbohydr. Res.* 66:145 (1978).
3. J. Boyd and J.R. Turvey, Structural studies of alginic acid using a bacterial poly (α -L-guluronate) lyase, *Carbohydr. Res.* 66:187 (1978).
4. M. Dubois, K.A. Gilles, J.K. Hamilton, P. A. Rebers and F. Smith. Colorometric determination of sugars and related compounds, *Anal. Chem.* 28:350 (1956).
5. D.J. Wedlock, B. A. Fasihuddin and G.O. Phillips, Comparison of molecular weight determination of sodium alginate by sedimentation diffusion and light scattering, *Int. J. Biol. Macromol.* 8:57 (1986).

6. D.J. Chapman and V.J. Chapman. "Seaweeds and Their Uses", Chapman and Hall Publishers, London and New York (1980).

7. D.J. Wedlock and B.A. Fasihuddin, Effect of formaldehyde pre-treatment on the intrinsic viscosity of alginate from various brown seaweeds, *Food Hydrocolloids,* 4:41 (1990).

RHEOLOGICAL PROPERTIES OF AQUEOUS SOLUTIONS OF AUBASIDAN

N.M. Ptitchkina, L.V. Novokreschonova, and A.G. Ishin

Research and Technological Institute of Agricultural Biotechnology
Saratov 410020, Russia

ABSTRACT

Influence of concentration, temperature, pH, and addition of KCl and NaCl on viscosity of aqueous solutions of microbial polysaccharide Aubasidan was studied by the rotation method (Cylinder-Cylinder) in the shear rate range 0.15-1312 s^{-1}.

INTRODUCTION

Aubasidan is an extracellular β-glucan secreted only by certain strains of yeast-like fungus *Pullularia pullulans*[1]. Aubasidan posesses several potential applications in cosmetic, medicine, and manufacturing industries as a floculant, adhesive, binder, etc. It can form biodegradable films and fibers with properties similar to synthetic polymers. The incorporation of aubasidan in food products results in saving various ingredients, reduction in calories, and improvement in taste[2].

This article is focused on study of rheological properties of aubasidan aqueous solutions.

MATERIALS AND METHODS

Aubasidan used was obtainad from Research Institute of Spirtbioproduction (Kiev, Ukraine). It is a cream, tasteless, odourless powder, hardly soluble in water and

Food Hydrocolloids: Structures, Properties, and Functions
Edited by K. Nishinari and E. Doi, Plenum Press, New York, 1994

practically not soluble in organic solvents. Polymer is not uniform in respect of molecular mass. The latter is in the interval from 100.000 to one million or even higher.

Viscosity measurements were made with Rheotest-2M (shear rate range 0.15-1312 s^{-1}) for aqueous solutions in the concentration range 1-5 wt% at the temperatures in the range 10-80° C.

RESULTS AND DISCUSSION

It can be seen from Fig.1 that aubasidan flow curves have concavity to the shear rate scale at low shear rates, i.e. the solutions are pseudoplastic fluids. At shear rates in the range from 1 s^{-1} to about 100 s^{-1} the dependence of viscosity η on shear rate $\dot{\gamma}$ can be expressed by the power law

$$\eta = K \dot{\gamma}^{n-1}. \tag{1}$$

The flow behavior index "n" decreases from 0.30 to 0.18, if solution concentration increases from 1 wt% to 5 wt%.

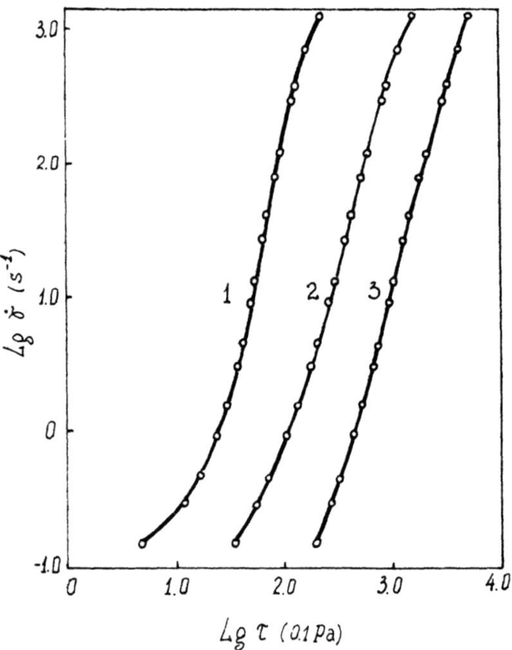

Figure 1. Flow curves at 25° C for aubasidan aqueous solutions at concentrations: 1.0 wt% (1), 3.0 wt% (2), and 5.0 wt% (3).

Aubasidan solution viscosity varies with concentration C accoding to relationship

$$\eta = A\ C^a.\tag{2}$$

Coefficient "a" decreases with increasing shear rate at a given temperature. With varying temperature, "a" exhibits a maximum at about 50° C at a given shear from the range of the power law (1) action. Values of "a" are in the range 1.5-2.3 for aubasidan solutions in the concentration range 1-5 wt%.

The viscosity of aubasidan solution are relatively insensitive to temperature in the range 10-80° C. Temperature dependence increases with decreasing concentration at a given shear rate and with decreasing shear rate at a given solution concentration, but effects are small.

At hihg shear rates (500-1300 s^{-1}) the aubasidan solutions have practically stable viscosity over a wide pH range (3-12). The pH effect increases with decreasing rate of shear, Fig.2.

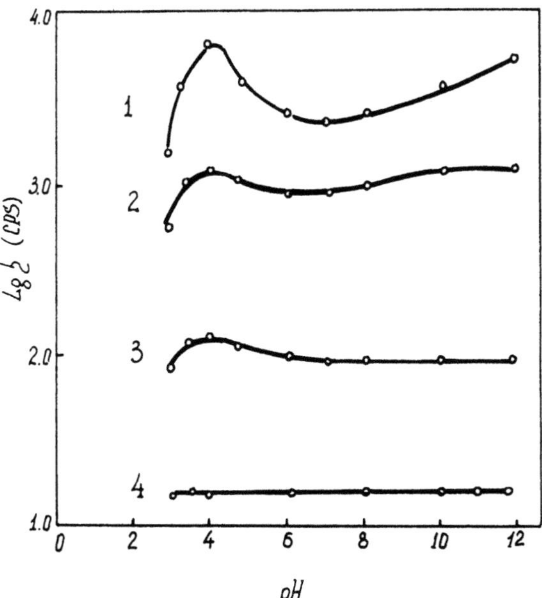

Figure 2. Dependence of 2% aubasidan solution viscosity on pH at 25° C and at shear rates: 0.9 s^{-1} (1), 9.0 s^{-1} (2), 121.5 s^{-1} (3), and 1312 s^{-1} (4).

The addition of salt (KCl or NaCl) gives rise to an decrease in viscosity of aubasidan solutions at all shear rates studied. It can be seen from Fig.3 that a reduction in viscosity in the presence of 0.1 M NaCl is the same as that in 0.1 M KCl. At lower and higher concentrations effects of NaCl and KCl on viscosity differ essentially.

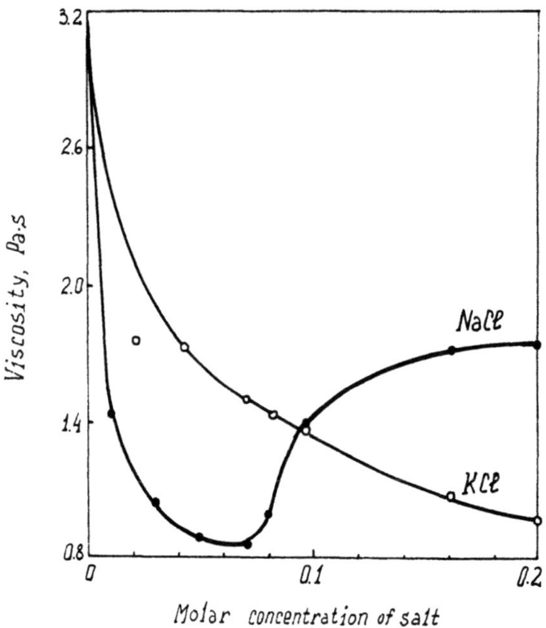

Figure 3. Dependence of 1% aubasidan solution viscosity on salt concentration at 25° C and at shear rate 121.5 s^{-1}.

The main rheological properties of aubasidan dispersions can thus be summarised as follows: high pseudoplasticity, low dependence of viscosity on temperature, stable viscosity over a wide pH range at high shear rates, and decreased viscosity in the presence of salts. The results given can be useful for diverse applications of aubasidan.

REFERENCES

1. N.I.Elinov. "Chemistry of Microbial Polysaccharides", High School Press, Moscow (1984).
2. A.F.Nevolnichenko,S.M.Shulga,N.V.Pritulskaja,I.A.Markovcova,and S.B.Varlamova. New sort of foodstuffs with use of aubasidan as a structure former, in: "Totals and Prospects of Use of Natural and Synthetic Polymers in Food Productions", - Proceedings of All-Union Conference on Food, 1-5 December 1991, Suzdal -, Moscow (1991).

INFLUENCE OF SMALL ADDITIVES OF AUBASIDAN
ON GEL FORMATION IN AQUEOUS SOLUTIONS
OF RED SEAWEED POLYSACCHARIDES

N.M.Ptitchkina[1], E.V.Karmanova[1], V.G.Artjuhov[2], and A.G.Ishin[1]

[1]Research and Technological Institute of Agricultural Biotechnology
(Russia)
[2]Research Institute of Spirtbioproduction (Ukraine)

ABSTRACT

Small (0.1-0.5 wt%) additives of microbial polysaccharide Aubasidan increase
viscosity and gel formation temperature of aqueous solutions of red seaweed
polysaccharides (agar, furcellaran, agaroid) and strengthen their gel structure.

INTRODUCTION

Agar, agaroid, fulcellaran, and carrageenan are the polysaccharides contained in
red seaweeds. The unique ability of low concentration aqueous solutions of the red
seaweed polysaccharides (RSPS) to form gels makes RSPS useful in food production,
microbiology, and in other fields.

In confectionary the aqueous RSPS solutions are used in combination with sugar.
For example, sugar content is up to 70% of the product weight in marmalade
production. But high sugar consumption promotes the progress of dental caries, the
numerous cases of the metabolism breach and cardio-vascular deseases. In this
connection, the problem of replacement of sugar by low-calorie sources of sweetness is
the urgent problem.

In this work we use Otison instead of sugar. Otison is the effective sugarsubstitu-
ent (potassium acetosulfame analogue) synthesized recently by Ukrainian scientist [1].

Food Hydrocolloids: Structures, Properties, and Functions
Edited by K. Nishinari and E. Doi, Plenum Press, New York, 1994

Small (0.2-0.4 wt%) additives of Otison ensure the sweet taste of food. Sweet jelling compositions on the bases of RSPS with Otison instead of sugar are the products of dietary and medical-prophylactic purpose.

Dietary and prophylactic action of the above mentioned compositions can be intensified by adding some bioactive polymers, for example glucans. The latter stimulate immunosystem, posses oncostatic action. In this study we use Aubasidan,an extracellular β-glucan elaborated by the yeast-like fungus *Pullularia pullulans* [2]. It is the purpose of this article to report the information about influence of small (up to 0.5 wt%) additives of Aubasidan on some characteristics of the RSPS-water-Otison system.

MATERIALS AND METHODS

Agar, agaroid, and furcellaran used were commercial products produced by firms "Serva" (USA), "Agaroid fabric" (Ukraine), and "Nacotne" (Latvia), respectively. Otison and Aubasidan were from Ukraine (Pilot plant of Research Institute of Spirtbioproduction).

Viscosity measurements were made by the falling ball method with Heppler viscosimeter. The temperatures of gel formation were obtained by the method of solution surface control in tube of small diameter. The gel strength was determined with Valenta device.

RESULTS AND DISCUSSION

As a rule, polysaccharide additives increase viscosity of aqueous solutions. Aubasidan is not an exception to the rule, Fig.1.

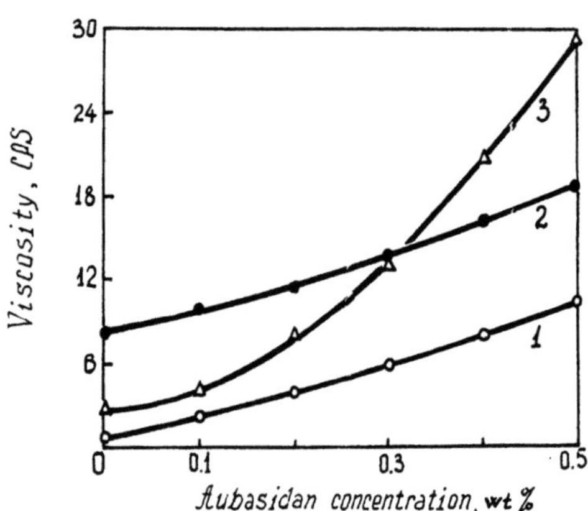

Figure.1. Aubasidan concentration dependence of viscosity for aqueous solutions of agar (1), agaroid (2), and furcellaran (3) at 80° C. Otison content - 0.3 wt%. Algal polysaccharide concentration: 0.85 wt% (1), 1.5 wt% (2), and 1.25 wt% (3).

In our experiments the RSPS concentration was varied in practically important range 0.5-2.5 wt%; sweetener content was 0.3 wt%. Viscosity was measured at 80° C. Under such conditions, viscosity of solutions at small (up to 0.5 wt%) Aubasidan additives was 2-20 times higher then that in Aubasidan absence.

Addition of Aubasidan to RSPS solutions has a insignificant effect on temperature of gelation, Tabl.1: increase in gelling temperature is 1-3° C.

Table 1. Aubasidan concentration dependence of gelling temperature (°C) of aqueous solutions of algal polysaccharides. Otison content - 0.3 wt%.

Aubasidan conc., wt%	Agar 0.85 wt%	Furcellaran 1.25 wt%	Agaroid 1.50 wt%
0	29	33	25
0.1	31	34	27
0.3	32	35	28
0.5	32	35	28

Effect of Aubasidan on gel strength of algae polysaccharides is more interesting. With varying Aubasidan concentration, the gel strength exhibits a maximum at about 0.3 wt%, Fig.2. The maximum position does not depend practically on both the character of polysaccharide and its concentration in solution. Unfortunatelly, we cannot explain now a nature of competitive processes that cause presence of this maximum. We need in complementary experiments to find the explanation of the matter.

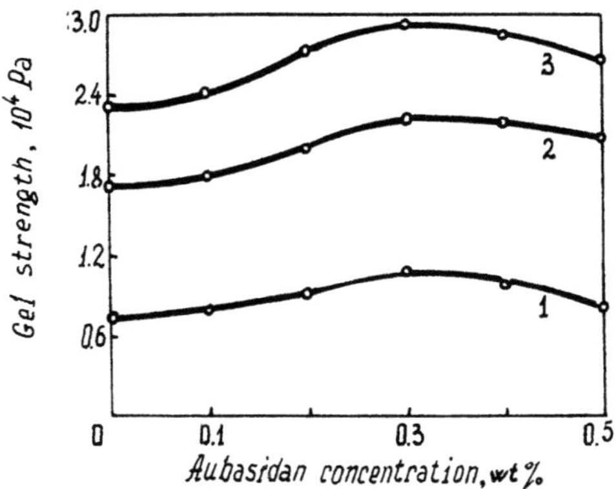

Figure 2. Aubasidan concentration dependence of gel strength of agar at 25° C and at concentrations 0.85 wt% (1), 1.5 wt% (2), and 1.8 wt% (3). Otison content - 0.3 wt%.

On the other hand, the experimental data obtained in this study give recomendations for practical use of Aubasidan. It is expedient to add Aubasidan to sweet jelling products on the basis of RSPS in the amount of 0.3 wt%, at wich the largest increase in gel strength is observed. It makes possible to save 18-24% of algae polysaccharides, to reduce the final calorie value, and to improve the product taste.

REFERENCES

1. V.V.Jaremenko, G.K.Drogovoz, V.G.Artjuhov, N.A.Nagurnaja, and S.M.Shulga. Modified way of synthesis and physico-chemical properties of Otison, in: "Cemistry and Technology of Food", Publ.H. of VNII Sug.Ind., Kiev (1988).
2. N.I.Elinov. "Chemistry of Microbial Polysaccharides", High School Press, Moscow (1984).

RHEOLOGICAL AND ORGANOLEPTIC PROPERTIES

OF FOOD HYDROCOLLOIDS

Edwin R. Morris

Cranfield Institute of Technology
Silsoe College
Silsoe
Bedford MK45 4DT, UK

ABSTRACT

Most polysaccharide thickeners exist in solution as disordered coils, and their maximum viscosity at low shear rate depends only on degree of space-occupancy, $c[\eta]$, the product of concentration (proportional to the number of coils present) and intrinsic viscosity (proportional to their size). Suppression of perceived flavour and taste also depends directly on degree of space occupancy. Reduction in viscosity at higher shear rates (shear thinning) can be fitted to a simple equation for all "random coil" thickeners, and their dynamic viscosity (η^*) from low-amplitude oscillatory measurements is identical to their steady-shear viscosity (η) from normal rotational measurements. The perceived "thickness" of these solutions correlates well with either η at a shear rate of 50 s^{-1} or η^* at a frequency of 50 rad s^{-1}. Ordered polysaccharides (such as xanthan), however, can form "weak gels", with η^* (characterising the unbroken network) higher than η (where the network is destroyed in making the measurement). It is the higher values that reflect thickness in the mouth. Perceived "sliminess" also depends directly on η^* at 50 rad s^{-1}, not on degree of shear-thinning as previously believed. The extreme shear-thinning of xanthan "weak gels", however, gives almost perfect flavour release. The stress and strain at which "true" gels break correlate with, respectively, perceived "hardness" and "brittleness". Flavour release from gels is independent of hardness, but correlates directly with "brittleness", with brittle gels giving better release than more elastic materials.

INTRODUCTION

Industrial hydrocolloids have a wide range of applications in the food industry including, for example, their use as process aids and for product stabilisation against migration of water or growth of ice crystals. Their main function, however, is to generate and control product texture.

For the product developer, there is an obvious advantage in using instrumental measurements of texture, rather than relying on his own, perhaps atypical, judgement of eating quality, or resorting to the cumbersome and time-consuming expedient of using a large taste panel. However, unless such measurements reflect the subjective assessment of the consumer, they are of little value, and may indeed be positively misleading. The aim of this article is to review the progress that has been made in relating the objective rheology of food hydrocolloids to their effect on specific subjective sensory attributes.

Food Hydrocolloids: Structures, Properties, and Functions
Edited by K. Nishinari and E. Doi, Plenum Press, New York, 1994

QUANTIFYING SUBJECTIVE PERCEPTION

The most reliable way of converting subjective sensory judgements to numerical form[1] is by the technique of "ratio scaling" or "magnitude estimation". For example, using this method to quantify perceived sweetness, a typical procedure[2] would be to select as standard a sample near the middle of the range to be assessed and assign it a "sweetness score" of 100. Panellists would then be asked to give appropriate scores to the other samples. For example, one perceived to be three times as sweet as the standard would be scored at 300, one half as sweet would be scored at 50, and so on.

Magnitude estimation has been applied in many areas of sensory analysis, such as loudness of sound or brightness of light, and it is found that the perceived intensity (P) is related to the objective intensity (I) by a "power law":

$$P = k\,I^n \tag{1}$$

Thus plotting log P against log I yields a straight line of gradient n and intercept log k.

$$\log P = \log k + n \log I \tag{2}$$

The value of k depends on the sample chosen as reference, and has therefore no fundamental significance. The exponent (n), however, depends directly on the specific attribute being evaluated, and varies widely between different types of sensory stimuli.[1]

OBJECTIVE VISCOSITY AND PERCEIVED "THICKNESS"

The textural attribute that has been most extensively investigated is the perceived "thickness" (T) of fluid foods and beverages in the mouth. The obvious choice of instrumental measurement for quantitative correlation with subjective thickness is, of course, viscosity.

Newtonian Liquids

For solutions of small molecules, such as sugar syrups, the rate of movement (shear rate, $\dot{\gamma}$) increases in direct proportion to the force applied (shear stress, τ), so that viscosity ($\eta = \tau/\dot{\gamma}$) remains constant at a single, fixed value. This is known as Newtonian behaviour.

As shown in Fig. 1, log T for such systems (obtained[3] by the magnitude estimation procedure described above) varies linearly with log η. Thus the perceived thickness and objective viscosity of Newtonian liquids comply with the general "power law" relationship (Eqn (1)) observed for other types of sensory perception.

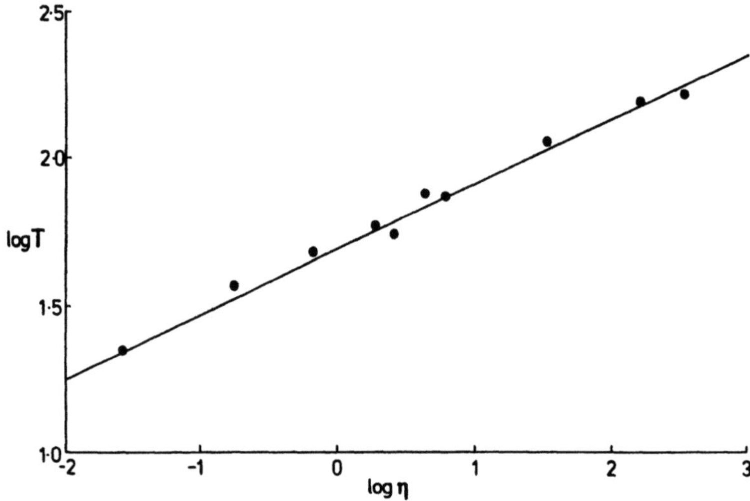

Fig. 1. Correlation of perceived thickness (T) with objective viscosity (η) for a range of Newtonian fluids. (From Cutler, Morris and Taylor,[3] with permission).

Non-Newtonian Materials

Most fluid foods (soups, sauces, dressings, etc.), particularly those thickened with hydrocolloids, do not show Newtonian behaviour. Increasing the applied stress normally produces more than a proportionate increase in shear rate, with consequent decrease in the measured viscosity ("shear thinning"). A central issue in attempting to relate the perceived thickness of such systems to their objective viscosity is therefore selection of an appropriate shear regime for the instrumental measurements.

The first approach to this problem was by using samples whose "flow curves" of η *vs.* $\dot{\gamma}$ crossed one another, and identifying conditions of shear-rate where the measured viscosities followed the same order as perceived thickness. Using this method, Wood concluded[4] that the effective shear rate in the mouth is about 50 s⁻¹.

More recently, the validity of this conclusion was tested[3] for a wide range of fluid foods and hydrocolloid solutions by using the linear relationship between log T and log η shown in Fig. 1 to identify for each sample the single viscosity value that would give perfect agreement with panel scores for in-mouth thickness. As shown in Fig. 2, viscosity measurements at 50 s⁻¹, as suggested by Wood[4], were in reasonable agreement with those derived from panel scores for most of the samples studied. There was, however, an obvious systematic tendency for viscosity at this shear rate to underestimate the perceived thickness of samples thickened with xanthan. The magnitude of the discrepancy increased with increasing concentration of polymer. To understand the origin of this behaviour we must first consider the molecular basis of non-Newtonian viscosity in hydrocolloid solutions.

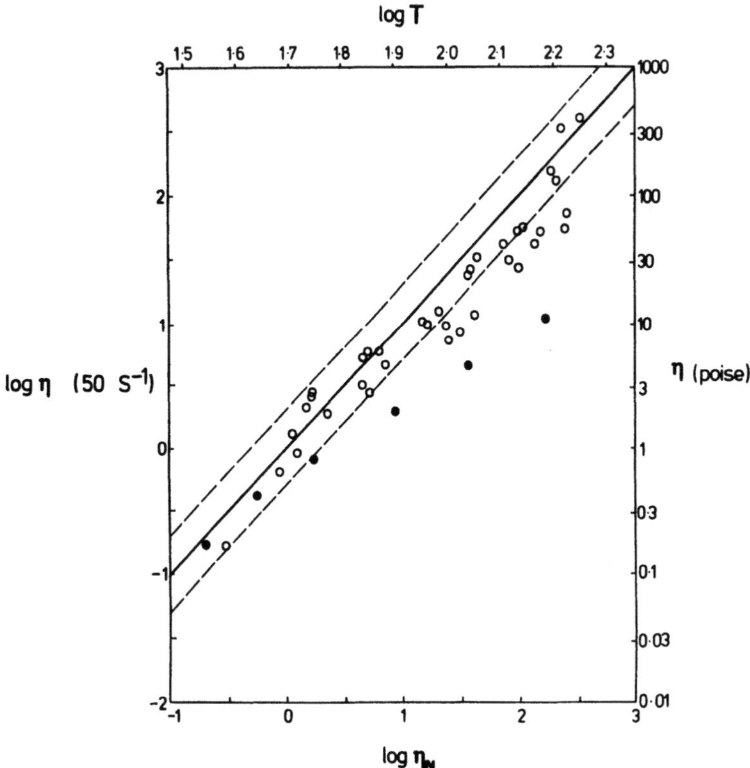

Fig. 2. Comparison of objective viscosity at 50 s⁻¹ for xanthan "weak gels" (●) and other fluids (O) with the "equivalent Newtonian viscosity" (η_N) derived from panel scores for thickness (T) using the relationship shown in Fig. 1. The solid line corresponds to perfect agreement, and the dashed lines indicate confidence limits (p = 0.95) of the panel scores. (From Cutler, Morris and Taylor,[3] with permission).

Origin and Form of Shear-Thinning

Most food hydrocolloids exist in solution as fluctuating, disordered coils.[2] In dilute solution, where the individual coils are free to move independently, their viscosity is essentially Newtonian (i.e. virtually independent of shear rate). With increasing concentration, however, the coils are forced to interpenetrate one another, to form an entangled network. The viscosity at the onset of coil-overlap is very low (normally close to 10 times that of water[5]), so that in most practical applications the chains are densely entangled.

The viscosity of such solutions remains Newtonian only at low shear rates, where there is sufficient time for the entanglements pulled apart to allow the solution to flow to be replaced by new entanglements between different chains. At higher shear rates, however, where the rate of re-entanglements lags behind the rate of forced disentanglement, the network is depleted and viscosity falls (often by two or three orders of magnitude[5] over the range of shear rates accessible on normal commercial viscometers). For most commercial hydrocolloids, the form of shear-thinning can be matched accurately[6] by the simple relationship:

$$\eta = \eta_0 \,/[1 + (\dot{\gamma}/\dot{\gamma}_{1/2})^p\,] \tag{3}$$

where η_0 is the maximum Newtonian viscosity at low shear rates, $\dot{\gamma}_{1/2}$ is the shear rate required to reduce the viscosity to $\eta_0/2$, and p is the maximum (negative) slope of $\log \eta$ vs. $\log \dot{\gamma}$ at high shear rates, which is found empirically[5,6] to have a constant value of $p = 0.76$ for samples of high polydispersity (as in all commercial thickeners).

Xanthan generates viscosity in an entirely different way. Under virtually all conditions of practical relevance for food use, the polymer exists in solution in a rigid, ordered chain conformation, and forms a tenuous three-dimensional network by weak, side-by-side association of the ordered chains. Rupture of the network to allow the solution to flow requires a finite stress. Thus, as indicated in Fig. 3, the viscosity at low shear rates does not level-out to a "Newtonian plateau", as in the case of entangled coils, but rises continuously with decreasing $\dot{\gamma}$.

Dynamic Viscosity as an Index of "Thickness"

A further consequence of the "weak gel" character of xanthan solutions is that the "dynamic viscosity", η^*, obtained from low-amplitude oscillatory measurements where the network remains intact, is higher than the normal "steady shear" viscosity, (η) from, for example, rotational measurements, where the network is broken down.

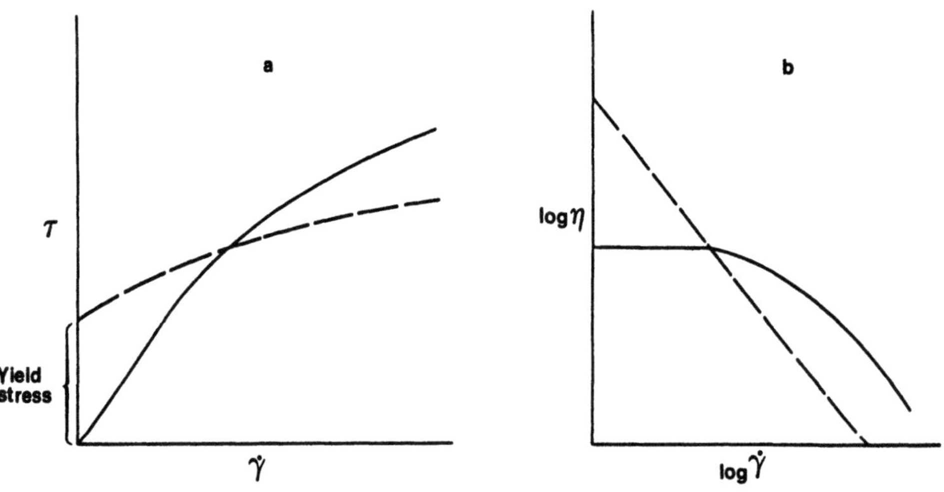

Fig. 3. Shear rate ($\dot{\gamma}$) dependence of: a) shear stress (τ) and b) viscosity ($\eta = \tau/\dot{\gamma}$) for "weak gels" (- - - -) and solutions of entangled coils (———).

As shown in Fig. 4, the magnitude of the discrepancy between η and η^* increases steadily with increasing concentration of xanthan,[8] paralleling closely the increasing divergence observed (Fig. 2) between the measured rotational viscosities of xanthan "weak gels" and the values anticipated from panel scores for perceived thickness.

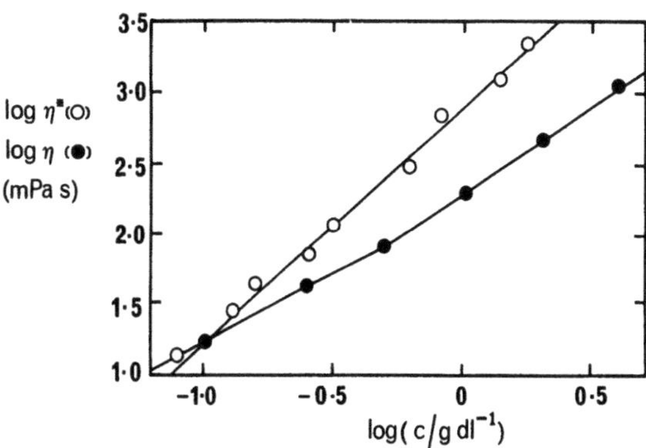

Fig 4. Comparison of dynamic viscosity (η^*) at 50 rad s^{-1} and steady shear viscosity (η) at 50 s^{-1} for xanthan. (From Baines and Morris,[8] with permission) .

For most other materials, including specifically solutions of disordered hydrocolloids, where there are no enthalpic associations between the chains, η and η^* are closely super-imposable[2] at equivalent numerical values of shear rate ($\dot{\gamma}$ /s^{-1}) and frequency (ω /rad s^{-1}). Thus by raising the values obtained for "weak gels" and similar structured fluids while leaving those of other materials unchanged, the use of oscillatory measurements of η^* at 50 rad s^{-1}, in place of rotational measurements of η at 50 s^{-1}, gives excellent correlation (Fig. 5) with subjective assessments of thickness.[9]

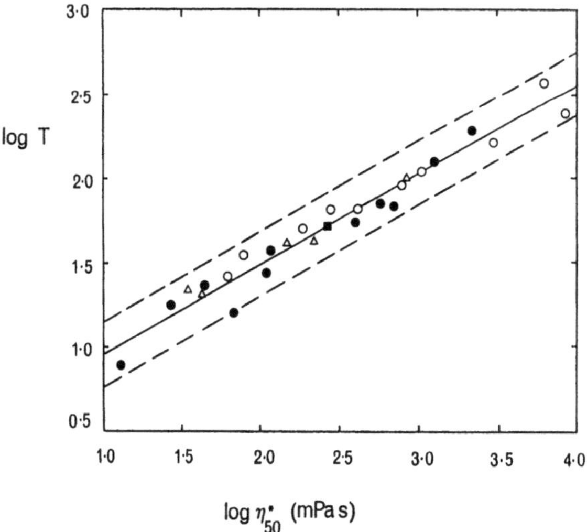

Fig. 5. Correlation of perceived thickness (T) of sample thickened with guar gum (O), starch (Δ), and xanthan (●) with dynamic viscosity (η^*) at 50 rad s^{-1}. Dashed lines show the confidence limit (p = 0.95) on panel scores. (From Richardson, Morris, Ross-Murphy, Taylor, and Dea,[9] with permission.).

Previous attempts to reconcile steady shear viscosity with perceived thickness, by using measurements at lower shear rate (e.g. 10 s-1) for all samples,[3] or by suggesting that thicker samples are evaluated at lower shear rates in the mouth,[10] probably reflect a forced compromise between the behaviour or "weak gels" and normal solutions since, as indicated in Fig. 3b, the viscosity of the structured systems becomes disproportionately higher as shear rate is decreased.

"STICKINESS", "SLIMINESS" AND "MOUTHFEEL"

In addition to "thickness", verbal descriptions[11] of the in-mouth texture ("mouthfeel") of hydrocolloid systems usually include other perceived attributes such as "stickiness" or "sliminess". Both of these also correlate directly[9] with measured values of η^* at 50 rad s-1. This is at first sight highly surprising, since there is a widespread belief in the food industry that "mouthfeel", and in particular "sliminess", depend on shear-thinning behaviour.

The origin of the apparent conflict is indicated in Fig. 6. In the original investigation that gave rise to the idea of an (inverse) correlation between perceived sliminess and degree of shear thinning,[12] a large number of different polysaccharides were used to prepare solutions with the same viscosity (1.2 Pa s) at low shear-rate (0.5 rpm on a Brookfield viscometer), and their perceived sliminess was found to decrease systematically with increasing (negative) slope of η vs. $\dot{\gamma}$. However, when the standardisation of viscosity at low shear rate is removed,[9] it is the viscosity at higher shear-rate (~50 s-1) that determines perceived sliminess, not the slope of the flow curve. In other words, samples with the same viscosity at 50 s-1 (or, more generally, the same value of η^* at ~50 rad s-1) will be perceived as equally slimy (and equally thick), irrespective of their degree of shear-thinning.

This does not imply that the earlier interpretation was unfounded. In systems where product-performance is based on viscosity characteristics at low shear-rate (e.g. for surface coating or particle suspension) there will indeed be a systematic dependence of viscosity at 50 s-1 (or η^* at 50 rad s-1), and hence of perceived sliminess, on degree of shear thinning. What it does imply is that, using food hydrocolloids, "thickness" and "mouthfeel" cannot be manipulated independently in product formulation.

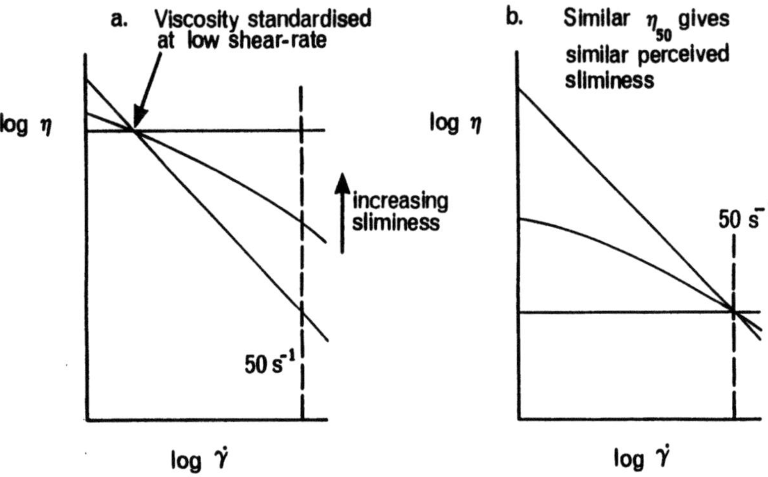

Fig. 6. Objective correlate of perceived "sliminess" .
a) Samples standardised to constant viscosity at low shear rate: η (50 s⁻¹) decreases with degree of shear thinning, with consequent decrease in sliminess.
b) Similar values of η (50 s⁻¹) give similar "sliminess" despite different low-shear viscosities. (From Baines and Morris,[8] with permission).

EFFECT OF TEXTURE ON FLAVOUR/TASTE INTENSITY

It is well known in the food industry that the amount of flavouring required to produce the same subjective flavour intensity is often much higher in thickened or structured products than in more fluid systems. A similar suppression of perceived intensity is also well established[13] for taste attributes (sweet, sour, bitter, salty). In a few specific cases, direct binding of flavour/taste components to hydrocolloid chains may be a significant factor, but it is evident from studies of model systems that there is a much more general dependence of flavour/taste suppression on texture.

"Random Coil" Thickeners

Fig. 7 shows panel scores from magnitude estimation of the perceived thickness, sweetness and flavour intensity for a range of solutions all containing the same objective concentration of sucrose and flavouring, but thickened with different concentrations of the same "random coil" (i.e. conformationally-disordered) hydrocolloid.[14] The sharp break in the concentration-dependence of perceived thickness is accompanied by a corresponding abrupt increase in concentration-dependence of objective viscosity. This is a well-understood phenomenon, and corresponds to the onset of coil-overlap and entanglement.[5]

The concentration at which overlap occurs is known as c^*, and is inversely proportional to the volume of the individual coils, as characterised by, for example, intrinsic viscosity, $[\eta]$, with $c^* \approx 4/[\eta]$ for most commercial hydrocolloids. Measured values of η_0 (the maximum viscosity at low shear-rate) for different "random coil" thickeners superimpose closely[5] when plotted as a function of the extent of space-occupancy, $c[\eta]$, rather than as a function of concentration alone.

At concentrations below c^*, where the individual coils are free to move independently, the presence of the thickener has no detectable effect on the perceived intensity of flavour or sweetness (Fig. 7), but at higher concentrations the panel scores decrease steeply, with both attributes showing the same concentration-dependence of suppression. Similar experiments with different "random coil" thickeners,[15] or with different molecular weights of the same material,[14] give curves of the same form, but offset to higher or lower concentration (for, respectively, smaller or larger coils).

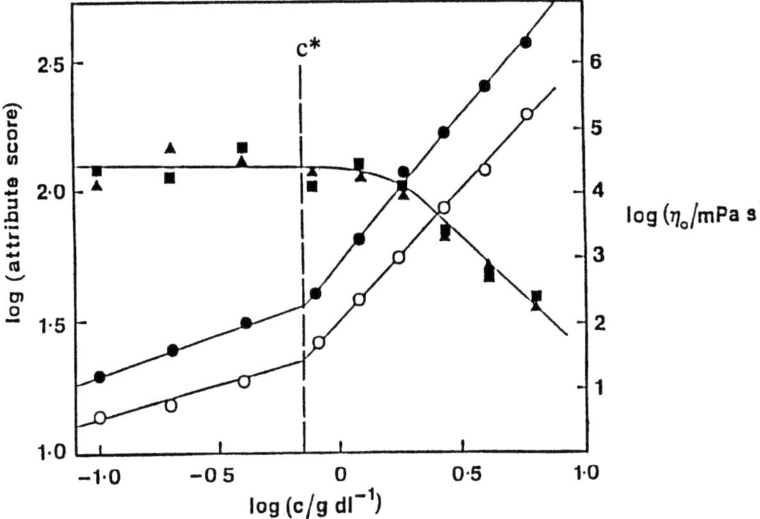

Fig. 7. Variation in perceived "thickness" (●), sweetness (▲), and flavour intensity (■), and in objective "zero-shear" viscosity (η_0, ○ , right-hand axis) for samples incorporating the same fixed concentrations of sucrose and flavouring, but different concentrations (c) of a "random coil" polysaccharide (guar gum). (From Baines and Morris,[14] with permission).

However, when concentrations are expressed as a fraction or multiple of the coil-overlap concentration (i.e. as c/c*) the results obtained for different samples converge[15] to a single "master curve". The extent of suppression is thus directly dependent on the degree of overlap of the individual polymer coils. Since objective viscosity is also directly dependent on the degree of space-occupancy by the polymer,[5] it seems likely that the dominant factor in suppression is increased viscosity reducing the rate of transport of flavour/taste components from the interior of the sample, where they cannot be perceived, to the surface, where they can be.

Polysaccharide Gels

As would be anticipated from the proposal that flavour/taste release from thickened systems depends on ease of mixing in the mouth, formation of a gel network (where no mixing can occur) causes extreme suppression. However, since gel samples are not perceived as totally lacking in flavour and taste, some limited release must be occurring by a different mechanism.

Diffusion of small molecules through polymer networks is extremely slow, and is therefore unlikely to contribute significantly. A more likely release mechanism is exposure of fresh surfaces by fracture of the gel on chewing. Gel fracture can be characterised by two objective instrumental parameters, the stress and strain at the point of rupture. Yield stress correlates well[17] with perceived "hardness" or "firmness", whereas yield strain is inversely related to perceived "brittleness".

As shown in Fig. 8, the perceived intensities of sweetness and flavour for polysaccharide gels with the same objective concentrations of sucrose and flavouring decrease systematically with increasing yield strain (M.G.E. Gothard and E.R. Morris, unpublished).

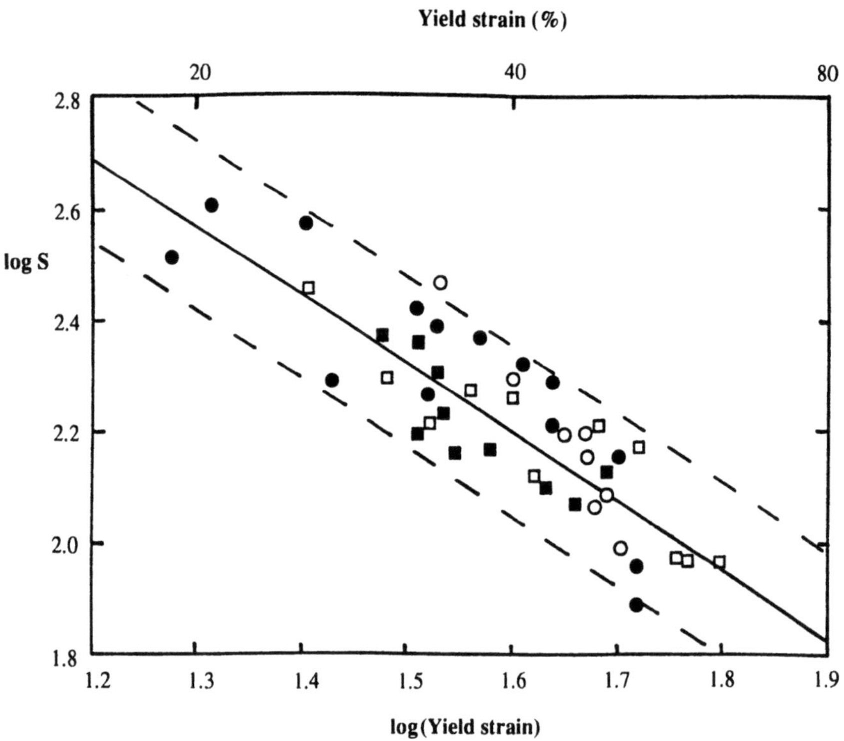

Fig. 8. Variation of perceived sweetness and flavour intensity (S; mean scores for both attributes) with yield strain for gels of gellan gum (●), kappa carrageenan (○), calcium alginate (■) and xanthan/locust bean gum (□). Dashed lines show the confidence limits (p = 0.95) on panel scores. (Gothard and Morris, unpublished).

The results presented in Fig. 8 were obtained for systems chosen to span the widest possible range of failure properties: a combination of xanthan and locust bean gum (LBG), which gives very elastic gels; gellan gum, which forms very brittle gels with calcium ions; and two systems of intermediate brittleness, calcium alginate and kappa carrageenan (K+ salt form). In each case, polysaccharide concentrations were varied over the full range at which measurable gels could be prepared. As in the previous investigations outlined above, flavour and sweetness were again suppressed to the same extent, and the panel scores shown are the mean of the individual values.

In addition to assessing sweetness and flavour, panellists were also asked to rate the gels for perceived firmness. As shown in Fig. 9, the flavour/taste intensities for samples of equivalent perceived firmness differed widely between the different gel systems, with gellan

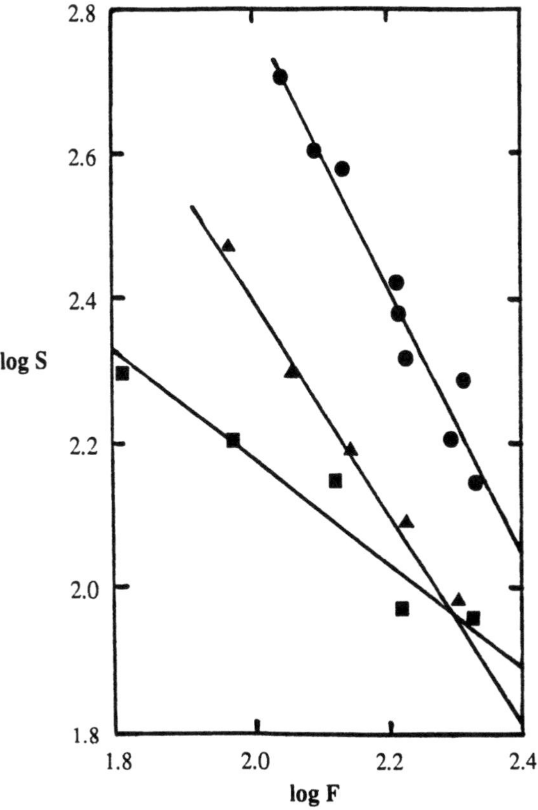

Fig. 9. Correlation of flavour/taste intensity (S; mean scores for both attributes) with perceived firmness (F) for samples gelled with various concentrations of gellan gum (●), kappa carrageenan (▲), and xanthan/locust bean gum (■). Results for calcium alginate were closely similar to those for carrageenan, and are omitted for clarity. (Gothard and Morris, unpublished).

giving much higher flavour/taste scores than xanthan/LBG, and the other samples showing intermediate release properties. As in previous studies, the subjective assessments of firmness correlated well with instrumental values of yield stress, so the same systematic differences between different gelling materials (Fig. 9) were evident when flavour/taste intensity was plotted against objective yield stress. It therefore appears that the principal determinant of flavour/taste suppression in gels is the amount of deformation required to break the network, not the force involved. Thus equivalent flavour/taste release can be achieved in gels of progressively firmer texture (objective and perceived) as the brittleness of the gels increases, with gellan gum having particularly good release characteristics.

Xanthan "Weak Gels"

In contrast to the pronounced suppression observed for "true" gel networks, xanthan solutions cause far less suppression of flavour and taste than disordered hydrocolloids at an equivalent degree of space-occupancy (i.e. at equivalent $c[\eta]$.) In terms of the proposed mechanism of suppression in thickened systems, the difference can be explained by the extreme shear-thinning of "weak gels" (Fig. 3) giving much lower viscosities, and therefore more efficient mixing, at the comparatively high shear rates that appear to operate in the mouth (~ 50 s^{-1}). For normal commercial xanthans, detectable suppression occurs[18] only at concentrations well above those used in any food applications ($>1\%$ w/v). Thus for all practical purposes xanthan gives perfect "flavour release" (i.e. indistinguishable from unthickened samples).

REFERENCES

1. S.S. Stevens, "Psychophysics" (G. Stevens, ed.), Wiley, New York (1975).
2. E.R. Morris, Polysaccharide solution properties: origin, rheological characterisation and implications for food systems, *in*: "Frontiers in Carbohydrate Research 1", R.P. Millane, J.N. BeMiller, and R. Chandrasekaran, eds., Elsevier, London (1989).
3. A.N. Cutler, E.R. Morris, and L.J. Taylor, Oral perception of viscosity in fluid foods and model systems, *J. Text. Stud.* 14: 377 (1983).
4. F.W. Wood, Psychophysical studies on the consistency of liquid foods, *in*: "Rheology and Texture of Foodstuffs", SCI Monograph, London (1968).
5. E.R. Morris, A.N. Cutler, S.B. Ross-Murphy, D.A. Rees, and J. Price, Concentration and shear rate dependence of viscosity in random coil polysaccharide solutions, *Carbohydr. Polym.* 1:5 (1981).
6. E.R. Morris, Shear-thinning of "random coil" polysaccharides: characterisation by two parameters from a simple linear plot, *Carbohydr. Polym.* 13: 85 (1990).
7. S.B. Ross-Murphy, V.J. Morris, and E.R. Morris, Molecular viscoelasticity of xanthan polysaccharide, *Faraday Symp. Chem. Soc.* 18: 115 (1983).
8. Z.V. Baines and E.R. Morris, Effect of polysaccharide thickeners on organoleptic attributes, *in*: "Gums and Stabilisers for the Food Industry 4", G.O. Phillips, D.J. Wedlock, and P.A. Williams, eds., IRL Press, Oxford (1988).
9. R.K. Richardson, E.R. Morris, S.B. Ross-Murphy, L.J. Taylor, and I.C.M. Dea, Characterisation of the perceived texture of thickened systems by dynamic viscosity measurements, *Food Hydrocolloids*, 3:175 (1989).
10. F. Shama and P. Sherman, Identification of stimuli controlling the sensory evaluation of viscosity, *J. Text. Stud.* 4:111 (1973).
11. A.S. Szczesniak, Clasification of mouthfeel characteristics of beverages, *in*: "Food Texture and Rheology", P, Sherman, ed., Academic Press, London (1979).
12. A.S. Szczesniak and E. Farkas, Objective characterisation of the mouthfeel of gum solutions, *J. Food Sci.* 27:381 (1962).
13. R.M. Pangborn, Z.M. Gibbs, and C. Tassan, Effect of hydrocolloids on apparent viscosity and sensory properties of selected beverages, *J. Text. Stud.* 9:415(1978).
14. Z.V. Baines and E.R. Morris, Flavour/taste perception in thickened systems: the effect of guar gum above and below c*, *Food hydrocolloids*, 1:197 (1987).
15. Z.V. Baines and E.R. Morris, Suppression of perceived flavour and taste by food hydrocolloids, *in*: "Food Colloids", R.D. Bee, P. Richmond, and J. Mingins, eds., Special publication No. 75, Royal Society of Chemistry, Cambridge (1989).
16. B.-O. Häglund, M. Elisson, and L.-O. Sundelöf, Diffusion permeability in concentrated polymer solutions, *Chemica Scripta*, 28:129 (1988).
17. F.W. Wood, Psychophysical studies on liquid foods and gels, *in*: "Food Texture and Rheology", P. Sherman, ed., Academic Press, London (1979).
18. E.R. Morris, Organoleptic properties of food polysaccharides in thickened systems, *in*: "Industrial Polysaccharides", M. Yalpani, ed., Elsevier, Amsterdam (1987).

FOOD HYDROCOLLOIDS IN THE DAIRY INDUSTRY

P.M.T. Hansen

Department of Food Science and Technology
Columbus Ohio, 43210, USA

ABSTRACT

Food hydrocolloids are used in a variety of manufactured dairy
products as stabilizers and thickening or gelling agents. A
more recent application relates to the use of some hydrocolloids
for increasing the soluble fiber content of milk-based, dietary
products. The current trend toward new dairy products with
lower fat and lower total solids content has created a need for
improved stabilization technology. A governing principle for
the stabilization of many dairy products is a recognition that
combinations of hydrocolloids can be more effective than the use
of a single stabilizer. This realization and decades of
industrial experience have led to the introduction and marketing
of a large number of blended hydrocolloids formulated for
specific dairy products. For ice cream and other frozen dairy
desserts, typical stabilizer blends involve the use of primary
colloidal stabilizers (CMC, alginate, and locust or guar bean
gum) in combination with the secondary stabilizer, carrageenan.
In these blends, carrageenan serves the purpose of preventing
serum separation of ice cream mix caused by the use of the
primary stabilizers. While the tendency for serum separation is
related to the concentration levels and polymer size of the
primary colloids, the effectiveness of carrageenan as a
balancing factor may be related to its specific milk protein
reactivity.

INTRODUCTION

The successful manufacture of shelf-stable dairy products
depends to a large extent upon the use of functional ingredients
to control technical and physical-chemical problems related to
the inherent instability of milk systems and to incorporate
desirable textural attributes with formulated products. In this
respect, food hydrocolloids are multifunctional ingredients
which are used in a variety of manufactured dairy products as
stabilizers, thickening or gelling agents, and in some cases as
a source of soluble fiber for dietary purposes. Food
hydrocolloids are associated with an extensive range of
functional properties and gum technology has become an important
element in food product development (1, 2). Many of the unique

Food Hydrocolloids: Structures, Properties, and Functions
Edited by K. Nishinari and E. Doi, Plenum Press, New York, 1994

properties of food hydrocolloids are related to their colloidal
dimensions and to interaction phenomena with other food
components (3,4). The role of hydrocolloids in stabilizing
particulate matter and emulsions has been examined from a
thermodynamic viewpoint (5). This presentation is concerned
with exploring approaches toward selecting hydrocolloid
stabilizers for dairy products.

FOOD REGULATIONS

The use in the USA of food hydrocolloids as additives to dairy
products is regulated by the Food and Drug Administration (FDA)
and limited by two provisions found in the Code of Federal
Regulations (CFR). The first provision relates to foods for
which definitions and standards of identity have been
established. For such foods, the addition of a food
hydrocolloid as an optional ingredient is permitted only if such
addition has been specifically included in the standard of
identity (Table 1). The second provision limits the use to only
those ingredients which are permitted food additives for direct
addition to foods (6) or to those ingredients which appear under
the listings *generally recognized as safe* (GRAS) (7) or *specific
substances affirmed as GRAS* as shown in Table 2.

Any approved ingredient must be used in accordance with current
Good Manufacturing Practices (GMP) which include the requirement
that a direct human food ingredient be of appropriate food
grade; that it is prepared and handled as a food ingredient; and
that the quantity of the ingredient added to food does not
exceed the amount reasonably required to accomplish the intended
physical, nutritional, or other technical effects in food (8).
In some cases, FDA has defined maximum levels of addition
consistent with GMP.

Milk-based infant formulas are not included under the
requirements for specific standardized milk and cream (Part
131). However, they are regulated by general provisions
stipulated under CFR Part 107, Infant Formula (9). Under the
provisions of this act it is mandatory for manufacturers to
notify the regulatory agency of any change in formulation.

The list of GRAS substances (Part 182) is not necessarily
inclusive of all options. For example, Konjac flour does not
appear in the listing but has GRAS standing. Unmodified food
starches are also GRAS but are not presently included in the
listing. Pending affirmation, these products will be listed
under CFR 184.1847.

Recently the Food and Drug Administration and the USDA Food
Safety and Inspection Service have proposed a series of new food
labeling regulations for the USA which may have far reaching
consequences for the food industry (11). One of the proposals
would amend the food standard regulations to allow defined
descriptors to be used on standardized food (e.g. light egg nog,
light butter, nonfat ice cream etc.). Thus, this new regulation
would improve the opportunity for manufacturers of dairy
products to develop new products with an improved nutritional
profile, yet retain the elements of the original standard of
identity.

Table 1. Dairy Products Defined Under FDA Standards of Identity (10).

CFR No.	Product	Stabilizer addition[1,2,3]
Milk and cream:		
131.110	Milk	none
131.110	Milk (flavored e.g chocolate)	stabilizers permitted
131.111	Acidified milk	stabilizers permitted
131.112	Cultured milk	stabilizers permitted
131.115	Concentrated milk	none
131.120	Sweetened condensed milk	none
131.122	Sweetened cond. sk. milk	none
131.123	Lowfat dry milk	stabilizers & emulsifiers
131.125	Nonfat dry milk	none
131.127	Nonfat dry milk,vit.fortified	none
131.130	Evaporated milk	stabilizers & emulsifiers
131.132	Evaporated skim milk	stabilizers & emulsifiers
131.135	Lowfat milk	none
131.136	Acidified lowfat milk	stabilizers permitted
131.138	Cultured lowfat milk	stabilizers permitted
131.143	Skim milk	none
131.144	Acidified skim milk	stabilizers permitted
131.146	Cultured skim milk	stabilizers permitted
131.147	Dry whole milk	stabilizers & emulsifiers
131.149	Dry cream	stabilizers & emulsifiers
131.150	Heavy cream	stabilizers & emulsifiers
131.155	Light cream	stabilizers & emulsifiers
131.157	Light whipping cream	stabilizers & emulsifiers
131.160	Sour cream	texture improvers
131.162	Acidified sour cream	texture improvers
131.170	Eggnog	stabilizers permitted
131.185	Sour half-and-half	texture improvers
131.200	Yogurt	stabilizers permitted
131.203	Lowfat yogurt	stabilizers permitted
131.206	Nonfat yogurt	stabilizers permitted
Selected cheese products:		
(see part 133 for other cheese varieties)		
133.128	Cottage cheese (creaming mixture)	safe & suitable ingr.
133.129	Dry curd cottage cheese	none
133.133	Cream cheese	stabilizers < 0.5%
133.147	Grated American cheese food	melting salts < 3%
133.162	Neufchatel cheese	stabilizers < 0.5%
133.167	Pasteurized process cheese	melting salts < 3%
Frozen desserts:		
135.110	Ice cream and frozen custard	safe & suitable ingr.
135.115	Goat's milk ice cream	safe & suitable ingr.
135.120	Ice milk	safe & suitable ingr.
135.125	Goat's milk ice milk	safe & suitable ingr.
135.130	Mellorine	safe & suitable ingr.
135.140	Sherbet	safe & suitable ingr.
135.160	Water ices	safe & suitable ingr.

[1]Stabilizers refer to approved food gums.
[2]Safe & suitable ingr. refer to approved non-milk derived ingredients, including stabilizers and emulsifying agents.
[3]Texture improvers refer to safe and suitable ingredients that improve texture, prevent syneresis, or extend the shelf life of the product.

Table 2. Hydrocolloids Permitted in Food in the USA (6,7)

CFR No.	Product	Designation

Food additives permitted for direct addition to food[1]:

CFR No.	Product	Designation
172.365	Kelp	special dietary additives
172.610	Arabinogalactan	gums and related substances
172.615	Chewing gum base	gums and related substances
172.620	Carrageenan	gums and related substances
172.623	Carrageenan/w. polysorbate 80	gums and related substances
172.626	Salts of carrageenan	gums and related substances
172.655	Furcellaran	gums and related substances
172.660	Salts of furcellaran	gums and related substances
172.665	Gellan	gums and related substances
172.695	Xanthan	gums and related substances
172.868	Ethyl cellulose	multipurpose additives
172.870	Hydroxypropyl cellulose	multipurpose additives
172.872	Methyl ethyl cellulose	multipurpose additives
172.874	Hydroxypropyl methylcellulose	multipurpose additives
172.892	Food starch-modified	multipurpose additives
172.898	Bakers yeast glycan	multipurpose additives

Substances generally recognized as safe (GRAS)[2]:

CFR No.	Product	Designation
182.1480	Methylcellulose	multipurpose GRAS food ingr.
182.1745	Sodium carboxymethylcellulose	multipurpose GRAS food ingr.
182.7255	Chondrus extract	stabilizers

Specific substances affirmed as GRAS [3]:

CFR No.	Product	Designation
184.1115	Agar-agar	affirmed as GRAS
184.1011	Alginic acid	affirmed as GRAS
184.1120	Brown algae	affirmed as GRAS
184.1121	Red algae	affirmed as GRAS
184.1133	Ammonium alginate	affirmed as GRAS
184.1187	Calcium alginate	affirmed as GRAS
184.1230	Calcium sulfate	affirmed as GRAS
184.1277	Dextrin	affirmed as GRAS
184.1330	Acacia (gum arabic)	affirmed as GRAS
184.1333	Gum ghatti	affirmed as GRAS
184.1339	Guar gum	affirmed as GRAS
184.1343	Locust (carob) bean gum	affirmed as GRAS
184.1349	Karaya gum (sterculia gum)	affirmed as GRAS
184.1351	Gum tragacanth	affirmed as GRAS
184.1445	Maltodextrin	affirmed as GRAS
184.1588	Pectins	affirmed as GRAS
184.1724	Sodium alginate	affirmed as GRAS
184.1847	Food starch, unmodified	pending affirmation

[1] Food additives permitted for direct addition to food have been approved following required testing for safety.

[2] Substances generally recognized as safe (GRAS) are additives for which the safe use in food has prior sanction.

[3] Affirmed direct food substances are GRAS-list substances which have been tested according to regulatory guidelines.

(Differences between [1]food additives, [2]GRAS substances and [3]affirmed GRAS substances are now primarily related to historical usage and fine points of law. Generalizations about differences in safety are probably not warranted. The permitted usages of [1]food additives are generally more restricted than for [2,3]GRAS substances).

STABILIZATION OF DAIRY PRODUCTS

Hydrocolloids are important functional ingredients in a wide
variety of dairy products where they serve as texture modifiers
and stabilizers to improve shelf life. Other uses have come to
light in recent years, which may have considerable potential for
food product development. For example, various hydrocolloids
can provide improved body in artificially sweetened foods and
mask bitter aftertastes (12). The masking of the bitter
aftertastes associated with calcium fortification of cottage
cheese dressing may be satisfactorily controlled with fairly
high levels of guar and locust bean gum (13). Stabilizer
applications are directed toward different categories
represented by:

- Sterilized milk concentrates
- Chocolate milk beverages
- Cultured milks
- Milk gels
- Frozen desserts

Carrageenans are in many cases the stabilizers of choice because
of attributes related to suspension power, milk gel formation
and a modifying influence on the action of other colloids as
shown in Table 3 (14).

Sterilized milk concentrates and chocolate milk

There are several instability features of concern in the
manufacture of sterilized milk concentrates, including infant
formulas, which may not be related, and which require different
control measures. Some of the stabilizing treatments are
physical in nature and do not rely on the use of stabilizing
additives. For example, before such products can be safely
sterilized, the milk must undergo a *forewarming* treatment for
the purpose of improving the coagulation stability of the milk.
Concentrated products may be conditioned by first forewarming
the milk in a vat (90-100°C for up to 10 min) or by ultra-high-
temperature sterilization (UHT), (130-150°C for 1-5 sec) (15).
During forewarming the majority of the whey proteins is
denatured, interactions take place between whey proteins and
casein micelles and calcium salts are precipitated. The
resulting protein complex is rendered stable to subsequent
sterilization without age gelation taking place during storage.
Heat stability *per se* of batch-sterilized, concentrated milks
may be controlled in part by the addition of small amounts of
phosphates to adjust the salt balance of the system.

The successful applications of different carrageenans for
stabilizing evaporated milk and infant formulas against
destructive creaming (4) and for suspending cocoa particles in
chocolate milk(16) reflect a specificity in their physical
function which may be related to unique gelling characteristics
of κ-and ι-carrageenan (17) or to their interactions with the
casein system (18, 19). The concentrations of carrageenan used
in these products are generally at such low level that the
effect on overall viscosity is minimal. Therefore, the
mechanism for stabilization must be accounted for

Table 3. Application of Carrageenan in Dairy Products (14).

Product	Type	Use level (%)	Function or target
<u>Chocolate milk</u>			
a. past. process	κ-carrageenan	.02 –.03	cocoa suspension
b. cold process	λ-carrageenan	.05	cocoa suspension
<u>Evaporated milk</u>	alk. mod. κ/λ	.01 –.02	creaming stability
<u>Infant formula</u>	alk. mod. κ/λ	.01 –.02	creaming stability
<u>Liq. coffee whiteners</u>	κ/λ-carrageenan +(CMC, alginate, galactomannan)	.01 –.03	protein stability
<u>Ice cream and sherbet</u>	κ-carrageenan +(CMC, alginate, galactomannan)	.01 –.05	secondary stab. prev. syneresis
<u>Milk shake</u> and inst. breakfast pwdrs.	λ-carrageenan	.03 –.05	thickening, stab. improved mouthfeel
<u>Milk puddings</u>			
a. cooked	κ-carrageenan	.10 –.30	milk gelation
b. instant pwdrs.	λ-carrageenan	.20 –1.0	cold milk gelation
<u>Whipped products</u>			
a. dessert mix	κ-carrageenan	.15 –.50	foam stability
b. frozen wh. topping	κ/λ-carrageenan	.03 –.05	red. syneresis
<u>Cream cheese</u>	κ-carrageenan (locust bean g.)	.01 –.05	prev. syneresis

through interaction of the hydrocolloid with other components of the system. It may be noted, that in accelerated tests for predicting long-term creaming stability of infant formula emulsion systems (20) results showed that the protein concentration and/or protein type were more important in stabilization than any other factor, including levels of lecithin or carrageenan.

<u>Cultured dairy products.</u> Stabilization is apparently best achieved through the use of blends of hydrocolloids, which may include xanthan, modified starch, guar and locust gum, sodium carboxymethylcellulose (CMC) as well as pectin and carrageenan. The function of the individual constituents in such blends may not be the same. Typically, some of the stabilizers take aim at improving the body and texture of the acid-coagulated milk or cream while others are directed toward preventing serum separation (syneresis).

In a successfully made cultured dairy product, the final pH will be established in the vicinity of 4.3-4.6. This pH-range

corresponds to the isoelectric point of the milk proteins, where
they attain their minimum solubility, and where the tendency for
syneresis is strongest. The use of negatively charged
hydrocolloids, such as xanthan, carboxymethylcellulose and
pectin provides a powerful means for minimizing whey expulsion.
The capacity for these stabilizers to undergo electrostatic
interactions with proteins by the mutual attraction of the
anionic groups on the stabilizers to positively charged sites on
the proteins will result in an overall shift in the isoelectric
point of the resulting complexes to lower values (21), with the
net effect of reduced syneresis.

In contrast, the neutral stabilizers (starch, guar and locust)
do not undergo pH-dependent interactions with the proteins and
these are, therefore, more likely to improve body and texture by
increasing the viscosity of the aqueous phase of the system.

Stirred-type yogurt and drinking-type yogurt are subject to
different considerations. The gel structure is no longer
maintained in these products but is deliberately destroyed
during pumping to provide for smoother and more liquid products.
The products are often flavored with fruit puree which may
create additional stability problems. The choice of stabilizers
would logically be toward anionic-type stabilizers, such as CMC
and pectin, which soften the curd and minimize whey separation.
The application of high-methoxy-pectin as a protein stabilizer
in long-life drinking yogurt has been described (16). This type
of fluid yogurt, with an expected shelf life of around three to
six months, has been heat treated after fermentation under pH
conditions which make the proteins highly heat sensitive.

The choice of an appropriate stabilizer system for cultured
dairy products must be viewed in light of the fact that
acceptable products can be produced with proper attention to
processing details without the need for any stabilizing
additives at all. For example, homogenization at about 20 MPa at
a temperature of 55-70° C improves stability and consistency and
imparts a better "body" to the products (22). Application of
high-heat treatment of the milk prior to innoculation is
particularly critical. Heat treatment at 90-95°C for up to 5
minutes will help ensure that the coagulum of a set-type yogurt
will be firm.

MILK PUDDINGS AND GELS

Ready-to-eat milk desserts with a creamy or gelled texture and
long shelflife have become popular in Europe and are growing in
popularity in the USA. Custard-type milk desserts may be
prepared using only starch as the gelling agent. Klose and
Glicksman(23) have described two types of puddings: a cooked
milk pudding in which raw starch is the sole gelling agent, and
instant puddings in which the gel is formed with cold milk by
the synergistic effect of a pregelatinized starch and a milk
protein-phosphate salt reaction. However, milk gels and
puddings can also be formed by cooking milk with alginate,
carrageenan or furcellaran to form a blanc mange or flan (23).
The concentration of sodium alginate to achieve the desired
setting of a cold milk pudding is about 1 - 1.5%, which may be
used in combination with 0.1 - 3% dicalcium phosphate and 0.6 -
1.2% of sodium hexametaphosphate (23). Cooked milk puddings may

be prepared from κ-carrageenan or furcellaran, used either alone or in combination with starch. According to Klose and Glicksman (23), levels of approximately 0.4% furcellaran and 1% starch give satisfactory products.

A study (24) on the application of UHT and aseptic packaging to prepare milk desserts with long life has shown that it is possible to obtain satisfactory products but that the resulting texture is highly dependent upon the type and the ratio of the thickening and gelling agents used. The results showed that a blend of acetylated di-starch waxy adipate (3%) with carrageenan (0.2%) was suitable for UHT processing of creamy desserts utilizing direct as well as indirect heating.

A study of the cold-setting mechanism of carrageenan-based milk gelation has suggested that this is a three-stage process which includes (a) increased viscosity and molecular alignment, (b) molecular interaction and (c) aggregation (25).

FROZEN DESSERTS

The primary role of hydrocolloid stabilizers in frozen dessert products is to bind water in the mix as water of hydration, or by entrapment in a gel structure (23). It is the water holding capacity which makes hydrocolloids such effective agents for producing satisfactory body, smooth texture, slow meltdown, and heat-shock resistance in the resulting products. The following factors are considered important in choosing a stabilizer (26):

- Ease of incorporation with the mix
- Effect on viscosity and whipping properties of the mix
- Type of body desired in the mix
- Effect on melt-down characteristics
- Ability to retard ice-crystal growth
- Amount required to produce the stabilization
- Stabilizer cost

To this list must be added the need to distinguish between stabilization of frozen, hardened products, and stabilization of soft-serve items, intended for consumption before hardening. Clearly, the stabilization of hardened ice cream must address the problems associated with ice crystal growth over time ("heat-shock") which can usually be controlled by the use of blends of primary stabilizers, such as alginate, CMC, guar and locust gum. The tendency for these hydrocolloids to cause serum separation of the mix can to some extent be controlled by incorporating carrageenan in the stabilizer blends (23) or small amounts of calcium sulfate (27).

In contrast, ice crystal growth is of no consequence for soft-served products which are typically consumed within minutes of being drawn from the freezer. In this case the main concern relates to the importance of the frozen product having an appropriate stiff body with slow meltdown characteristics. This may be achieved by selecting blends of primary stabilizers which will maximize the viscosity without causing serum separation.

Since the mix for soft-serve freezers is kept in the liquid
state until drawn from the freezer, it is particularly
important, that any tendency for serum separation is controlled
by, for instance, including carrageenan in the stabilizer blend.

The growing interest in the USA for frozen desserts of lower fat
and total solids content has resulted in the need for better
knowledge for selecting effective stabilizing additives and, in
particular, choosing the appropriate concentrations for use. As
total solids levels are reduced it appears necessary to increase
the stabilizer levels to compensate for the weaker body and
texture. In *light* dairy desserts, there is an increased use of
hydrocolloids, such as polydextrose and microcrystalline
cellulose, which are effective as bulking agents and fat
replacers (28).

SERUM SEPARATION BY PRIMARY STABILIZERS

The cause for serum separation of milk by the use of a wide
range of hydrocolloids appears to be directly related to the
concentration level used and to the polymer size of the colloids
in question (29). This is shown in Figure 1, where serum
separation for two milk/colloid systems has been measured by the
protein depletion following centrifugation of the samples (29,
30). For both systems, serum separation is delayed as the
polymer size of the colloids is reduced as indicated by the
shift of the critical concentration (C*) to higher values as the
intrinsic viscosity is decreased.

The dependency of the onset of serum separation on polymer size
is shown clearly by the plot in Figure 2 which demonstrates that
serum separation in skimmed milk may be predicted from knowledge
of the intrinsic viscosity.

CARRAGEENAN AS A SECONDARY STABILIZER

In general, commercial carrageenans possess the ability to gel
or to thicken milk systems and to react with milk proteins.
Selection and blending of the raw material is important for
controlling the specific properties required. In addition,
commercial carrageenans are usually standardized with dextrose
to insure uniform performance. Various testing procedures are
used as a guide for uniformity. Gel strength and viscosity
measurements are used to measure gelling and thickening
properties of carrageenan/milk and carrageenan/water systems.
Testing procedures may vary throughout the industry and
commercial carrageenans may consequently be described by
different sets of specifications although the functional
properties may be similar.

Stabilization by carrageenans of α_s-casein (31) or of rennet-
treated milk proteins (32) against calcium-induced precipitation
points to a specific functional property of carrageenans which
may be attributed to the strongly acidic ester sulfate groups
associated with these hydrocolloids (18). These findings have
been adapted for routine analysis in a test which may possibly
be used as a performance test to demonstrate differences among
carrageenans and blends of carrageenan. The test procedure is

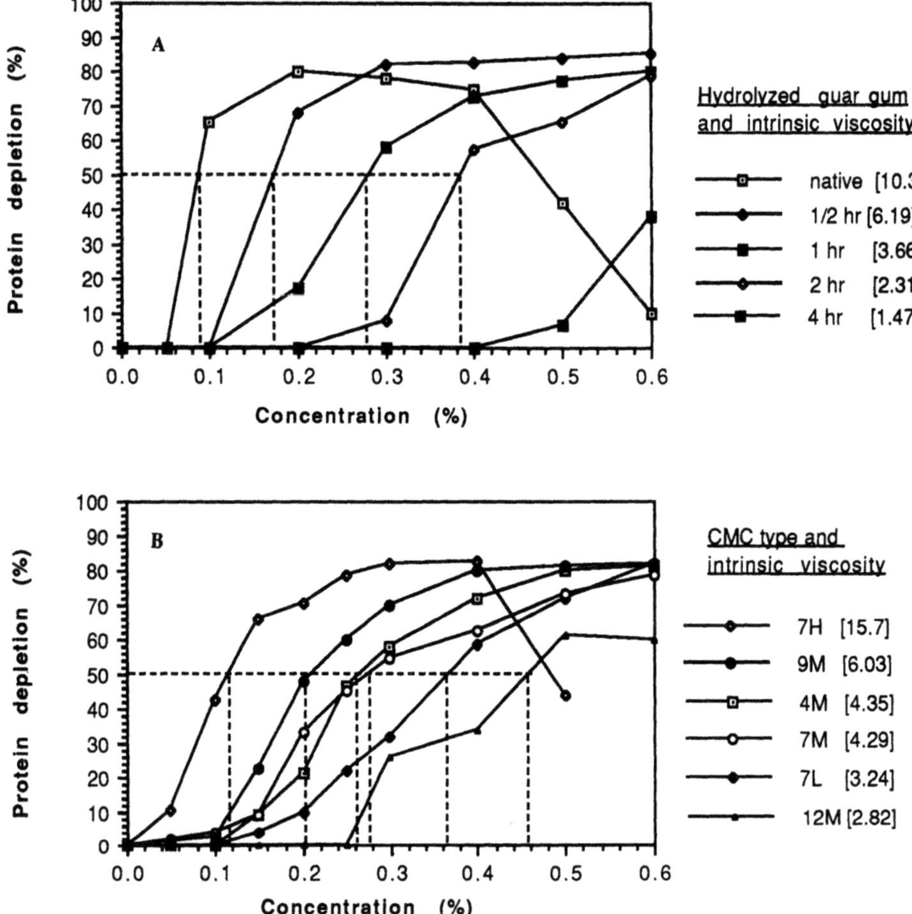

Fig. 1.   Serum separation in skimmed milk induced by addition of
hydrocolloids of different polymer size (intrinsic
viscosity indicated in brackets). A: native and
depolymerized guar gum. B. different types of
carboxymethylcellulose. (Serum separation was measured
as depletion of protein in the supernatant after aging
samples overnight in the refrigerator followed by
centrifugation at 3000 rpm in a clinical centrifuge.
The critical concentration for serum separation (C*)
was determined as the concentration required to produce
50% protein depletion).The dependency of the onset of
serum separation on polymer size is shown clearly by
the plot in Fig. 2 which demonstrates that serum
separation in skimmed milk may be predicted from
knowledge to the intrinsic viscosity.

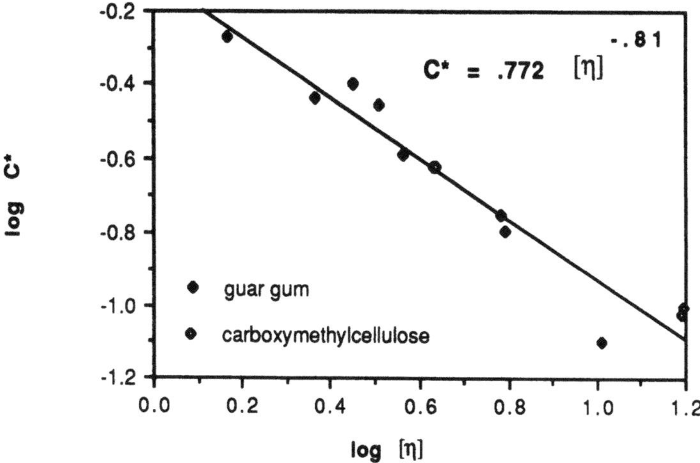

Fig. 2.   Hydrocolloid-induced serum separation of skimmed milk
as a function of polymer size. C* is the critical
concentration of the colloid to cause 50% serum
separation; [η] is the intrinsic viscosity, which is
used here as a measure of polymer size (29, 30).

outlined in Table 4 and has been found to be specific for
carrageenan, furcellaran and other sulfated polysaccharides. No
stabilization reaction has been detected for any of the common
neutral or carboxylated hydrocolloids.

The test is based upon the variable ability of carrageenans to
inhibit calcium aggregation of rennet-treated sodium caseinate.
A diluted solution of sodium caseinate contained in a series of
test tubes is mixed cold with increasing proportions of the test
material at a specified calcium chloride concentration. The
slightly milky solutions are then treated with rennet enzyme for
sufficient time to cause the caseinate to aggregate under the
influence of calcium ions. The turbidity (%T) in all sample
tubes is then measured relative to an untreated blank of sodium
caseinate containing only 0.02 M $CaCl_2$. Turbidity is gradually
cleared by the addition of carrageenan but not by other common
food hydrocolloids. The stabilization profile is produced by
plotting turbidity (%T) versus the ratio of stabilizer-to-
protein. In some cases the measured stabilization ability may
exceed one hundred percent and indicates that some carrageenan
stabilizers effectively clear the inherent turbidity of the
reference calcium caseinate solution.

The results in Figure 3A show the comparative ability of four
commercial samples to stabilize the calcium-sensitive system.
In this particular case it was found that the unspecified
industrial sample exhibit stabilizing properties which appear to
be intermediate to the samples designated as lambda- and kappa-
carrageenan. The sample designated as iota-carrageenan show a
profile of comparatively low stabilization capacity at low

concentration but increasing strength at higher concentrations, where the maximum stabilization has increased to values beyond one hundred percent of the reference solution.

The stabilization profiles in Figure 3B show results obtained with different mixtures of iota-and lambda-carrageenan. For the 35/35 ratio the maximum stabilization capacity remains high while the steepness of curve for iota-carrageenan has been moderated. Finally, the curves in Figure 3C show that alkaline modified chondrus extract (lambda/kappa-carrageenan) displays a greater stabilization capacity than either of the unmodified constituents. The curves for moderately and strongly alkaline-modified lambda-carrageenan show evidence for the modification having caused a drop in the maximum stabilization capacity at the higher concentration but also having enhanced the stabilization at the low concentration region.

While the described test relates to measurement of a unique functional behaviour of carrageenan, it is still uncertain whether the measured values reflect in any substantial way the stabilizing function of these particular hydrocolloids. Nevertheless, the stabilization test profiles may provide a useful means for detecting differences among a variety of available products.

Table 4. Stabilization of rennet treated casein against calcium precipitation (33).

___Reagents___

Solution A:	0.5% of sodium caseinate in water (wt./wt.)
Solution B:	0.025% solution of carrageenan in water(wt./wt.)
Solution C:	water
Solution D:	$CaCl_2$-solution (.2 M)
Solution E:	sgl. strength rennet diluted 2 ml in 100 ml.

___Schedule___

	Tube #	A Casein (ml)	B Test sol. (ml)	C H_2O (ml)	D $CaCl_2$ (ml)	E Rennet (drop)	Stab/ prot. ratio	Turbidity (%T)
Reference	1	1.00	0.00	8.00	1.00	0	0.000	100
	2	1.00	0.00	8.00	1.00	1	0.000	---
	3	1.00	1.00	7.00	1.00	1	0.050	---
	4	1.00	2.00	6.00	1.00	1	0.100	---
	5	1.00	3.00	5.00	1.00	1	0.150	---
	6	1.00	4.00	4.00	1.00	1	0.200	---
	7	1.00	5.00	3.00	1.00	1	0.250	---
	8	1.00	6.00	2.00	1.00	1	0.300	---
	9	1.00	7.00	1.00	1.00	1	0.350	---
	10	1.00	8.00	0.00	1.00	1	0.400	---

___Method___

1. Combine all reagents except for rennet in test tubes at 0-2°C.
2. Add rennet in sequence to the samples tubes (#2-10), then transfer tubes to 30°C water bath and incubate for 20 minutes.
3. Measure the turbidity (%T) of all samples at 620 nm against water as a blank. Calculate stability from turbidity values relative to the reference sample (tube #1=100% stability). Plot stability (%T) vs. stabilizer/protein ratio.

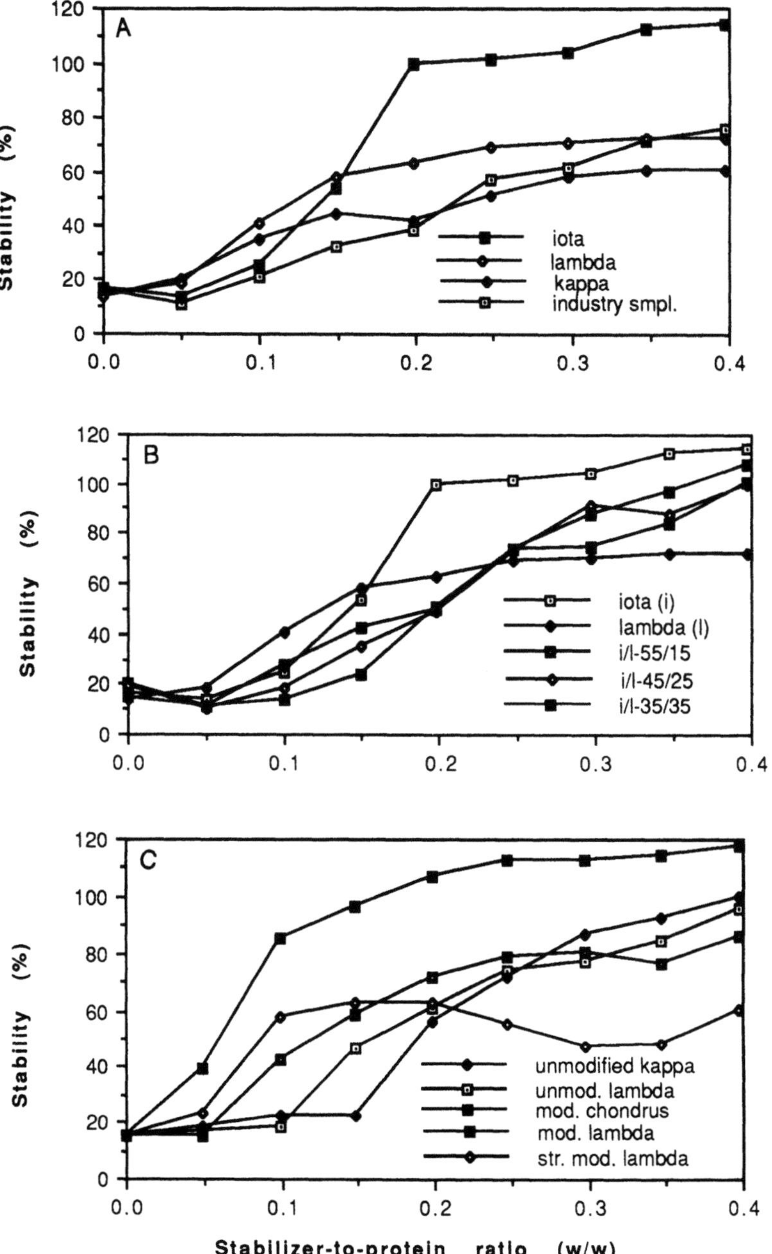

Fig. 3 Stabilization by carrageenans of rennet-treated
casein against calcium precipitation.
A: Comparison of commercial carrageenan products. B:
Comparison of blends of iota (i) and lambda (l)
carrageenan blends. C: Effect of moderate and strong
alkali modification on stabilization capacity of
selected carrageenans (33).

REFERENCES

1. Glicksman, M. (1982). Food Hydrocolloids, Vol 1, CRC Press, Florida.
2. Dickenson, E. and Stainsby, G. (1982). Colloids in Food, Applied
     Science, London.
3. Mitchell, J.R. and Ledward, D.A. (eds.) (1986). Functional Properties
     of Food Macromolecules, Elsevier Applied Science, London.
4. Hansen, P.M.T. (1982).Prog. Fd. Nutr. Sci., 6: 127-138.
5. Dickenson, E.Proc.(1987). 4th Int. Conference on Gums and
     Stabilisers for the Food Industry. Wrexham, Clwyd, Wales.
6. Code of Federal Regulations. (1992). 21 CFR 172. U.S. Government
     Printing Office, Washington DC.
7. Code of Federal Regulations. (1992). 21 CFR 182 and 184. ibid
8. Code of Federal Regulations. (1992). 21 CFR 184.1(b). ibid
9. Code of Federal Regulations. (1992). 21 CFR 106 and 107. ibid
10. Code of Federal Regulations. (1992). 21 CFR 131 and 135. ibid
11. Mermelstein, N.H. 1992. Food Technology, (January): 56.
12. Glicksman, M. and Farkas, E. 1966. Food Technol. 20(2):156
13. Puspitasari, N.L., Lee, K. and Greger, J.L. 1990. J. Dairy Sci.
     74(1):1.
14. Towle, G.A. (1973). Carrageenan, Ch. V in Industrial Gums, 2nd ed.
     (R.L. Whistler and J.N. BeMiller eds.) Academic Press New York.
15. Harwalkar, V.R. 1983. Age gelation of sterilized milks, Ch. 7 in
     Developments in Dairy Chemistry-1 (P.F. Fox, ed.) Applied Science
     Publishers, London.
16. Carnell, J. 1992. Dairy Ind. Int. 57(1):25.
17. Rees, D.A. 1969. Adv. Carbohydrate Chem. 24:267.
18. Lin, C.F. and Hansen, P.M.T. 1970. Macromolecules, 3: 269.
19. Snoeren, T.H.M., Payens, T.A.J., Jeunink,J. and Both, 1975.
     Milchwissenschaft 30:393.
20. Fligner, K.L., Fligner, M.A. and Mangino, M.E. 1991. Food
     Hydrocolloids 5(3):269.
21. Hidalgo, J.E. and Hansen, P.M.T.. 1969. J. Agric. and Food Chem. 17:
     1089.
22. Anon. (no year). Dairy Handbook. Ch. 10, Cultured-Milk Products. Alfa-
     Laval Food Engineering AB, Lund, Sweden.
23. Klose, R.E. and Glicksman, M. (1972). Gums, Ch. 7 in Handbook of Food
     Additives, 2nd edition.(T.E. Furia ed.) CRC Press, Cleveland Ohio.
24. Rapaille,A., Vanhemelrijck, J. and Mottar, J. 1988. Dairy Ind. Int.
     53(9):21.
25. Xu, S.Y., Stanley, D.W., Goff, H.D., Davidson, V.J. and Le Maguer, M.
     1992. J. Food Sci. 57(1):96.
26. Potter, F.E. and Williams, D.H. 1950. Milk Plant Monthly, 39(4):6
27. Bassett, H.J. 1988. Dairy Field. 171(3):18.
28. Tharp, B.W. and Gottemoller, T.V. 1990. Food Technology (October):86.
29. Oen, P., Hansen, P.M.T. and Chism, G.W. 48th Annual IFT Meeting, New
     Orleans, Louisiana. June 19-22, 1988. Abstract # 535.
30. Lin, P., Desai, N. and Hansen, P.M.T. 1986. Relationship betwwen
     polymer size of guar gum and critical concentration for protein
     precipitation (abstract). 46th Annual IFT Meeting, Dallas, TX, June
     15, 1986.
31. Hansen, P.M.T. 1968. J. Dairy Sci. 51:192.
32. O'Loughlin, K. 1973. Stabilization of rennet-treated milk proteins by
     carrageenan, MS Thesis, The Ohio State University, Columbus.
33. Lee, S.- C. 1992. Stabilization of rennet-treated casein by
     carrageenan, MS Thesis, The Ohio State University, Columbus.

RHEOLOGICAL STUDIES OF FISH PROTEINS

D. D. Hamann

Food Science Department
North Carolina State University
Raleigh, NC 27695-7624 USA

ABSTRACT

Fish protein is highly functional in the sense that it will bind water and form a strong deformable gel in a suitable environment of ionic strength, pH, and temperature conditions. Specific gel properties are dependent on factors including species, fish quality, and degree of protein degradation during processing and storage. Rheology reveals species differences, fish handling and storage effects, changes during thermal processing, and final product quality. Rheology serves as a research tool in the study of fish protein gelation mechanisms. The role of chemical bonds stabilizing fish protein gels may be studied via rheological comparison with gels of better known chemistry and structure.

INTRODUCTION

Rheological studies of fish proteins have become important because of the internationalization of the surimi industry initially developed in Japan. Surimi is a wet concentrate of the proteins of fish muscle produced by water washing of minced flesh. With the addition of cryoprotectants (commonly 4% sucrose + 4% sorbitol) surimi can be stored frozen for over a year with little loss of its gel-forming functionality. Analog food products based on surimi have become popular in the U.S. and Europe although the variety of products is much less than in Japan where surimi has a long history.

Our knowledge of how fish protein functions in surimi products is based on early work by Japanese researchers. A brief history is given by Okada[1] in a recent book on surimi[2]. Niwa[3] summarizes our understanding of fish protein chemistry, and Matsumoto and Noguchi[4] cover the very important subject of cryostabilization.

Fish protein paste (with NaCl) is characterized rheologically as being both viscous and elastic. During thermal processing the protein paste becomes a gel with only a small proportion of viscous character remaining. Rheological studies of fish protein are separated into two categories, small strain (deformation) and large strain (fracture). Small strain studies can monitor changes occurring during thermal processing revealing the timing, rates,

Food Hydrocolloids: Structures, Properties, and Functions
Edited by K. Nishinari and E. Doi, Plenum Press, New York, 1994

and significance of gelation events. Variation in the profile of a rheology property as a function of temperature during gelation is due to factors including fish species, processing and surimi history. Both thermal scanning and isothermal testing are valuable. Scanning can mimic an industrial process. Determining the effect of a specific temperature on a paste with only a cold temperature history requires a step change isothermal test.

Large strain results reveal textural properties of the food product; something small strain properties frequently fail to do[5]. Properties measured relate to gel hardness and gel deformability (closely related to sensory texture profile analysis cohesiveness[5]). Fundamentally, hardness is measured as fracture stress (Pa units) and deformation is measured as fracture strain (dimensionless). Fracture strain is particularly important in evaluating surimi quality because deformability is an important textural property of crab analogs in the U.S. and kamaboko quality in Japan. Fracture strain is a very good measure of gelling quality because it is primarily dependent on surimi gelling quality and is not strongly influenced by small differences in protein concentration, filler ingredients, or thermal processing history which strongly affect fracture stress[6].

When a paste (or gel) is deformed by external force, the magnitude of the force required is the sum of the resistance of the paste due to its viscous nature plus that due to its elastic nature. Viscous resistance is proportional to the <u>rate of deformation</u>, and the work of deformation (force x distance) is lost as heat energy to the paste. Elastic resistance is directly proportional to the <u>deformation</u>, and the work of deformation is recoverable. If the material is incompressible, volume changes do not occur so only shear (shape changes) are considered. Force applied tangentially to a unit area is the shear stress and shear strain is the trigonometric tangent of the angle of rotation of a plane perpendicular to the stressed plane as a result of the shear stress. The ratio of shear stress to shear strain is the shear modulus, G, and is a measure of the material overall rigidity or stiffness (resistance to change in shape). Small strain rheology instruments often produce cyclic deformation in the form of sinusoidal motion. A shear modulus designated G' is the rigidity due to elastic response of the material and a modulus G" is the rigidity due to viscous response of the material. It is important when testing muscle pastes/gels that the gel not be strained too rapidly, otherwise the usefulness of the test decreases since the viscous response does not occur. Surimi pastes/gels tested near 1 Hz and above will show very little viscous character[6]. The test deformation can also be in the form of a sawtooth function (triangular deformation/time graph). Hamann and MacDonald[6] give detailed coverage of several rheology and mechanical texture evaluation methods. Fish pastes (usually surimi pastes) subjected to small strain triangular or sinusoidal deformation patterns will be considered.

GEL FORMATION RHEOLOGY

Species Differences

Triangular deformation testing of Alaska pollack (*Theragra chalcogramma*) and Atlantic croaker (*Micropogan undulatus)* surimi pastes (3% NaCl; 5:1 water to protein ratio) shows species differences. Figure 1[7] shows G as it changed due to thermal processing at 1° C/min. Figure 2[7] shows the percent of the total energy input per cycle lost as heat due to the material viscosity. Two treatments applied to each species were 1) the control: loading the test fixture and testing immediately after paste preparation, and 2) the setting treatment: loading the paste in the test fixture and holding for 24 hr at 4° C prior to testing. At 10° C, before any heating, the set gels are about twice as rigid as the control (Fig. 1). This difference is partially lost during heating, with more being retained by the pollack than the croaker (very little is retained by the croaker). The shear modulus at 70° C, after thermal processing, can be lower than values earlier in the thermal process history. Most of the rigidity based on hydrogen bonding is not present at 70° C. Cooling the cooked gel back to room temperature and determining the shear modulus will yield a value approximately double that determined at 70° C.

The cause of the increase in shear modulus due to the setting treatment is different in the two species. The bonds formed during the setting treatment are thermally stable in pollack. The bonds formed in the croaker are not thermally stable, so after reaching 45° C in the thermal scan, the set and control profiles are nearly the same. This suggests that 4° C setting of croaker paste produces increased hydrogen bonding, but in pollack paste also produces covalent bonds and hydrophobic associations. The result for pollack is consistent with low temperature setting effects reported by several researchers[3, 8, 9, 10]. Non-disulfide covalent cross-linking of myosin in surimi is reportedly produced by the enzyme transglutaminase[11].

Room temperature fracture values of shear stress and strain are given at the termination of each plot in Fig. 1 to show the effect of the setting treatment on texture properties. The 4° C setting increased the fracture stress of pollack to about 1.5 times the control value, but 4° C setting had no effect on croaker shear stress. Shear strains were not influenced by setting for either species. The fracture data confirm that 4° C setting produced stable changes in pollack but not in croaker. Pastes made from a number of species inhabiting moderate and warm waters, including croaker, require higher temperatures to set irreversibly due to higher thermal stability of their proteins[12].

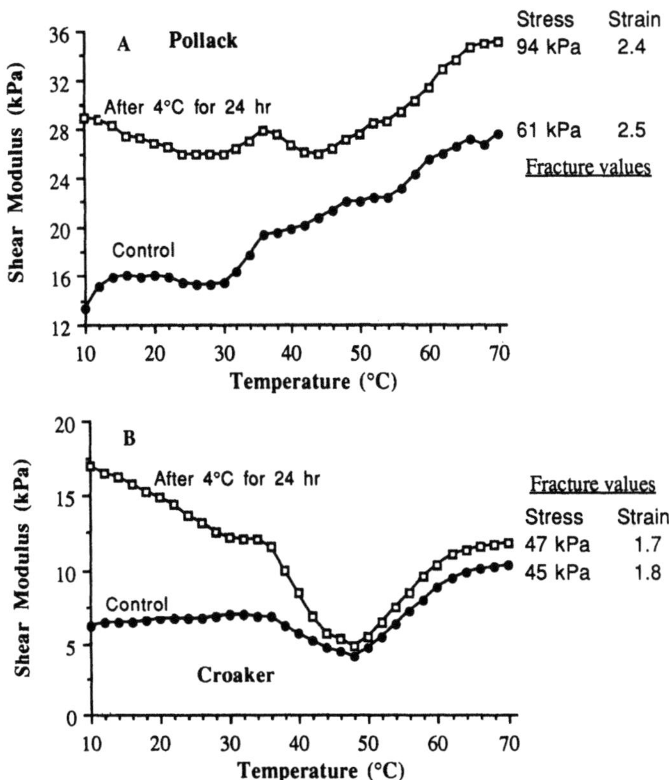

Figure 1. Shear modulus vs temperature for surimi pastes (1/5 = water/protein; NaCl = 3%) heated at 1° C / min.; (A) is for Alaska Pollack and (B) for Atlantic croaker. Open squares are for pastes subjected to a setting treatment prior to heating. Test motion was sawtooth at 0.04 Hz, strain amplitude 0.02. For comparison, fracture values are room temperature shear stress and shear strain at fracture in torsion. From Kim[7].

Differences in thermal stability between pollack and croaker sols can be seen by inspecting the viscous energy loss (EL) curves (Fig. 2). A gel is more elastic as energy loss decreases. The pollack control was least elastic initially, but rapidly became more elastic by 25° C. The set pollack was much more elastic initially (10° C), but showed a similar response to temperature as the control with a rapid decrease in EL by 25° C. The pollack gels gained most of their increase in elasticity by 45° C. The transition from viscous to elastic character occurred at higher temperatures and over a narrower temperature range for croaker. Initiation was at about 38° C and the transition was fairly complete by 50° C. The transition occurred over about 12 degrees in croaker, but required about 20 degrees in pollack. Transitions seen in Fig. 2 could have been seen equally well using a sinusoidal deformation function and plotting the phase angle between stress and strain (δ) or G" versus temperature.

Below 20° C in Figs. 1 & 2, the set pastes decrease in G and increase in EL as temperature increases. Some of the structure developed in setting is lost, probably due to a decrease in hydrogen bonds, so the pastes become less stiff and more viscous. This suggests that the 4° C setting treatment exposes sites for hydrogen bonds. The control pastes are much less stiff at 10° C, but show a slight increase in G as temperature increases.

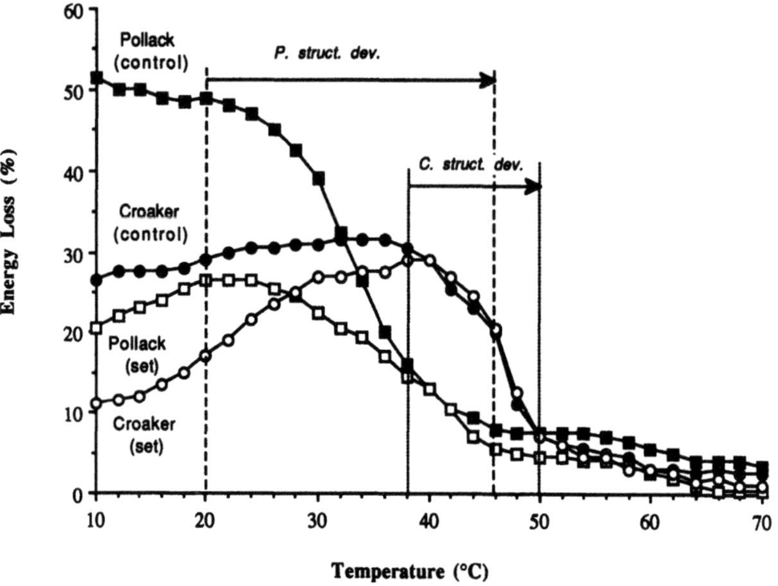

Figure 2. Energy loss vs temperature for the scans of Fig. 1; <u>P. struct dev.</u> indicates the critical temperature range over which elasticity increases in pollock pastes; <u>C. struct dev.</u> indicates this range for croaker.

Quality Differences

Several changes in slope in Fig 1 are significant, being dependent on fish species and the muscle protein history. Figure 3 for pollack paste[13] demonstrates an effect of protein history. The control surimi was of high quality as indicated by the final gel fracture strains listed in Fig. 3. The treatment subjecting the surimi to 6 freeze thaw cycles lowered quality so set gel fracture stress dropped by 27%. Frozen and thawed surimi gels not set were also weaker and less cohesive (lower fracture strain). High quality pollack surimi pastes do not usually produce large modulus peaks during a thermal scan. However, the freeze/thaw denaturation produced the large 38° C peak shown in Fig. 3. The cause of this peak is not completely clear, but gelation kinetics are involved.

Wu et al.[14] give gelation rate constants (rate of increase in G) for chicken myosin (10 mg/mL) based on isothermal testing. Chicken myosin will gel at temperatures slightly below 44° C, but at 44° C the rate constant was 1/10 of the rate at 46° C and only 0.003 of the highest rate which was at 52° C. The rate then dropped off rapidly at 54° C and was about 1/3 the 52° C value.

Although pollack proteins are less thermally stable than chicken proteins, and the protein concentration is higher in surimi pastes than the solutions of Wu et al.[14], qualitative comparisons can be made. In Fig. 3 slow unfolding and aggregation increase structural stiffness at temperatures below 15° C. This is followed by a near zero slope section to about 28° C. The zero slope section is likely produced by a combination of conformation loss and structure building so that little change in structural rigidity occurs. Rapid structure building starts at about 28° C as indicated by the increase in slope in Fig. 3 and the steep negative slope in Fig. 2. By comparison with Wu et al.[14] one can estimate the gelation rate constant is maximum near 40° C, a temperature associated with the negative slope for the freeze/thawed surimi and slower increase in G for the control surimi. The reason for the negative slope is that the structure already developed via myosin head groups at temperatures below 38°C loosens up due to additional loss of protein native conformation. Clark and Ross-Murphy[15] write of "growth of G' prior to tail-section unfolding, then a sharp drop in G' as tail unfolding takes place." The rate constant would be expected to be maximum at a temperature that produces loss of protein native conformation quite rapidly, but the temperature cannot be so high that the aggregation rate severely limits molecular alignment. Wu et al.[14] show that, for chicken myosin, isothermal G development at the temperature of highest rate constant (52° C) did not produce the highest equilibrium G value. The equilibrium G values at 48 and 50° C were about 1.7 times higher.

The basic elastic framework in pollack surimi gels seems to be formed by about 45° C when heating at 1° / min. (Fig. 2). At this temperature the slope in Fig. 3 for freeze/thaw surimi again becomes positive and the two gels stiffen in similar fashions as temperature increases. The reason that the control pollack surimi paste does not exhibit the strong 38° C peak is because much of the protein is nearer the low entropy native state so more energy is required to unfold it prior to aggregation. A more gradual structural change results.

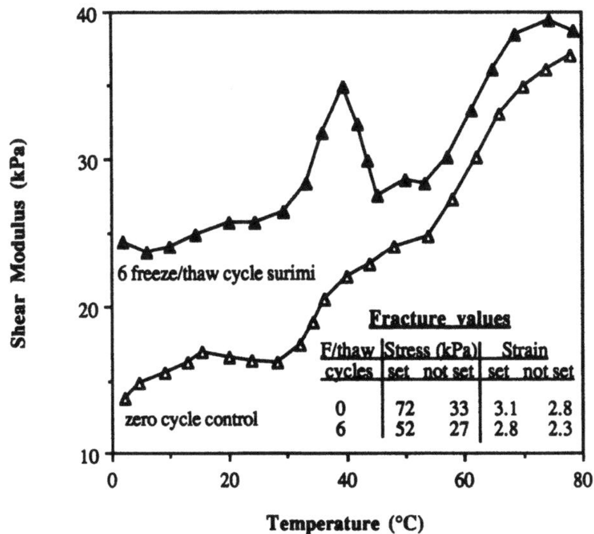

F/thaw cycles	Stress (kPa)		Strain	
	set	not set	set	not set
0	72	33	3.1	2.8
6	52	27	2.8	2.3

Figure 3. Shear modulus curves for ramp heating of Alaska pollack surimi pastes (water = 80% ; NaCl = 3%) heated at 1° C / min.; solid triangles are for pastes made from surimi subjected to 6 freeze - thaw cycles; open triangles are for the control. Test motion was sawtooth at 0.04 Hz. For comparison, fracture property values are room temperature shear stress and shear strain at fracture in torsion. Data from Kim[13].

Possibly the most important observation of Fig. 3 is that the final small strain shear modulus values did not correlate with fracture stress values (gel hardness). The freeze denatured surimi actually produced a higher G than the higher quality surimi. Fracture stress and strain, however, did reveal the lowering of surimi quality. Shear moduli, G or G', by themselves are not reliable predictors of gel texture.

RHEOLOGY OF PROCESSED SURIMI GELS

Small Strain

Thermally processed surimi gels are stable in that the gelation process is irreversible. However, cooling the gels shown in Fig. 1 or Fig. 3 will produce stiffer (higher G) gels and reheating will return the values to near the values before cooling. Increasing and decreasing hydrogen bonding is the dominant cause of this reversible change. Figure 4 compares a cooling G' curve for a surimi gel[16] with a similar curve for a 0.45% high methoxy pectin gel[17]. Pectin is thought to be stabilized by a combination of hydrogen bonds and hydrophobic associations[18]. Qualitative changes in G' below about 65° C are similar for the two gels. Both surimi and pectin G' values increase with decreased temperature below about 50° C where hydrogen bond strength is the dominant factor. Surimi gel G' values decrease slightly from 50° to 85° C, whereas the pectin gel G' values increase from 65 to 85° C. Pectin G' has been shown to decrease at temperatures much above 85° C[17] suggesting that the increase in G' from 65° to 85° C is caused by strengthening of hydrophobic associations which weaken at higher temperatures. It is thought that hydrophobic association strength is maximum near 60° C[3], but values of G' for methyl cellulose (hydrophobically stabilized) also increase as temperature is raised to 85° C[17]. This further strengthens the argument that the rise in G' for pectin gel up to 85° C is the result of hydrophobic associations. It is likely that the formation of new hydrophobic associations continues up to 85° C in the cases of pectin and methyl cellulose although the strength of each association is less than if the temperature were lower.

The rise in pectin gel G' from 65 - 85° C is partly due to rubber (entropy) elastic character so that G or G' increase with test temperature. Case[19] has demonstrated that glucomannan gels (konjac mannan) exibit dominant rubber elastic character although they are not considered to have any covalent bonds. This is because these are entangled systems having physical junction zones[20] so that the structure responds to short time rheological testing in a manner similar to gels formed by covalent cross links. Long time rheological testing (e.g., stress relaxation) reveals the noncovalent structure[19].

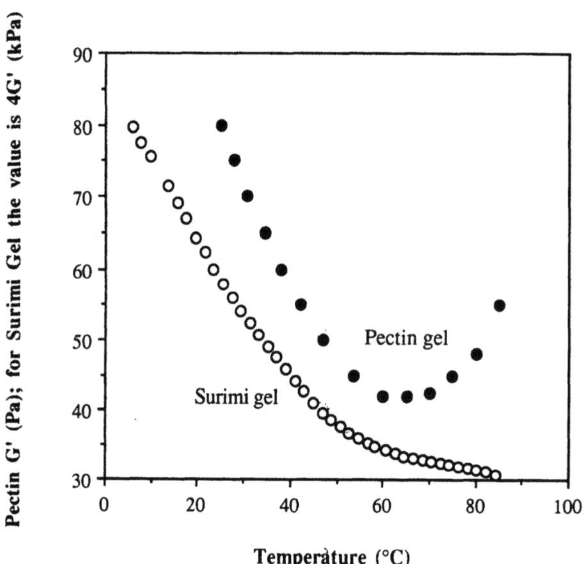

Figure 4. The effect of gel temperature (cooling) on the storage moduli (G') of a cooked (90° C) Alaska pollack surimi gel (H_2O = 78%; NaCl = 2%; pH = 7)[16] and a 0.45% high methoxy pectin gel[17]. Instrument settings were: frequency = 0.05 Hz; shear strain amplitude = 0.02.

At temperatures above 65° C hydrogen bonds in surimi gel decrease to the point that the entropy elastic effects probably can be seen. Niwa et al.[21] demonstrated that the entropy elastic nature of surimi gels is revealed after the influence of low energy bonds is reduced by stress relaxation which presumedly allows entanglements to slip producing an equilibrium situation dominated by the strong bonds. Then, G' increases as temperature increases due to more rapid thermal motion of molecular chains. For surimi gel (Fig. 4) the magnitude of G' as temperature increases above 50° C (without stress relaxation) is the result of at least 3 simultaneous events: 1) entropy elasticity causes G to increase, 2) hydrophobic associations change in strength or number causing G to change accordingly, and 3) further reductions in hydrogen bond strength cause G to decrease. The net result depends on which factors dominate.

Large Strain

Fracture data can assist in understanding protein gel structure and chemistry as well as serve as food texture information. For reference we will consider a gel of known structure. Polyacrylamide gels are covalently bonded gels known to exhibit rubber elastic behavior under small strain so G increases linearly with increased temperature. Figure 5[22] shows fracture shear stress and fracture shear strain as functions of test temperature. Fracture stress does not change significantly, but fracture strain decreases linearly as temperature increases. If fracture stress is divided by fracture strain to obtain G at fracture, this fracture G and the small strain G are not significantly different over the test temperature range[22].

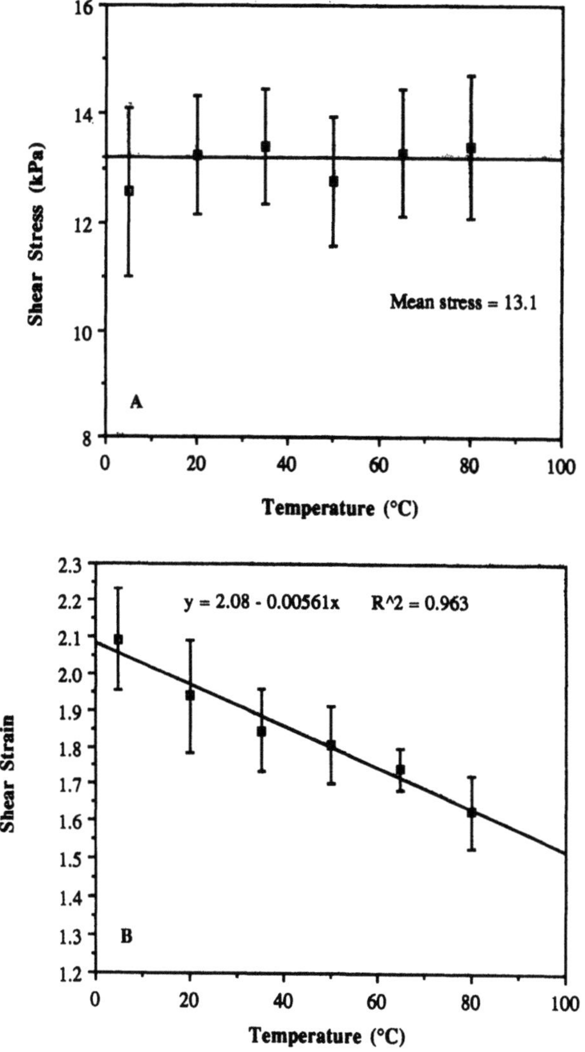

Figure 5. Fracture shear stress (A) and fracture shear strain (B) of a polyacrylamide gel as functions of temperature. Bars indicate the standard deviations. From Gonzalez[22].

Figure 6[23] shows fracture results for Alaska pollack surimi gels. The set gels were held at 25° C for 1 hr prior to a final cook at 90° C for 20 min. The control gels were cooked at 90° C for 20 min. soon after paste preparation. Surimi gel fracture stress responded to test temperature differently when the setting treatment was used. The control gels were less strong overall, and below 35° C fracture stress did not change much with test temperature. Fracture stress for the set gels increased linearly as temperature decreased over the whole range. Fracture strain showed a quadratic response to temperature with the set gels decreasing less in fracture strain at the high and low temperatures. Results similar to those for the control gels in Fig. 6, but done with a penetration test have been reported by Niwa et al.[21].

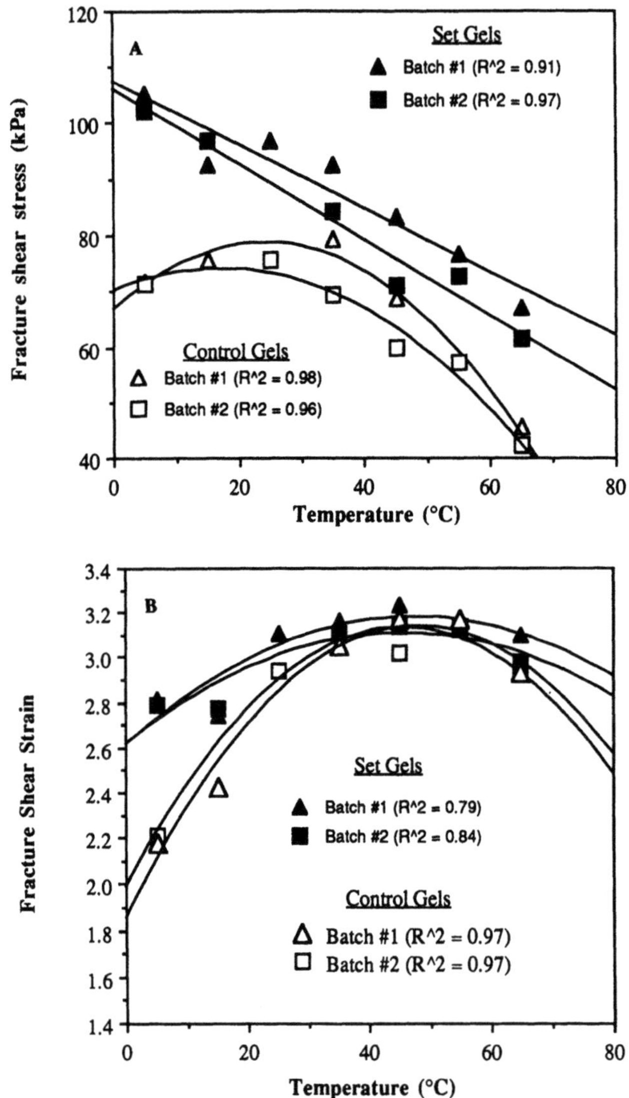

Figure 6. Fracture shear stress (A) and fracture shear strain (B) of Alaska pollack surimi gels (H_2O = 74%; NaCl = 2%) as functions of temperature. Results for duplicate preparations (Batches) are shown . Points are mean values of 5 or more specimens. Set gels were held at 25° C for 1 hr followed by cooking at 90° C for 20 min.; Control gels were cooked soon after paste preparation. From Howe[23].

Set pollack gels have more covalent crosslinking[10, 11] which can explain some of the differences seen. Covalent bonds produce a stable foundational fracture stress as seen for polyacrylamide in Fig. 5. The linear increase in set surimi gel fracture stress as temperature is reduced is likely due to increased hydrogen bond strength superimposed on the foundational covalent strength and less stable hydrophobic associatons. Hydrophobic associations would be quite stable from 65° C to about 40° C and weaken at colder temperatures where hydrogen bonding would be greatest. In the case of the control gels weakening of hydrophobic associations offsets the strengthening of hydrogen bonds.

The fracture strain change with temperature is also consistent with the logic that control gels are more dependent on hydrophobic associations than set gels. The decreasing fracture strain measured as temperature decreases below about 50° C is likely due to weakening of hydrophobic associations which evidently are dominant enough to strongly influence fracture strain. The hydrogen bonding causing the increase in fracture stress of set gels as temperature decreases does not seem to influence fracture strain strongly. Set surimi gel fracture strain does not decrease as much with temperature because of the greater influence of covalent bonds. At the high temperature end (above 60° C) the greater decrease in fracture strain for control gels could also be due their greater dependence on hydrophobic associations which are weakening with increased temperature.

APPLICATION

The data and discussion presented are intended to demonstrate that rheology is a valuable approach to understanding gelation of fish proteins as well as other food ingredients. Comparing results from a variety of materials improves understanding of structure - rheology relationships. However, this understanding is fallible and should be confirmed or invalidated by complimentary methods. Much of the discussion presented here confirms the work of researchers in Japan and other countries upon whose foundation we build.

Most of the fish protein data presented was for Alaska pollack surimi using a very limited set of storage and processing treatments. The temperature dependent fracture properties of other species of commercial importance must also be examined. For example, it was seen that the cohesiveness (shear strain) of cold pollack surimi gel is lower than gel served at room temperature. Cohesiveness is maximum near 45° C which is often the temperature of an entree. Serving temperature dependence of cohesiveness is reduced by using a setting treatment on Alaska pollack surimi, but we would not expect this to be general over all commercial species. Consideration of these points suggests that more work is needed and that compilation of results in a data bank would be helpful. Textural hardness is equally important but easier to adjust by changing protein concentration, using starch, using a setting treatment, or otherwise changing the thermal process. An appropriate data bank for a variety of species would reduce dependence on trial and error.

REFERENCES

1. M. Okada, History of surimi technology in Japan, in: "Surimi Technology," T. C. Lanier and C. M. Lee, eds., Marcel Dekker, Inc. New York (1992).

2. T. C. Lanier and C. M. Lee, "Surimi Technology," Marcel Dekker, Inc. New York (1992).

3. E. Niwa, Chemistry of surimi gelation, in: "Surimi Technology," T. C. Lanier and C. M. Lee, eds., Marcel Dekker, Inc. New York (1992).

4. J. J. Matsumoto and S. F. Noguchi, Cryostabilization of protein in surimi, in: "Surimi Technology," T. C. Lanier and C. M. Lee, eds., Marcel Dekker, Inc. New York (1992).

5. J. G. Montejano, D. D. Hamann, and T. C. Lanier, Comparison of two instrumental methods with sensory texture of protein gels, *J. Texture Studies*, 16: 403 (1985).

6. D. D. Hamann and G. A. MacDonald, Rheology and texture properties of surimi and surimi-based foods, in: "Surimi Technology," T. C. Lanier and C. M. Lee, eds., Marcel Dekker, Inc. New York (1992).

7. B. Y. Kim, Rheological investigation of gel structure formation by fish proteins during setting and heat processing, Ph. D. dissertation, North Carolina State University, Raleigh, NC (1987).

8. T. Numakura, N. Seki, I. Kimura, K. Toyoda, T. Fujita, K. Takama, and K. Arai, Cross-linking reaction of myosin in fish paste during setting (suwari), *Nippon Suisan Gakkaishi*, 51: 1559 (1985).

9. S. Nishimoto, A. Hashimoto, N. Seki, I. Kimura, K. Toyoda, T. Fujita, and K. Arai, Influencing factors on changes in myosin heavy chain and jelly strength of salted meat paste from Alaska pollack during setting, *Nippon Suisan Gakkaishi*, 53: 2011 (1987).

10. G. G. Kamath, Investigation of physico-chemical basis for the unique "setting" phenomenon of Alaska pollack and Atlantic croaker surimi, Ph. D. dissertation, North Carolina State University, Raleigh, NC (1990).

11. N. Seki, H. Uno, N-H. Lee, I. Kimura, K. Toyoda, T. Fujita, and K. Arai, Transglutaminase activity in Alaska pollack muscle and surimi, and its reaction with myosin B, *Nippon Suisan Gakkaishi*, 56: 125 (1990).

12. A. A. Hashimoto, A. Kobayashi, and K. Arai, Thermostability of fish myofibrillar Ca-ATPase and adaption to environmental temperature, *Bull. Japan Soc. Sci. Fish.*, 48:671 (1982).

13. B. Y. Kim, D. D. Hamann, T. C. Lanier, and M. C. Wu, Effects of freeze-thaw abuse on the viscosity and gel-forming ability of surimi from two species. *J. Food Sci.*, 51: 951 (1986).

14. J. Q. Wu, D. D. Hamann, and E. A. Foegeding, Myosin gelation kinetic study based on rheological measurements, J. Agric. & Food Chem., 39: 229 (1991).

15. A. H. Clark and S. B. Ross-Murphy, Structural and mechanical properties of biopolymer gels. *Adv. Polymer Sci.*, 83: 57 (1987).

16. D. D. Hamann, Viscoelastic properties of surimi seafood products, in "Viscoelastic Properties of Food," M. A. Rao and J. F. Steffe , eds., Elsevier Sci. Pub. LTD, Barking, Essex, England (1992).

17. S. E. Case, J. A. Knopp, D. D. Hamann, and S. J. Schwartz, Characterization of gelation of konjac mannan using lyotropic salts and rheological measurements, in: "Gums and Stabilizers for the Food Industry Vol. 6," P.A. Williams and D. J. Wedlock, eds., Oxford University Press, New York (1992).

18. D. Oakenfull and A. Scott, Hydrophobic interaction in the gelation of high methoxyl pectin, *J. Food Science*, 49: 1093 (1984).

19. S. E. Case, Rheology and chemistry of konjac mannan, Ph. D. dissertation, North Carolina State University, Raleigh, NC (1992).

20. A. H. Clark, S. B. Ross-Murphy, K. Nishinari, and M. Watase, Shear modulus - concentration relationships for biopolymer gels. Comparison of independent and cooperative crosslink descriptions, in: "Physical Networks," W. Burchard and S. B. Ross-Murphy, eds., Elsevier Sci. Pub. LTD, Barking, Essex, England (1990).

21. E. Niwa, E. Chen, T. Wang, S. Kanoh, and T. Nakayama, Extrordinarity in the temperature-dependence of physical parameters of kamaboko, *Nippon Suisan Gakkaishi*, 54: 1789 (1988).

22. C. Gonzalez., E. A. Foegeding, S. E. Case, and D. D. Hamann (to be published).

23. J. Howe, Surimi gel fracture properties; the influence of test temperature and strain rate, M.S. thesis, North Carolina State University, Raleigh, NC (1991).

CRITICAL BEHAVIOR OF AGAROSE NEAR THE SOL-GEL TRANSITION POINT

Tomoyuki Fujii, Hitoshi Kumagai, and Toshimasa Yano

Department of Agricultural Chemistry, The University of Tokyo
1-1-1 Yayoi, Bunkyo-ku, Tokyo 113
Japan

ABSTRACT

Not only the macroscopic property of the sol-gel transition system, such as viscosity or elastic modulus, but also the microscopic property, such as the molecular weight distribution of clusters or the correlation length, should be measured to learn whether the sol-gel transition can be described by the percolation theory. In this work, dynamic light scattering (DLS) method was carried out to determine the cluster radius of agarose sol. The z-average cluster radius showed a tendency to diverge when the concentration increased and approached the sol-gel transition point. It was possible to fit the data with scaling law, and the sol-gel transition concentration and the critical exponent were determined to be 4.5×10^{-3} wt% and 0.92 at the temperature of 27.0 °C, respectively. When the temperature was lowered toward the sol-gel transition temperature, the cluster radius became larger. The sol-gel transition temperature and the critical exponent were determined to be 32.6 °C and 1.42 for 0.15 wt% agarose sol, respectively. From the above results, scaling law was shown to be effective when observing the physical property changes near the sol-gel transition point.

INTRODUCTION

Recently it has been noted that sol-gel transition is analogous to critical phenomena[1], that is, the percolation theory may be applicable to the sol-gel transition. In this connection, the elastic shear modulus E of gel and the exponent t ($E \propto \varepsilon^{t}$) have been measured near the sol-gel transition point[2-4]. However, not only the macroscopic property of a sol-gel transition system, such as viscosity or elastic modulus, but also the microscopic property, such as the molecular weight distribution of clusters or the correlation length, should be measured to learn whether the sol-gel transition can be described by the percolation theory. In this work, dynamic light scattering (DLS) method was carried out to observe the sol-gel transition behavior. The distribution of the hydrodynamic radii of clusters of agarose, which forms clusters by non-covalent bonds, was investigated. Then the z-average of hydrodynamic radius of the cluster, which behaves similarly to the correlation length, is discussed from the viewpoint of scaling concept.

Food Hydrocolloids: Structures, Properties, and Functions
Edited by K. Nishinari and E. Doi, Plenum Press, New York, 1994

THEORY

Percolation theory can be explained as follows[5,6]: On a lattice model, each site of a lattice is occupied randomly with probability p, independent of its neighbors. Then a cluster is defined as a group of neighboring occupied sites. Fig. 1 shows the percolation networks obtained with this procedure. Thus for small p a collection of finite clusters is present, whereas for p close to $p=1$, one infinite network exists. Therefore, there is a sharp phase transition at some intermediate critical point $p=p_c$, where an infinite cluster starts to appear. In gelation, monomers are thought to occupy the sites of a lattice, and a p implies an initial concentration of a monomer. A polymer corresponds to a cluster, a gel an infinitely large cluster. A sol-gel transition point is defined as a point of $p=p_c$. The scaling law, in which the value of physical property is described by the power form near the critical point, holds near the sol-gel transition point.

The correlation length, ξ, is defined as some average distance of two sites belonging to the same cluster:

$$\xi^2 = \frac{\Sigma_r r^2 g(r)}{\Sigma_r g(r)} \tag{1}$$

where $g(r)$ is the correlation function defined as the probability that a site at distance r from an occupied site in a finite cluster is also occupied and belongs to the same cluster. The correlation length is related to a 'radius of gyration' R_s:

$$\xi^2 = \frac{\Sigma_s R_s^2 s^2 n_s}{\Sigma_s s^2 n_s} \tag{2}$$

The expression of the right side of Eq. (2) is called a z-average of the squared cluster radius. This correlation length diverges at the critical point.

Percolation theory states that the z-average of the cluster radius of gyration behaves similarly to the correlation length ξ, which is described near the critical point as follows:

$$\xi \propto |p-p_c|^{-\nu} \tag{3}$$

where ν is its critical exponent. The critical exponent of the correlation length in percolation theory has been predicted to be about 0.9 by the computer simulation.

Fig. 1. Site percolation on the square lattice, illustrating various clusters. (p : the fraction of filled sites)

MATERIALS AND METHODS

Material

Agarose (Nacalai Tesque, Tokyo) was dissolved in distilled water filtered through a Millipore filter of 0.025 μm pore diameter.

Dynamic light scattering (DLS) method[7]

The DLS method has been used to measure the hydrodynamic radius of colloidal particles. The time dependence of the fluctuations of scattered intensity is caused by the Brownian movement of the particles. Therefore, particle size can be determined by the DLS method from the intensity of the fluctuations that occur over short time intervals. For this method the values of the intensity (photocount) $I(t)$ at some arbitrary time t and at some later time $I(t+\tau)$, where τ is the length of the interval, are required. From these two photocounts, the autocorrelation function, $C(\tau)$, and the normalized autocorrelation function, $g^1(\tau)$, are calculated as follows:

$$C(\tau) = <I(t)*I(t+\tau)> \tag{4}$$

$$g^1(\tau) = \frac{C(\tau) - C(\infty)}{C(0) - C(\infty)} \tag{5}$$

If particles are monodisperse, $g^1(\tau)$ can be written as a single exponential function

$$g^1(\tau) = \exp(-\Gamma\tau) \tag{6}$$

Γ is the relaxation rate, and related to the translational diffusion coefficient D of the particle by

$$\Gamma = D\,Q^2 \tag{7}$$

where Q is the magnitude of the scattering vector,

$$Q = \frac{4\pi n}{\lambda} \sin\frac{\theta}{2} \tag{8}$$

λ being the wavelength of the incident light, n the refractive index of the solution, and θ the scattering angle. D is related to the hydrodynamic radius R_h by the Stokes-Einstein equation:

$$D = \frac{k_B T}{6\pi\eta R_h} \tag{9}$$

where k_B is Boltzmann's constant, T the absolute temperature, and η the solvent viscosity. In general, a sample is polydisperse. Therefore, a correlation function should be written with a distribution function, as follows:

$$g^1(\tau) = \int_0^\infty G(\Gamma)\,\exp(-\Gamma\tau)\,d\Gamma \tag{10}$$

where $G(\Gamma)$ is the normalized distribution function of G. The normalized size distribution function $G(R_h)$ is obtained from $G(\Gamma)$ by using Eqs. (7), (8), and (9). A z-average value of the hydrodynamic radius $<R_h>_z$ is given by

$$\langle R_h \rangle_z = \int_0^\infty R_h \, G(R_h) \, dR_h \qquad (11)$$

The scattered light intensity from a sample was measured with the Malvern 4700 photon correlation spectroscopy system (Malvern, UK), utilizing a 30 mW, 633 nm He-Ne laser (NEC, Tokyo). The scattering angle was fixed to 90°.

RESULTS AND DISCUSSION

Agarose forms clusters by non-covalent bonds. Since the correlation length is expressed in the form of Eq. (2), we should measure a distribution of gyration radii of a polymer and apply scaling theory to its behavior near the sol-gel transition point. However, in order to determine a distribution of gyration radii, we must fractionate a sample in advance and then determine a gyration radius of the each fraction. It is impossible to determine a distribution of gyration radii of an agarose cluster since the non-covalent bonds by which biopolymer forms cluster is so weak as to be destroyed during fractionation. On the other hand, hydrodynamic radii of biopolymer clusters and its distribution can be measured by using DLS method without destroying the clusters. Fig. 2 shows the typical distribution of the hydrodynamic radii of the clusters of agarose by using DLS method. The agarose concentration was 3.0×10^{-3} wt% at the temperature of 27.0 °C. It was possible to measure the cluster radius of agarose and determine its distribution. The distribution curve has a single peak with a gradual decrease toward the larger value of radius. This result indicates that the DLS method is useful for investigating the sol state of biopolymer such as agarose. In this condition the z-average radius was determined to be 116 nm from the distribution curve in Fig. 2.

The z-average of cluster radius showed a tendency to diverge when the concentration increased and approached the sol-gel transition point. Fig. 3 shows the double logarithmic plot of the z-average of hydrodynamic radius of the cluster vs. the deviation in concentration from the sol-gel transition point. The z-average of cluster radius ranged from about 100 nm up to 1000 nm, thus indicating a tendency of agarose clusters to form an infinite cluster as concentration increases. This shows that the value of the cluster radius obtained by using DLS method is important for the sol-gel transition. Furthermore, since the agarose is thought to have a wide molecular weight distribution, the cluster distribution curve indicates the effect not only of clusterization but also of polydispersity of agarose molecular weight. However, the clusters formed were considered to be sufficiently larger than the largest molecule of agarose. In addition, larger particles have more influence over the z-average compared to smaller particles. Therefore, the effect of polydispersity of agarose molecules

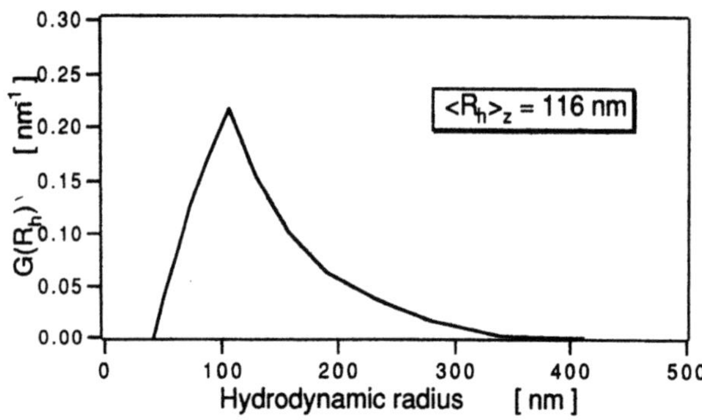

Fig. 2. Typical distribution of the hydrodynamic radii of the clusters of agarose. (C = 3.0×10^{-3} wt%)

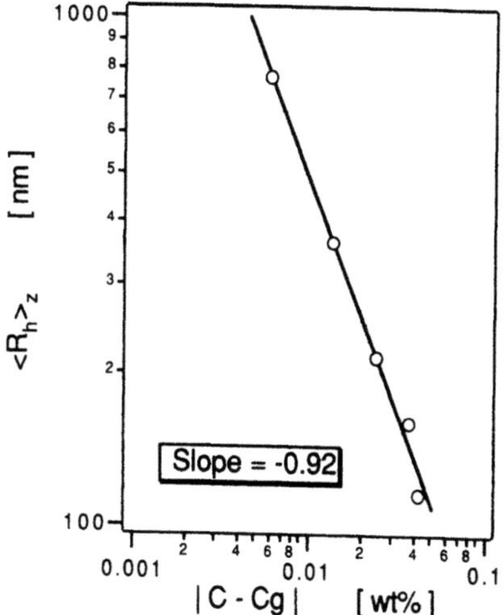

Fig. 3. Double logarithmic plot of the z-average of hydrodynamic radius of the cluster vs. the deviation in concentration from the sol-gel transition point.

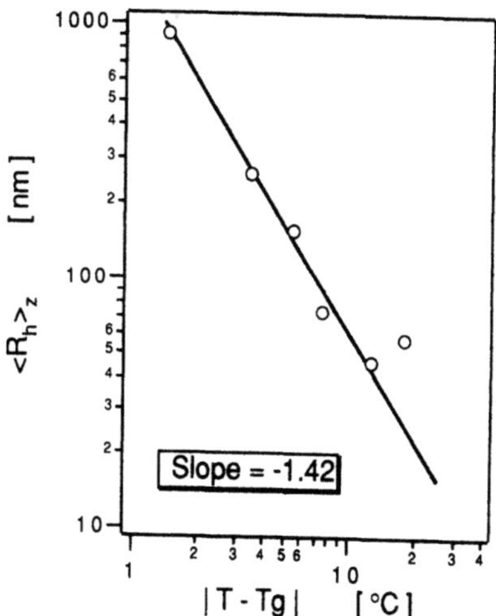

Fig. 4. Double logarithmic plot of the z-average of hydrodynamic radius of the cluster vs. the deviation in temperature from the sol-gel transition point.

on the z-average radius is founf to be negligible. It was possible to fit the data with scaling law:

$$\langle R_h \rangle_z \propto |\, C - Cg\, |^{-\nu_C} \tag{12}$$

and the sol-gel transition concentration and the critical exponent were determined to be 4.5×10^{-2} wt% and 0.92 at the temperature of 27.0 °C, respectively. The value of the critical exponent was found to be close to the theoretical value of that of the correlation length. In this work, the z-average of the hydrodynamic radius, not the gyration radius, was compared with the correlation length. Since the critical exponent of z-average of hydrodynamic radius was close to that of the correlation length, a hydrodynamic radius of the cluster of agarose behaved similarly to the gyration radius. In other words, the z-average hydrodynamic radius of agarose cluster, one of the microscopic properties in sol-gel transition system, was described by the percolation theory.

Fig. 4 shows the double logarithmic plot of the z-average of hydrodynamic radius of the cluster vs. the deviation in temperature from the sol-gel transition point. The z-average of hydrodynamic radius of an agarose cluster did not depend on the temperature at high temperature condition. This result suggested that molecular motion was so vigorous that bonds could not form at high temperature condition. When the temperature was lowered toward the sol-gel transition temperature, the cluster radius became larger. By fitting the data with scaling law:

$$\langle R_h \rangle_z \propto |\, T - Tg\, |^{-\nu_T} \tag{13}$$

the sol-gel transition temperature and the critical exponent were determined to be 32.6 °C, and 1.42, for 0.15 wt% agarose sol, respectively. Although the data was described by scaling law, the critical exponent thus obtained was larger than the theoretical value.

Considering the sol-gel transition of agarose, agarose molecule plays the role of a unit which occupies a site of the lattice for percolation theory. Since the radius of agarose molecule is thought to depend on its stiffness, distribution of cluster radii is also influenced by the stiffness of the polymer. When the temperature is constant, the stiffness is constant even when the concentration is various. On the other hand, when the temperature is not constant, the stiffness of polymer is thought to be different. The result in Fig. 4 suggests that scaling law of the stiffness of polymer chain must be considered.

In conclusion, percolation theory was found to be applicable to the behavior of the z-average of the hydrodynamic cluster radius near the sol-gel transition point, and the concept of 'scaling' was shown to be effective to understand the sol-gel transition on molecular grounds.

Acknowledgements

We wish to thank Prof. Shozo Ishizaka of The University of Tsukuba for the useful discussion.

REFERENCES

1. P.G. de Gennes, "Scaling Concept in Polymer Physics," Cornell University Press, Ithaca, NY (1979).
2. M. Adam, M. Delsanti, and D. Durand, Mechanical measurement in the reaction bath during the polycondensation reaction near the gelation threshold, *Macromolecules*, 18:2285 (1985).
3. M. Tokita, and K. Hikichi, Mechanical studies of sol-gel transition: Universal behavior of elastic modulus, *Phys. Rev.*, A35:4329 (1987).
4. M. Tokita, Gelation mechanism and percolation, *Food Hydrocolloids*, 3:263 (1989).
5. D. Stauffer, A. Coniglio, and M. Adam, Gelation and critical phenomena, *Adv. Polym. Sci.*, 44:103 (1982).
6. D. Stauffer, "Introduction to Percolation Theory," Taylor & Francis, London (1985).
7. B. Chu, "Dynamic Light Scattering," Academic Press, London (1975).

SMALL-ANGLE NEUTRON SCATTERING STUDIES OF PROTEOGLYCAN FROM SHARK FIN CARTILAGE

Yoh Sano

National Food Research Institute, Tsukuba City
Ibaraki 305, Japan

ABSTRACT

The gross structural properties, size and shape of the proteoglycan complex from shark fin cartilage in native state were determined by the small-angle neutron scattering (SANS) method in different D_2O/H_2O ratio. A contrast matching point of the complex was estimated to be 53% D_2O/H_2O. The dependence of radius of gyration on the solvent contrast was positive, indicating that the mucopolysaccharide is on the surface of the core protein. The proteoglycan complex in native state was approximated as the two uniform concentric ellipsoidal shell model having the axial ratio of about 3 in SANS method.

INTRODUCTION

Chondroitin sulfate, one of natural acid mucopolysaccharides, composes the primary part of the cartilage, together with collagenlike protein, and is present as proteoglycan in connective tissues and in biological membranes, giving the charge and the anisotropic structure of biological membranes, and controlling the transport of the drug through the membrane.[1-2] The maintenance of the integrity connective tissue is, in part, dependent on the macromolecular properties of

Food Hydrocolloids: Structures, Properties, and Functions
Edited by K. Nishinari and E. Doi, Plenum Press, New York, 1994

its constituent proteoglycan. Shark fin cartilage proteoglycan complex consists of a protein core with covalently bound
mucopolysaccharide chains of chondroitin sulfate and keratan
sulfate. These mucopolysaccharides are important pharmacolog-
ically as a blood anticoagulant and also as a rejuvenator.
The anticoagulant activity increases with increasing molecular weight and is related to structural features, e.g.,
molecular shape and size.

In the present paper, the gross structural property of
mucopolysaccharide and the neutron scattering length density,
the size and shape of the isolated proteoglycan complex in
the native state were determined by the small-angle neutron
scattering(SANS) method in aqueous solution of different $D_2O/$
H_2O ratio, i.e., 100%, 70%, 20% and 0% D_2O/H_2O, where a unit
of the ratio is mol%.

MATERIALS AND METHOD

The complex of proteoglycan in native state was isolated
from shark fin cartilage by means of cold ethanol fractionation method and gel filtration method. The molecular weight
of the present sample in 10mM phosphate buffer (ionic
strength is 0.02) determined by Zimm plot with laser light
scattering method was 3.7×10^6.

Neutron scattering experiment was performed with SANS-U
(JRR-3M) of the Japan Atomic Energy Research Institute
(JAERI).[3,4] Sample concentration was mainly 10.4 mg/ml and pH
of the solution is kept at 7.2 with using 10mM phosphate
buffer.

RESULTS AND DISCUSSION

SANS MEASUREMENTS OF PROTEOGLYCAN COMPLEX

The scattering function I(Q) as a function of scattering
vector Q ($=4\pi \sin\theta/\lambda$, where 2θ is the scattering angle, λ
is the wavelength) obtained in different D_2O/H_2O ratio are
shown in Fig.1. At this concentration of the proteoglycan
complex (10.4 mg/ml) the scatter of the data was relatively
small and I(Q) monotonously increased with decreasing the
scattering angle at the smallest Q-range.

The example of Guinier plot derived from the above I(Q)
is shown in Fig.2. The Guinier plot in all the cases with
different D_2O/H_2O ratios also showed a good straight line.
The radius of gyration, Rg, and the intensity extrapolated to
Q=0, I(0), were obtain from this straight line using the

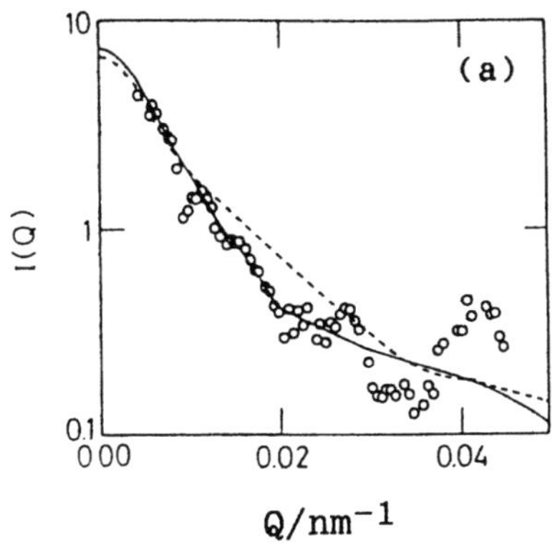
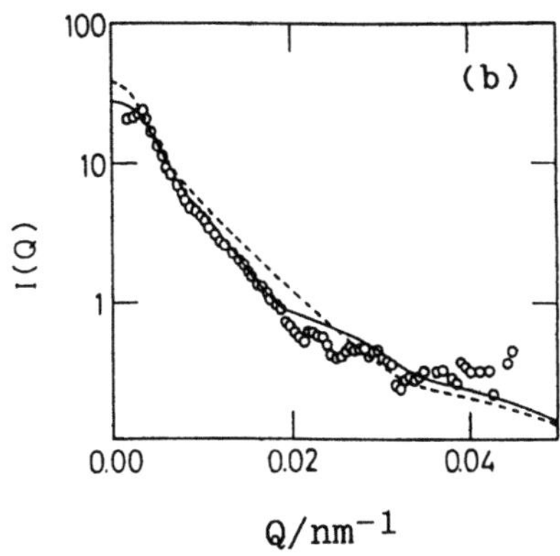

Figure. 1. Scattering functions of proteoglycan complex (10. 4 mg/ml) in aqueous solution at 70%(a) and 0%(b) D_2O/H_2O ratio. The real and dotted lines show the theoretical scattering curves calculated as a model of the two uniform concentric ellipsoidal shells having the axial ratio of 3(real) and 6(dotted line), respectively.

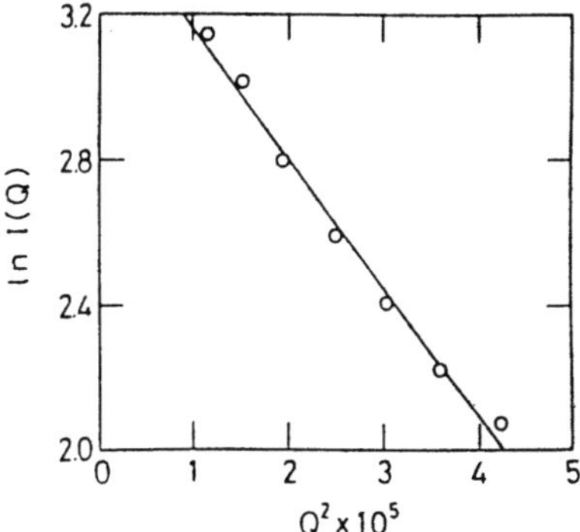

Figure 2. Guinier plot of proteoglycan complex in 0% D_2O/H_2O ratio.

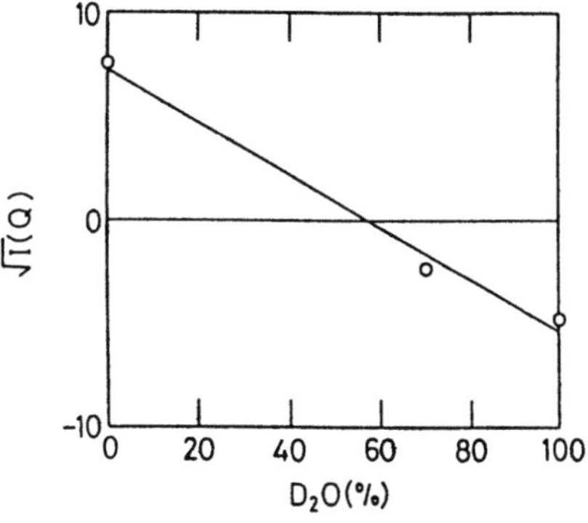

Figure 3. Square root of I(0) vs D_2O/H_2O ratio of proteoglycan complex

least square fitting method. The results of I(0) and R_g were shown in Fig.3 and Fig.4, respectively.

A contrast matching point of the complex, where the overall shape of the proteoglycan complex disappears, was estimated to be 53% D_2O/H_2O by using square root of I(0) of Guinier plot, plotted against D_2O/H_2O ratio as shown in Fig. 3. The mean excess scattering density $\bar{\rho}$, defined by the difference between the mean scattering density of the proteoglycan complex and the scattering density of the solvent, was determined as 3.4×10^{-10} cm^{-2}. The matching point calculated

from the crystal data by assuming full H-D exchange is 55.3%
D_2O/H_2O for ionized chondroitin sulfate and 46.3% D_2O/H_2O for
ionized keratan sulfate. The 53% D_2O/H_2O in the present paper
is, therefore, very much reasonable from the dissociation
constant of carboxyl and sulfate groups in the crystal data.

R_g^2 is correlated with $\bar{p}$ as (so-called Stuhrmann plot[5])

$$R_g^2 = R_{gc}^2 + \alpha(1/\bar{p}) - \beta(1/\bar{p})^2$$

where R_{gc}, α and β are the radius of gyration of the
particle at infinite contrast($\bar{p}$=0), a radial second moment
of the scattering density distribution and a displacement of
the center of the scattering density distribution with respect to the center of the geometrical shape, respectively.
The dependence of Rg on the solvent contrast, as shown in
Fig.4, showed that a radial second moment of the scattering
density distribution was positive, indicating that the mucopolysaccharide such as chondroitin sulfate is on the surface
of the molecule with a higher scattering density than that of
the core protein.

The proteoglycan complex in native state can be approximated as a model of two uniform concentric ellipsoidal shells
as mentioned above. The scattering function calculated by a
spherical shell model gives periodically harmonic scattering
patterns indicating that the model is unsuitable in the
present case. The comparison between the data (open circles
in Fig.1) and the theoretical scattering curves calculated by
changing the various axial ratios, in which the real line
calculated as a model of the two uniform concentric ellipsoidal shells having the axial ratio of 3 agrees quite well with
the experimental data, indicates that the proteoglycan complex in SANS method is approximated as the two uniform concentric ellipsoidal shell model having the axial ratio of
about 3.

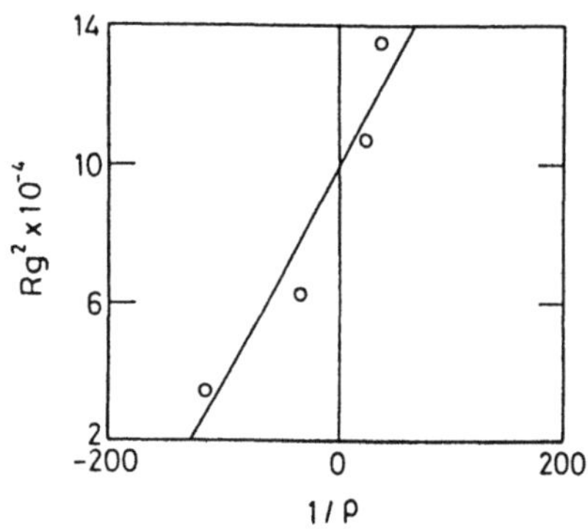

Figure 4. Stuhrmann plot of proteoglycan complex.

The ratio of the geometric radius Rg to the hydrodynamic radius RH is defined as the δ-parameter which is strongly dependent on the molecular architecture. The experimental δ-parameter is about 1.1, indicating the star-like molecule calculated theoretically. The model obtained by laser-light scattering (Zimm plot method) and quasi-elastic light scattering method quite agrees well with that of SANS method.

However, an equivalent spherical Stokes radius from a diffusion coefficient and also the radius of gyration determined by laser-light scattering method were quite larger than the molecular size from SANS method. The polyanionic mucopolysaccharide chains of the concentric binding region are approximated as the semi-flexible rods that extend radially from the protein core into the solvent. Such a structure of the polyanionic mucopolysaccharide would trap much water and hence lead to a Stokes radius much larger than that predicted on the basis of the SANS experiment.

REFERENCES

1. Sano,Y.,1986, Equilibrium constants for the association of chondroitin sulfate with bovine serum albumin, Bull.Natl.Inst.Agrobiol.Resour., 2: 1-12.
2. Sano,Y.,1985, Viscometric method for determination of Michaelis constant of hyaluronidase on chondroitin sulfate, Bull. Natl. Inst. Agrobiol. Resour., 1: 53-62.
3. Sano,Y., Niimura,N., and Tanaka,I.,1992, The 4th International Conference on Biophysics and Synchrotron Radiation, Tsukuba, Abstract p. 355.
4. Sano,Y., Niimura,N., Hiragi,Y., Hirai,Y., and Inoue,H.,1990, KENS Report-VIII, Small angle neutron scattering of TMV particle, 1989/ 1990: 203.
5. Ibel,K., and Stuhrmann,H.B.,1975, Comparison of neutron and x-ray scattering of dilute myoglobin solutions, J.Mol.Biol.,93: 255-265.

PREPARATION OF A POLYELECTROLYTE COMPLEX GEL AND ITS pH-DEPENDENT SWELLING BEHAVIOR

Chia-Hong Chu, Takaharu Sakiyama, Tomoyuki Fujii, and Toshimasa Yano

Department of Agricultural Chemistry
Faculty of Agriculture
The University of Tokyo
1-1-1 Yayoi, Bunkyo-ku, Tokyo 113, Japan

ABSTRACT

A polyelectrolyte complex gel was made by mixing chitosan and xanthan solutions under the presence of NaCl. To study the effect of environmental pH on the swelling behavior, the gel in cylindrical shape was immersed in aqueous NaOH or HCl solution, and the diameter of the gel was measured at appropriate intervals. In the range of pH 10-12, the diameter increased with time, and the gel reached to the swelling equilibrium not more than 7 days. The gel also swelled at pH below 1.5, though it dissolved in the range of pH 0.2-1. No swelling was observed at pH from 2 to 9. The effect of salt concentration on the pH-dependent swelling of the gel was also investigated.

INTRODUCTION

It is known that polyelectrolyte complexes[1] are formed by the electrostatic interaction between oppositely charged polyelectrolytes in aqueous solution. The formation and structure of polyelectrolyte complexes have been studied[2] using various polyelectrolytes, including polysaccharides.[3]

In a polyelectrolyte complex, oppositely charged polyelectrolytes are linked to each other by electrostatic bonds.[1,2] When weak polyelectrolytes are involved in the complex, the number of the electrostatic bonds should be affected by the environmental pH because of the change in degree of dissociation of polyelectrolytes. This may lead to pH dependence of the net charge fixed on the complex. Since the ionization of the functional group plays an important role on the swelling of such polymer gels as polyacrylate gel,[4] the degree of the swelling of the polyelectrolyte complexes may depend on environmental pH. To study the swelling behavior, a bulky polyelectrolyte complex gel is necessary. However, the polyelectrolyte complexes are usually obtained as precipitates since the complex formation is very rapid because of the strong electrostatic attraction between polyanion and polycation. Little work on the preparation of a bulky polyelectrolyte complex gel has been reported.

Food Hydrocolloids: Structures, Properties, and Functions
Edited by K. Nishinari and E. Doi, Plenum Press, New York, 1994

In this study, a bulky polyelectrolyte complex gel in cylindrical shape is prepared from oppositely charged polysaccharides. Xanthan, which has carboxy groups, was chosen as a weak polyanion component, and chitosan, which has amino groups, as a weak polycation component. For the cylindrical complex gel, the effect of environmental pH on the swelling behavior is studied. The swelling experiments are also performed under the presence of NaCl at various pH since the salt concentration may affect the swelling behavior of the complex gel.

MATERIALS AND METHODS

Preparation of Polyelectrolyte Complex Gel

Chitosan and xanthan were obtained from Fuji Shoji Co. Ltd. and Kimitsu Chemical Industry Co. Ltd., respectively. These were used without further purification.

A chitosan solution was prepared by dissolving 0.2g of chitosan in 4.8g of 1% acetic acid, followed by adding 0.3g of NaCl. A xanthan solution was prepared similarly except for the use of distilled water instead of 1% acetic acid. Equal volumes of the chitosan and xanthan solutions were mixed on a boiling water bath. After adding a cylindrical frame (4mm in inner diameter), the mixture was centrifuged to remove air bubbles, and kept at 5°C for one day. Gelation of the mixture occured upon washing with distilled water. A cylindrical gel was obtained by taking the gel out of the frame, and stored in distilled water at 5°C.

Swelling Behavior of Polyelectrolyte Complex Gel

Swelling experiments for the cylindrical complex gel were performed as follows. Under a nitrogen atmosphere, a piece of cylindrical gel of initial diameter Di (about 4mm) was immersed in 150ml of NaOH or HCl solution of a specified concentration at 5°C. The diameter of the gel was measured at adequate time intervals until no change in the diameter was detected. The equilibrium swelling ratio was calculated as the ratio of the equilibrium diameter, De, to Di. To study the effect of salt concentration on the swelling of the gel, NaCl was added to the solution in which the gel was immersed. In all the swelling experiments, the change in solution pH before and after each immersion was not more than 0.2 in pH unit.

RESULTS AND DISCUSSION

The complex gel prepared in this study is shown in Fig.1 (a). Without the addition of NaCl in the gel preparation, aggregation of the polyelectrolytes occured immediately on mixing, and such a bulky gel as shown in Fig.1 was not obtained. This may be because of immediate formation of polyelectrolyte complex since xanthan and chitosan have oppositely charged functional groups, namely carboxy and amino groups. The addition of NaCl made it possible to obtain a homogeneous viscous mixture of polyelectrolytes, which set to gel after washing. The presence of NaCl in the mixture may suppress the binding of polyelectrolytes since the electrostatic force between oppositely charged groups is screened at a relatively high ionic strength.[5] By removing NaCl from the mixture, the electrostatic attraction may be restored, resulting in the formation of a polyelectrolyte complex gel.

Figure 1 (b) shows a swelled complex gel in equilibrium. In all the experiments, it took not more than 7 days until the swelling equilibrium was reached. The equilibrium swelling ratio is shown in Fig.2 as a function of environmental pH. In alkaline solutions, swelling of the gel was observed in the range of pH 10.0-12.0. The maximal equilibrium swelling ratio was 3.6 at pH 10.0. In acid solutions, swelling of the gel was also observed in the range of pH 0-1.5, though the gel dissolved in a couple of days at pH from 0.2 to 1.0.

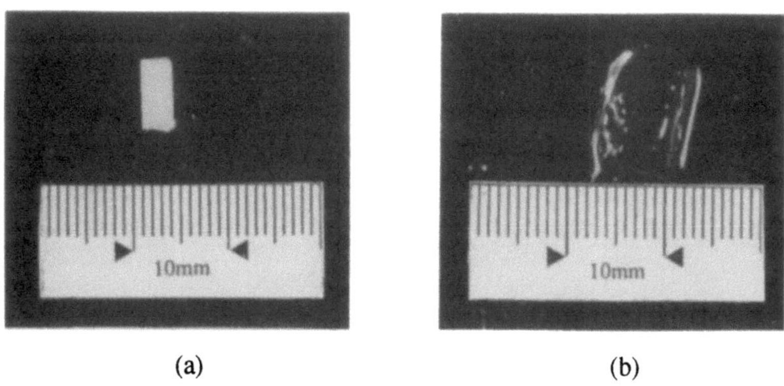

(a)   (b)

Figure 1. Xanthan-chitosan complex gel.(a):before swelling (b):swelled in NaOH solution of pH 10.5.

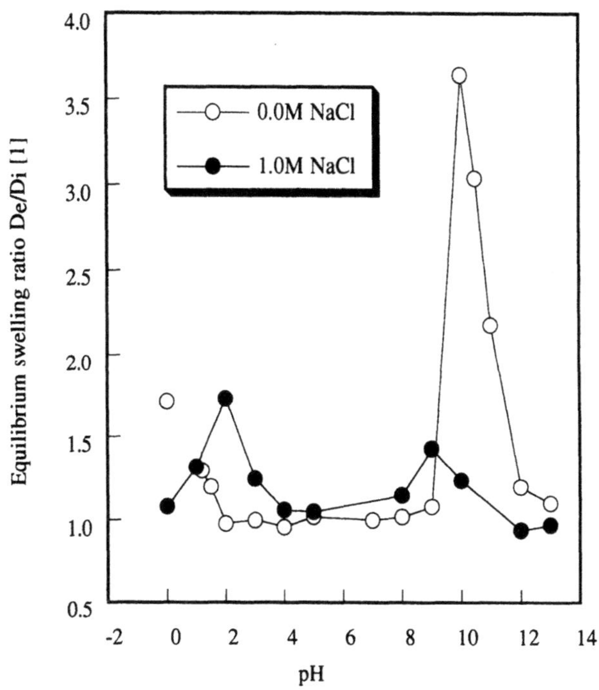

Figure 2. Equilibrium swelling ratios of xanthan-chitosan complex gels in HCl or NaOH solutions.

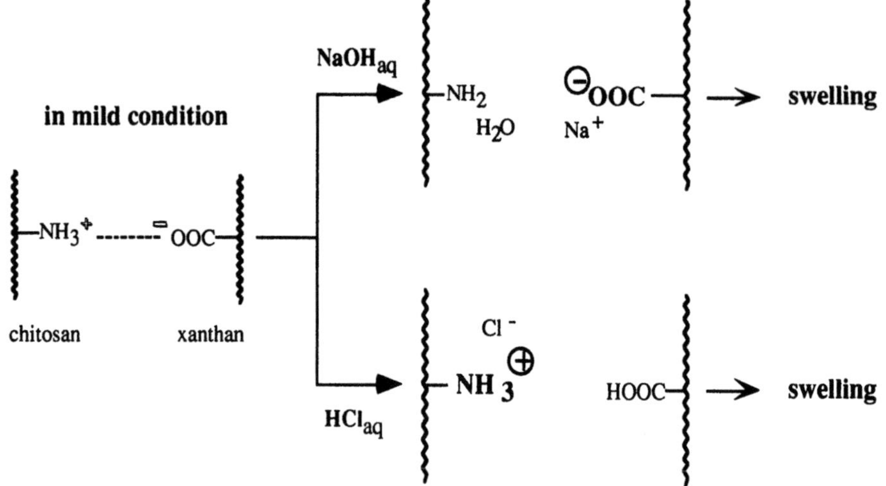

Figure 3. Swelling mechanism of polyelectrolyte complex gels.

The equilibrium swelling ratio under the presence of 1M NaCl is also shown in Fig.2 as a function of environmental pH. By the presence of NaCl, the equilibrium swelling ratio was significantly lowered at pH 10. At pH 1.0, dissolution of the gel were not observed under the presence of NaCl. The pH values of the maximal equilibrium swelling in alkaline and acidic solutions were shifted to neutral side by the presence of NaCl.

The complex gel prepared in this study swelled at specific ranges of pH. A possible mechanism of the swelling is illustrated in Fig.3. In a mild acidic condition, the amino group of chitosan can interact with the carboxy group of xanthan since they have opposite charges. When environmental pH is increased beyond 10.0, the amino group may lose its positive charge though the carboxy group may remain in a dissociated state. This means disappearance of the interaction between these two functional groups and appearance of repulsion between negatively charged carboxy groups, which may be the reason for the swelling of the gel in alkaline solutions. The swelling mechanism of the gel in acidic solutions may be similar to that in alkaline solutions. The low equilibrium swelling ratios under the presence of NaCl may be ascribed to the decrease in repulsion between the charged groups.

REFERENCES

1. R.M. Fuoss and H. Sadek, Mutual interaction of polyelectrolytes, Science., 110:552(1949).
2. B. Philipp, H. Dautzenberg, K.-J. Linow, J. Kötz, and W. Dawydoff, Polyelectrolyte complexes-recent developments and open problems, Prog. Polym. Sci. 14:91(1989).
3. A. Nakajima and K. Shinoda, Complex formation between oppositely charged polysaccharides, J. Colloid Interface Sci. 55:126(1976).
4. T. Tanaka, Gels, Sci. Am., 244:110(1981).
5. K. Abe, H. Ohno, and E. Tsuchida, Phase changes of polyion complex between poly(methacrylic acid) and a polycation carrying charges in the chain backbone, Makromol. Chem. 178:2285(1977).

MECHANICAL PROPERTIES OF FOODS
IN THE VICINITY OF GELATION POINT

Hitoshi Kumagai, Tetsuo Inukai, Tomoyuki Fujii, and Toshimasa Yano

Department of Agricultural Chemistry, The University of Tokyo
Bunkyo-ku, Tokyo 113, Japan

ABSTRACT

The dependences of mechanical properties for gelatin and κ-carrageenan on temperature (T) and concentration (ϕ) were measured in the vicinity of gelation point, and analyzed by the scaling equations derived from the percolation theory. The values of ϕ or T at the gelation point, ϕg or Tg, were estimated by viscosity measurement or dynamic light scattering method. The values of the critical exponent for sol viscosity, s, and those for gel elasticity, t, were evaluated by fitting the experimental data to the scaling equations. The values of s for gelatin were estimated to be about 1, this value of s being larger than that predicted by the percolation theory assuming the superconductor/normal mixture model. The values of t for gelatin and κ-carrageenan were 1.7-2.0, these values of t being similar to that predicted by the percolation theory.

INTRODUCTION

Mechanical properties near gelation point have been studied since de Gennes[1] suggested that the percolation theory was applicable to gelation phenomena. Such works have been mainly carried out on synthetic polymers, not so many works having been done on natural polymers like food gels. Therefore, it is important to confirm that the percolation theory is applicable to food gels with complicated gelation mechanism, not only for controlling the mechanical properties of food gels but also for examining the universality of the percolation theory for describing the gelation phenomena.

In this study, the dependences of mechanical properties for gelatin and κ-carrageenan on temperature and concentration were measured in the vicinity of gelation point, and analyzed by the scaling equations derived from the percolation theory.

THEORETICAL

Sol viscosity, η, and gel elasticity, G, are expressed by the following scaling equations derived from the percolation theory, near gelation point.

$$\eta = C \ |Xg\text{-}X|^{-s} \tag{1}$$

$$G = C' \ |X\text{-}Xg|^{t} \tag{2}$$

Food Hydrocolloids: Structures, Properties, and Functions
Edited by K. Nishinari and E. Doi, Plenum Press, New York, 1994

where C and C' are positive constants, s is the critical exponent for sol viscosity, t is the critical exponent for gel elasticity, X is a parameter like temperature (T) and monomer concentration (ϕ), Xg is X at the gelation point.

EXPERIMENTAL

Measurements of sol viscosity and gel elasticity

Sol viscosity was measured with a cone and plate viscometer (Tokyo Keiki, ELD type, Tokyo). Gel elasticity was measured with a Rheolograph Sol (Toyo Seiki Seisakusho, Tokyo) at a frequency of 2 Hz.

Sample Preparation

Gelatin. Gelatin powder was the alkali pig skin type (Kanto Chemical Co., Inc., Tokyo). A specified amount of the gelatin powder was suspended in distilled water, and swollen for about ten minutes at room temperature. After the sample was dissolved at 60°C, the sample temperature was lowered to 40°C. The sample was aged in the sample holders of the experimental apparatuses for 18hrs at specified temperatures.

κ-Carrageenan. A specified amount of κ-carrageenan (Sigma Chemical, typeI) and KCl were suspended in distilled water, and dissolved at 60°C. Thereafter, the sample temperature was lowered to 40°C. The sample was aged in the sample holders of the experimental apparatuses for 90 min.

Estimation of the parameters, ϕg, s, and t

Viscosity or correlation length of the bonded monomer cluster diverges at gelation point. Therefore, Xg was estimated from the measurement of viscosity or correalation length. The parameter Xg for gelatin was estimated by extrapolating $1/\eta$ to zero. As for κ-carrageenan, Xg was taken as X where the correlation length diverged. The correlation length was measured with the Malvern 4700 photon correlation spectroscopy system (Malvern, UK), utilizing a 30 mW, 633 nm He-Ne laser (NEC, Tokyo).

The critical exponent s was estimated from the slope of the double logarithmic plot of η vs. IXg-XI. The critical exponent t was estimated from the slope of the double logarithmic plot of G vs. IX-Xgl.

RESULTS

Figure 1 shows the double logarithmic plots of the reduced concentration (ϕg - ϕ) vs. η. A linear relationship was observed, the critical exponent, s, being about 1.1 from the slope.

Figure 2 shows the double logarithmic plot of (ϕ- ϕg) vs. G. A linear relationship was observed, the critical exponent, t, being about 2 from the slope.

Table 1 shows the values of the critical exponents s and t of gelatin. The values of s were larger than that calculated from the percolation theory based on the superconductor-normalconductor random mixture model, 0.7-0.8. On the other hand, the values of t were similar to that calculated by the computer simulation, 1.7-2.0.

Table 2 shows the values of the critical exponent t of κ-carrageenan. The values of t were similar to that predicted by the percolation theory.

DISCUSSION

As shown in Table1, the values of s were larger than that calculated from the percolation theory based on the superconductor-normalconductor random mixture model.[1] On the other hand, we investigated the concentration dependence of sol viscosity for

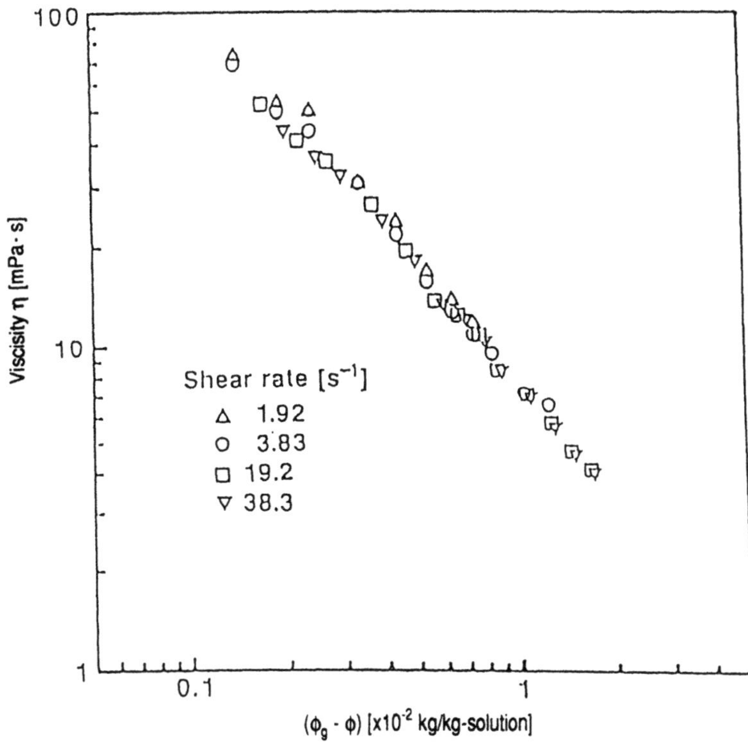

Fig. 1   Double logarithmic plot of η vs. (ϕ_g-ϕ) for gelatin sol. Temperature was 20°C.

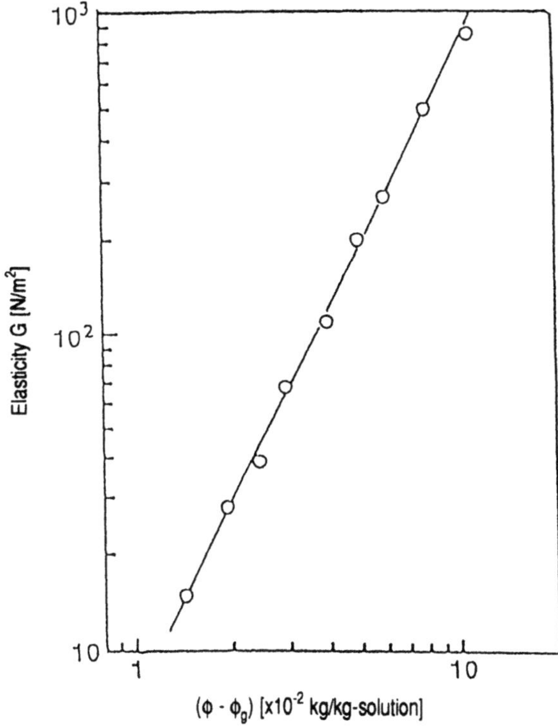

Fig. 2   Double logarithmic plot of G vs. (ϕ-ϕ_g) for gelatin gel. Temperature was 20°C.

polyacrylamide, and estimated the critical exponent s to be about 0.7 [2], this value of s being similar to the calculated value. The reason for this difference would be caused by the following two assumptions on which such calculation was based[3].

(a) The entanglement between adjacent molecules is neglected.
(b) All the effects of hydrodynamic interaction are omitted.

During the gelation of gelatin, coil-helix transition occurs and the formation of helices implies an aggregation of the chain. Therefore, as for gelatin, the two assumptions above, especially (a), would not be satisfied, causing the discrepancy between the calculated and experimental value of s. On the other hand, polyacrylamide is a copolymer, and the bond is relatively simple. Then the above assumptions may be satisfied.

As shown in Tables 1 and 2, all the values of t were similar to that calculated by the computer simulation. This indicates that the elasticity of gelatin and κ-carrageenan are described by the percolation theory.

Table 1. Critical Exponents for Viscosity and Elasticity of Gelatin.

	Parameter X	
	Concentration ϕ	Temperature T
s	1.1 ± 0.1	0.89 ± 0.09
t	2.05 ± 0.06	1.9 ± 0.1

Table 2. Critical Exponents for Elasticity of κ-Carrageenan.

	Parameter X	
	Concentration ϕ	Temperature T
t	1.74 ± 0.07	1.8 ± 0.3

CONCLUSION

1) The values of the critical exponent for sol viscosity s of gelatin were larger than that calculated from the percolation theory based on the superconductor-normalconductor random mixture model.

2) The values of the critical exponent for gel elasticity t of getatin and κ-carrageenan were similar to that predicted by the percolation theory.

ACKNOWLEDGEMENT

The authors express their thanks to Prof. A. Shimada, Assoc. Prof. K. Hatae and Dr. K. Sasaki in Ochanomizu University for their help in Rheolograph Sol measurement.

REFERENCES

1. P.G. de Gennes, "Scaling Concepts in Polymer Physics," Cornell University Press, Ithaca and London, 1979.
2. T. Yano, H. Kumagai, T. Fujii and T. Inukai, The concentrationn dependence of mechanical properties of polyacrylamide near sol - gel transition point, *Biosci, Biotech, Biochem.*, in press.
3. P.G. de Gennes, Viscosité - viscosité près d'une transition so-gel, *C.R. Acad. Sc. Paris*, B286: 131 (1978).

INTERACTION BETWEEN POLYELECTROLYTE INHIBITORS OF CALCIUM PHOSPHATE FORMATION AND CALCIUM ION

Kazutaka Yamamoto, Hitoshi Kumagai, Takaharu Sakiyama, Hiroyuki Ogawa, and Toshimasa Yano

Department of Agricultural Chemistry, The University of Tokyo, Bunkyo-ku, Tokyo 113, Japan

ABSTRACT

The interaction between the inhibitors of calcium phosphate formation and calcium ion was investigated. Firstly, the induction time, when the transition of amorphous calcium phosphate to crystalline is known to occur, was measured in base titration experiments as an index of the inhibitory activity against calcium phosphate formation. Polyelectrolyte inhibitors delayed the induction time. The double logarithmic plots of the induction time vs. polyelectrolyte concentration were sigmoidal, and there existed a concentration threshold for the inhibitory activity of each inhibitor. Secondly, calcium ion activity was measured in the presence of the inhibitors, using ion meter connected to calcium ion selective electrode. The calcium ion activity gradually decreased near the concentration threshold for the inhibitory activity.

INTRODUCTION

The formation of calcium phosphate often causes problems in food industry. Calcium phosphate precipitate on the plate of heat exchangers reduces heat transfer and lowers efficiency. During ultrafiltration of milk and whey, calcium phosphate precipitation leads to clogging of the pores of the membrane, resulting in a lower flux and a reduced efficiency. Moreover, the formation of calcium phosphate is also important from a physiological point of view, e.g. biological calcification of bones and teeth, and some inhibitors of calcium phosphate formation are known to affect the biological calcification. However, it is not clearly understood how the inhibitor behaves in the inhibition of calcium phosphate formation. To understand the behavior of the inhibitor, we evaluated the inhibitory activity against calcium phosphate formation and also measured the calcium ion activity in the presence of inhibitors to investigate the interaction between the inhibitors and calcium ion.

Food Hydrocolloids: Structures, Properties, and Functions
Edited by K. Nishinari and E. Doi, Plenum Press, New York, 1994

MATERIALS AND METHODS

Materials

Poly-L-glutamate (hereafter, PLGA) of a weight-average molecular weight 13,300 (Sigma Chemical Co.) and alginate (G1-L type, Kimitsu Chem. Ind., Tokyo) were used as polyelectrolyte inhibitors and citrate (Kanto Chemicals Co. Ltd., Tokyo), as a low molecular weight inhibitor.

Inhibitory activity against the calcium phosphate formation

The inhibitory activity was evaluated using the induction time as an index, obtained from the base titration curve. Alginate, PLGA or citrate were dissolved into 95.0ml of water at desired concentrations in the reaction vessel at $25.0 \pm 0.5°C$. After adding 2.5ml of 120mM KH_2PO_4 , the solution pH was adjusted to 6.00 with 1N KOH or 1N HCl. Then, 2.5ml of 200mM $CaCl_2$ was gradually added to the solution. The pH value was kept at 5.9 $\pm$ 0.1 with 1N KOH during the addition of $CaCl_2$, and reached 6.00 at the end of the addition. Thereafter, solution pH was adjusted to 7.40 with 1N KOH, in order to induce a

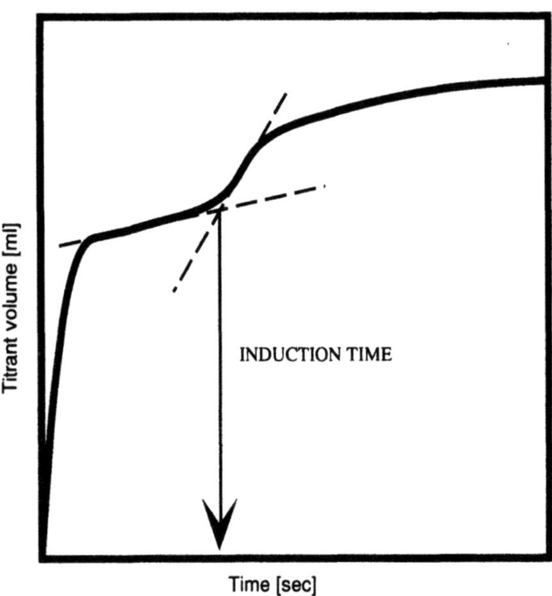

Fig. 1   Schematic diagram to evaluate the induction time from a base titration curve.

spontaneous precipitation. The mixture was maintained at the pH 7.40 throughout the reaction using a pH-stat, and the amount of 1N KOH added was continuously recorded. During the experiment, nitrogen gas was bubbled into the solution to exclude CO_2. All solutions were prepared with deionized and distilled water. All the chemicals were of reagent grade.

The induction time, when the transition of amorphous calcium phosphate to crystalline is known to occur,[1] was estimated by the method of Meyer and Eanes[2] from a base titration curve, as illustrated in Fig. 1.

Calcium ion activity

Effect of the inhibitors on the calcium ion activity in the solution was investigated to use the ion meter connected to calcium ion selective electrode and reference electrode (double junction type). Solutions of 1.8mM calcium, containing inhibitors at desired concentrations, were prepared with 0.1M Tris-HCl (pH 7.4) buffer solution. All solutions were prepared with deionized and distilled water. The solution was stirred by magnetic stirrer and kept at $25.0 \pm 0.1°C$ throughout the experiment. All the chemicals were of reagent grade.

RESULTS AND DISCUSSION

Figure 2 shows the effect of polyelectrolyte inhibitors concentration on the induction time. The concentrations of polyelectrolyte inhibitors were expressed as monomer concentration. Figure 2 also shows the effect of citrate, a well-known inhibitor of low molecular weight, on the induction time. The double logarithmic plots of polyelectrolyte concentration vs. the induction time were sigmoidal. There existed a concentration threshold for the inhibitory activity against calcium phosphate formation. On the other hand, as the citrate concentration increased, the induction time increased, the plot being concave upwards. These results suggest that the inhibitory mechanism of polyelectrolyte inhibitor is different from that of low molecular weight inhibitor.

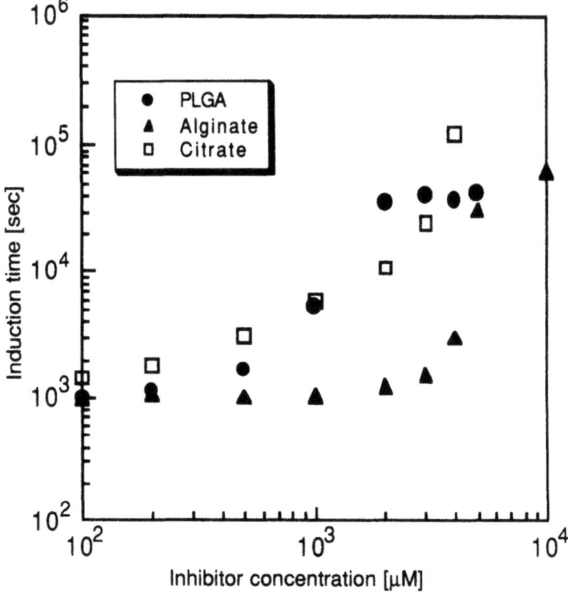

Fig. 2   Effect of inhibitor concentration on the induction time when the transition of amorphous calcium phosphate to crystalline hydroxyapatite.

Figure 3 shows the effect of inhibitor concentrations on the calcium ion activity detected by ion meter connected to calcium ion selective electrode. The concentrations of polyelectrolyte inhibitors were expressed as monomer concentration. As shown in Fig. 3, calcium ion activity decreased as each inhibitor was added to the system. Citrate, alginate, PLGA reduced the calcium ion activity in that order. In the case of PLGA, the calcium ion activity was effectively reduced above the concentration which is electrostatically equivalent to the total calcium concentration in the system. These results indicate that the inhibitors interact with calcium ions.

As shown in Figs. 2 and 3, the calcium ion activity decreased in the same order of concentration as the concentration range tested in the evaluation of the inhibitory activity. In the case of PLGA, the calcium ion activity effectively decreased near the concentration threshold of the inhibitory activity. It was suggested that the interaction between the inhibitors and calcium ion plays an important role on the inhibition of calcium phosphate formation.

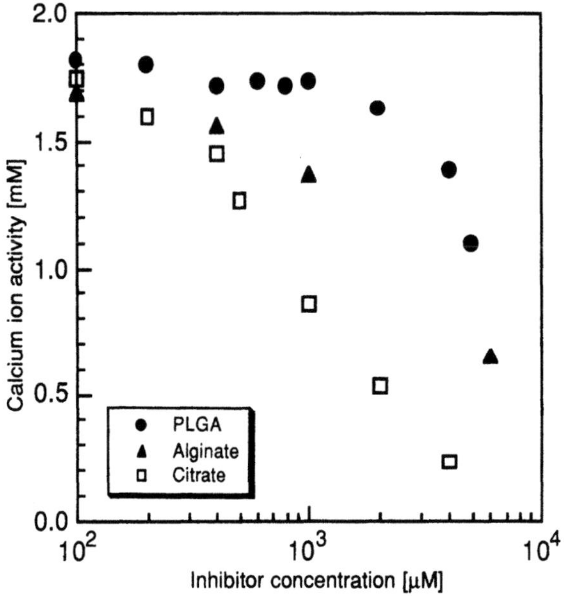

Fig. 3   Effect of inhibitor concentration on the calcium ion activity

CONCLUSIONS

It was suggested that the inhibitory mechanism of polyelectrolyte inhibitor is different from that of low molecular weight inhibitor.

Alginate, PLGA, and citrate reduced the calcium ion activity in the same concentration range as tested in the evaluation of inhibitory activity against the formation of calcium phosphate. The result suggests that there exists the interaction between the inhibitors and calcium ions and that the interaction plays an important role on the inhibition of calcium phosphate formation.

REFERENCES

1 A. L. Boskey and A. S. Posner, Conversion of amorphous calcium phosphate to microcrystalline hydroxyapatite. A pH-dependent, solution mediated, solid-solid conversion, *J. Phys. Chem.* 77: 2313 (1973).
2. J. L. Meyer and E. D. Eanes, A thermodynamic analysis of the amorphous to crystalline calcium phosphate transformation, *Calcif. Tissue Res.* 25: 59 (1978).

THE SOL-GEL TRANSITION OF FOOD MACROMOLECULES UNDER HIGH PRESSURE

Kunihiko Gekko

Department of Food Science and Technology, Faculty of Agriculture
Nagoya University, Nagoya 464-01, Japan

ABSTRACT

The stability of thermoreversible gels of some food macromolecules were studied under high pressures up to 300 MPa. The melting temperature of gelatin and agarose gels increased with increasing pressure, but carrageenan gels were oppositely destabilized by pressure as well as the heat-setting gels of soy protein and ovalbmin. The volume changes due to sol-gel transition were estimated by thermodynamic analyses and discussed in terms of the characteristic hydration modes of cross-linking junctions in relation to the mechanism of gelation.

INTRODUCTION

Various types of experiments have been carried out to elucidate the mechanism of gelation of polysaccharides and proteins.[1] However, only limited data have been reported on the pressure effects or volume changes on the gel systems,[2] probably due to the technical difficulties. Since the sol-gel transition is always accompanied by the modification of hydration around the polymer molecules, a high pressure study should be a diagnostic test for the structural characteristics of the cross-linking junctions such as the stabilizing force, size and lateral aggregation, which have not been definitely understood. Such data would be also useful for utilization of high pressure in biotechnology and food processing.[3] From these viewpoints, we examined the pressure effects on the sol-gel trasition of some food macromolecules and discussed the mechanism of gelation in terms of the polymer-water interactions.

Food Hydrocolloids: Structures, Properties, and Functions
Edited by K. Nishinari and E. Doi, Plenum Press, New York, 1994

METHODOLOGY

The aqueous solutions of carrageenans, agarose, gelatin, soy protein and ovalbumin were cooled or heated to form gels in a small test-tube containing a stainless steel ball. The test-tube was sealed with a Parafilm and set bottom up in a light path of the high pressure vessel with sapphire windows (Kobe Steel Ltd, Kobe). After the system was equilibrated at the working temperature, the pressure of the sample was elevated at a rate of 10 MPa/min. In some cases (agarose and gelatin), the temperature was increased at a constant pressure. The pressure or temperature at which the transmitted light was intercepted by the ball dropping down was defined as the melting pressure or melting temperature of the gel. Thus, the stability of gels was determined as a function of temperature and pressure.

For the reversible sol-gel transition, the thermodynamic approach would be essentially available although this transition is not a real thermodynamic equilibrium as shown by hysteresis. According to the Eldridge and Ferry's equation,[4] we can estimate the enthalpy of cross-links formation, ΔH, from the polymer concentration dependence of melting temperature of gels,

$$\log C = \Delta H/2.303RT + \text{constant} \tag{1}$$

where C is the polymer concentration in g/l, R the gas constant, and T the absolute temperature of gel melting. If the sol-gel transition is assumed to be a phase transition, the volume change, ΔV, and entropy change, ΔS, of gelation could be evaluated by using the Clausius-Clapeyron equation

$$(dT/dP)_m = \Delta V/\Delta S = T\Delta V/\Delta H \tag{2}$$

Here, it should be noted that the thermodynamic parameters thus obtained refer to the quantities per mol of cross-linking junction.

RESULTS AND DISCUSSION

Carrageenan

Fig.1 shows the typical sol-gel phase diagram of κ-carrageenan at various polymer concentrations.[5] The similar diagram was also observed for ι-carrageenan. Evidently, the melting temperature of the gels decreases linearly with increasing pressure, indicating the destabilization of the gels under high pressure. The slopes of the linear lines, $(dT/dP)_m$, were almost independent of concentrations of polymer and KCl added, and the averaged values were -0.057 and -0.040 K/MPa for κ- and ι-carrageenans, respectively. As shown in the inset of Fig.1, there were linear relationships between the logarithmic polymer concentration and the reciprocal absolute temperature of gel melting. Thus, the equation (1) holds at any pressure as well as atmospheric pressure. The ΔH values estimated from the slopes of the linear lines are listed in Table 1 together with the corresponding entropy change, ΔS, and volume change, ΔV, which were calculated at atmospheric pressure and 25 °C.

The negative values of ΔH and ΔS indicate that the stabilization of the gel is predominantly brought about by a large exothermic effect overcoming the loss of conformational entropy due to the polymer-polymer hydrogen bonds, which may be formed between the OH groups at O(2) and O(6) positions of different D-galactose residues. Such an

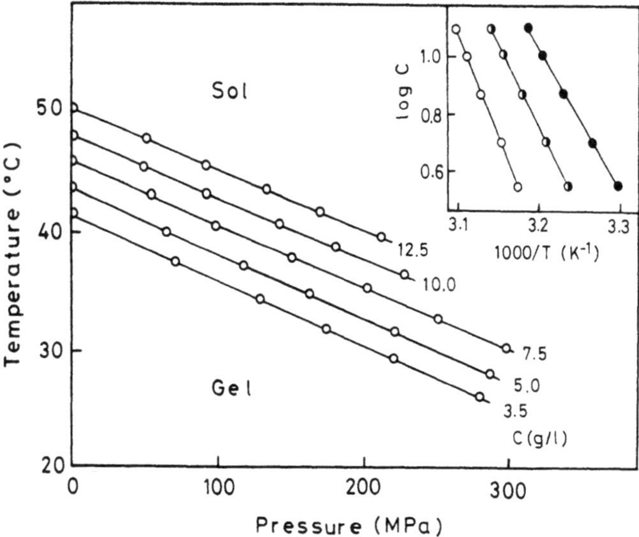

Fig. 1. Typical sol-gel phase diagram of κ-carrageenan in 0.025 M KCl at various carrageenan concentrations. The inset in the figure shows the typical plots of the logarithmic carrageenan concentration as a function of the reciprocal absolute temperature of gel melting at 0.1 MPa ($\bigcirc$), 100 MPa ($\oct 0.1$), and 200 MPa ($\bullet$). Taken from Gekko and Kasuya.[5]

intermolecular hydrogen bond formation should release the water of hydration from the OH groups of galactose residues, resulting in the increase in volume of the system since water of hydration is considerably contracted comparing with free water.[6] Assuming the double-helix model for the cross-linking junctions, the ΔV values observed correspond to the volume change of 2.5-4.9 and 1.7-3.4 ml/mol of unit cross-link (two disaccharide residues) for κ- and ι-carrageenans, respectively. These values are close to the volume change due to dehydration, 2.7 ml/mol of OH groups of carrageenans, which was determined by the density measurements.[7] These findings suggest that a double-helix structure of carrageenan gels may persist in solution as well as in the solid state, and that the pressure-induced destabilization of carrageenan gels could be essentially explained in terms of the positive volume change due to dehydration upon the coil-double helix transition of polymer molecules.

Agarose

The melting temperature of agarose gel increased linearly with increasing pressure opposite to carrageenan gels.[8] The slope, $(dT/dP)_m$, averaged for agarose concentrations is listed in Table 1 together with the thermodynamic parameters, ΔH, ΔS, and ΔV. The ΔH and ΔS values are comparable with those for carrageenan gels, but the ΔV value is negative for agarose. Since the cross-links of agarose gel are also considered to consist of the intermolecular double helix , the negative ΔV value cannot be explained by such a simple dehydration model as assumed for carrageenans. There might be some specific hydration modes in the cross-linking junctions of agarose gel as proposed below for gelatin gel.

Table 1. Thermodynamic parameters for sol-gel transition of food macromolecules in water at 1 atm and 25 °C.

Sample	$(dT/dP)_m$ (K/100MPa)	ΔH (kJ/mol)	ΔS (J/K·mol)	ΔV (ml/mol)
κ-carrageenan[5]	-5.7			
0.015 M KCl		-71	-238	13.5
0.025 M KCl		-136	-456	23.7
0.035 M KCl		-185	-619	36.3
ι-carrageenan[5]	-4.0			
0.03 M KCl		-59	-197	8.5
0.05 M KCl		-76	-255	9.9
0.10 M KCl		-38	-126	5.3
Agarose[8]	2.3	-34	-115	-2.6
Gelatin[9]				
Bloom No. 60	3.9	-196	-660	-25.7
Bloom No. 225	3.2	-195	-656	-20.8
Bloom No. 304	2.9	-187	-627	-18.3
Ovalbumin (pH 7)	-12.1	-0.94	-3.5	0.37
Soy Protein (pH 7.8)	-17.7	-1.7	-5.5	0.98

ΔH, ΔS, and ΔV are expressed in units per mol of cross-links.

Gelatin

As shown in Fig. 2a, the melting temperature of gelatin gels increased linearly with increasing pressure,[9] indicating the pressure-induced stabilization of the gels. The slope appeared to be independent of gelatin concentration while it decreased with increasing Bloom No. of gelatin samples (Table 1). The $(dT/dP)_m$ values obtained, 0.03-0.04 K/MPa, are close to the pressure dependence of denaturation temperature of native collagen.[10] This coincidence supports the general conclusion that the cross-linking junctions of gelatin gels consist of the intermolecular hydrogen bonds between polypeptide chains similar to the triple helix of native collagen. The negative ΔH and ΔS values are consistent with such hydrogen-bonding mechanism for cross-linking junctions but ΔV is negative although the amount of hydration of collagen increases by about 10% on melting of triple helix. However, it is probable that during gelation of gelatin, some water molecules are bound within the cross-linking junctions since the triple helix of native collagen is known to be stabilized by the intermolecular bridging water molecules, amounting to 0.15 g/g of collagen, one-third of the total amount of hydration.[11] Since their movement would be largely restricted, such a bridging hydration could probably produce a large volume contraction enough to compensate for the positive volume change due to a decrease in total amount of hydration upon gelation. Assuming the volume contraction of -2.9 ml/mol of water for such bridging hydration, which is comparable with the electrostriction of water around ionic groups, we can tentatively predict the volume change observed for the sol-gel trasition of gelatin. Thus, it is highly possible that the

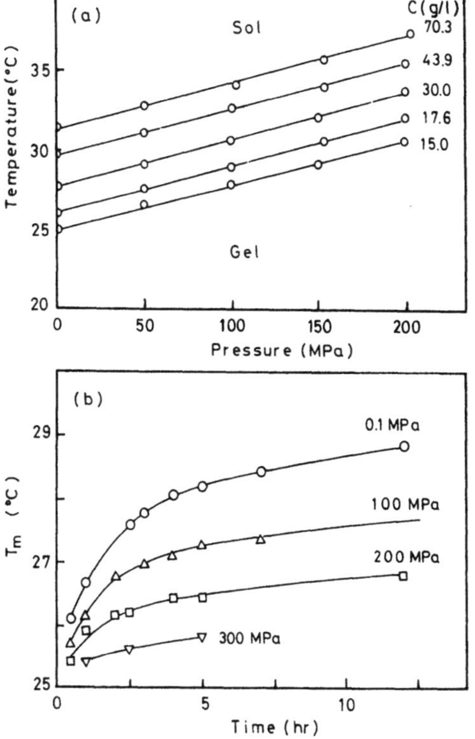

Fig. 2. (a) Typical sol-gel phase diagram of gelatin (Bloom No. 304) at various gelatin concentrations. (b) Melting temperature of 7% gelatin gel (Bloom No. 60) as a function of gel setting time under various pressures. Taken from Gekko and Fukamizu.[9]

pressure-induced stabilization of gelatin gels could be mainly attributed to these specific hydration structures of the cross-linking junctions. In this respect, it is noteworthy that DNA double helix, which involves the bridging water molecules, is stabilized by pressure.[12] Similar bridging hydration may be also expected for agarose gels since they are stabilized under high pressure.

Fig. 2b shows the melting temperature of gelatin gel as a function of gel-setting time under various pressures at 20 °C. Although gelatin gels were stabilized by pressure, the kinetic process of gelation was suppressed under high pressure, indicating the positive activation volume of gelation. This would be possible if random association of polypeptide chains occurs to release some water of hydration before the interchain triple helix is formed to involve the bridging water molecules. The resulting large activation volume for gel to sol transition would be responsible for stabilizing the gel state under high pressure.

Soy Protein and Ovalbumin

The thermoreversible gels of soy protein and ovalbumin were prepared by heat denaturation and 0 °C setting. The melting temperature of these gels decreased linearly with increasing pressure (Gekko et al., to be published). As shown in Table 1, the absolute values of $(dT/dP)_m$ are considerably high for both globular proteins as compared with other fibrous polymers as shown above. The ΔH and ΔS values are negative and ΔV is positive, but their absolute values are very small. These results suggest that different from the gels of fibrous polymers, the cross-linking junctions of globular protein gels may be very small as regarded as an entanglement of protein molecules polymerized by heat denaturation.

As demonstrated above, the sol-gel transition of food macromolecules is sensitively affected by pressure. It is interesting that the stability of these gels shows the opposite response to pressure nevertheless their cross-linking junctions commonly melt when heated. Although the origin of the volume changes observed should be analyzed in more detail on a molecular level, these pressure effects could be essentially explained in terms of the different hydration modes of polymers at the sol and gel states. As an example of irreversible and enzyme-treated gels, we also examined the pressure effects on milk curdling.[13,14] The milk curdling was accelerated under pressures up to 20-40 MPa, but it was depressed at higher pressures and the curd formation was completely inhibited at the pressures above 130 MPa. This result might be explained in terms of a compensation of pressure effects on the two processes, the aggregation of casein micelles (positive activation volume) and the growing up or synerethis of the micelle aggregates (negative activation volume). Further advanced high pressure studies on the sol-gel transition of biological macromolecules would promise to give an insight into the mechanism of gelation and utilization of high pressure in biotechnology and food processing.

REFERENCES

1. J. R. Mitchell and D. A. Ledward, "Functional Properties of Food Macromolecules," Elsevier Applied Science Pub., London (1986).

2. K. Suzuki, Y. Taniguchi, and T. Enomoto, The effect of pressure on the sol-gel transitions of macromolecules, *Bull. Chem. Soc. Japan* 45:336 (1972).

3. C. Balny, R. Hayashi, K. Heremans, and P. Masson, "High Pressure and Biotechnology," John Libbey Eurotext, Paris (1992).

4. J. E. Eldridge and J. D. Ferry, Studies of the cross-linking process in gelatin gels--(III) dependence of melting point on concentration and molecular weight, *J. Phys. Chem.* 58:992 (1954).

5. K. Gekko and K. Kasuya, Effect of pressure on the sol-gel transition of carrageenans, *Int. J. Biol. Macromol.* 7:299 (1985).

6. K. Gekko and Y. Hasegawa, Compressibility-structure relationship of globular proteins, *Biochemistry* 25:6563 (1986).

7. K. Gekko, H. Mugishima, and S. Koga, Compressibility, densimetric and calorimetric studies on hydration of carrageenans in the random form, *Int. J. Biol. Macromol.* 7:57 (1985).

8. K. Gekko and M. Fukamizu, Effect of pressure on the sol-gel transition of agarose, *Agric. Biol. Chem.* 55:2427 (1991).

9. K. Gekko and M. Fukamizu, Effect of pressure on the sol-gel transition of gelatin, *Int. J. Biol. Macromol.* 13:295 (1991).

10. K. Gekko and S. Koga, The effect of pressure on thermal stability and in vitro fibril formation of collagen, *Agric. Biol. Chem.* 47:1027 (1983).

11. C. A. J. Hoeve and P. C. Lue, The structure of water absorbed in collagen--(I) dielectric properties, *Biopolymers* 13:1661 (1974).

12. S. A. Hawley and R. M. Macleod, Pressure-temperature stability of DNA in neutral salt solutions, *Biopolymers* 13:1417 (1974).

13. K. Ohmiya, K. Fukami, S. Shimizu, and K. Gekko, Milk curdling by rennet under high pressure, *J. Food Sci.* 52:84 (1987).

14. K. Ohmiya, T. Kajino, S. Shimizu, and K. Gekko, Effect of pressure on the association states of enzyme-treated caseins, *Agric. Biol. Chem.* 53:1 (1989).

UTILIZATION OF HIGH HYDROSTATIC PRESSURE
TO MAKE ALGINATE GELS

Toshiaki Shioya, Ryogo Hirano, Atsumi Tobitani

Technical Research Institute, Snow Brand Milk Products Co., Ltd.
1-1-2, Minamidai, Kawagoe, Saitama 350, Japan

ABSTRACT

Alginate gel formation under high hydrostatic pressure was studied. Sodium alginate was mixed with various calcium salts and water was added to make slurry. An hydrostatic pressure of 900 MPa was applied. Calcium salts with different solubility were used, i.e. calcium chloride, calcium citrate, calcium lactate, calcium gluconate, and calcium carbonate.

The texture of resultant gels varied with the calcium salts and their concentration. Both soluble and insoluble calcium salts could be used to make the gels. Calcium chloride provided an opaque, homogeneous, and firm gel. Calcium citrate provided transparent, homogeneous, and soft gel. Gel containing 15% alginate was also prepared, the concentration of this gel was 4 times higher than that of conventional alginate gels. These results suggest that insoluble calcium salts dissociate and that the hydration of sodium alginate is accelerated under high hydrostatic pressure.

It was found that high hydrostatic pressure treatment could be applied to make alginate gels. The gel can be utilized for calcium-enriched gel foods with unique textures and also as durable carriers for immobilization of microorganisms.

INTRODUCTION

High hydrostatic pressure has recently become the subject of renewed interest in the food industry[1]. When high pressure is applied to foods, volume-increasing reactions will tend to be inhibited by pressure, while reactions leading to a decrease in volume will tend to be promoted. The disruption of hydrophobic and ion-pair bonds, and the formation of hydrogen bonds are considered to occur under such high hydrostatic pressure. Such non-covalent interaction causes the aggregation or gel formation of proteins and polysaccharides. The effect of pressure on the sol-gel transition have been reported[2,3].

Alginate is well known to form gels in the presence of polyvalent ions such as calcium. The gel can be prepared by several ways[4] such as diffusion setting, internal setting or bulk setting, and dropping method. The dropping method is used to make alginate capsules and to immobilize living cells, in which a calcium solution is dropped into an alginate solution through a nozzle. Calcium chloride and calcium lactate are commonly used to make alginate gels. We intend to make alginate gel by using both soluble and insoluble calcium salts under high hydrostatic pressure.

Food Hydrocolloids: Structures, Properties, and Functions
Edited by K. Nishinari and E. Doi, Plenum Press, New York, 1994

MATERIALS AND METHODS

Sodium alginate (350G, M/G ratio=0.2) was purchased from Kibun Food Chemifa (Japan). Analytical grade calcium salts, i.e. calcium chloride, calcium citrate, calcium lactate, calcium gluconate, and calcium carbonate, were used.

A given amount of sodium alginate and calcium salt was mixed, then water was added to make slurry. The slurry was packed into a polyethylene bag. Then a hydrostatic pressure ranging from 100 MPa to 990 MPa was applied to the bag by the high pressure apparatus (MCT -150,Mitsubishi Heavy Ind.) at room temperature. The pressure was maintained for 30 minutes, then released.

The dynamic viscoelasticity of pressure-induced gel samples was measured by a viscoelastic meter (VEM-800D,Yokohama System,Japan). A gel sample measuring 1x1x1cm was cut out and set on the parallel plate of the viscoelastic meter. The sample was covered with liquid paraffin to prevent evaporation of water during measurement. The frequency dependence from 1 to 100 Hz was measured in compression mode at 25 °C.

RESULTS AND DISCUSSION

1. Alginate Gels Prepared with Various Kinds of Calcium Salts

Alginate gel formation with calcium salts, it can be assumed that two carboxyl groups of alginate react with one calcium ion in stoichiometry[4]. When one mole of alginate reacts with 0.5 mole calcium, it can be defined as a 100% calcium equivalent concentration. Various calcium salts were mixed with sodium alginate (12%) as a 100% calcium equivalent concentration, with water added to make slurry. Then, a hydrostatic pressure of 900 MPa was applied for 30 minutes at room temperature. The appearance of resultant alginate gels

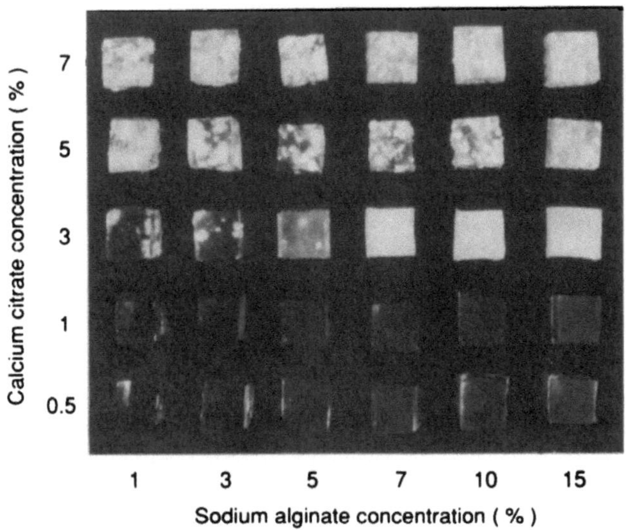

Fig.1 Pressure induced-alginate gels prepared with various concentration of sodium alginate and calcium citrate. The gels were formed under a hydrostatic pressure of 900 MPa for 30 min.

266

were as follows:

[calcium carbonate] : white,cloudy,very soft gel that dissolves in hot water.
[calcium gluconate] : transparent, high elastic gel.
[calcium lactate]   : stiff, transparent gel with many white spots.
[calcium chloride]   : very stiff, cloudy gel.

These results suggested that alginate dissolved and simultaneously calcium salts dissociated under such high hydrostatic pressure. The carboxyl groups of alginate may reacted with calcium ions during depressurizing process, that is, a transparent and homogenous gel may have been formed.

Alginate gels with various textures ranging from soft to hard can be prepared by choosing appropriate calcium salts. The texture differences in the gels may be caused by the solubility of calcium salts and the dissociation state under high hydrostatic pressure.

2. Alginate Gels with Calcium Citrate

Alginate gels were prepared with various concentrations of calcium citrate. A hydrostatic pressure of 900 MPa was applied to make gels. The appearance of the gels is shown in Fig.1. Transparent, homogeneous alginate gels were obtained in the range of 1% to 10%

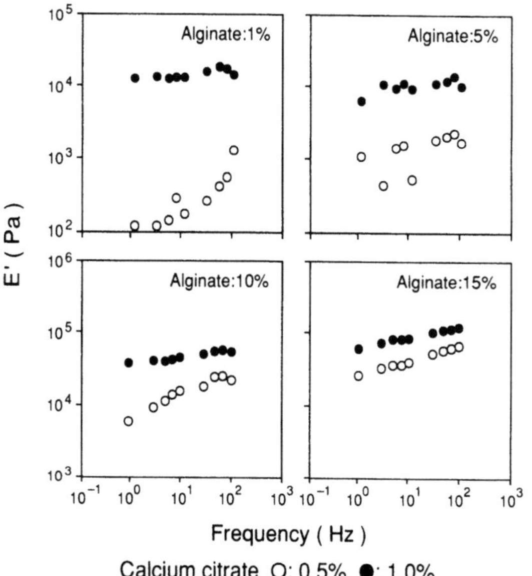

Fig.2 Frequency dependency of dynamic elasiticity (E'). Sodium alginate concentration was in the range of 1% to 15%. The gels were prepared with 0.5% and 1% calcium citrate.

alginate and 0.5% to 1.0% calcium citrate. The other gels had white spots scattered in the transparent gel phase or were cloudy.

The calcium equivalent concentration was calculated to estimate why the gel became white or cloudy. The calculated calcium equivalent concentration was in the range of 2.3% to 486%. A gel [15%alginate+3%calcium citrate] had a 13.9% calcium equivalent concentration.

The calcium concentration may not indicate that excess amount of calcium was added, however the gel became white. It can be assumed that water in the mixture of alginate and calcium salt was utilized to dissolve the alginate and to dissociate the calcium salt. The dissociation of calcium salts under hydrostatic pressure might require some amount of water. So the water amount per calcium salt was calculated. The amount of water needed for dissolving alginate was about 1g per 1g alginate from DSC measurements. The amount of available water, taking into account the amount of water for alginate, was calculated. The white gel started to form [1% alginate + 3% calcium citrate] (calcium equivalent concentration : 208%, available water for dissociation of calcium : 32g/g), and [3% alginate + 3% calcium citrate] (calcium equivalent concentration : 69.4%, available water for dissociation of calcium : 30g/g). It may be concluded that the former gel became white due to an excess of calcium, and that the latter gel became white due to insufficient water. The amount of water needed for the dissociation of calcium citrate was estimated at about 32g/g under a hydrostatic pressure of 900 MPa.

3. Dynamic Viscoelasticity of Alginate Gels

Fig.2 shows the frequency dependency of dynamic elasticity(E') of various kinds of pressure-induced alginate gels. E' increased with alginate and calcium concentration. Alginate gels prepared with 1% calcium citrate showed higher E' than the gels prepared with 0.5% calcium citrate over the applied frequency range. E' formed a nearly horizontal line over the frequency range in [1% alginate + 1%calcium citrate] gel. On the other hand, frequency dependency was observed in [1% alginate+0.5% calcium citrate] gel and in higher alginate gel such as [15% alginate + 1% calcium citrate], which showed an increase in E' with frequency increases.

The frequency dependency of these gels may be related to the gel structure, that is, the number of close link points, the strength of these close link points and the entanglement of alginate molecules. An entanglement network system shows the frequency dependence, while a gel shows no frequency dependency [5]. The different frequency dependency of alginate gels may correspond to the calcium equivalent concentration, which affects the number of close link points and these strength of the close link points. Alginate gels with higher calcium equivalent concentrations give a higher density of close link points and show no frequency dependency.

CONCLUSION

High hydrostatic pressure treatment was found to be applied to make alginate gels using both soluble and insoluble calcium salts without preparing solutions. This pressure-induced gel can be made with much higher concentrations of alginate and calcium salts. It was found that gels are stable in 10% sodium chloride solution for 3 days. The gels can be utilized for calcium-enriched gel foods with unique textures and also as durable carriers for the immobilization of microorganisms and enzymes.

REFERENCES

1.D.Farr, High pressure technology in the food industry, Trend in Food Sci. & Tech., 1:14(1990)
2.K.Suzuki,Y.Taniguchi,T.Emoto, The effect of pressure on the sol-gel transition of macromol, Bulletin of chemical society of Japan, 45:336(1972)
3.K.Gekko,K.Kasuya, Effect of pressure on the sol-gel transition of carrageenans, Int.J.Biol.Macromol.,7:299(1985)
4.W.J.Sime, Alginates, "Food Gels", P. Harris ed., Elsevier Applied Sci., London, (1990)
5.A.H.Clark,S.B.Ross-Murphy, Structure and mechanical properties of biopolymer gels, Advance in Polymer Sci.,83:85(1987)

COAGULATION BEHAVIOR OF EGG PC VESICLES WITH DIFFERENT PARTICLE SIZES

Kunio Furusawa[1], Ken-ichi Watanabe[1], and Hideo Matsumura[2]

[1]Dept. of Chemistry, Univ. of Tsukuba, Tsukuba 305, Japan
[2]Electrotechnical Laboratory, AIST, MITI, Tsukuba 305, Japan

INTRODUCTION

The importance of the interaction between amphoteric phospholipid bilayer membranes has been pointed out by many authors to understand the membrane stability against coagulation and membrane fusion.[1-3] Long-range interaction force, felt during the initial mutual approach of two membranes, is relatively well understood theoretically and experimentally. However, short-range forces occurring near contact between membranes have not realized fully, which come from specific nature of chemical groups. It has been realized that the membrane made of hygroscopic phospholipid experiences a strong repulsive force at close contact. The force is called "hydration force" or "steric force", which originates from the work of removing water molecules from the membrane surface or is caused by the local thermal motion of the membrane.[4] Below 2-3 nm separation, the hydration force dominates over Van der Waals attractive force and poses a significant energetic barrier to membrane contact. As the result, colloid particles consisted of those phospholipid membranes can be stably dispersed in electrolyte solution.

In this study, coagulation behavior of egg phosphatidylcholine (PC), -phosphatidylethanolamine (PE), and -phosphatidylglycerol (PG) vesicles has been studied to clarify the differences in short-range interaction between various lipid membranes and the nature of coagulation of PC vesicles.

EXPERIMENTAL

Phospholipid Vesicles; Egg PC, PE, and PG were purchased from Sigma Chemical Co. Ltd. and used for vesicle preparation without further purification. Vesicles with different sizes (d=80-500 nm) were prepared by extrusion method using a polycarbonate filter having respective pore size. Large size vesicles with d=1-2 μm were prepared by the Vortex mixing method with filtration to exclude lager sizes and dialysis to exclude smaller ones.

Coagulation behavior; Coagulation of the vesicles was detected by the change in light transmittance (at 500 nm wave length) of the vesicle dispersion using usual spectrophotometer and/or by the dynamic light scattering measurement (Otsuka elect. ELS-800) after adding various concentrated $CaCl_2$ aqueous solutions.

Electrophoresis; The electrophoretic mobility of large vesicles were measured in various salt solutions by a micro-electrophoretic apparatus

Food Hydrocolloids: Structures, Properties, and Functions
Edited by K. Nishinari and E. Doi, Plenum Press, New York, 1994

(Rank Brothers MK-2) using a rectangular glass cell. Zeta potentials
were calculated by Smoluchowski-equation.

RESULTS AND DISCUSSION

Generally speaking, colloid particles like silica, latexes, and
lipid particles (vesicles or liposomes) are stably dispersed in aqueous
solution by electrostatic repulsive force when it dominates over
attractive Van der Waals force. To reduce the electrostatic repulsive
force, surface charges on particle are neutralized or screened by adding
metal salts to the aqueous solution. When sufficient small repulsive
potential barrier is realized, these colloid particles begin to aggregate
each other and usually make large flocs. To examine the colloidal
stability of phospholipid vesicles, zeta potential measurements by which
we can estimate the extent of electrostatic potential, and light
transmittance measurements/dynamic light scattering measurements which
give us an information about the sizes of the phospholipid vesicles, are
conducted. The data of these experiments are shown, consecutively.

Figure 1 shows the zeta potentials of vesicles made of each kind of
phospholipid bathed in various concentration of $CaCl_2$ aqueous solution.
Egg PC vesicles have zero zeta potential around 1×10^{-3} M $CaCl_2$, but PE
and PG vesicles still have net negative charges judging from the zeta
potential. So, the electrostatic repulsive force between PC vesicles can
be diminished around this $CaCl_2$ concentration. Different from PC
vesicles, however, PE and PG vesicles will be affected by electrostatic
repulsion when they encounter each other in the same $CaCl_2$ concentration
range.

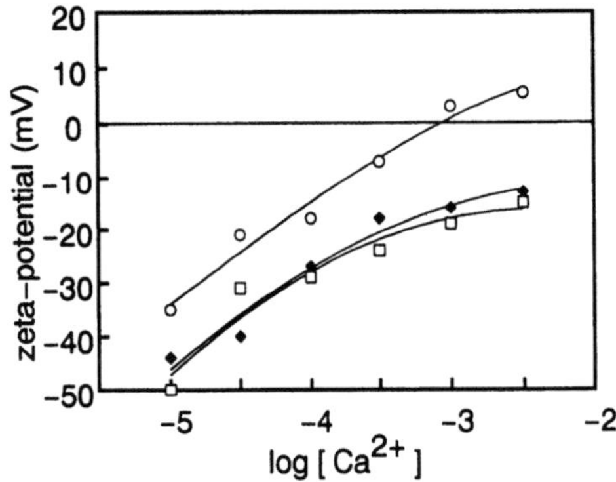

Figure 1. Zeta potentials of Egg-PC (O), -PE (□), and -PG (◆) vesicles bathed in various concentration
of $CaCl_2$ aqueous solution.

Figure 2 shows the transmittance change for three kinds of phospholipid vesicles (size; diameter d=100-200 nm) bathed in 1×10^{-3} M $CaCl_2$
aqueous solution. The transmittances of PE and PG vesicle dispersions
decrease with time, which indicates that the coagulation is induced by
the strong attractive Van der Waals force over electrostatic repulsion.
This means that the primary potential barrier of total potential is not
so high around that $CaCl_2$ concentration. On the other hand, PC vesicle
does not cause any aggregation, which means that rather strong repulsive
force acts on encounter of two PC vesicles despite zero zeta potential.

270

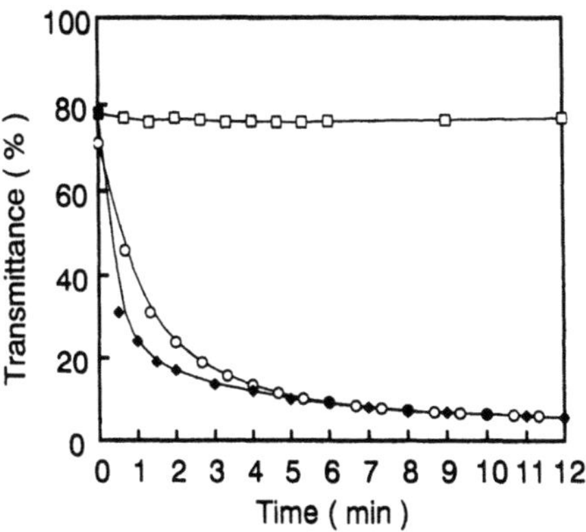

Figure 2. Change in light transmittance vs. elapsed time: PC (□), PE (◆), and PG (○) -vesicles bathed in 1×10^{-3} M $CaCl_2$ aqueous solution.

This repulsive force is due to so-called "hydration force" and originates from the hygroscopic nature of choline residue of PC molecules.

Figure 3 shows transmittance change for the vesicle dispersions for various sizes (d=100, 200, 350 nm, and 2μm) bathed in 1×10^{-3} M $CaCl_2$ aqueous solution. These data indicates that the coagulation can occur when the diameter of vesicles exceeds some value. This size-effect on coagulation behavior suggests that the coagulation may be caused not by primary minimum but by the presence of secondary minimum of inter-vesicle potential curve. This speculation can be verified by the peptizing experiments of flocs.

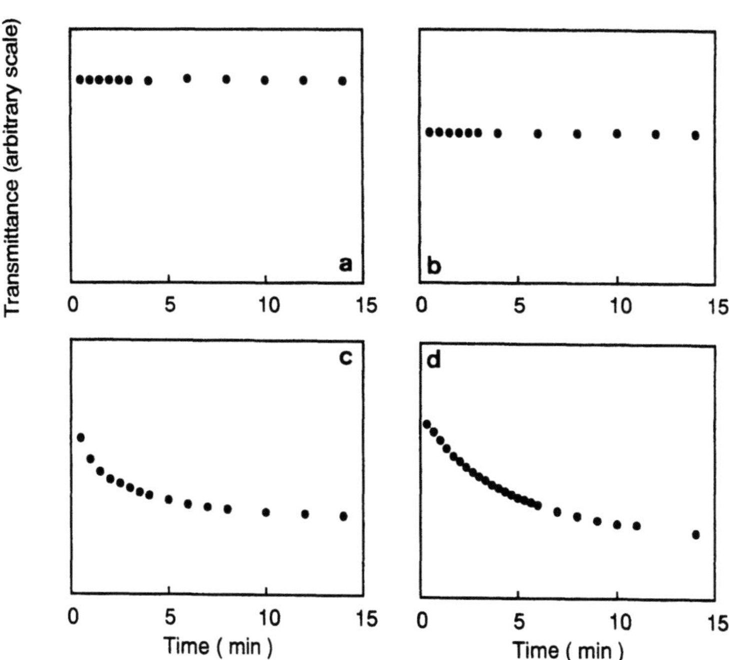

Figure 3. Effect of vesicle size on light transmittance: a: d=100nm, b: d=200 nm, c: d=350 nm, and d: d=2 μm.

Figure 4 shows the change in hydrodynamic diameter of vesicles with the change in the concentration of Ca^{2+} ions. Here, the relative value of diameter against initial value of diameter of vesicles is plotted as a function of elapsed time. Dilution of Ca^{2+} ion from $2x10^{-4}$ M to $4x10^{-5}$ M brings the change in the relative diameter of vesicles from 1.7 to 1. This means that the aggregated vesicles can be re-dispersed by the dilution of the salt concentration in the solution, that is, PC vesicles form a loose and reversibly aggregated flocs.

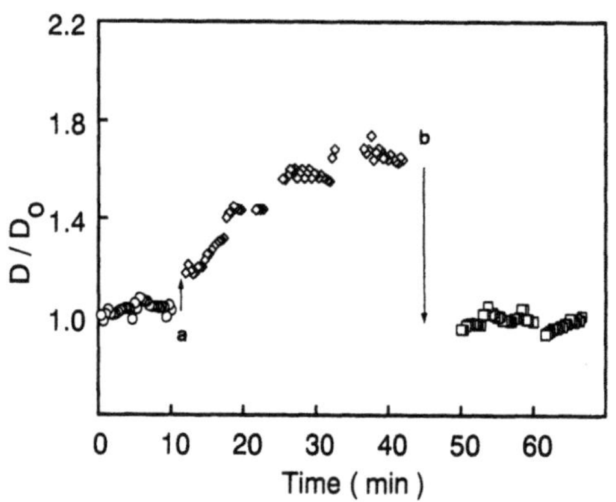

Figure 4. Change in relative diameter of aggregated bodies by the change in salt condition vs. time. D_0 is diameter of original vesicles.
(a) $CaCl_2$: from 0 to $2x10^{-4}$ M, (b) $CaCl_2$: from $2x10^{-4}$ to $4x10^{-5}$.

All these results indicate that the hydration force prevents PC vesicles from coagulation into primary minimum of interaction potential curves. However, when the vesicles have large diameter, a shallow potential minimum is generated rather apart from the surface by the combination of repulsive hydration force and attractive Van der Waals force, which can cause a reversible coagulation.

REFERENCES

1. S.Nir and J.Bentz, On the forces between Phospholipid Bilayers, J. Colloid Interface Sci., 65: 399 (1978).
2. L.J.Lis, M.McAlister, N. Fuller, R.P.Rand, and V.A. Parsegian, Interactions between Neutral Phospholipid Bilayer Membranes, Biophysical J., 37: 657 (1982).
3. B.L. Gamon, J.W. Virden and J.C. Berg, The Aggregation Kinetics of an Electrostatically Stabilized Dipalmitoyl Phosphatidylcholine Vesicle System, J. Colloid Interface Sci., 132: 125 (1989).
4. J.N. Israelachvili, "Intermolecular and Surface Forces," Academic Press, New York (1992).

FUNCTIONALITY OF DIFFERENT PROTEIN PRODUCTS IN MEAT EMULSIONS WITH REDUCED NaCl CONTENT

Gerald Muschiolik

WIP Research Group
Functionality and Application of Proteins
University of Potsdam
D-1505 Bergholz-Rehbrücke

ABSTRACT. Meat emulsions (ME) were prepared using different sodium chloride content and with different whey and plant protein products (soya, faba bean). Rehydrated protein products (PP) were added to ME. The influence of protein functionality parameters (solubility, emulsification, gelation, water absorption) on ME properties (consistency of raw and cooked emulsion, cook loss) was calculated by rank correlation and regression analysis for a range of solubilities of the protein products (PP). Rank correlation shows only a low dependence. Regression analysis gives more information on interactions if performed for highly or moderately soluble PP. Tests of other applications demonstrate that the kind of preparation of ME is of importance for improving the techno-functionality. ME with reduced NaCl content shows better fat binding (emulsification) if liquid meat fat is emulsified with PP solution and then mixed with meat. For improving the water and fat binding of ME with reduced NaCl content it is important to take into consideration the functionality mechanisms of added proteins (surface activity, gelation, swelling) in order to select a suitable application technology (preparing pre-emulsion, adding protein powder or solutions to ME).

INTRODUCTION

The influence of different protein products (PP) on meat emulsions (ME) has been well documented.[1,2] The main aims of protein applications in such cases are to reduce cook loss, fat content and to improve comminuted meat texture. Our main intention besides investigating the above mentioned characteristics is to substitute the loss of salt soluble meat proteins in NaCl reduced ME with non-meat proteins of good emulsifiying and gelation properties. Comparable investigations although with different NaCl contents are familiar from previous studies.[2-4]

The objectives of this research are to study: (i) the influence of salt content on the functionality parameters of PP, (ii) the influence of PP on the quality of ME with reduced NaCl content, (iii) the connections between protein parameters and the parameters of meat products and (iv) the influence of the kind of application on ME properties.

Food Hydrocolloids: Structures, Properties, and Functions
Edited by K. Nishinari and E. Doi, Plenum Press, New York, 1994

MATERIAL AND METHODS

Protein Preparates

Faba bean protein isolates (FBPI 1, FBPI 2, protein 77% and 78% resp.) were prepared from flour by extraction at pH 7.0, precipitating the extracted protein at pH 4.2, neutralization to pH 7.0 followed by spray drying (FBPI 1).[5] FBPI 2 was acetylated with acetic anhydride (degree of acetylation 97%, in terms of % blocked epsilon amino-groups in relation to the original).

Whey protein concentrates (WPC 1 to WPC 3, protein 74.3 to 78.0%) were obtained from Danmark Protein A/S, Soy protein isolate (SPI, protein 83%) was obtained from Protein Technologies Int. and soy protein concentrates (SPC 1 to 5, protein 61.1 to 67.0%) were donated by Central Soya Overseas.

Characterization of Protein Preparates

Nitrogen solubility index (NSI, 2,5% w/v PP solution, pH 6.0, centrifugation at 5800g for 10min), emulsifying activity (EA, 2.5% w/v PP solution, 50 ml sunflower oil, 47.4% water, centrifugation at 1000g for 10min), emulsion stability (ES, heated EA emulsion, 80°C 15min, centrifugation see EA), gelling or penetration (PU, 14% w/v PP dispersion, 90°C, 15 min, penetration with cone-shaped probe attachement), water absorption (WA, capillary suction method[6]) were characterized with different NaCl content (1.3 and 2.0% w/v).

Preparation of Meat Emulsions

The frozen meat and fat were thawed to 0°C, the model systems were prepared by chopping and emulsifying with the kitchen mixer `multiboy LZ 2000` (Elbtalwerk,Heidenau, Germany). The temperature of the meat emulsions was not higher than +8°C.

ME was prepared in different ways according to the form in which PP was added. Preparation A (PP solution to ME) was used for runs with statistical analyses. Preparations B to D were used for testing the influence of PP application on ME properties (B, PP powder to ME; C, pre-emulsion of PP solution and warmed meat fat to homogenized meat; D, cold fat dispersed in PP solution and added to meat before homogenisation).

Characterization of Non-heated Meat Emulsions

The adhesion (ADH, stickiness) and the cohesion (COH, stretching of batters) were measured using a stainless steel circular plate of 30 mm diameter.[7] ADH represents the maximum measured tension force and COH was expressed as the maximum extension reached by the stretched sample under tension between dish and plate before separation.

Characterization of Heated Meat Emulsions

Cook loss of fat (CLF) and cook loss of water (CLW) were determined by packing 30 g of comminuted meat into open end glass tubes of 17 mm diameter. Duplicate tubes were maintained at an internal temperature of 80°C for 15 min. These were than cooled to 25°C, the liquid phase separated by centrifugation at 200g for 3 min and finally dried at 104°C for 12 hrs so as to estimate fat and water content.[8]

Firmness (N) of heated emulsion (prepared as for CLF) was estimated after 12 hrs storage at +6°C by pressing the sample through a 0.5 mm slot. Penetration analysis (PUP) was performed using a Penetrometer (see PU).

Statistical Analyses

Spearman rank correlation and multiple regression analysis with reduction of variables were preformed.[9]

The influence of proteins on ME properties was calculated for functionality parameters with the same NaCl content (see Table 3).

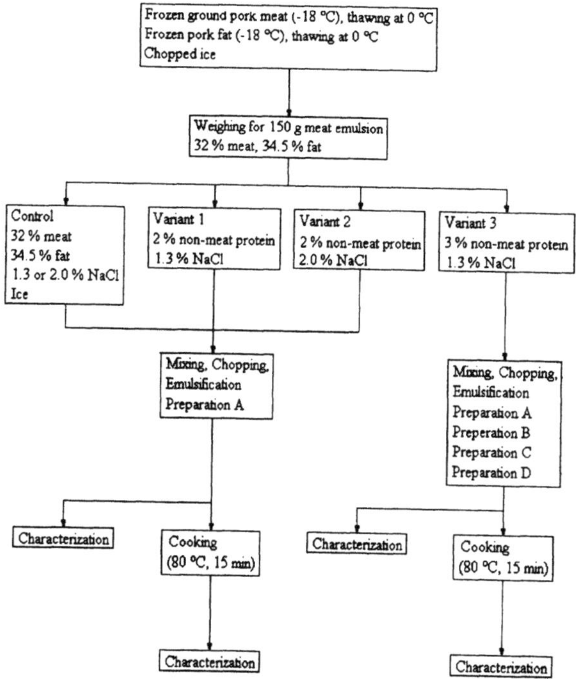

Fig. 1 Flow chart and variants of frankfurter model meat emulsion preparations.

RESULTS

The functional properties of PP depending on salt content are summarised in Table 1. It is seen that very different properties are observed depending on the NaCl content.

The rank correlations as summarized in Table 2 demonstrate high dependence of the statistic or the actual thing that depends on solubility. No correlation is found for calculations with preparates having NSI over 44% (correlation factors below 0.50).

From regression analysis the effects of protein functionality parameters on the ME properties are summarized in Table 3.

It is shown that no clear interactions are found for all PP and ME properties. These results depend on the type of applied proteins (protein solubility) and are partly non-significant (P> 0.05).

PP of varying NSI values influence the ME in different ways. So for example the cohesion or stretching of emulsions increases (P>0.05) with EA (for NSI>44%) or decreases (P<0.05) with EA (for NSI<44%). A contrary effect of ES on cohesion and PU with cook loss of water (P>0.05) is also observed.

PP with NSI over 44% show an increase of the cohesion of unheated ME with rising WA, EA and pH and an increase in the firmness of heated ME with WA, pH and low gel penetration. Under conditions of high PU (P>0.05), we observe an expected cook loss of water whereas cook loss of fat is increased with rising pH, WA and reduced with rising ES.

Under conditions of protein application with low NSI values, the adhesion is reduced significantly with falling ES.

The method of producing the pre-emulsions clearly influences certain batter properties e.g. the decrease of cooking loss (preparation C, Table 4). This is also the case when preparing pre-emulsions, with liquid fat leading to lower CLF or use of dry powder (SPC 5) showing a reduced CLW. Consistency (N and PUP) of emulsion D (mixing meat fat and protein solution) shows the lowest firmness of all preparations A to C.

Table 1. Functional properties of the protein preparates depending on NaCl content.

NSI, %	Protein functionality	Meat emulsion protein	coefficient
8 to 97	NSI	ADH	-0.65
	WA	ADH	0.57
	ES	ADH	-0.53
	WA	PUP	-0.50
<44	EA	COH	-0.61
	EA	CLW	0.52
	ES	CLW	0.57
>44	Coefficients lower than		0.50

Table 2. Rank correlations between functionality parameters of protein preparates and properties of meat emulsions with different NaCl content.

No.	Protein-sample	NSI, %		EA, %		ES, %		PU, 0.1 mm		WA, ml/g sample	
		1.3	2.0	1.3	2.0	1.3	2.0	1.3	2.0	1.3	2.0
1	FBPI 1	57	70	71	69	72	71	250*	250*	1.9	2.0
2	FBPI 2	78	77	84	90	90	92	33	25	3.2	2.9
3	WPC 1	89	94	70	69	83	84	13	10	2.8	2.6
4	WPC 2	91	91	65	69	79	79	31	40	2.1	1.8
5	WPC 3	95	96	72	69	87	91	11	13	2.4	2.2
6	SPI	8	9	62	60	70	70	31	24	5.6	5.1
7	SPC 1	11	8	57	59	57	59	250*	250*	4.1	3.8
8	SPC 2	17	16	58	57	59	57	250*	250*	5.8	5.5
9	SPC 3	10	11	57	58	59	61	66	74	5.6	5.5
10	SPC 4	18	18	58	59	57	57	92	69	5.3	5.1
11	SPC 5	17	16	60	61	60	59	53	63	4.7	4.8

DISCUSSION

The results show a difference in correlation and regression depending on the solubility range of non-meat proteins. One can be distinguish here between soluble high surface active and less soluble high water binding (swelling) PP, both these parameters are influenced by salt but rather not as expected or as known for salt soluble myofibrillar proteins (NaCl increases water binding[10]).

Thus to select a suitable applicable protein it is desirable to classify proteins in solubility ranges (highly or moderately soluble) and to make the final choice with reference to it's functionality.

Table 3. Influence of protein functionality parameters on the properties of meat emulsions (results of regression analysis, ***P<0.05, *P>0.05, - decreasing effect, + increasing effect).

Properties of meat batters	NSI of protein preparates					
	8 to 97 %		> 44		< 44	
	Parameters	Effect	Parameters	Effect	Parameters	Effect
Adhesion, kPa					ES	-
	pH	-				
	NSI	-	NSI	-		
			WA	+		
Cohesion, mm					NSI	+
	WA	-	WA	+		
	EA	-	EA	+	EA	-
			ES	-	ES	+
			pH	+		
Cook loss, % Water	PU, Gel	-	PU, Gel	+	PU, Gel	-
	ES	+			ES	+
					pH	+
Cook loss, % Fat	PU, Gel	+				
			pH	+		
			ES	-		
					WA	+
Firmness, N	pH	+	pH	+		
	WA	+	WA	+		
			NSI	-	NSI	-
			ES	-	ES	-
			PU, Gel	-	PU, Gel	-
					EA	+
Consistency, PUP	WA	-	WA	-	WA	-
					NSI	+

Table 4. Influence of different preparations on the properties of meat emulsions (3% non-meat protein, 1.3% NaCl).

Nr.	Protein-product	Prepa-ration		non heated		heated			
						Cook loss			
				Adhesion	Cohesion	Water	Fat	Firmness	Penetration
			pH	kPa	mm	%	%	N	PUP
1	FBPI 2	A	6.5	1.57	5.5	11.1	7.8	41.8	37
2		B	6.4	1.53	4.0	15.6	4.1	32.3	65
3		C	6.4	1.50	8.1	15.6	1.6	35.5	41
4		D	6.4	1.61	9.8	12.3	13.2	24.7	37
5	WPC 2	A	6.4	0.96	12.9	20.0	18.0	46.9	75
6		B	6.4	1.20	13.9	18.1	8.6	49.4	54
7		C	6.4	1.29	12.9	19.6	4.0	38.0	46
8		D	6.4	1.39	11.7	17.4	11.0	30.4	86
9	SPC 5	A	6.7	1.48	7.7	12.2	9.3	50.0	37
10		B	6.5	1.58	9.3	3.2	16.8	43.8	47
11		C	6.5	1.46	6.9	11.9	1.4	33.5	37
12		D	6.5	1.55	9.8	15.8	13.2	23.5	44

For the stability of ME (fat binding) a stable interface membrane is of importance.[10] The formation of a stable membrane can be limited by competition between protein fractions if different soluble and surface active meat and non-meat proteins are acting at the fat globule.

High soluble and surface active non-meat proteins interfere with the adsorption of soluble meat proteins at the fat and therefore the preparation of pre-emulsions[11,12] using non-meat protein and warmed liquid meat fat could be the preferred application for soluble and surface active proteins. This is demonstrated by the results of Table 3.

Low soluble proteins with high EA increases CLW (Table 2, NSI<44%). The general conclusion could be that meat emulsions will be reduced in CLW with proteins of lower solublility and high WA. Proteins with low NSI resp. high WA improve firmness (higher N or lower PUP, Table 3)

We can conclude that for influencing CLW and consistency it is of advantage to mix dry or dissolved high swelling PP with meat.

The results allow us to conclude that for the reduction of salt content in meat emulsions the role of emulsifying meat proteins can be substituted by other proteins, but this depends on the solubility, surface activity and water binding of the PP and it is important to find the optimal application technology for non-meat proteins.

For improving the reduced water binding and consistency high swelling PP should be added to the ME. The fat binding of ME should be influenced by preparing pre-emulsions with highly soluble surface active PP.

Investigations on combinations of such applications are under progress.

ACKNOWLEDGEMENT

The author wishes to express his appreciation to Karin Lengfeld for skillful technical assistance and Helga Rubbert who carried out the statistical analyses.

REFERENCES

1. C.Lacroix and F.Castaigne, Meat emulsification, effects of some ingredients: vegetable protein, salt, sugar and glycerol, Sciences des Aliments 4:505(1984).
2. A.M.Hermansson, Functional properties of added proteins correlated with properties of meat systems, J.Food Sci. 40:611(1975).
3. L.Ojalainen and E. Poulanne, Der Einfluß von geringem Salzgehalt und niedrigem pH-Wert auf die Wasserbindung bei Brühwurst, Fleischwirtschaft 68:482(1988).
4. O.Chevalier and Y.Timmermans, Possible applications of pea protein isolate in meat products, ZFL 43:283(1992).
5. Ch.Schneider,M.Schultz and H.Schmandke, Preparation of broad bean (Vicia faba L. minor) products, Nahrung 29:785(1985).
6. L.Heinevetter and J.Kroll, Zur Bestimmung der Wasserbindung pulverförmiger, quellfähiger Substanzen mittels einer Kapillarsaugmethode, Nahrung 26:K17 (1982).
7. G.Muschiolik, Behaviour of low-calorie spreads based on protein-stabilized oil-in-water systems, in:"Food
Polymers Gels and Colloids," E. Dickinson,ed.,The Royal Soc.Chem., Cambridge (1991).
8. G.Muschiolik and H. Schmandke, Einfluß von Fremdproteinen auf die Eigenschaften von erhitztem Fleischbrät, Nahrung 26:65(1982).
9. N.H.Nie "SPSS Statistical Package for the Social Sciences," McGraw Hill Bock, N.Y.(1975).
10. A.Gordon and S.Barbut, The effect of chloride salts on the texture, microstructure and stability of meat batters, Food Microstructure 8:171(1989).
11. P.A.Morrissey,D.M.Mulvihill and E.M.O'Neill, Functional properties of muscle proteins, in:"Developments
in Food Proteins-5," B.J.F.Hudson,ed.,Elsevier Appl.Sci, London,N.Y.(1987).
12. D.Scheid, Einflüsse auf die Emulsionsbildung bei der Wurstherstellung, Fleischwirtschaft, 57:1761(1977).

GELATION OF LOW DENSITY LIPOPROTEIN (LDL)FROM HEN EGG YOLK DURING FREEZING AND THAWING

Toshio Wakamatsu

Research Institute, Q.P. Corp.
No.13-1, 5-Chome,Sumiyoshi-cho, Fuchu-shi
Tokyo, Japan

ABSTRACT

Gelation of 40% LDL solution with 1-10% NaCl was inhibited during frozen storage at higher than the eutectic temperature of sodium chloride (-21.13°C). Frozen storage of LDL solutions with more than 4% NaCl at lower than the eutectic temperature induced the gelation, whereas gelation was inhibited by addition of 1% and 2% NaCl even at lower than the eutectic temperature. Differential scanning calorimetry analyses revealed that when NaCl acts as an inhibitor of gelation, it increased the unfrozen water in the LDL solutions through formation of LDL-water-NaCl complex where the water is not frozen even below -55°C; and when it acts as an accelerator of gelation, it promoted removal of water from the complex. It was concluded that he key factor for the gelation is not salt concentration in unfrozen phase but dehydration to less than the critical moisture of LDL.

INTRODUCTION

Frozen storage of hen egg yolk induces the loss of its fluidity and solubility. This phenomenon is called gelation. The key component which participates in the gelation is low density lipoprotein (LDL), which constitutes about 65% of yolk proteins[1]. Several hypotheses on the cause of LDL damage during frozen storage are proposed: (1)salt concentration in unfrozen phases, (2)pH change in an unfrozen phase[2], (3)dehydration due to transition of water to ice.[3] However, experimental evidences that verify each of them have seldom been demonstrated.

The purpose of this study is to clarify the gelation mechanism in LDL-water and the LDL-water-sodium chloride system by tracing the transition of water to ice quantitatively with the aid of a differential scanning calorimeter(DSC).

Food Hydrocolloids: Structures, Properties, and Functions
Edited by K. Nishinari and E. Doi, Plenum Press, New York, 1994

MATERIALS AND METHODS

Low density lipoprotein (LDL)

LDL was prepared as described in a previous study,[1] and then exhaustively dialysed against distilled water for 3 days at 5°C.

Freezing treatment

Sodium chloride was added to 40% LDL in its concentration range of 1-10%(w/w) and stored overnight at 5°C. Fifteen grams of the sample containing a given amount of sodium chloride were frozen and stored in the range from -20°C to -60°C for 72 hrs after being sealed in a stainless steel container (3 x 3 x 3 cm) and then thawed at 25°C for 30 min.

Apparent viscosity and turbidity measurement

The apparent viscosity was measured on a model RVT Brookfield synchro-lectric viscometer (Brookfield Engineering Laboratories, Inc., Massachusetts) at 25°C. The absorbance at 660 nm of a diluted sample (1% solid) with 1 M sodium chloride solution was measured by an electrophotometer (model 200-20, Hitachi Seisakusho, Tokyo).

Determination of unfrozen water

DSC analyses were performed on a Model SSC-540 differential scanning calorimeter (Seiko Electronics Ltd., Tokyo). Transition heats of a 40-60 mg of sample in an aluminum container (diameter 4 mm, height 8 mm) during freezing and thawing were measured. A sample was scanned from room temperature to -55°C at the rate of 0.6°C/min, and then from -55°C to room temperature at the same rate. The amounts of frozen water were calculated by dividing the measured exothermic and endothermic heat by transition heat of pure water. The unfrozen water was calculated by subtracting the frozen water from the whole water in a sample.

RESULTS AND DISCUSSION

1. Prerequisite for egg yolk gelation

The results obtained from determination of the unfrozen water by DSC and measurement of the extent of gelation at various freezing temperatures are summarized as follows[4]:(1)super-cooling even at -10°C did not cause gelation, (2)yolk gelation was suppressed under the condition that the ice content is less than approximately 80% and (3)the viscosity of egg yolk did not alter during frozen storage at -60°C, but when it was transferred to a -20°C freezer, the viscosity increased with time. These results suggest that the gelation is induced by at least two steps: (1)the unfrozen water is reduced to less than the critical moisture, and (2)the micro-ice crystals, which may be formed between LDL particles during freezing period, are removed by merging with macro-ice crystals during frozen storage and thawing.

2. The critical moisture of LDL

It was assumed, as described above, that the gelation process is initiated by removal of water around LDL through formation of macro-ice crystals. To confirm this assumption ,the water in LDL solution was gradually removed in a low relative humidity chamber at 4°C and the relationship was plotted between the water content and solubility of LDL[5]. The solubility of LDL was lost when the water content decreased to 0.11-0.16 g/g LDL. This means that the amount of water is indispensable for stabilizing LDL and it was defined as critical moisture. The unfreezable water which was defined as the unfrozen water at -55°C of LDL in the present study was 0.10g/g LDL. In addition, the monolayer and multilayer water estimated from desorption isotherm of LDL were 0.04g and 0.06g/g LDL, respectively. The thickness of unfreezable water layer was estimated as 7-8 Å by dividing the amount of unfreezable water by the surface area of LDL (124 m^2/g LDL) estimated by the BET equation. The diameter of a water molecule is 2.8Å, so the the thickness of the unfreezable water corresponds to 2-3 water molecules. The postulated model of water phases around LDL was shown in Fig 1. It is assumed that the removal of a definite amount of water corresponding to the difference between critical moisture and unfreezable water is responsible for aggregation of LDL.

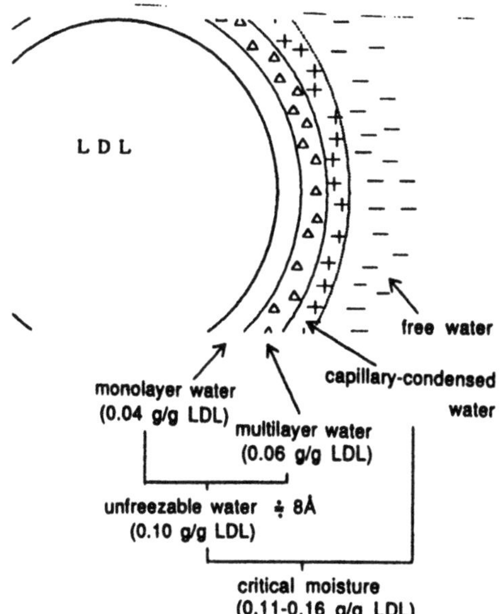

Fig.1 Postulated model of water phases around LDL

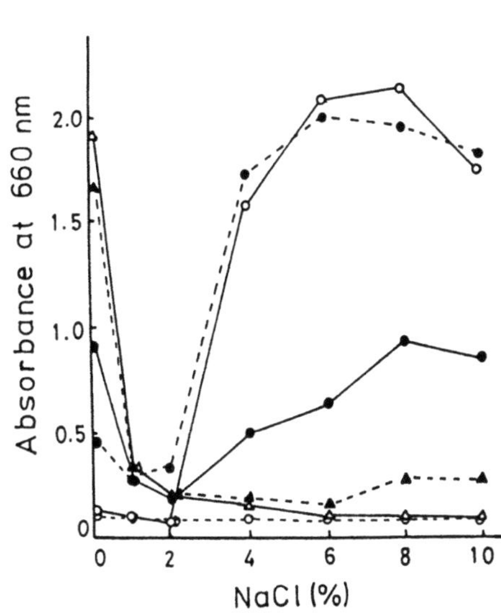

Fig.2 Sodium chloride effect on turbidity of frozen-thawed LDL. Freezing Temp.: O-O; unfrozen , △—△;-20°C, ▲- -▲ ; -25°C, ●—● ; -30°C, ●- -●;-40°C, O—O ;-60°C

3. The action of sodium chloride in gelation process

The turbidity change of LDL with 0-10% sodium chloride after frozen storage at -10°C--60°C, which is almost in accord with viscosity change, was shown in Fig.2. The turbidty change decreased with the increase of salt concentrations at -20°C. In contrast, turbidity increased with the increase of salt concentrations in the range below -30°C. This phenomenon is reasonably explained by DSC analyses of LDL solutions with sodium chloride[6].

(1)In the case of LDL without sodium chloride, only one peak due to transformation of free water to ice was detected(Fig. 3 A). Turbidity change was the highest at -20°C(Fig.2). At this temperature, micro-ice crystals may be easily removed between LDL particles during frozen-storage, and this dehydration may cause LDL damage, thereby induce aggregation(Fig.4 A).

(2)When LDL solutions with less than 2% sodium chloride were frozen at -20°C--60°C, gelation was inhibited. In addition, only one transition peak due to transformation of free water to ice was detected(Fig. 3 B). In contrast, in the case of sodium chloride solutions with less than 24% solute, two peaks caused by transformation of free water to ice and transformation of bound water of ions to ice(formation of eutectic mixture),were detected[7]. These results suggest that free sodium or chloride ions may not exist, that is, almost all of sodium or chloride ions are bound to LDL and the water which consist of LDL-water-NaCl complex is not frozen even at -60°C(Fig.4 B). This water may protect LDL from freezing damage and prevent the mutual contact of LDL particles.

(3)When LDL solutions with more than 4% sodium chloride were scanned from -55°C to room temperature, two endothermic transitions occurred, and the amount of water involved in the eutectic mixture rose with the increase of sodium chloride amount(Fig.3 C-F). In addition, gelation of these solutions occurred during frozen storage below -30°C. These results suggest that the sodium or chloride ions are adsorbed onto LDL almost to the limit and a portion of sodium chloride may exist as free ions(Fig.4 C). When these LDL solutions are frozen below -30°C, which is lower than the eutectic temperature of sodium chloride (-21.13°C), the water that binds to free ions may transform to ice (formation of eutectic mixture) after the freezing of free water. This may accelerate the water migration from the LDL-water-NaCl complex to the eutectic mixture; finally, the water in the complex decreases to the unfreezable level, where the protective effect is lost.

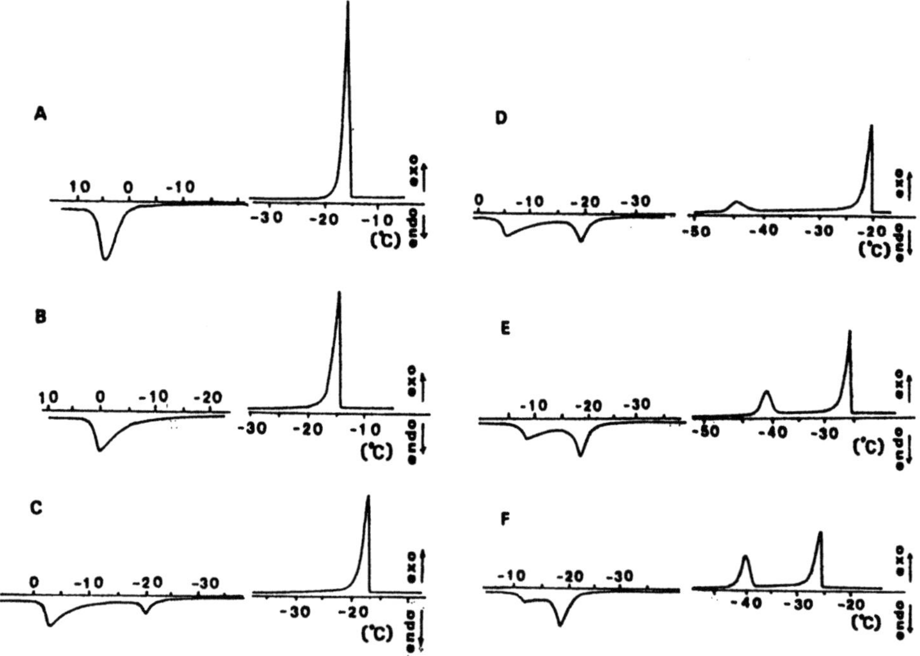

Fig. 3 DSC thermograms of LDL solutions with sodium chloride at room temperature to -55°C. NaCl conc.:(A),0%; (B), 2%; (C), 4%; (D), 6%; (E), 8%; (F),10%.

When LDL solutions with 1%-10% of sodium chloride was frozen at -20°C and -25°C, gelation was inhibited(Fig.2). DSC analyses from room temperature to -21°C revealed that only one transition had occurred, and that unfrozen water increased with the rise of sodium chloride amount. These results indicate that only free water is frozen and bound water of free ions is not, even if free ions may exist. Consequently, the disruption of the complex may not be induced.

REFERENCES

1. T. Wakamatsu, Y. Sato and Y. Saito, Identification of the components responsible for the gelation of egg yolk during freezing, *Agric. Biol. Chem.* 46:1495 (1982).
2. F.S. Soliman and L. van den Berg, Factors affecting freeze aggregation of lipoprotein, *Cryobiology* 8:265 (1971).

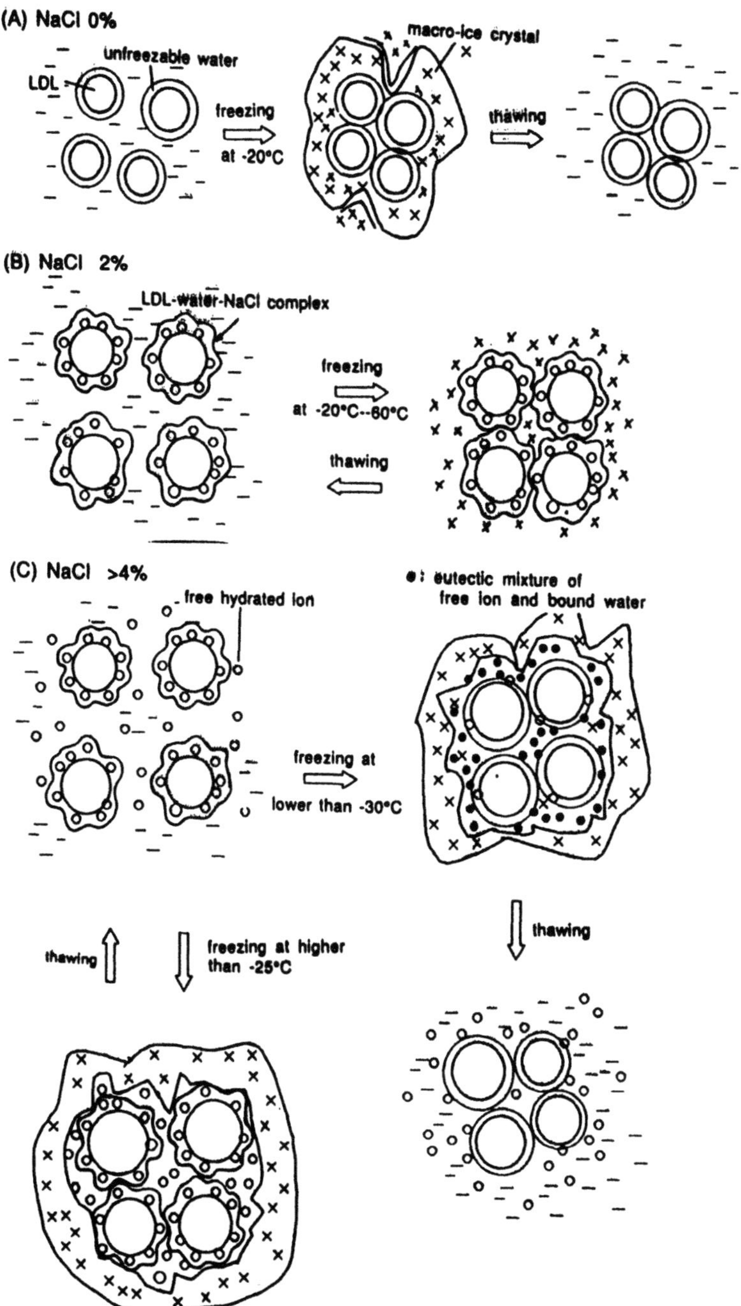

Fig. 4 Postulated model of the action of NaCl in the gelation process of LDL during freezing and thawing

3. Y. Sato and T. Aoki, Influences of various salts on gelation of low density lipoprotein (egg yolk) during its freezing and thawing, *Agric. Biol. Chem.* 39:29 (1975).

4. T. Wakamatsu, Y. Sato and Y. Saito, Effects of freezing temperature and storage time on gelation and quantity of unfrozen water of hen egg yolk, *Nippon Nogei Kagaku Kaishi.* 55:699 (1981).

5. T. Wakamatsu, Y. Sato and Y. Saito, Relationship between solubility change during dehydration and unfreezable water in egg yolk and low density lipoprotein, *Nippon Nogeikagaku Kaishi.* 56:117 (1982)

6. T. Wakamatsu, Y. Sato and Y. Saito, On sodium chloride action in the gelation process of low density lipoprotein(LDL) from hen egg yolk, *J. Food Sci.* 48:507 (1983).

7. T. Wakamatsu, Y. Sato and Y. Saito, Determination of unfreezable water in sucrose, sodium chloride and protein solutions by differential scanning calorimeter, *Nippon Nogeikagaku Kaishi* 53:415 (1979).

EFFECT OF STARCH ON THE EVALUATION OF SURIMI GELS MADE FROM VARIOUS KINDS OF FISH SPECIES

Satoshi F. Noguchi and Keiichi Nakagawa

Central Research Institute
Taiyo Fishery Co., Ltd. Tsukuba, Japan

ABSTRACT

The present study is concerned with the rheological properties of cooked surimi gels made from various kind of fish species and starch from the point of view of quality measuring system of surimi. The textural changes of cooked gels samples were observed with punch test and torsion test. The values of stress were increased by addition of starch in any tested surimi gel samples,but the values of strain varied with fish species and grades. Composite characteristics of cooked surimi gels containing various amount of starch were assessed by examining the microstructure under cryo-scanning microscope. The microstructures of tested gels suggest that the gel-strengthening effect of starch is closely related with a kind of competition between water holding capacity of surimi gel matrix and swelling power of starch globules by heating.The results of present study show that the effect of starch is varied with fish species and grades complicatedly,and suggested that the difficulties of use of starch for the practical quality measurement system of widely varied surimi samples.

INTRODUCTION

The ingredient behavior of textural properties of surimi or kamaboko gel have been studied in relation to their composite characteristics such as starch and egg albumin like other food gels. In particular, Japanese surimi scientits have interest in the textural changes of surimi gels by addition of starch from the point of view of the quality

Food Hydrocolloids: Structures, Properties, and Functions
Edited by K. Nishinari and E. Doi, Plenum Press, New York, 1994

measurement system of frozen surimi. Alaska pollock surimi dominates the market today,and the success of surimi industry world wide has made possible the marketing of local or underutilized species. However, some surimies made from new and underutilized fish species have very low elastic properties and have involving difficulties in preparering test gel sample. The other hand, it is known that starch inclusion will greatly increase the jelly strength by the conventional experiences of Japanese kamaboko makers, and the optional articles of including potato starch is given for the low grade Alaska pollock surimi in Japanese standard quality measurement system. The present study is concerned with the rheological properties of the thermal induced surimi gels made from various kind of fish species and potato starch from the point of view of quality measuring system of surimi.

MATERIALS AND METHODS

High and low grade frozen surimies prepared from Alaska pollock (Theragra chalcogramma) and Atka mackerel (Pleurogrammus azounus), Low grade blackmouth croaker (Atrobucca nibe) surimi from China, surimi prepared Threadfin bream (Nemipterus tambuloides) from Thailand, and surimi prepared from Hoki (Macruronus navaezelandiae) from New Zealand were obtained from Taiyo Fishery Co.Ltd. (Tokyo, Japan). Surimies prepared from Blackmouth croaker and Largehead hairtail (Trichiurus lepturus) and Sardine (Sardinopus melanosticus) were obtained from Nagasaki suisankako kyodokumiai (Nagasaki Japan). The preparation of surimi gel samples for the punch test and the measurement of the punch force and deformation were applied by the methods of Japanese surimi quality measurement system. The 25mm log gel sample is centered under the 5mm diameter plunger of rheometer and measured the stress and strain at the point of breaking . The plunger speed was set at 60mm/min. For the preparation of gel samples for the torsion test, the specified vacuum equipped cutter/mixture machine was used and the followed procedures after grinding were same as the punch test. Dumbbell shaped specimen 2.87cm in length and 2.0cm end diameter at the center for torsion test were prepared. The torsion test apparatus using Brookfield viscometer was used to measure the shear stress and strain. The sample was rotated at 2.5rpm. Composite characteristics of cooked surimi gels containing various amounts of starch were assessed by examining the microstructure under a cryo-scanning microscope.

RESULTS

The textural changes of cooked surimi gels with and without 3% potato starch were observed by the punch test and torsion test. The results of both test are shown in Fig.1 by a kind of textural map suggested by Hamann. The gel-strengthening effect of varying level of starch on textural properties of high and low grade surimi are observed by punch test and shown in Fig.2. Composite characteristics of cooked surimi gels containing various amounts of starch were assessed by examining the microstructure under a cryo-scanning microscope and the electron micrographs of surimi gels with 3% starch are shown in Fig.3.

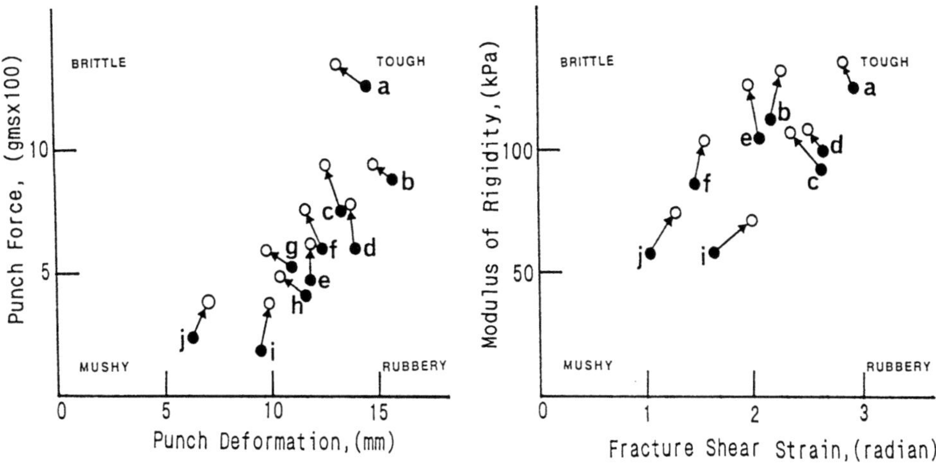

Fig.1.The gel-strengthening effect of 3% starch on the textural properties of cooked surimi gels made from various kinds of fish species and grades. Closed circles indicate starch free surimi gels and open circle indicate 3% starch added gels . Alphabets indicate   the used surimies; a:high grade Alaska pollock surimi (moisture content 73.8%), b:Threadfin bream(77.9%), c:Hoki(74.1%),   d:Croaker(74.5%), e:Largehead hairtail(78.2%), f:Sardine (76.8%), g:Low grade croaker(75.4%), h:Alaska pollock(78.4%), i:Low grade Alaska pollock(79.2%), j:Atka mackerel(77.5%).

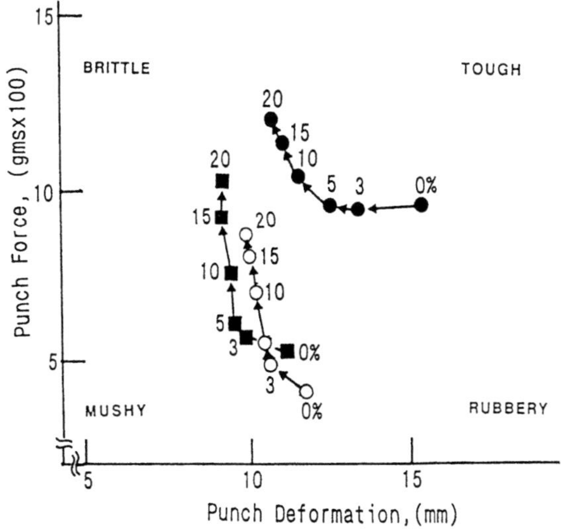

Fig.2. Effect of varying level of starch on the textural properties of cooked gels made from three kinds of surimies. Numbers indicate % of added starch. Closed circle:high grade Alaska pollock surimi(moisture content;75.4%). Open circle:Alaska pollock(78.4%). Closed square:Low grade croaker(75.4%).

DISCUSSION

The values of stress, punch force and modulus of rigidity increased by addition of starch in any tested surimi gel samples, though the rates of increasing are varied with fish species and grades complicatedly (Fig.1). In general, the strain values of cooked gels made from low grade surimies were increased by addition of starch, but the values of high grade surimies were decreased (Fig.1). The rates of change of strain values of tested three kind of surimies are large and marked by addition of 3 and 5% starch. But, over 10% starch, the changes of values of punch deformation of all the tested gel samples are a little, and the values of each tested surimies seemed to converge around the value of 10mm(Fig.2). The present rheological study shows the gel strengthening effect of starch is varied with fish species and grades complicatedly. The microstructures of tested gels show that starch globules are broken and swelled in the gels made from low grade surimi, but are not swelled in the gels of high grade surimi (Fig.3). The typical structural feature is observed in the boundary layer between starch globules and surimi gel matrix. The present observation suggest that the gel-strengthening effect of starch is closely related with a kind of competition between water holding capacity of surimi gel matrix and swelling power of starch globules by heating or the pulling of water against each other.The results suggest the difficulties of use of starch for the practical quality measurement system of varied surimi samples.

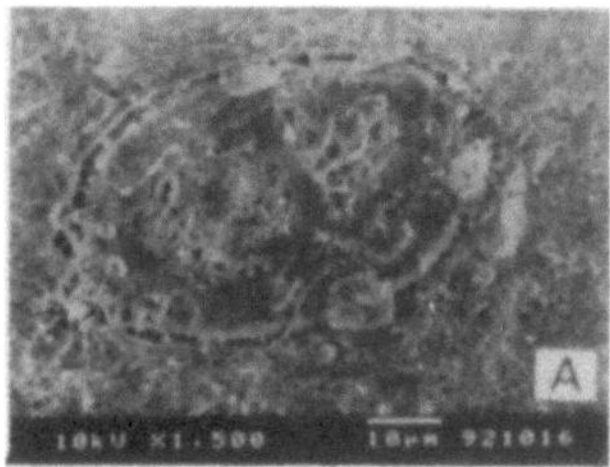

Fig.3. Electron micrographs of 3% starch added surimi gel. A indicates high grade Alaska pollock surimi. B indicates Alaska pollock surimi. C indicates low grade Blackmouth croaker surimi. Used gel samples were same as Fig.2.

REFERENCES

1.C.M.Lee and J.M.Kim, The relationship of composite characteristics to rheological properties of surimi sol and gel, in "Food Engineering and Process" Vol.1 p63-79, N.LeMaguer and P.Jelen ed., Elsevier Applied Science Publishers,London (1986).

2.C.M.Lee, M.C.Wu and M.Okada, Ingredient and formulation technology for surimi based products in "Surimi Technology" P273-30,T.C.Lanier and C.M.Lee ed., Marcel Dekker Inc. New York (1992).

3.D.D,Hamann and G.A.MacDonald, Rheology and texture properties of surimi and surimi based foods in "Surimi Technology" p429-500 ,T.C.Lanier and C.M.Lee ed., Marcel Dekker Inc. New York (1992) .

THE USE AND CONTROL OF CHEMICAL REACTIONS TO ENHANCE GELATION OF MACROMOLECULES IN HEAT PROCESSED FOODS

Sandra E. Hill, Helen J. Armstrong and John R. Mitchell

Department of Applied Biochemistry and Food Science
University of Nottingham
Sutton Bonington Campus
Nr. Loughborough
Leics. LE12 5RD, U.K.

INTRODUCTION

Often the study of food hydrocolloids involves extensive research into the macromolecular conformation and a colloid science approach to functionality. In this paper two pieces of work are discussed that demonstrate the effect that low molecular weight materials can have when co processed at retort temperatures with hydrocolloids.

The two areas of work are:

a) The control of oxidative reactions to prevent viscosity and gel strength loss on heating galactomannan solutions.

b) The use of the Maillard reaction to enhance protein gelation.

METHODS AND MATERIALS

The hydrocolloids and proteins were supplied by Sigma Chemical Company and were used without further purification. Other chemicals were supplied by Fisons PLC or Sigma.

Preparation of Solutions and Gels. The gums were prepared in mixed phosphate buffer at pH 7.0. When the antioxidants (sodium sulphite and n-propyl gallate) were

Food Hydrocolloids: Structures, Properties, and Functions
Edited by K. Nishinari and E. Doi, Plenum Press, New York, 1994

incorporated they were added to the buffer prior to the addition of the gums. Guar gum was added to the buffer at ambient temperature while carrageenan, xanthan and locust bean gum were added at 80° C. Cans were filled with the solutions and seamed leaving no head space. Cans were retorted at 120° C for up to 60 minutes.

Proteins were dissolved in distilled water in the presence or absence of reducing sugars and glucono-δ-lactone (GDL). The gels were heated in glass bottles in a water bath or under retort conditions. In some cases to facilitate subsequent rheological testing, gels were prepared by heating solutions contained in dialysis tubing (21mm diameter) covered with aluminium foil.

Viscosity and Gel Strength. Viscosity measurements were determined using a Deer rheometer equipped with a cone (4°) and plate geometry. Gel properties were determined by recording the peak penetration force made directly in cans and bottles or by compressing cylinders (2cm high by 2cm diameter) using a TAXT2 Texture Analyser (Stable Microsystems Ltd. Surrey, U.K.) All readings were taken at ambient temperature after ageing the gels for about 18 hours. A 1.20 cm (0.6 cm into bottles) diameter plunger was used at a speed of 2cm/sec to a distance of 25mm.

THE CONTROL OF OXIDATION REACTIONS TO PREVENT THE VISCOSITY AND GEL STRENGTH LOSS ON HEATING GALACTOMANNAN SOLUTIONS

It is well established that heat treatment can result in substantial decreases in viscosity and gel strengths, the decrease being strongly dependent on pH. However, even at neutral pH values polysaccharides will degrade as a result of oxidative reductive depolymerisation (ORD) reactions. This form of degradation can be controlled by the addition of antioxidant systems. The work that has been carried out has mainly been concerned with the stabilisation of polymers used in oil field applications (e.g. Wellington, 1983) and little attempt has been made to extend this approach to food systems using non toxic antioxidants. Here we demonstrate that a combination of low levels of the additives sodium sulphite and propyl gallate can substantially reduce the viscosity and gel strength loss which occurs on heat sterilisation of systems containing galactomannans.

RESULTS AND DISCUSSIONS

Initial results indicated that there was a marked drop in the viscosity of galactomannans when they were heated to 100° C and the viscosity drop was even higher at retort temperatures. For example the relative viscosity for a 0.2% solution of guar at pH 7.0 was 5.66 after heating at 100° C for 10 minutes this being reduced to 3.21 after retorting. If the binary antioxidants described in this paper were used the viscosity without autoclaving was 10.75 and 10.15 after heating at 121° C for 30 minutes. Inclusion of propyl gallate alone had little effect on the stability of the galactomannans to retorting. It has been reported that the addition of sulphite alone enhances the thermal stability of guar gum (Rodrigez, 1985) but, as

Figure 1 demonstrates the binary system is very much more effective. The figure displays the viscosity of the 0.8% guar solutions following retorting as a function of the antioxidant ratio at a total concentration of 200ppm (0.02%). The data shows the very strong synergism between the two additives and the optimum ratio is about 3:1 sulphite to gallate. Maximum stabilising effect occurs at about 300ppm and no further thermal stability occurs at higher levels of antioxidant.

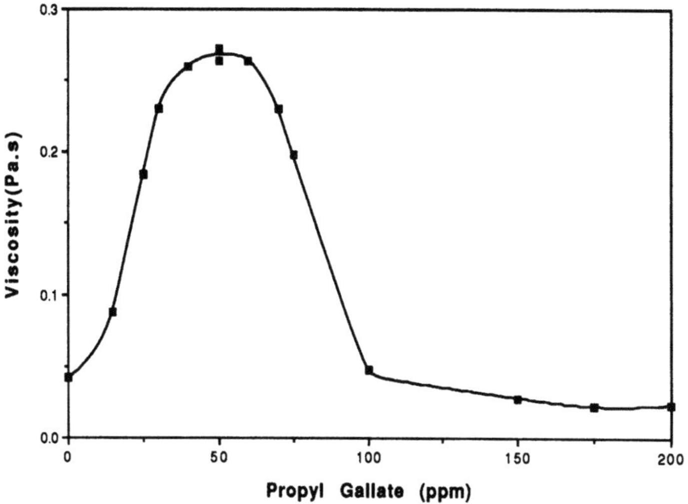

Figure 1. Effect of antioxidant ratio on the viscosity (measured at 20° C and $50s^{-1}$) of a 0.8% (w/v) guar gum solution at pH 7.0. The total antioxidant concentration (gallate plus sulphite) was 200ppm and numbers on the abscissa are the propyl gallate concentration. The viscosity of an unretorted control containing no additives was 0.5 Pa.s.

It is interesting to compare the effect of the antioxidants on the thermal stability of guar with their ability to prevent the breakstrength loss which occurs on autoclaving mixed carrageenan- locust bean gum gels. A somewhat similar dependence on additive ratio was a observed to that obtained for the viscosity of guar gum solutions. There was some suggestion that the sulphite: gallate ratio required for optimum gel strength is higher than that required to control the guar viscosity. It was found that the gel strength could be enhanced by a factor of about 4 compared with the system containing no additives although the strength of the non autoclaved system was not maintained.

The effect of total additive concentration at ratios 17:3 gallate to sulphite is displayed in Figure 2. This set of data implies that retorted gel strength does not reach a maximum until the additive level exceeds 300ppm although substantial improvement is obtained at far lower levels.

It is well recognised that the strength of mixed carrageenan gel systems are strongly influenced by pH and K^+ levels. However, it was found that the binary antioxidant system had a beneficial effects in the absence or presence of potassium ions.

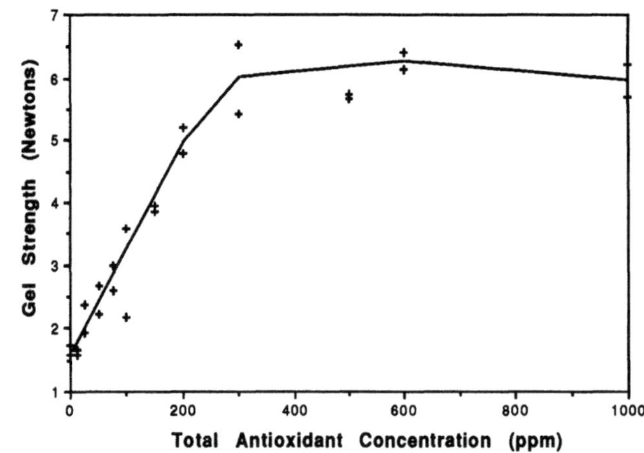

Figure 2. Effect of total antioxidant concentration at a sulphite: gallate ratio of 17:3 on the strength of retorted gels at pH 7.0 containing locust bean gum and 0.4% carrageenan. The average strength of non retorted control containing no antioxidants was 7.0N.

There was also an increase in gel strength throughout the pH range except at very low pH values (<5) were a gel did not form presumably because at these low pH values hydrolysis becomes far more important as a degradation mechanism compared with ORD reactions. The protection from a molecular weight decrease due to heat treatment should result in less brittle gels being formed (Mitchell, 1980) and this was clearly seen in the mixed carrageenan-locust bean gum gels.

Neither xanthan gum nor locust bean gum form gels when they are the only polysaccharide present. However, a mixture of the two gums form "elastic" gels of low modulus. This implies that the crosslink density is low and the distance between the junction zones along the locust bean gum chain will be large. A reduction in locust bean gum molecular weight will rapidly reduce the functionality (number of regions of the chain participating in crosslinks) below the critical value of two required for gel formation. As expected, autoclaving had a dramatic effect on the mixed locust bean gum- xanthan gel and gels were not formed after the heat treatment. However, in the presence of the binary antioxidants, which protect the locust bean gum and xanthan in aqueous solutions, an elastic gel was achieved.

THE USE OF THE MAILLARD REACTION TO ENHANCE PROTEIN GELATION

In investigating the role of antioxidants in preventing viscosity loss in some heat sterilised food systems containing galactomannans, proteins and reducing sugars we observed an increase in viscosity as a result of the heat treatment rather than a decrease. This occurred even in the absence of the galactomannan but not when the reducing sugar was omitted. It seemed probable that the Maillard reaction was important in promoting gelation. This phenomena was subsequently studied more systematically with bovine serum albumin as the protein and xylose as the reduciong sugar.

RESULTS AND DISCUSSION

When 2% BSA solution was heated in the presence of xylose at 121° C for 1 hour a gel was formed. There was correlation between the extent of the Maillard reaction, as evidenced by the change of colour, and the strength of the gels. At these concentrations of BSA gelation does not normally occur in the absence of sugar or in the presence of the non reducing sugar, sucrose. The effectiveness of the reducing sugars in promoting gelation follows the previously reported series for the reactivity of reducing sugars for the Maillard reaction (Hurrel, 1979). When the solutions were heated for 1 hour at 100° C no gelation was observed for 2% bovine serum albumin in the presence of 2% sugars, however, when the protein concentration was raised to 3% gelation did occur with xylose but not with the other sugars. We have investigated other globular proteins and the promotion of gelation in the presence of a reducing sugar appears a fairly general phenomena.

At constant xylose concentration the breakstrength and modulus of the BSA gel was approximately linearly related to the BSA concentration. It would appear that the xylose concentration required for maximum gel strength increases with the concentration of bovine serum albumin.

The most obvious mechanism for gelation is a pH reduction which is the result of the development of acidic products due to sugar degradation mechanisms which is part of the Maillard reaction. In the absence of reducing sugars the initial BSA solution pH was 6.5 on autoclaving (121° C for 30 min) this decreased by about 0.2 units. In the presence of the reducing sugars, however, a substantial pH reduction was observed. Autoclaving 3% BSA and 2% xylose resulted in a gel of pH 4.2. There is a strong correlation between gel strength and the final pH for the BSA Maillard gels prepared under a range of conditions (Figure 3).

We have found that gels at low protein concentrations can also be formed using glucono-δ-lactone (GDL) to produce a pH fall in conjunction with high temperature. There are however, important differences between the two systems. The lysine and amino groups are modified in the presence of the reducing sugars and there is evidence that there is a net increase in charge on the protein with a lowering of the isoelectric point.

A gel prepared by autoclaving 3% BSA and 0.75% GDL at 121°C for 30 minutes to give a final pH of 5.6 was completely soluable in 1% sodium dodecyl sulphate (SDS) and 1% β mercaptoethanol (Sheard et al., 1984) where as a "Maillard" gel with a pH of 5.1 prepared with 3% BSA and 2% xylose retained its structural integrity in this solvent system. This indicates that the GDL gel was held together by conventional crosslinks (disulphide and non-covalent bonds) while the "Maillard" gels have additional crosslinks.

CONCLUSION

The work described demonstrates that the performance of macromolecules can be dramatically affected by the presence of other material. It shows that the use (as in Maillard

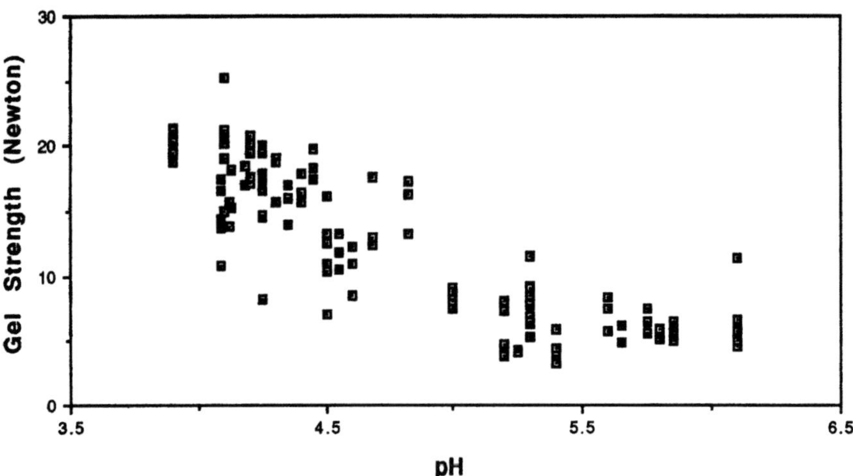

Figure 3. The breakstrength of retorted (121° C for 60 min) gels formed from 3% BSA versus the final pH of the gel. The graph combines results from a range of sugars at various concentrations (0 to 10%).

reaction) and control (as in the limiting of depolymerization) of chemical reactions can enhance the functionality of hydrocolloids. We have also carried out work using Maillard induced pH fall to release calcium thus gelling alginates at specific times and temperatures.

REFERENCES

Hurrel, R.F., Lama, P. and Carpenter, K.J. 1970, Reactive lysine in foodstuffs as measured by rapid dye-binding procedure, *J.Fd.Sci.*44:1221-1227.

Rodriguez, P. 1985, US Patent 4514318

Sheard, P. R., Ledward, D. A. and Mitchell, J. R. 1984, Role of carbohydrates in soya extrusion. *J. Food Technol.* 19:475-483

Wellington, S.L. 1983, Biopolymer solution stabilisation-polymer degradation and antioxidant use, *Soc. Pet. Eng. J.* 9296: 901-912

MECHANISM OF WALL EFFECT ON FLUIDITY OF MILK IN A CAPILLARY

Teruko Nakamura, Akemi Yamamoto, Kishiko Sakanishi
and Takeshi Mineshita

Department of Food Science
Tezukayama College
Nara, Japan

INTRODUCTION

Milk is one of the most important and popular biofluids
containing various nutritients. It is significant to investigate rheological properties of milk in a capillary
to explain the milk flow on tongue and through throat for
the consumer preference. In the previous papers[1]-[5], flow
properties of milk in a capillary have been studied by
viscosity measurements with various bore-sizes of capillary viscometer and photo-microscopic observation. In
this paper, effects of silicone coated glass capillary
tube on flow properties of milk were elucidated in comparison with the result obtained in a non-coated glass
capillary tube.

EXPERIMENTAL

Fig.1 is the schematic view of the experimental method.
The apparatus used throughout
this experiment was a Maron-
Belner type low shear capillary viscometer combined
with a photomicroscope.
Viscosity measurements were
carried out with various
bore sizes of capillary
viscometer with continuous
varying pressure head of
shear stress range from 0.2
to 30dynes/cm^2.

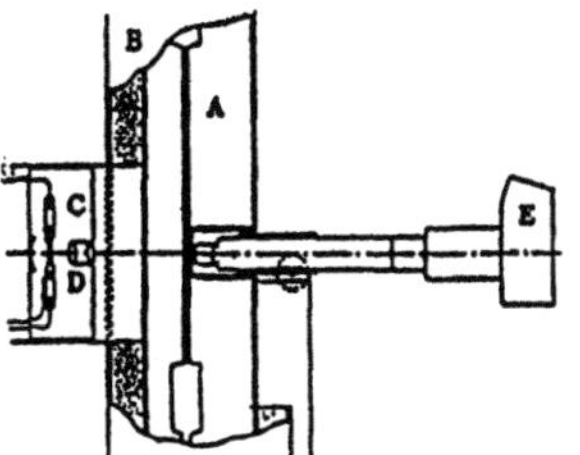

Fig.1 Vertical section of photomicroscopic
system combining with a viscomerer.

A. Capillary portion B. Water bath
C. Xenon lamp D. Condenser lens
E. Photomicroscopic camera

Food Hydrocolloids: Structures, Properties, and Functions
Edited by K. Nishinari and E. Doi, Plenum Press, New York, 1994

To take a photograph of the image of milk fat droplets flowing in a capillary, photomicroscope combined with a capillary viscometer was used. In this experiment, silicone coated glass capillary tube was used to

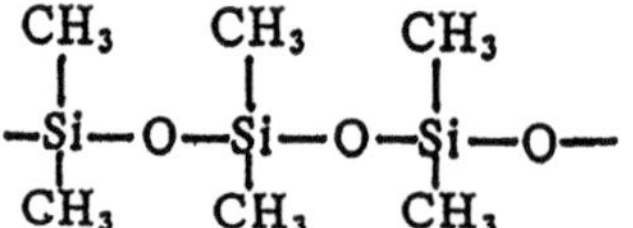

Fig.2 Chemical components of dimethyl-polysiloxane.

compare with non-coated glass capillary tube. Chemical component of used silicone coating agent is a polymer of dimethyl-poly-siloxane as shown in Fig.2. This polymer chain stretches its polar oxygen binding toward the capillary wall, and hydrophobic binding of methyl radical stretches toward the center of the capillary tube. The thickness of the coated polymer layer is about 100Å.

RESULTS AND DISCUSSION

The viscosity (F/G) versus shear stress(F) plots of milks of 1% fat are shown in Fig.3. These figures (a),(b) and (c) show human fresh milk, cow's fresh milk and homogenized milk, respectively. With the sample of this concent-

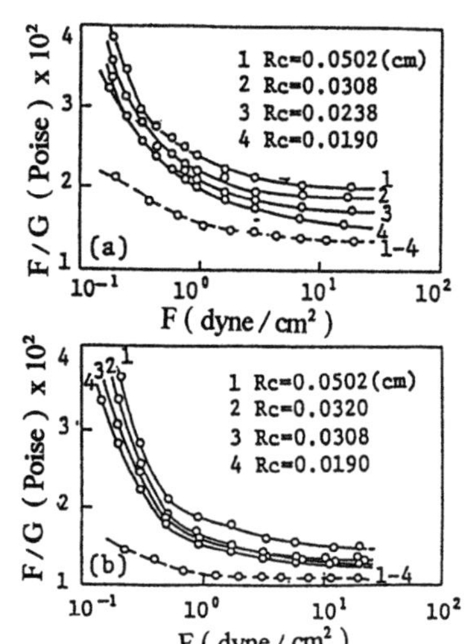

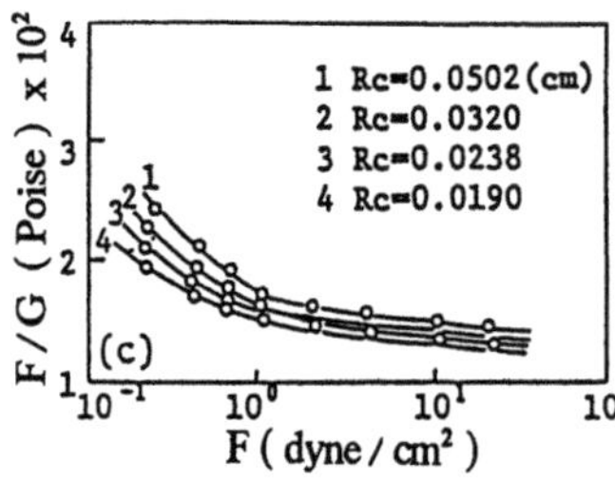

Fig.3 Viscosity versus shear stress plots of milk. (20C)

a: Human fresh milk (fat : 1.0%)
b: Cow's fresh milk (fat : 1.0%)
c: Homogenized milk (fat : 1.0%)

ration of milk fat, a non-Newtonian flow is observed in all ranges of shear stress, and a curve shown by dotted line obtained with skim milk is observed to be slightly non-Newtonian flow behavior. Except for this suspension of the skim milk, the viscosity of the emulsion decreases apparently 10-30% or more between the smallest capillary bore and the largest one at a given shear stress. The flow behaviors of the other sample of cow's fresh milk in various capillary bore sizes are shown in this figure. The shear dependence on the viscosity is observed marked-ly at a lower shear stress than at a higher stress, while the difference in viscosity due to the size of the capil-lary bore is marked at a high shear stress. The result of viscosity versus shear stress plots of homogenized milk is shown in this figure. In this case, both shear stress dependence and capillary bore size dependence on the viscosity are more easily seen at a low shear stress rather than at a higher one. Fig.4 shows the concentration dependence of homogenized milk and human milk on viscosity with silicone coated capillary tube and non-coated glass capillary tube. Fig.4 (a) is the results obtained by narrower capillary bore size, and 4(b) is the results of wider capillary bore-size. From this figure, linear relationship between viscosity and concentration

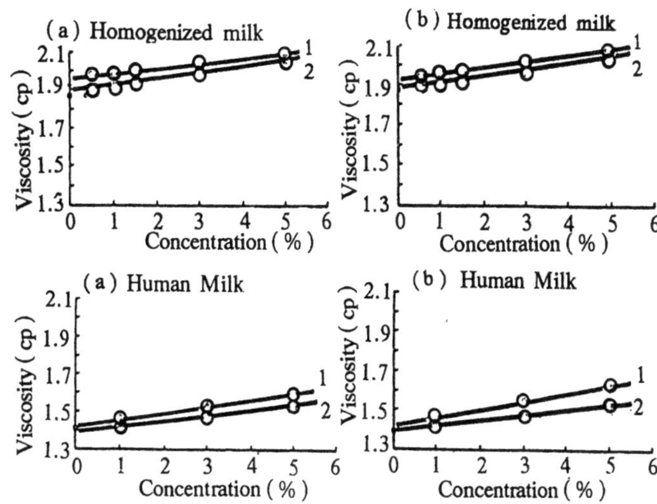

Fig.4 Relationship between viscosity and concentration of milk fat globules in each milk (20C°)

1 : Non-coated glass   2 : Silicone coated glass

of homogenized milk was observed. In this case, dependence of a capillary bore-size and silicone coated effect on viscosity could be observed slightly. This figure also shows the concentration dependence on viscosity of human milk obtained with silicone coated capillary tube and non-coated glass capillary tube. Also, in this case, linear relationship between viscosity and concentration of milk fat globules was observed in both capillaries. And, viscosity values measured by a viscometer having a narrower silicone coated capillary tube appeared to decrease more than those measured with a non-coated capillary tube. This difference between homogenized milk and human milk will depend on the dispersion states of milk in a capillary. Photomicroscopic observation of dispersion states of each milk was shown in Fig.5. Fig.5(A) and 5(B) show the photomicrogram of dispersion states of homogenized milk and human milk in a capillary observed with various radii of glass capillary tube. It is obviously shown that human milk formed aggregate structures at the near side of the center of the capillary tube, and less aggregates were observed at the nearest side of the

capillary wall. In the case of homogenized milk, because
of its homogenization, smaller aggregates were observed
at the center of the tube, but they were too small to
identify clearly. Consequently, concentration distribution of milk fat globules is calculated as shown in the
next figure.

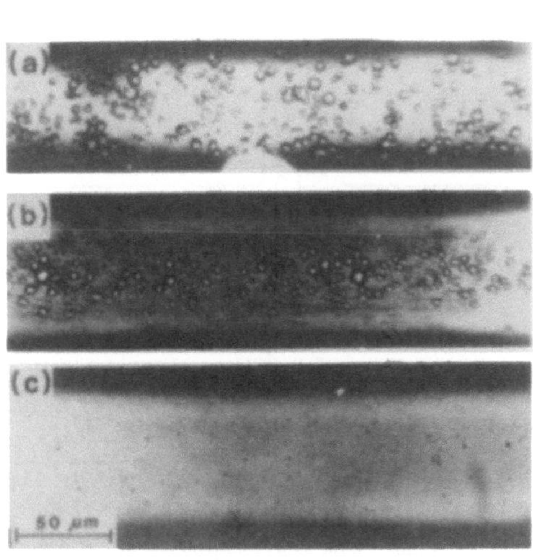

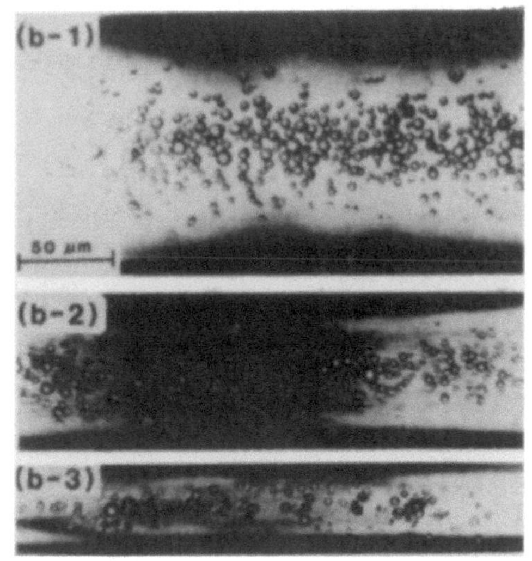

Fig.5-A, Photomicrograph of milk fat droplets flowing in a capillary
a) Cow's fresh milk b) Human milk
c) Homogenized milk

Fig.5-B Photomicrograph of human milk
flowing in a capillary
(b-1) 120μm   (b-2) 80μm
(b-3) 50μm

Fig.6(a) and 6(b) show the profile of concentration distribution of milk fat globules in a capillary tube, and
these results were calculated from the photomicrogram as
shown above. The ratio (ΔN/N) means the numbers of fat
globules occupied on a certain area against the total
numbers of them on a unit area. As shown in Fig.6(a), in
the case of human milk compared with the non-coated
capillary tube, this ratio in a slicone coated capillary
tube decreased at the near side of the center of the tube
and increased at the near side of the wall. This fact
means that a milk fat globule of human milk has the
hydrophilic binding surface and, as a result, a repulsion
force will occur at the wall in a silicone coated capillary
tube. Therefore, aggregates of fat globules could not be
formed at the nearest side of the capillary wall, but
formed at a certain distance from the wall. The concentration distribution of fat globules of homogenized milk in
a capillary, calculated under the same conditions as above
was shown in Fig.6(b). In this case, the ratio of the
numbers in a silicone coated capillary tube changed remarkably in both the center of the tube and the near side

of the wall
compared
with the
non-coated
glass capil-
lary tube
It means
that protein
molecules
like casein
surrounding
milk fat
globules were

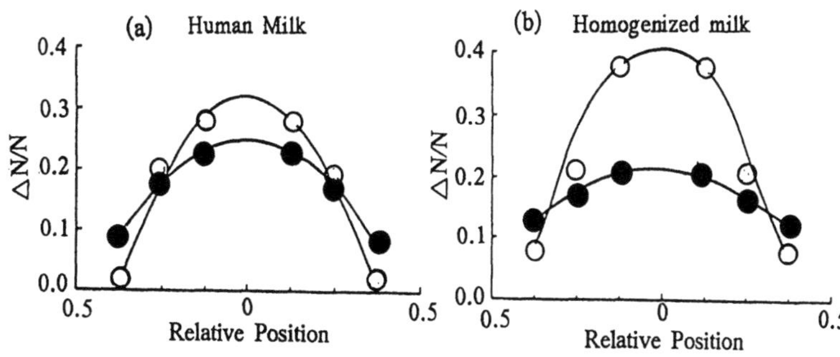

Fig.6 Ratio of partial number of milk fat globules against
total number versus relative position in a capillary.
○ Non-coated glass   ● Silicone coated glass

cracked by treatment of homogenization and, as a result,
an attractive force occurred between hydrophobic bindings
of milk fat surface and capillary wall. Thus, the flow
properties of milk will be changed with the formation
change of aggregation of milk fat globules in a capillary.
Consequently, the dependence of the particle size and size
distribution of fat globules of each milk will be one of
the important factors to determine the formation change of
aggregates of milk fat globules in a capillary.

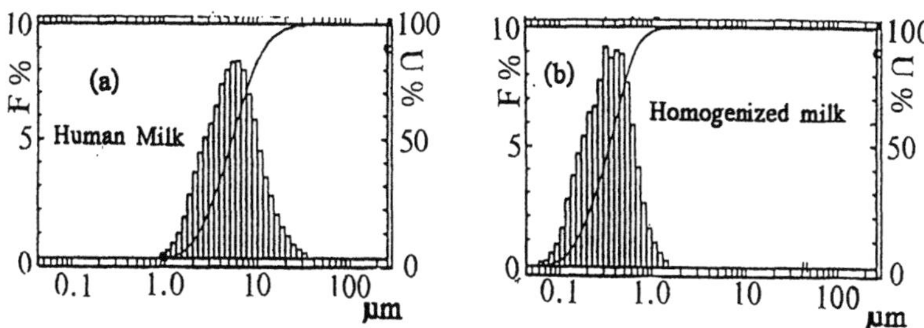

Fig.7 Percent distribution versus particle size distribution of milk fat globules.

Fig.7(a)and 7(b) show the particle size and the size distribution of milk fat globules of human milk and homogenized milk. Both of them show smooth distribution curve
and histogram, and those of human milk show a larger
particle size, whose average diameter is about 5.37 micron
meters. On the other hand, milk fat globules of homogenized milk show a very small particle size and their
average diameters are about 0.342 micron meters. In this
case, milk fat globules of homogenized milk will become
smaller due to homogenization, and in addition to this, a
protein layer will be cracked by homogenization; therefore, hydrophobic interaction occurred at the capillary
wall. A milk fat globule is composed of four main layers.
That is, innermost of the layer is composed of liquid fat
layer, and next to the inside layer included globulin, and

outermost of the layer is surrounded by a binding water
layer. A milk fat globule of human milk is surrounded by
the protein layer and a binding water layer, therefore,
larger fat globules were observed. However, in the case of
homogenized milk, they were removed by treatment of homogenization, so in this case, very small fat globules were
observed as shown in this experiment.

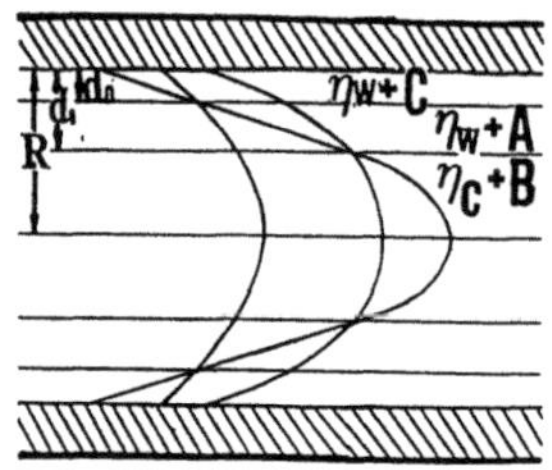

Fig.8 Schematic model of
flow in the presence of wall
layer.

As the results of it, non-Newtonian flow behavior and the
dependence of the capillary bore-size on viscosity of milk
in a capillary were caused by the formation change of
aggregate structure of the milk fat globules in a capillary. Consequently, a flow model of milk in a capillary is
suggested as shown in Fig.8. It is composed of three
layers, one is the layer of the center core of a tube, and
other one is the wall layer. However, the wall layer must
be composed of double layers in a flowing liquid, one is
the nearest side of the capillary wall, and another one is
the near side of the wall, and this layer is at a certain
distance from the wall, and it will be influenced by the
wall. From this flow model of double layers, the thickness
of the wall layers (do and di) will be calculated.

REFERENCES

1. T.Mineshita et al, J.Food Processing and
   Preservation 4, p.1 (1980).
2. T.Mineshita and A.Yamamoto, Proceedings of 5th
   Int. Congress on Biorheology p.89 (1981).
3. T.Mineshita et al, Biorheology 19:376 (1982).
4. A.Yamamoto et al, Biorheology 20:623 (1983).
5. T.Mineshita et al, J.Texture studies 17:205(1986).

MEASUREMENT OF FLUID VISCOSITY AND GEL
SETTING POINT BY A HOT-WIRE METHOD

Tomoshige Hori and Kensuke Itoh

Snow Brand Milk Products Co. Ltd.
Technical Research Institute
Kawagoe, Saitama 350, Japan

ABSTRACT

The theoretical considerations and application to the measurement
of food gel setting are reported for a new hot-wire technique that
applies the phenomenon of free-convection heat transfer to the
surrounding fluid from a line heat source. With fluids of different
viscosity, the rate of convection heat transfer to each fluid at the
surface of the heat source differs, heat being more rapidly
transferred the lower the viscosity of the fluid is, so that the
temperature of the heat source in an equilibrium state, when the heat
being generated by the source is constant, will be correspondingly
lower. This is because a lower viscosity fluid possesses a thinner
boundary layer around the surface of the heat source in the
equilibrium state. The temperature of a hot-wire probe designed for
laboratory and in-line use proved to be a vital measurement during the
process of gel setting. The inherent simplicity and precision of
this hot-wire technique provide a sound basis for practical
applications.

Food Hydrocolloids: Structures, Properties, and Functions
Edited by K. Nishinari and E. Doi, Plenum Press, New York, 1994

INTRODUCTION

True automation in the cheese industry had still not been applied by 1973 as was reported in the IDF Annual Bulletin.[1] This was because an objective method for detecting milk clotting during the process had not then been established, the secret of cheesemaking being in the most timely cutting of the milk coagulum which is referred to as curd.[2]

Three types of method for the objective measurement of milk clotting have already been outlined: 1) mechanical, 2) spectroscopic, and 3) ultrasonic methods.[3] Each of the mechanical methods suffers from the basic problem that the curd is destroyed. The spectroscopic and ultrasonic methods, while theoretically sound, still have some drawbacks.

The objective of this research has been to establish a new technique[4] that will detect the onset of gel formation accurately and in line.

THEORETICAL CONSIDERATIONS

Equation for Free-Convection Heat Transfer

In an equilibrium state of laminar free-convection heat transfer, the following functional relationship exists:[5]

$$f(Nu, Gr, Pr) = 0 \tag{1}$$

where

$$Nu = \alpha L / \lambda \tag{2}$$

$$Gr = L^3 g \, \beta \Delta \theta_s / \nu^2 \tag{3}$$

$$Pr = \nu / a \tag{4}$$

$$Q = \alpha \, S \Delta \theta_s \tag{5}$$

$$\Delta \theta_s = \theta_s - \theta_0 \tag{6}$$

When characteristic length L, gravitational acceleration g and surface area S in Equations 3-5 are assumed as constant, substitution of Equations 2-6 into Equation 1 yields the following relationship:

$$f(\nu , \lambda , a, \beta , \Delta \theta_s , Q) = 0 \qquad (7)$$

This basic expression (Eq. 7) indicates that the first four physical properties of a fluid can be correlated with the other two heat-transfer characteristics: i.e., the temperature difference between the surface of the heat source and the surrounding fluid, and the heat generated by the source.

Measurement of Food Gel Formation

Of the four physical properties in Equation 7, the thermal conductivity, thermal diffusivity and coefficient of volumetric expansion of foodstuffs are practically constant. This is because foodstuffs such as milk and gelatin solution consist principally of water, and the solid content is maintained constant during the gel-setting process. Therefore, the heat-transfer relationship in Equation 7 for processing a fluid food can be obtained as follows:

$$\nu = f(\Delta \theta_s , Q) \qquad (8)$$

Surface Temperature Measurement

The equation for steady-state heat conduction from a vertical cylinder in which heat is being generated,

$$\{ d^2/dr^2 +(1/r)(d/dr) \} \ \theta (r) = - \ W/ \lambda_w \qquad (9)$$

leads to the analytical solution

$$\theta_w - \theta_s = C (Q / \ell) \qquad (10)$$

where θ_w is the integrated mean temperature of the cylinder, defined by

$$\theta_w = (4 / \pi d^2) \int_0^{d/2} 2 \pi r \ \theta (r) \ dr \qquad (11)$$

This would be the case for both a single thin hot wire made of a metal such as platinum and a hot-wire probe comprising a stainless steel tube with a built-in resistor.

Therefore, provided heat Q generated in the platinum wire or built-in resistor is constant, the heat transfer relationship in Equation 8 for processing the fluid can be developed as follows:

$$\nu = f(\Delta \theta_w) \tag{12}$$

$$\Delta \theta_w = \theta_w - \theta_0 \tag{13}$$

Equations 12-13 indicate conclusively that the temperature of the built-in hot wire can detect the onset of gel setting directly as the change in fluid viscosity.

MATERIALS & METHOD

Pasteurized whole milk was coagulated with rennet in the curdmaking tank of a cheese plant, and commercially available sterilized whole milk was treated with a starter culture in the laboratory. A gelatin solution was also tested by cooling in individual vessels exposed to a constant temperature. The gelation time and/or temperature of each of the three solutions were then measured with the hot-wire probe to detect gelling or clotting.

RESULTS & DISCUSSION

The temperature of the probe with a built-in resistor proved to be a vital measurement during the process of gel formation. The temperature of the thin hot wire detected both the gel setting and gel melting points with little influence from the cooling or heating rate of the sample. The hot wire covered with a stainless steel tube could also detect the clotting time, which is tentatively defined as the time at the point of inflection in the time vs. temperature difference curve(Figure 1). In addition, hot-wire measurements during the curdmaking process at a cheese plant proved the practical potential of the hot-wire technique. The hot-wire temperature could detect the onset of milk clotting resulting from an enzymatic reaction in line and in real time without disturbing the milk coagulum.

CONCLUSION

These results amply verify the validity of the present hot-wire
method, which is a unique technique developed by the authors.   The
non-rheological hot-wire probe can detect the change in viscosity of a
fluid containing a high proportion of water by straightforward
measurement of the hot-wire temperature, which provides high potential
for practical applications.

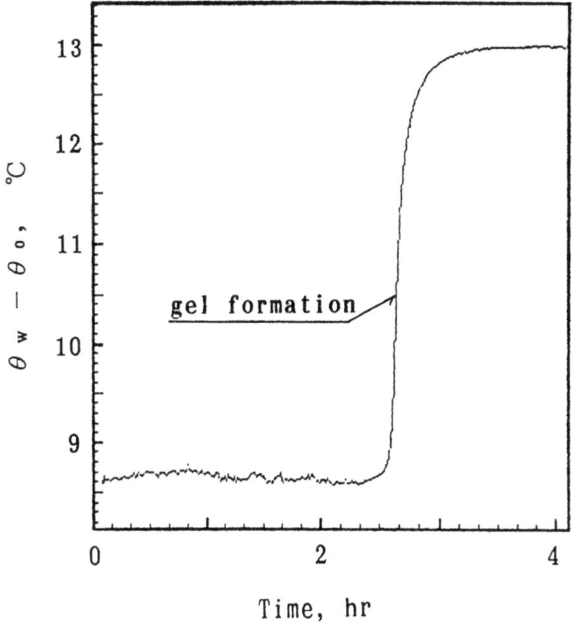

Figure 1. Measurement of the process of gel
formation during the starter-culture treatment of
sterilized milk by a hot-wire probe (2mm diameter,
100mm length)

NOMENCLATURE

a	Thermal diffusivity, m^2/s
C	Numerical constant in Equation 10, mK/J
g	Gravitational acceleration, m/s^2
Gr	Grashof number, dimensionless

L	Characteristic length, m
ℓ	Length of the hot-wire probe, m
Nu	Nusselt number, dimensionless
Pr	Prandtl number, dimensionless
Q	Total heat flux, J
r	Coordinate in a radial direction, m
S	Surface area, m^2
W	Heat generated in a unit volume, J/m^3
α	Heat transfer coefficient, W/m^2K
β	Coefficient of volumetric expansion, 1/K
$\Delta\theta_s$	Temperature difference between θ_s and θ_o, $^\circ C$
$\Delta\theta_w$	Temperature difference between θ_w and θ_o, $^\circ C$
θ	Temperature, K
θ_o	Temperature of the surrounding fluid, $^\circ C$
θ_s	Surface temperature, $^\circ C$
θ_w	Hot-wire temperature, $^\circ C$
λ	Thermal conductivity, W/mK
λ_w	Thermal conductivity of the hot wire, W/mK
ν	Kinematic viscosity, m^2/s

REFERENCES

1. C.H. Brissenden, Automation in the cheese industry, International Dairy Federation Bulletin, 72:1 (1973).

2. T. Hori, Measurement of fluid viscosity by a hot-wire method and its application to the cheese industry, in "Advances in Food Engineering," ed. by P. Singh and A. Wirakartakusmah, CRC Press, Florida (1992).

3. J. Tomathow and E. Voss, Methods for the determination of the firmness of milk coagulum, International Dairy Federation Bulletin, 99:1 (1977).

4. T. Hori, Objective measurements of the process of curd formation during rennet treatment of milk by the hot-wire method, J. Food Sci., 59(4):991 (1985).

5. W. Nusselt, Das Grundgesetz des Warmeubergangs, Gesundh Ing., 38(42):477,490 (1915).

EFFECT OF THE ADDITION OF HIGH METHOXY PECTIN ON THE RHEOLOGY AND COLLOIDAL STABILITY OF ACID MILK DRINKS

Alan Parker, Patrick Boulenguer and Thierry P. Kravtchenko

Sanofi Bio-Industries
Baupte
50500 Carentan
France

ABSTRACT

The effect of high methoxy pectin on the behaviour of model acid milk drinks has been investigated using viscometry, laser diffraction size measurement and micro-electrophoresis. The results of these different tests are interpreted using the theory of colloidal stability. With increasing pectin concentration, the behaviour changes from flocculated with few particles below 0.2µm diameter and shear thinning, time-dependent rheology to non-flocculated, many sub 0.2µm particles and low viscosity with Newtonian rheology. With even higher pectin concentrations, the acid milk drink remains non-flocculated, but with increasing viscosity and close to Newtonian rheology.

It is shown that pectin adsorbs to the surface of the casein aggregates and acts as a dispersing agent. Its stabilising effect cannot be explained by electrostatic repulsion, but presumably by steric stabilisation. The pectin is probably anchored to the casein surface by electrostatic attraction.

INTRODUCTION

When yoghurt is converted into a drinkable form, the resulting product suffers from sedimentation problems. It has been known since the late 50's (1) that high-methoxy pectin can "stabilise" such products, i.e. prevent sediment formation. Although such acid milk drinks are very widespread, particularly in the Far East, little has been written about the mode of action of the pectin for this application. In the early 80's, Glahn (2) and van Hooydonk et al. (3) looked at the factors affecting the performance of pectin in acid milk drinks.

Yoghurt consists of a gel made up of aggregates of casein. The structure of such acid milk gels is highly complex and depends critically on the method and rate of acidification (4). Model gels can be prepared at room temperature by using the decomposition of glucano-delta lactone as a source of hydrogen ions. Glahn (2) used gels prepared in this way to study the effect of the stabilisation of milk proteins by pectin and other hydrocolloids. In the present study, the starting material was yoghurt produced by fermentation with lactic acid bacteria, which was sheared before addition of pectin. This approach is closer to the manufacturing process, but obviously presents potential problems of reproducibility.

The effect of increasing amounts of pectin on the physical behaviour of acid milk drinks is described. These results are then compared with those from particle size

Food Hydrocolloids: Structures, Properties, and Functions
Edited by K. Nishinari and E. Doi, Plenum Press, New York, 1994

measurement and consumer acceptability tests. The theory of colloidal stability is used to correlate these results.

MATERIALS AND METHODS

The pectin used throughout this study was an experimental sample of lemon pectin produced by Sanofi Bio-Industries (SBI), France. Its stabilising properties are similar to those of SBI commercial products. Its degree of methoxylation, determined by the titration method (5), was 73%. The weight- average molecular weight was 450,000, as determined by HPLC, using PEO calibration (6).

The yoghurt used was a commercial low-fat product "Taille-Fine" made by Gervais-Danone, France.

Model acid milk drinks were prepared by diluting yoghurt in the ratio 2:1 with pH 4 Na-citrate buffer, and then shearing, using a homogeniser (Rannie Mini-Lab) at 200 bar. The diluted yoghurt was immediately mixed with the pectin solution.

The particle size distribution was measured by low angle laser diffraction using a Malvern MasterSizer (45mm focal length lens, code for optical properties of the particles, "presentation", 0403), after two weeks storage at 4°C. The dispersing medium was pH 4 Na-citrate buffer. It was found that gentle handling of the samples was necessary to obtain reproducible results, as violent mechanical action destroyed the flocs.

The viscosity of acid milk drinks was measured using a Contraves LS30 viscometer. Measurements were made at shear rates close to $5s^{-1}$, $20s^{-1}$ and $120s^{-1}$. For very time-dependent samples, the viscosity was taken after two minutes, if a steady reading had not been reached by this time.

The degree of flocculation was determined after two weeks storage at 4°C by the naked eye from the graininess and homogeneity of a drained film of product on the wall of a glass beaker.

The electrophoretic mobility of acid milk drink particles was measured in pH 4 Na-citrate buffer, using a Coulter DELSA. Electrophoretic mobilities were converted to zeta potentials using the Smoluchovski equation (7).

RESULTS AND DISCUSSION

A series of samples was prepared with constant yoghurt concentration and increasing concentrations of pectin. For each sample, the particle size distribution, viscosity at three shear rates and degree of flocculation were determined. Two types of behaviour could be distinguished as the pectin concentration increased: flocculated at low pectin concentrations and stabilised at higher concentrations. Results typical of the evolution of a sample with increasing pectin concentration are shown in fig. 1. The left-hand column shows the particle size distribution and the right-hand column the viscosity at increasing shear rates going from left to right.

The results typical of zero or low (≤ 250mgl^{-1}) pectin concentrations are shown in fig. 1a and 1b. Such samples were judged to be very flocculated by the naked eye. The particle size distribution is trimodal, this feature being very reproducible both for repeat measurements of a particular sample and for other flocculated samples. The population of large particles is close to 10μm in diameter. Rheologically, such samples were very viscous, extremely shear thinning and also very time-dependent.

Fig. 1c and 1d show results typical of a sample containing 750mgl^{-1} of pectin, judged to be moderately flocculated by the naked eye. The size distribution shows a much reduced peak at large size, and a much larger quantity of small particles, now with diameters stretching down to 0.1μm. With the addition of pectin, the shear thinning decreases as well as the mean viscosity.

Fig. 1e and 1f show results at the minimum concentration of pectin for which the product was unflocculated (1250mgl^{-1}), as judged by eye. The rheology is Newtonian and the viscosity is at a minimum. The size distribution shows that all of the large particles have disappeared, to be replaced by small ones.

At even higher concentrations of pectin (data not shown) the viscosity increases once again, but the rheology remains practically Newtonian and non-time-dependent. The

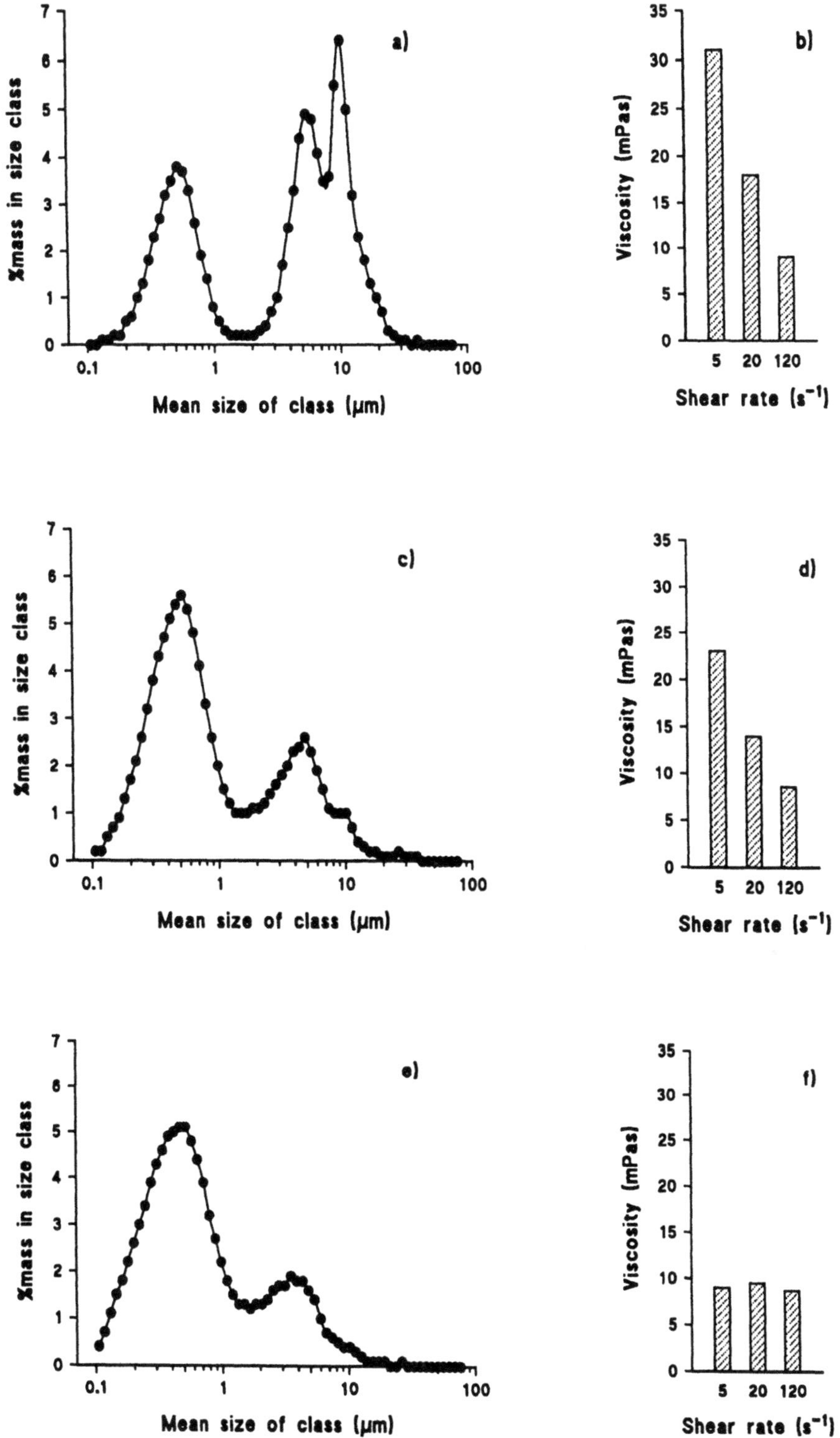

Figure 1. Comparison of particle size distributions (left hand column) and viscosity (right hand column) for very flocculated (a and b), moderately flocculated (c and d), and unflocculated (e and f) acid milk drinks.

size distribution is not changed from that shown in fig. 1e, neither does the degree of flocculation increase.

This sequence of rheological behaviour was found to be universal for high methoxy pectins, although the minimum viscosity, and the pectin concentration at which it occured, varied with the pectin sample, as did the sharpness of the viscosity rise at high pectin concentrations.

The different kinds of behaviour can be understood in terms of the theory of colloidal stability (7), which sums the attractive and repulsive forces acting on particles to predict their stability. A colloidal dispersion is said to be stable if its particle size distribution does not evolve, which implies that there is no net attraction between the particles. In contrast, a dispersion is said to be colloidally unstable if the particle size increases with time, implying net attraction between the particles. This definition differs from that widely used in industry, where stability can simply imply a lack of visible sedimentation, or even more broadly, an unchanged appearance after storage.

The interaction forces between particles, their "stickiness", explains all of the correlations seen between measurements when passing from a pectin-free sample to a stable one with Newtonian rheology. A net attraction between particles causes them to flocculate, which creates structure in the sample. This structure is broken down when shear is applied to the sample, i.e. a colloidally unstable dispersion is shear thinning. Breaking down the structure takes time, so such dispersions are also time-dependent. Obviously, the attraction between particles creates large flocs, and can lead to sedimentation problems. On the other hand, if the concentration of particles is not too high, repulsion between particles leads to a Newtonian rheology and no time-dependence, since there is no aggregated structure, i.e. flocculation.

It is therefore clear that as the concentration of pectin increases, the acid milk drink passes from a colloidally unstable state into a colloidally stable one.

In the particle size distribution of the flocculated sample, fig. 1a), the population of large particles, centred on a diameter of 10µm is interpreted as the flocs which form a sediment. The population of medium sized particles, centred on a diameter of 5µm, which is always present, is probably due to fat globules (about 1%). The population of small size is composed of non-flocculated particles. Note that in the very flocculated sample, fig. 1a, there are no particles below about 0.2µm.

Up to this point, we have only discussed experimental observations, and a way in which they can be unified using the theory of colloidal stability. The exact mechanism of by which the pectin acts has not been discussed.

The existing literature (2, 3, 8) suggests that adsorption of pectin onto casein aggregates occurs, although no data to check this idea has ever been presented. This point has been verified by measuring the apparent adsorption isotherm of pectin onto acid milk drink particles. The isotherm can only be apparent, because addition of pectin increases the surface area of casein available. For a true isotherm the adsorbed amount would be expressed as the quantity of pectin per unit of surface. Acid milk drinks were ultracentrifuged and the concentration of pectin in the supernatant measured, so that the quantity adsorbed onto the casein surface could be calculated by difference. Fig. 2 shows the resulting apparent adsorption isotherm, which confirms that pectin is actually adsorbed onto the surface of the casein aggregates. However, even at the highest pectin concentration used, there is no sign of a plateau, due to saturation of the surface. This phenomenon is expected, due to the increase in the quantity of surface with increasing pectin concentration: as more pectin is added, the particle size decreases, exposing more surface. The true adsorption isotherm can only be derived if the surface area is known, which is not possible for the irregular aggregates of casein.

All authors (2, 3, 8) agree that the stabilising effect of the pectin is "probably" due to interparticle repulsion caused by the negative charge acquired by the particles on adsorption of pectin. However, to our knowledge, no evidence has been presented to support this point of view.

To check this hypothesis, the zeta potential of a series of acid milk drinks containing increasing concentrations of pectin was measured. Without pectin, a potential of +1mV was found, which is expected for proteins below their isoelectric point. The zeta

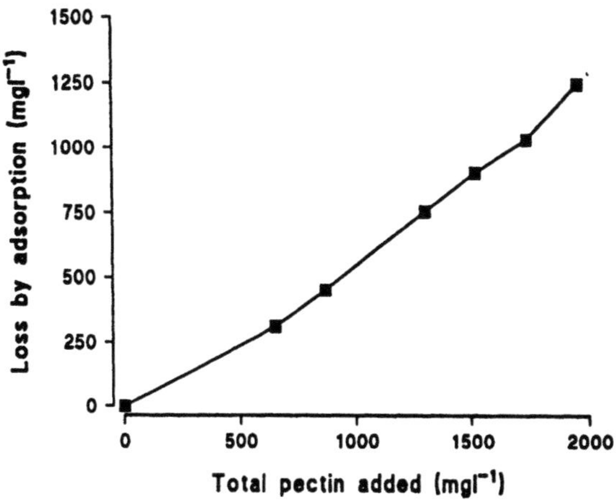

Figure 2. Amount of pectin adsorbed onto casein aggregates as a function of the total concentration of pectin added.

potential with pectin concentrations between 250 and 2000mgl^{-1} was -11±1 mV. The value did not vary with pectin concentration.

By analogy with casein micelles at pH 6 (9), it can be concluded that the electrostatic charge of the pectin covered casein particles at pH 4 is insufficient to stabilise them. In fact, a zeta potential of the order of -10mV does not provide sufficient repulsion to overcome the van der Waals attraction. The stability of casein micelles at pH 6 is explained (9) by the presence of soluble polymer chains (κ casein) attached to the surface. In the case of acid milk drinks, the stability may be explained by pectin molecules attached to the surface of the casein aggregates. It is well known (7) that such chains can "protect" colloids from flocculation. When two particles covered with soluble polymer layers approach too closely, the coats of adsorbed polymer interpenetrate and create a repulsion between the particles, so that flocculation is prevented.

Pectins which stabilise acid milk drinks at pH 4 cause phase separation when mixed with milk casein at its natural pH, close to 6 (10). The obvious difference between the system at pH 6 and pH 4 is that the electrostatic charge of the casein has changed from negative to positive. This difference changes the interaction of casein with pectin from repulsion, leading to phase separation, to attraction, leading to adsorption. This phenomenon strongly suggests that the interaction between pectin and casein is electrostatic, and not hydrophobic, since hydrophobic interactions are not pH dependent.

Polymeric dispersing agents must possess two contradictory properties: on the one hand, they must adsorb strongly, implying strong polymer/surface interactions, and on the other hand they must form a thick adsorbed layer in solution, implying weak polymer/surface interactions. Efficient polymeric dispersing agents are therefore always co-polymers, one (strongly interacting) molecular species serving to anchor the chains to the interface and the other (weakly interacting) serving to form the adsorbed layer. It is suggested that, in this application, high methoxy pectin conforms to this specification, since the free carboxylic acid groups form strong electrostatic bonds with the net positively charged surface of the casein aggregates and the methoxylated galacturonic acid units, which do not interact with the surface, form a stabilising layer in solution.

There are close analogies between the stabilisation of acid milk drinks by pectin and emulsification. In both cases, mechanical energy must be applied to the system to create small particles, and the presence of the dispersing agent serves to prevent their subsequent reaggreagation. In the model presented here, homogenisation of acid aggregated casein breaks electrostatic bonds, which have a substantial role in such gels (11). The pectin present in solution can then adsorb by interacting with the newly exposed positively charged groups. Unpublished measurements showed that, like emulsification, the action of pectin is not spontaneous: without a vigorous homogenisation, the pectin has little effect on the behaviour of the acid milk drink.

The theory of colloidal stability explains the physical changes occuring in acid milk drinks up to the viscosity minimum, but it is not clear why the viscosity starts to rise at pectin concentrations above that necessary for the minimum viscosity. The rheology remains close to Newtonian, excluding as explanations both pectin gel formation and exclusion flocculation, which occurs when sufficient non-interacting polymer is added to a colloidal dispersion (7). Neither can the viscosity rise be explained by a thickening effect of the pectin. Further work is being carried out in order to elucidate this peculiar phenomenon.

CONCLUSIONS

The picture of the effect of pectin on casein at pH close to 4 which emerges from this study is as follows:

1) - The dilute yoghurt consists of flocs formed from small casein aggregates held together by hydrophobic and electrostatic forces. These forces are relatively weak. The rheology is typical of a weakly flocculated dispersion, the viscosity is highly shear rate and time dependant.
2) Pectin can adsorb to the flocs of casein by electrostatic interactions, via its negative carboxylate groups, which interact with the positive groups on the surface of the flocs. The adsorbed pectin forms a barrier to interaction between casein particles and protects them from flocculation, principally by steric repulsion, since the electrostatic charge of the particles is insufficient to stabilise them. The reduced attraction between particles changes the rheology of the acid milk drinks: the shear thinning and time-dependence disappear if sufficient pectin is added.

On the basis of the results presented here, a simple and rapid test to evaluate the stabilising power of pectins has been developed, based on the sediment volume of acid milk drinks after stabilisation. This work will be the object of a following paper.

ACKNOWLEDGEMENTS

Many people contributed to this work, but we would like to especially thank Eric Aubrée, SBI Food Development Centre, France and Kimiaki Abe, Sanofi Bio-Industries, Japan. We are grateful to the management of SBI for permission to publish this work.

REFERENCES

1. Doesberg, J. J. and De Vos, L. (1959) *5th Internat. Fruit Juice Cong.*, 320-37.
2. Glahn, P. E. (1982) *Prog. Food Nutr. Sci.*, 6, 171-7.
3 van Hooydonk, A. C. M., Smalbrink, L. and Hagedoorn, H. G. (1982) *NIZO Bulletin* nos. 7 and 8.
4. Roefs, S. P. F. M., De Groot, A. E. A. and van Vliet, T. (1990) *Colloids Surfaces*, 50, 141-159.
5. Food Chemical Codex 1972. 2nd ed., pub. National Research Council, Washington DC, 580.
6. Brigand, G., Denis, A., Grall, M. and Lecacheux, D. (1990) *Carbohydr. Polym.*, 12, 61-77.
7. Hunter, R. J. (1986) *The foundations of colloid science, vol. 1*, pub. Oxford University Press, Oxford.
8. Amberg Pedersen, H. C. and Jørgensen, B. B., (1991) *Food Hydrocolloids*, 5, 323-328.
9. Payens, T. A. J. (1977) *Biophys. Chem.*, 6, 263-70.
10. Antonov, Yu. A., Grinberg, V. Ya., Zhuravskya, N. A. Tolstoguzov, V. B. (1982) *Carbohydr. Polym.*, 2, 81-90.
11. Roefs, S. P. F. M. and van Vliet, T. (1990) *Colloids Surfaces*, 50, 161-175.

CHANGES IN FIRMNESS OF STRAWBERRY FRUIT
DURING GROWTH AND PARTIAL PURIFICATION
OF ITS POLYGALACTURONASE

Yoichi Nogata, Koh-Ichi Yoza, and Hideaki Ohta

Chugoku National Agricultural Experiment Station
Ministry of Agriculture, Forestry and Fisheries
Fukuyama, Hiroshima 721, Japan

ABSTRACT

The correlation between firmness change in strawberry fruit (*Fragaria ananassa*, Duch. CV. Toyonoka) and molecular weight distribution of water soluble polyuronides (WSP) during softening was investigated. Lower-molecular-size polyuronides increased slightly with fruit development. A low level of polygalacturonase (PG) activity was found in this fruit throughout the developing stage, and it was shown to consist of three enzymes (PG1, 2, and 3) that could be separated by cation-exchange chromatography. PG1 released a small amount of oligomers in addition to galacturonic acid (GA). On the other hand, PG2 released only GA, indicating an exo-type PG.

INTRODUCTION

The economic value of fresh strawberry fruit strongly depends on the consumer's perception of its quality, in particular firmness. It is generally accepted that endo-PG plays a crucial role in converting protopectin in cell walls to a soluble form during ripening. In tomato there is evidence of a decrease in the average size of the soluble polyuronide chains late in ripening[1]. In addition, there is a correlation between the amount of PG, the rate of softening, and pectin degradation in this fruit[2, 3, 4]. Strawberry also showed an increased level of soluble polyuronides during softening[5, 6], although molecular size features of soluble polyuronides showed little consistent change during development[7], and little information relating to the amount of PG and the rate of softening is available. In this study, we investigated the relationship between firmness change and molecular size distribution of WSP, demonstrated the occurrence of PG, and evaluated the mode of its action.

MATERIALS AND METHODS

Strawberries were classified into five groups according to their developing stage; small green (SG), large green (LG), reddish (W/R), red ripe (R), and over ripe (OR). Each experiment was performed immediately after harvest.

Firmness was measured by a rheometer (Fudoh, NRM-3003D, Tokyo, Japan), using a 2 mm diameter disk shaped plunger, at a load speed of 300 mm/min.

Food Hydrocolloids: Structures, Properties, and Functions
Edited by K. Nishinari and E. Doi, Plenum Press, New York, 1994

The molecular weight distribution curve of the water soluble polyuronide fraction at the SG, LG and R stage was obtained by gel-filtration chromatography. An aliquot (0.2 ml) of each sample (1 mg of anhydro uronic acid/ml) was applied to a Superose 6 HR 10/30 column (10 x 300 mm, Pharmacia Fine Chemicals, Uppsala, Sweden) equilibrated with 50 mM sodium acetate containing 20 mM EDTA (pH 5.0) and eluted with the same buffer at a flow rate of 0.25 ml/min. Fractions of 0.4 ml were collected and analyzed for uronic acid by the hydroxydiphenyl reagent[8].

Enzymes were extracted by homogenizing 1 kg of R stage tissue in 2 l of cold water with 100 g of polyamide C-200 (Wako Pure Chemicals, Osaka, Japan). After adjustment to pH 6.0 with 1 N NaOH, the homogenate was centrifuged at 10 000 x g for 30 min. The insoluble fraction was washed with 1 l of cold water (pH 6.0) and centrifuged again. The enzyme was extracted by suspending the pellet in 1 l of 50 mM sodium acetate containing 1 M sodium chloride (pH 6.0). After standing for 12 hr at 0-4 °C, with occasional stirring, the suspension was centrifuged and the supernatant was dialyzed against 50 mM sodium acetate (pH 6.0), then used as the crude enzyme.

The enzyme activity assay was done as follows. The reaction mixture consisted of 50 μl of 0.2 M sodium acetate (pH 5.5), 100 μl of 0.4 % polygalacturonic acid (PGA), and 50 μl of enzyme solution. After 18 hr at 37 °C, the solutions were analyzed for reducing groups by the cyanoacetamide reagent[9].

Separation of R stage crude enzyme was performed by applying solubilized 40-80 % ammonium sulfate precipitate to a CM-Sepharose column (2.5 x 18 cm) equilibrated with 20 mM sodium acetate (pH 6.0) using a sodium chloride gradient (0-250 mM).

Identification of mode of action and reaction products of PG was carried out as follows. Reaction mixtures containing 0.5 ml of 0.4 % purified PGA, 0.25 ml of 0.2 M sodium acetate (pH 5.5), and 0.25 ml of concentrated enzyme preparation (PG1 or 2) from CM-Sepharose were incubated at 37 °C for 48 hr and heated at 100 °C for 5 min to stop the reaction. The control was heat-inactivated before incubation. Then, the mixture was passed through a Toyopak IC-SP M (H^+ form, Tosoh, Tokyo, Japan) and concentrated five times by centrifugal evaporation. Aliquots (50 μl) of reaction products released were separated on a A-2H^+ column (10 x 300 mm, Bio-Rad, CA, USA) at 85 °C using 5 mM sulfuric acid as eluent at a flow rate of 0.3 ml/min, monitored by an refractive index detector (Shimadzu RID-6A, Kyoto, Japan), and represented as RIU (refractive index unit).

RESULTS AND DISCUSSION

Firmness as measured by a rheometer decreased markedly during development from SG to LG stage (55 % loss), and continued to decrease slowly thereafter (Figure 1). Finally at the OR stage, the firmness decreased to approximately 10 % of the initial level.

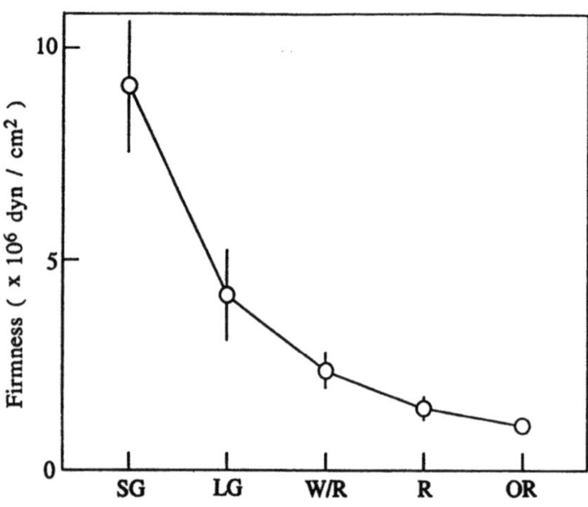

Figure 1. Changes in firmness during development of strawberry fruit.

Figure 2 shows the molecular weight distribution. Although little molecular weight change was observed, the amount of higher-molecular-size polyuronides decreased and that of lower-molecular-size polyuronides increased at the LG stage compared to the SG stage. There was little difference between the elution profile at the LG and R stage. These findings indicated the possibility of a correlation between the firmness change and molecular weight distribution during SG to LG stage.

Figure 3 presents the elution profile of strawberry PG at R stage on a CM-Sepharose column. Three PG-isoenzymes existed in the crude enzyme preparation of R stage strawberry. PG1 was eluted in the non-absorbed fraction while PG2 and 3 were bound enzymes and released by the sodium chloride linear gradient.

PG1 released a small amount of oligomers (8, 7, 6, 5 mer) in addition to a monomer (Figure 4a). Whereas PG2 released only a monomer (Figure 4b). These findings implied that the PG2 in strawberry was of the exo-type.

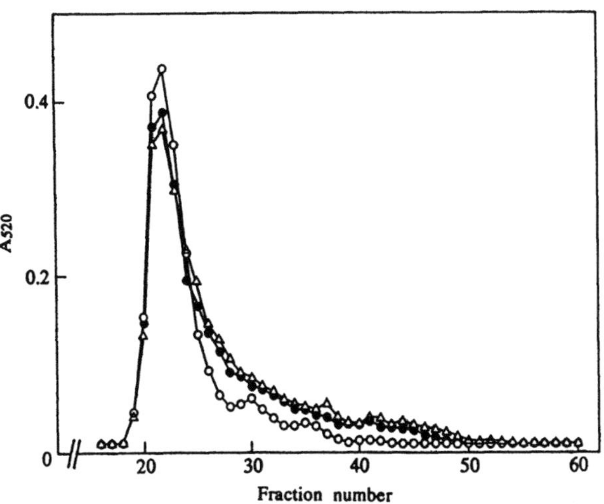

Figure 2. Gel filtration chromatography on a Superose 6 HR 10/30 of strawberry water soluble polyuronides. (O) SG, (●) LG, (△) R

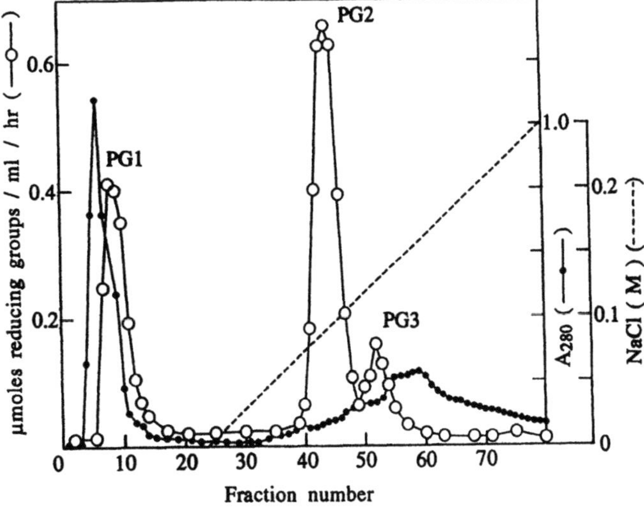

Figure 3. Elution profile of strawberry PG at R stage from a CM-Sepharose column. Fractions of 8.0 ml were collected. Activities were designated PG1, 2, and 3 in the order of their elution from this column.

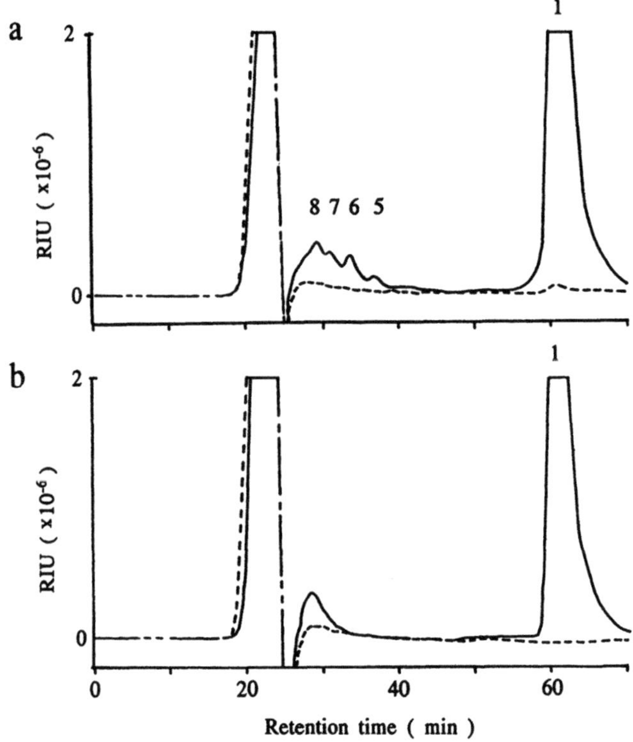

Figure 4. HPLC analysis on an A-2H$^+$ column of reaction products released by PG1 (a) or PG2 (b) from polygalacturonic acid after incubation for 48 hr. (—) activity, (---) control

Exo-PG has been suggested to play a role in the complete hydrolysis of the pectic fraction solubilized by the action of endo-PG[10], or to take part in the initiation of climacteric ethylene[11]. In this study, the activity levels of exo- and endo-PG during softening of strawberry fruit is still obscure, although as a result of the enzyme reaction, some of the WSP shifted to a lower-molecular-size during development from SG to LG stage.

REFERENCES

1. D.J. Huber, Polyuronide degradation and hemicellulose modifications in ripening tomato fruit, *J. Am. Soc. Hortic. Sci.* 108:405 (1983).
2. C.J. Brady, G. MacAlpine, W.B. McGlason, and Y. Ueda, Polygalacturonase in tomato fruits and the induction of ripening, *Aust. J. Plant Physiol.* 9:171 (1982).
3. C.J. Brady, W.B. McGlason, J.A. Pearson, S.K. Meldrum, and E. Kopeliovitch, Interactions between the amount and molecular forms of polygalacturonase, calcium and firmness in tomato fruit, *J. Am. Soc. Hortic. Sci.* 110:254 (1985).
4. C.J. Brady, S.K. Meldrum, W.B. McGlasson, and Z.M. Ali, Differential accumulation of the molecular forms of polygalacturonase in tomato mutants, *J. Food Biochem.* 7:7 (1983).
5. J.R. Woodward, Physical and chemical changes in developing strawberry fruits, *J. Sci. Food Agr.* 23:465 (1972).
6. M. Knee, J.A. Sargent, and D.J. Osborne, Cell wall metabolism in developing strawberry fruits, *J. Expt. Bot.* 28:377 (1977).
7. D.J. Huber, Strawberry fruit softening, the potential roles of polyuronides and hemicelluloses, *J. Food Sci.* 49:1310 (1984).
8. N. Blumenkrantz, and G. Asboe-Hansen, New method for quantitative determination of uronic acids, *Analyt. Biochem.* 54:484 (1973).
9. K.C. Gross, A rapid and sensitive spectrophotometric method for assaying polygalacturonase using 2-cyanoacetamide, *HortSci.* 17:933 (1982).
10. R. Pressy and JK Avants, Pear polygalacturonases, Phytochemistry, 15:1349 (1976).
11. E.A. Baldwin and R. Pressy, Treatment of tomatoes with an exo-enzyme increases ethylene and accelerates ripening, *Proc. Fla. State Hort. Soc.* 101:215 (1988).

HEAT-INDUCED TRANSPARENT GELS OF GLOBULAR PROTEINS

Etsushiro Doi,[1] Fumito Tani,[1] Michiyo Murata,[2] Taihei Koseki,[3] and
Naofumi Kitabatake[1]

[1]Research Institute for Food Science, Kyoto University
 Uji, Kyoto 611, Japan
[2]Kacho Jr. College, Higashiyama-Ku, Kyoto 605, Japan
[3]Department of Food Science, Mukogawa Women's University
 Nishinomiya, Hyogo 663, Japan

INTRODUCTION

Most globular protein solutions will form gels when heated and will not melt with
further heating. Globular protein gels are more nutritious than polysaccharide gels and are
also suitable for food. Globular protein gels differ from polysaccharide gels and gelatin gel
(Table 1) in several important ways. Most globular protein gels have a turbid appearance
similar to that of boiled egg white. The critical concentrations of gelling for globular
proteins are high (5 to 10%). However, these characteristics of globular protein gels can be
changed under certain conditions.

Table 1. General properties of globular protein gels and polysaccharide
or gelatin gels.

	Globular protein gel	Polysaccharide* or gelatin gel
Unit molecule	globular	fibrous
Appearance of gel	turbid	transparent
Gel formation	heat-set gel	cold-set gel
Gel-sol transformation	irreversible	reversible
Critical concentration for gelling	high	low

*The characteristics of agarose gel are taken as an example.

Food Hydrocolloids: Structures, Properties, and Functions
Edited by K. Nishinari and E. Doi, Plenum Press, New York, 1994

The properties of globular protein gels change as a function of variations in pH, variations in the ionic strength of the solution, and variations in the heating procedure. We have successfully manipulate these variables to create transparent gels from ovalbumin, bovine serum albumin, and hen egg lysozyme. Transparent ovalbumin gel has some properties in common with polysaccharide gels and gelatin gel, in terms of both gel structure as well as the gelling process.

"STRING-OF-BEADS" POLYMERS IN GLOBULAR PROTEIN GELS

When globular proteins are denatured by urea or guanidine hydrochloride in the presence of a reducing reagent, their peptide chains unfold into random coils. Normally, these peptide chains do not form gel networks. In the case of heating, the gel of globular proteins is most likely formed because the globular proteins do not completely unfold. These proteins do not completely denature and may retain their globular forms. These compact, partially denatured proteins may then form gel networks.

There are two extreme types of globular protein gels. One is composed of highly oriented "string-of-beads" polymers, while the other is composed of random aggregates of proteins.[1] The "string-of-beads" gels have been observed in heat-set gels of serum albumin, lysozyme, and ribonuclease. [2,3]

A typical string-of-beads polymer has been detected in the heat-set gel of soybean glycinin.[4-6] A glycinin molecule contains twelve subunits.[7] Whether these subunits dissociate during heat treatment and reassociate into gel strands is not clear.[4,5]

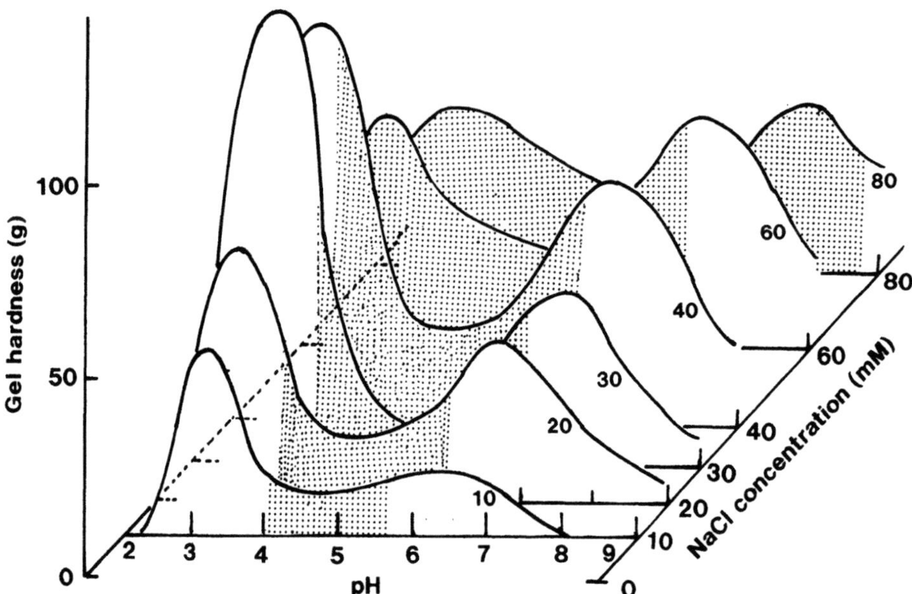

Figure 1. Hardness and turbidity of heat-induced ovalbumin gels at various pH and NaCl concentrations. A 5% ovalbumin solution containing 20 mM NaCl was heated at 80°C for 1 hr. The figures incorporate data from reference 8. Turbid gels are obtained inside the stippled area. Numerals under curves indicate the NaCl concentrations for each curve.

Ovalbumin is known to form a transparent gel under proper condition of the medium, such as pH and ionic strength. Further, it is known that this transparent gel is composed of string-of-beads polymers.

TRANSPARENT GELS OF OVALBUMIN

Figure 1 shows the effect of pH and NaCl concentration on the turbidity and gel hardness of 5% ovalbumin.[8] Ovalbumin solutions do not gel when heated, unless salt is added. Turbid suspensions containing coagulum are obtained near the isoelectric point (pI) of ovalbumin (pH 4.7), while transparent solutions are obtained at other pH values except for extreme acidic or alkaline pH regions..

At low NaCl concentrations (10-20 mM), turbid and soft gels are obtained near the pI, while transparent gels are obtained on either side of the pI, at approximately pH 3 and pH 7. At higher NaCl concentrations (30-80 mM), turbidity and gel hardness near the pI do not change significantly. The gel hardness at pH 3.5 and 7.5 is maximum at 30 mM and 50 mM NaCl, respectively, and decrease as NaCl concentrations increase. Gel turbidity increases as NaCl concentrations increase on either side of the pI. Hard gels are obtained at the critical pH and salt concentrations for turbidity, where translucent gels are obtained. Transparent and hard gels are obtained in a narrow range of pH and salt concentrations. This limitation can be overcome by varying the heating procedure.

Table 2. Critical concentrations for gelling of ovalbumin by one-step or two-step heating method at pH 7.0.

Exp. no.	Heating method	Buffer used for		Critical conc.(%)
		1st heating	2nd heating	
1	one-step	10 mM Na-phosphate		6.0
2	one-step	10 mM Na-phosphate 20 mM NaCl		3.5
3	two-step	10 mM Na-phosphate	+200 mM NaCl	1.0
4	two-step	10 mM Na-phosphate 10 mM NaCl	+200 mM NaCl	1.0
5	two-step	10 mM Na-phosphate 20 mM NaCl	+200 mM NaCl	0.9
6	two-step	10 mM Na-phosphate 30 mM NaCl	+200 mM NaCl	0.9

First and second heatings were conducted in glass tubes (0.6 X 10 cm) for 20 min at 80°C.
The formation of gel was judged by placing a steel ball (ϕ=4 mm) on the top of the gel.

TWO-STEP HEATING METHOD FOR TRANSPARENT OVALBUMIN GEL

A heated 5% ovalbumin solution (pH 7.5) will yield a transparent solution in the absence of NaCl, a transparent gel at 10-20 mM NaCl, a translucent gels at 30-40 mM NaCl and a turbid gel at higher NaCl concentrations.

The transparent solution obtained by heating ovalbumin in the absence of NaCl differs from the native solution in important ways. The heated solution contains heat-denatured ovalbumin, and has a higher viscosity than that of the native solution. In the two-step heating method, the solution is allowed to cool at room temperature. After cooling, NaCl is added to various concentrations and the mixture is reheated. This two-step heating method yields transparent gels at 10-200 mM NaCl, and translucent gels at 300-500 mM.[9] Gels produced by this method are harder than those produced at high salt concentrations by the one-step heating method.

Critical gel concentrations are also very low for gels prepared by the two-step heating method (Table 2). The lowest concentration for gelling is 0.9 %. This concentration is compatible to the critical concentration of agarose gel.

STRUCTURE OF TRANSPARENT OVALBUMIN GEL

The two-step heating method is useful both as a preparative method and as a means to elucidate the structure and mechanism of gelling. The transparent and viscous solution obtained by the first heating at low ionic strength contains a basic structure necessary to produce gel networks. This transparent solution is more easily analyzed by physical methods than are the turbid and solid gels.

The physicochemical properties of the transparent solutions obtained from ovalbumin at pH 7.0 in a 20 mM phosphate buffer by heating have been thoroughly examined.[10,11] Examination by circular dichroic spectrum analysis has revealed that the secondary structure of a heated ovalbumin molecule is not much different from that of a native molecule. Long

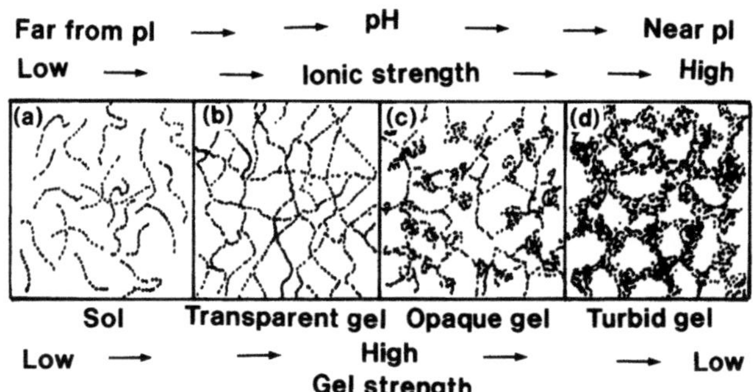

Figure 2. Model for the formation of gel networks by heated ovalbumin.

linear polymers (string-of-beads) in a heated 0.5% ovalbumin solution have been observed by transmission electron microscopy. A turbid suspension obtained after heating in the presence of 50 mM NaCl was found to contain random aggregates, while a transparent solution obtained by the two-step heating method, where 50 mM NaCl was added after the first heating, was found to contain cross-linked fibrous polymers.[10]

The weight-average molecular weight of the linear polymers formed from 0.5% ovalbumin in pH 7.0, with a 20 mM phosphate buffer at 80°C for 6 hr, was approximately 7,000,000.[11] The molecular dimensions of the polymers were analyzed by several methods including static light scattering, viscometry, and gel-permeation high pressure liquid

The weight-average molecular weight of the linear polymers formed from 0.5% ovalbumin in pH 7.0, with a 20 mM phosphate buffer at 80°C for 6 hr, was approximately 7,000,000.[11] The molecular dimensions of the polymers were analyzed by several methods including static light scattering, viscometry, and gel-permeation high pressure liquid chromatography (HPLC).[11] A linear polymer may be described in terms of a wormlike cylinder model as follows: M_L=1,600 Å^{-1}, d=120 Å, q=230 Å, where M_L, d, and q represent the molecular weight per contour length, diameter and persistence length, respectively.

The following models for the gelling process of ovalbumin[12] can explain the experimental results. The conformation of heat-denatured ovalbumin is not much different from that of the native molecule, except that some of the hydrophobic areas that were previously buried in the molecule become exposed on the surface. At pHs near the pI or at high ionic strength, denatured proteins randomly aggregate by hydrophobic interactions. At pHs far from the pI and at low ionic strengths, electrostatic repulsive forces hinder the formation of random aggregates, and linear polymers are formed.

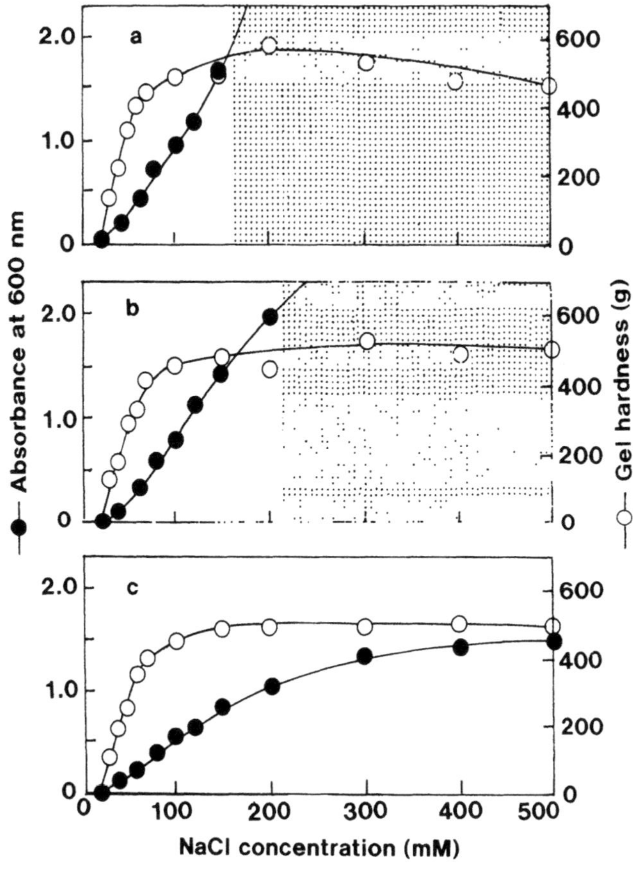

Figure 3. Effects of NaCl concentrations on the turbidity and hardness of 6% bovine serum albumin at pH 7.5 by one-step and two-step heating methods.[13] The first and second heating were conducted for 20 min at 85°C. (a) one-step heating method; (b) two-step heating method I (the first heating was conducted in the absence of NaCl); (c) two step heating method II (the first heating was conducted in the presence of 10 mM NaCl). Turbid gels are obtained inside the stippled area.

Figure 2 presents a model for the formation of random aggregate gels or string-of-beads gels, which depend on pH and ionic strength, as well as their relation to gel turbidity and gel hardness. Linear polymers are formed at pHs far from the pI and at low ionic strengths (Fig. 2 (a)). As electrostatic repulsive forces decrease (ionic strength increases or pH changes), three-dimensional networks begin to form a transparent gel (Fig. 2(b)). At high ionic strengths or at pHs near the pI, proteins polymerize to form a turbid gel composed of random aggregates (Fig. 2(d)). At intermediate ionic strengths or pHs, linear polymers and random aggregates are mixed together in the same gel, or linear polymers ball by intra-polymer interaction to form a translucent or opaque gel (Fig. 2(c)). Gel strength is maximum at the condition between that shown in Fig. 2(b) and that in Fig. 2(c).

In the two-step heating method, the first heating forms linear polymers (Fig. 2(a)) and the second heating causes a phase transition from Fig. 2(a) to Fig. 2(b), i.e., gel formation.

TRANSPARENT GELS OF BOVINE SERUM ALBUMIN AND LYSOZYME

The general applicability of the model descried above to proteins other than ovalbumin: i.e., bovine serum albumin[13] and hen egg lysozyme,[14] has been examined.

Heated 6% bovine serum albumin solutions give transparent gels or turbid gels depending on the pH and ionic strength of the medium as well as ovalbumin solutions.[13] The effects of salt concentration on the properties of gels prepared by the one-step or two-step heating methods are shown in Fig. 3. The two-step heating method, in which the initial heating occurs in the presence of 10 mM NaCl, yields transparent or translucent gels at wide range of NaCl concentrations.

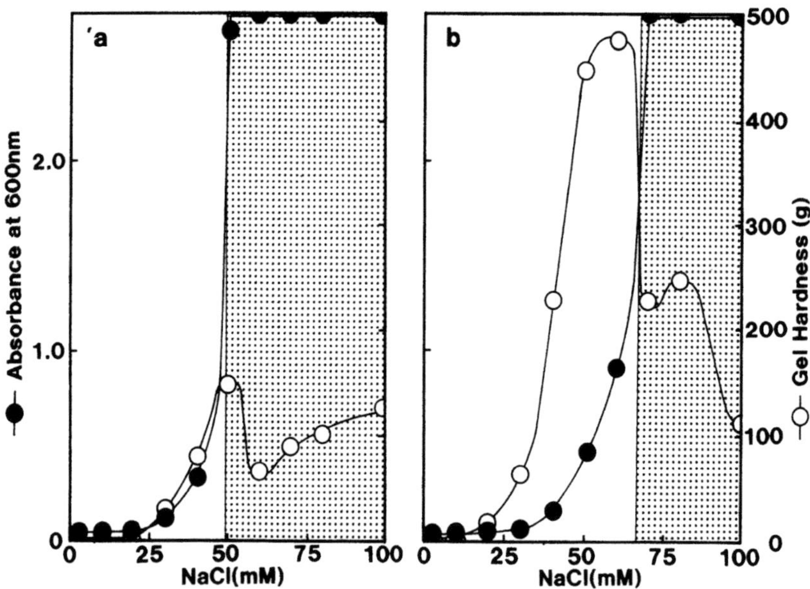

Figure 4. Effect of NaCl concentrations on the turbidity and hardness of 5% lysozyme at pH 7.0 in the presence of 7.5 mM dithiothreitol by (a) one-step and (b) two-step heating methods.[14] The first and second heating were conducted at 80°C for 20 min. Turbid gels are obtained inside the stippled area.

Electron microscopic observations of the heated serum albumin at different NaCl concentrations at pH 7.5, has documented the existence of the following: separated short polymers in the absence of NaCl, long linear polymers at 10 mM NaCl, linear polymers linked to each other at 20 mM NaCl, and random aggregates at 500 mM NaCl.[13] These results indicate that the gelling model for ovalbumin (Fig.2) can also be applied to serum albumin.

In the case of lysozyme, the presence of 7.5 mM dithiothreitol is necessary to produce gel.[15] A transparent gel is obtained in the presence of 40 mM NaCl, at pH 7.0, by the one-step heating method (Fig. 4(a)).[14] The two-step heating method allows transparent gels to be obtained over a slightly wider range of NaCl concentrations than is possible using the one-step heating method (Fig. 4(b)). The two-step heating method affects gel hardness more than any other characteristic. Gel which has been reheated in the presence of 50 mM NaCl is 4 times harder than gel obtained by the one-step heating method using the same concentrations

If the results obtained for the lysozyme gel can be explained in terms of the model in Fig. 2, then the linear polymers which were produced by the first heating (Fig. 2(a)) would form tight gel networks by the second heating, and would show high gel hardness. The presence of linear polymers has, in fact, been documented in a heated lysozyme solution using transmission electron microscopy.[14]

As an example of practical application of the two-step heating method, this method has been successfully used to prepare transparent gel from whole egg white.[16]

REVERSIBILITY OF GELLING PROCESS

The gelling process of ovalbumin involves several successive steps, as shown in Fig. 5. The first step involves the denaturation of the ovalbumin molecule by heat (Fig. 5(a)). This step is completely irreversible under normal gelling conditions,[10] because the denatured molecules immediately associate to form polymers. This polymerization reaction hinder the

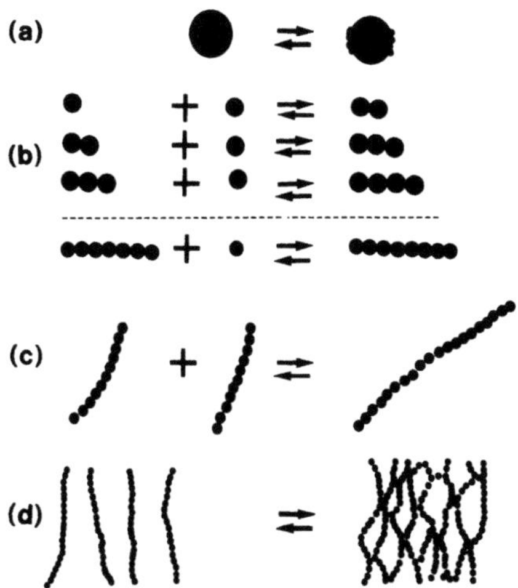

Figure 5. The intermediate steps for gel formation of ovalbumin by heating.

renaturation of the molecules after cooling. Recently, however, we have discovered conditions under which polymerization does not occur, i.e., under conditions of very low protein concentration and low ionic strength, at pH 7.5. Under these conditions, the heat denatured molecules remain as monomers, and then, slowly renature as they reach room temperature. At 37°C, the half time of renaturation is approximately 3 hr.

The second step (Fig. 5(b)) of the gelling process involves the association of the monomers to form linear polymers. This step can be partly reversed under high pressure (600 MPa).[17]

The third step of the gelling process involves the elongation of polymers by the combination of two separate polymers (Fig. 5(c)). This step can be reversed at 80°C, as indicated by the observed decrease in molecular weight of polymers diluted at this temperature.[11]. Pressure treatment at low ionic strength also causes the dissociation of polymers.[17]

The final step of the gelling process involves the formation of a network by cross-linking of linear polymers (Fig. 5(d)). A transparent gel prepared according to the one-step heating method (3.75% ovalbumin in a 10 mM Na-phosphate buffer, pH 7.0, heated for 20 min. at 80°C) melts when heated to a temperature below 100°C.[18] The melted sample (sol) forms a gel when cooled. This gel-sol transition can be observed with repeated heating and cooling. A similar gel-sol transition can be observed during pressure treatment.[17]

These results indicate that ovalbumin gel, which is a heat-set gel, behaves like a cold-set gel under certain conditions. Conditions of low protein concentration, close to the critical gel concentration, are suitable for transparent gels. At higher protein concentration than the critical concentration, the melting point is above 100°C, and no melting can be detected under normal heating conditions. Heating for gel formation in globular protein solutions is necessary in order to denature the protein molecules. Once the gel has formed, the ease of melting of the gel is determined by the bond strength between the networks in the gel. Since cold-set gels, polysaccharides and gelatin gels, are formed by hydrogen bonds, their bond strength is not greatly affected by heat. However, increasing molecular motion with increasing temperature can easily break these hydrogen bonds and melt the gels. Ovalbumin gels are thought to be held together mainly by hydrophobic interactions. The strength of hydrophobic interactions increases with increasing temperature, up to 110°C.[19] Therefore, higher temperatures are required to break the hydrophobic interactions in ovalbumin gels than are required to break the hydrogen bonds in polysaccharide gels.

Table 3. Properties of transparent ovalbumin gels and polysaccharide or gelatin gels.

	Ovalbumin gel	Polysaccharide or gelatin gel
Unit molecule	globular	fibrous
Appearance of gel	transparent	transparent
Gel formation	heat-set gel (cold-set gel)*	cold-set gel
Gel-sol transformation	reversible	reversible
Critical concentration for gelling	low	low

*A transparent ovalbumin gel is melted by heating and gelled by cooling as described in the text.

CONCLUSIONS

By controlling pH, ionic strength, protein concentration, and heating procedures, we can create transparent, translucent, opaque or turbid gel from globular proteins. Transparent gels are similar, in some respects, to polysaccharide gels or gelatin gel, as shown in Table 3.

The string-of-beads networks in transparent gel give rise to many of its properties. The molecular mechanism that accounts for the formation of the string-of-beads arrangement in denatured protein molecules is not known. One theory to explain the arrangement (a dimer model) will be soon published elsewhere.[20]

In accounting for these molecular mechanisms, it must be assumed that the denatured proteins do not completely unfold. The denatured proteins having a compact globular conformation or molten globule state are currently of great interest to protein chemists.[21,22] The molten globule-like state is important for the understanding of ordered structures formed by denatured proteins in food processing.

REFERENCES

1. M.P. Tombs, Gelation of globular proteins, *Faraday Discuss. Chem. Soc.* 57:158 (1974).
2. E. Barbu, and M. Joly, The globular-fibrous protein transformation, *Discuss.Faraday Soc.* 13:77 (1953).
3. A.H. Clark, F.J. Judge, J.B. Richards, J.M. Stubbs, and A. Suggett, Electron microscopy of network structures in thermally-induced globular protein gels, *Int. J. Peptide Res.* 17:380 (1981).
4. T. Nakamura, S. Utsumi, and T. Mori, Network structure formation in thermally induced gelation of glycinin, *J. Agric. Food Chem.* 32:349 (1984).
5. A.-M. Hermansson, Structure of soya glycinin and conglycinin gels, *J. Sci. Food Agric.* 36:822 (1985).
6. T. Mori, T. Nakamura, and S. Utsumi, Behavior of intermolecular bond formation in the late stage of heat induced gelation of glycinin, *J. Agric. Food Chem.* 34:33 (1986).
7. B.A. Badley, D. Atkinson, H. Hauser, D. Oldani, J.P. Green, and J.M. Stubbs, The structure, physical and chemical properties of the soy bean protein glycinin, *Biochim. Biophys. Acta*, 412: 214 (1975).
8. H. Hatta, N. Kitabatake, and E. Doi, Turbidity and hardnes of a heat-induced gel of hen egg ovalbumin, *Agric. Biol. Chem.* 50:2083 (1986).
9. N. Kitabatake, H. Hatta, and E. Doi, Heat-induced and transparent gel preparred from hen egg ovalbumin in the presence of salt by a two-step heating method, *Agric. Biol. Chem.* 51:771 (1987).
10. T. Koseki, N. Kitabatake, and E. Doi, Irreversible thermal denaturation and formation of linear aggregates of ovalbumin, *Food Hydrocolloids*, 3:123 (1989).
11. T. Koseki, T. Fukuda, N. Kitabatake and E. Doi, Characterization of linear polymers induced by thermal denaturation of ovalbumin, *Food Hydrocolloids*, 3:135 (1989).
12. E. Doi, and N. Kitabatake, Structure of glycinin and ovalbumin gels, *Food Hydrocolloids*, 3:327 (1989).
13. M. Murata, F. Tani, T. Higasa, N. Kitabatake and E. Doi, Heat-induced transparent gel formation of bovine serum albumin, *Biosci. Biotech. Biochem.* (Japan) (in press).
14. F. Tani, M. Murata, T, Higasa, M. Goto, N. Kitabatake, and E. Doi, Heat-induced transparent gel of hen egg lysozyme by a two-step heating method, *Biosci. Biotech. Bichem.* (Japan) (in press).
15. S. Hayakawa, and R. Nakamura, Optimization approaches to thermally induced egg white lysozyme gel, *Agric. Biol. Chem.* 50:2039 (1986).
16. N. Kitabatake, A. Shimizu, and E. Doi, Preparation of transparent egg white gel with salt by two-step heating method, *J. Food Sci.* 53:738 (1989).
17. E. Doi, A. Shimizu, H. Oe, and N. Kitabatake, Melting of heat-induced ovalbumin gel by pressure, *Food Hydrocolloids*, 5:409 (1991).
18. A. Shimizu, N. Kitabatake, T. Higasa, and E. Doi, Melting of the ovalbumin gels by heating, *Nippon Shokuhin Kogyo Gakkaishi*, 38:1050 (1991).
19. P.L. Privalov, and S.J.Gill, Stability of protein structure and hydrophobic interaction, *Adv. Protein Chem.* 39:191 (1988).
20. N. Nemoto, A. Koike, K, Osaki, T. Koseki, and E. Doi, Dynamic light scattering of aqueous solution of linear aggregates induced by thermal denaturation of ovalbumin, *Biopolymers,* (in press).
21. K. Kuwajima, The molten globular state as a clue for understanding the folding and cooperativity of globular protein structure, *Proteins: Structure, Function, Genetics,* 6:87 (1989).
22. D.O.V. Alonso, K.A. Dill, and D. Stiger, The three states of globular proteins: Acid denaturation, *Biopolymers*, 31:1631 (1991).

THERMODYNAMIC ASPECTS OF FOOD PROTEIN FUNCTIONALITY

V. B. Tolstoguzov

Institute of Food Substances of the Russia
Academy of Sciences, Vavilov Street 28, Moscow
117813, GSP-1, Russia

ABSTRACT

The term "functional properties" of food proteins may be defined as a
set of physico-chemical characteristics of a protein that affects its use
and contributes to the desired structural, mechanical and other important
physico-chemical properties of both the food processing systems and the
final food products. Many food systems and processing methods are of a
non-equilibrium nature. However, thermodynamic approach can provide
valuable information on the possible state and potential behavior of
proteins at interfaces and in bulk multicomponent food systems as well as
about the effects of the main thermodynamic variables (such as
temperature, pH, salt and biopolymer concentrations) upon a direction of
a system evolution. Of great importance to the functional properties of
food macromolecular substances is the phenomenon of incompatibility or
more precisely limited compatibility between proteins and polysaccahrides
in aqueous media. This research line is necessary for a deeper
understanding of solubility, gelforming, thickening, foaming and
emulsifying properties of various proteins. From thermodynamic point of
view the structure and properties of some conventional and novel food
systems are considered.

1 Introduction

1. 1. Conformational potential of proteins.

The term "functional properties" describes the ability of a protein
to perform its structural functions in foods. At the molecular level
this term implies the possibility of the use of a so-called
"conformational potential" of a food protein.[1] From physico-chemical
point of view the processing of a food protein into a food product may be

Food Hydrocolloids: Structures, Properties, and Functions
Edited by K. Nishinari and E. Doi, Plenum Press, New York, 1994

regarded as the formation of intermacromolecular bonds and supermacromolecular structural elements that mainly contribute to the quality of a given food product. The term "conformational potential" characterizes an ability of protein to form intermolecular junction zones that are required for the formation of desired structural, rheological and other physico-chemical properties of foods. As a measure of the conformational potential of a protein may serve a minimum threshold protein concentration above which a protein gel or a stable O/W emulsion (stabilized against coalescence by a given protein) can be obtained. Normally, the formation of intermolecular junction zones (or the establishing a balance between intra- and intermolecular interactions) in food systems is a time-depending, many stage process. Therefore, the kinetic approach is widely used for the study and development of food technologies.

1.2. Thermodynamic and kinetic aspects of protein functionality

In practice, many foods are of a non-equilibrium nature. This is due to slow rate processes of structure formation in concentrated, highly viscous systems containing various interacting macromolecules and colloidal particles strongly differing in nature, composition, size and shape. For instance, such processes are mass- and heat transfer in viscous heterogeneous, heterophase systems, conformational rearrangements of macromolecules, the aggregation and formation of three dimensional networks, specific phase transitions (phase separations, the crystallization and recrystallization of starches, the denaturation-renaturation of proteins), the orientation-disorientation of food macromolecules and their aggregates in a shear field and at rest. Accordingly, food system properties are usually dependent on time and preparative conditions. Since very high viscosity of food systems (especially below the temperatures of glass transition) a non-equilibrium food system from the thermodynamic point of view may be kinetically stable.

Kinetic approach underlies the majority of protein functionality tests, including solubility, dispersibility, swelling, gelation. foaming, emulsification, water-holding and fat-holding etc. These empirical characteristics of the performance of a particular protein in a given model food system are determined under the specific empirical set of constant test conditions. Different techniques, various instruments and operating conditions are used for evaluating functional properties of proteins. They may only provide, however, an approximate ranking of proteins at given test conditions. Main reasons are as follows. There are no standard, generally accepted methods, simple appropriate model systems and conditions utilized for measuring functional properties of proteins which could provide information on the performance of a given protein in a real food system. Proteins may fulfill one or more interrelated structural functions in a food product simultaneously. Functional properties of a protein are dependent upon its composition, structure and other intrinsic properties as well as upon the environment. Protein functional properties are related to the protein interaction with other components of a food system. They can be significantly altered with system composition. In many cases, it is impossible to consider the functional properties of proteins without taking into account the role of polysaccharide components of a given food system. Nevertheless, the kinetic approach attracts incomparably more attention than the thermodynamic one.

We will try to show, however, that thermodynamic approach deserves more attention. Thermodynamic approach takes account of the multicomponent nature of food systems. It provides information on the possible state and potential behavior of proteins at interfaces and in bulk multicomponent food systems as well as about the effects of the main thermodynamic variables (such as temperature, pH, salt and biopolymer concentrations) upon a direction of a system evolution. [2] We will neither discuss the molecular basis of the functional properties for individual proteins and no compare the functional properties of different proteins. We will concentrate on the common features of food proteins and polysaccharides and their behaviour in food system. These two types of macromolecular food ingredients and additives can be thermodynamically incompatible and carry out the most important structural functions in foods. This paper is devoted to the thermodynamic approach to protein functionality, more precisely, to the incompatibility phenomenon and its role in affecting the functional properties of food proteins and the structure formation in food systems.

1.3. Thermodynamic incompatibility of food macromolecules

Presumably, one of the most interesting features of protein-polysaccharide mixtures is the immiscibility or thermodynamic incompatibility of these biopolymers in aqueous solutions [1, 2]. This phenomenon strongly enlarges the application area of the thermodynamic approach to food system studies. Thermodynamic incompatibility in mixed solution of proteins and polysaccharides shows up at a sufficiently high total biopolymer concentration. The value of phase separation threshold, i.e. minimal total concentration of biopolymers at which phase separation occurs, mainly depends on excluded volume of biopolymers. Its value is normally over 12% for solution mixtures of various globular proteins. For globular protein-polysaccharide mixtures it usually exceeds 4%. Limited miscibility or limited thermodynamic compatibility is characteristic for structurally and/or chemically different biopolymers in mixed solutions and melts. [1-4] This phenomenon is of general nature and typical of food systems. In many cases, however, the total biopolymer concentration is not high enough and conditions are not sufficient for the phase separation of a given system. In the other cases a food system separates into two aqueous phases and its structure may be altered. In the former cases the functional properties of a protein can be modified. In the latter cases the system structure can be significantly changed. We will consider how this phenomenon can modify the functional properties of food this phenomenon can modify the functional properties of food hydrocolloids and govern the structure of food systems.

2. Phase diagrams for mixed solutions of biopolymers

The incompatibility phenomenon is illustrated in Fig. 1. This is the phase diagram typical for protein-polysaccharide-water systems. [2, 5-7] It describes the phase behaviour and phase equilibrium in mixed solutions of proteins and polysaccharides. Binodal curve separates the regions of compositions of the single- and two-phase state of the system. Compositions lying under the binodal curve correspond to single-phase aqueous solutions of biopolymers. In this region solutions of different biopolymers are completely miscible.

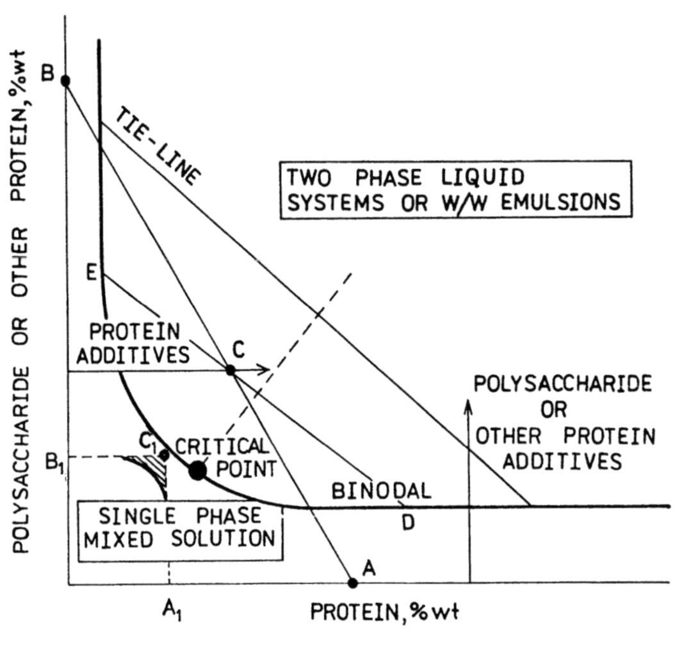

Fig. 1. Phase behaviour of mixed aqueous solutions of biopolymers.

The region lying above the binodal curve represents compositions of two-phase systems. The thin lines are the tie-lines. They connect the points representing the compositions of the co-existing phases. The two macromolecular components concentrate mainly in different phases. The upper and lower binodal branches are met at the critical point, where the system phases are of the same composition and volume. As a system composition moves away from the critical point the difference in composition between co-existing phases rapidly increases.

When we mix aqueous solutions of a protein, e.g. of composition A, and a polysaccharide solution whose composition corresponds to point B, and provide sufficiently high total biopolymer concentration exceeding the phase separation threshold, the mixture of composition C can be obtained. This mixed solution breaks down spontaneously into two liquid phases forming the water-in-water (W/W) emulsion. The two phases of W/W

emulsion are phase D and phase E. One of which, D, is rich in protein
and another, E, is rich in the polysaccharide. These phases can be
separated by a centrifuge. The phase volume ratio corresponds to the
ratio of the tie-line segments: EC/CD.

3. Membraneless osmosis

Besides the distribution of biopolymers the partition of water between the phases of a food system is important for protein functionality.
When the angle made by the tie-line with the concentration axis of protein
is increased the difference in water concentration between co-existing
phases rises. This means that the phase diagram asymmetry is increased.
Presumably, the ratio of the critical point coordinates and the angle made
by the tie-line with one of the concentration axes can be used as a measure of the relative hydrophilicity of biopolymers. Fig. 2 gives some

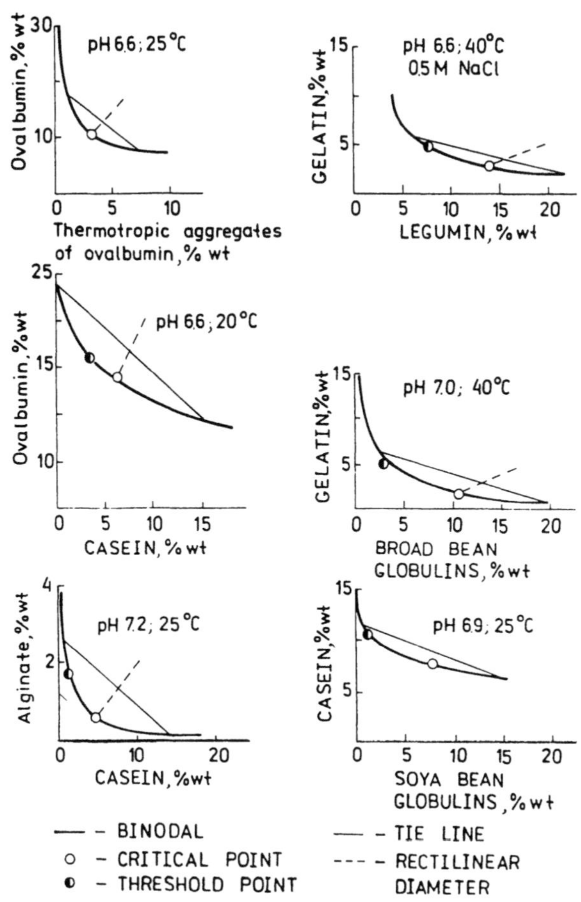

Fig. 2. Phase diagrams.

examples of phase diagrams for mixtures of casein with alginate, or
ovalbumin, or soybean globulins, for mixtures of gelatin with the 11S
globulin and the mixtures of broad bean globulins, and, finally, for
mixed solutions of native and soluble denatured forms of ovalbumin. The

co-existing equilibrium phases may differ strongly in concentration. This underlies a new method for concentrating protein solutions. This method has been called membraneless osmosis.[5] Its main idea is clarified in Fig. 3. Unlike conventional osmosis that is the transfer of

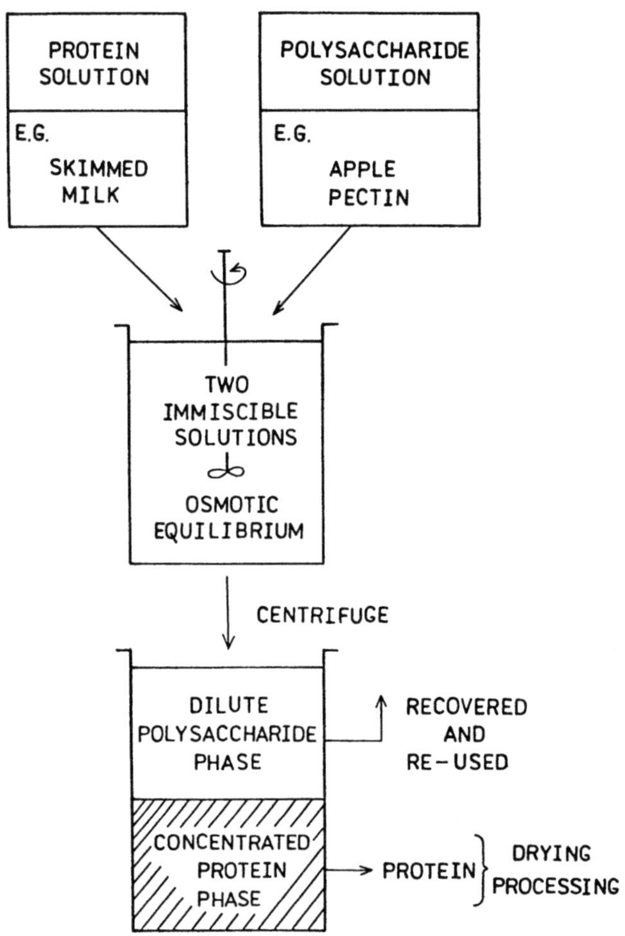

Fig. 3. General scheme of membraneless osmosis process.

solvent molecules to a solution through a semipermeable membrane. The membraneless osmosis process involves two immiscible solutions of biopolymers. A semipermeable membrane is replaced by the interface between the two immiscible membrane is replaced by the interface between the two immiscible aqueous solutions of biopolymers. The membraneless osmosis is the process of the diffusion transfer of water from one aqueous solution to another through the interface between the two immiscible solutions which are phases of water-in-water emulsion.

For instance, on adding 0.5-1% pectin to skimmed milk the mixture is separated into two liquid phases.[5] It takes several seconds only to complete the water transfer between the phases. Then two phases can be separated by sedimentation. Casein-rich phase with concentration of 20 - 30% contains 80 - 90% of milk proteins. The advantages of membraneless osmosis process are relatively low energy requirements and high efficiency due to the developed interface between the two aqueous phases.

This process governs the water partition between phases of many food systems. It affects use of the conformational potential of biopolymers.

It should be noted that the high value of phase diagram asymmetry is more typical of the protein-protein-water and protein-salt-water systems. This seems to be due to self-ordering structure of concentrated protein solutions. [1,7]

4. Phase behaviour of protein-polysaccharide mixtures. Effect of sugars

The effect of pH values, various salts and the composition, structure and conformation of biopolymers on the phase behaviour of protein-protein and protein-polysaccharide mixed solutions has been recently discussed. [1,2,5-7] Other interesting point is that the behaviour of a protein in mixed solutions with carbohydrates is strongly changed when we go over from carbohydrates of low molecular weight to high molecular weight polysaccharides, e.g., from sucrose to dextran. The solubility (in the vicinity of the isoelectric point) of food proteins studied such as casein and legumin as well as the co-solubility of different protein pairs (e.g., legumin and ovalbumin) are markedly increased with the concentration of (purified) sucrose from 10 to 50%. But dextran is immiscible with these proteins at the ionic strength higher than 0.1. On adding sucrose to the two phase legumin-dextran-water or casein-sodium alginate-water systems differences between co-existing phases in water content and volume fraction ratio are strongly reduced.

Presumably, this effect is of importance to the structure and properties of ice cream mix. The concentration of polysaccharide stabilizers commonly used in ice cream mixes varies from 0.1 to 0.5% wt. As was mentioned above, the addition of 0.5% or more apple pectin to skimmed milk results in phase separation of the mixture. This means that stabilizing agents and milk proteins may be only partially co-soluble. Stabilizers are presumably able to induce phase separation in the ice cream mixtures. Presumably, this also may result in a specific bimodal size distribution of both ice crystals and lipid dispersed particles in ice cream mix. Interfacial surfaces may affect the growth rate of ice crystals, both by mechanical obstruction and by a decreased rate of water transfer to the surface of growing crystals located closely to these interfacial surfaces. The propagation path of ice crystal growth in the medium under consideration may be limited by the size and concentration of water dispersed particles and by the distance between two dispersed particles within the continuous phase of the emulsion. The deformation of the dispersed particles under the shear stresses which arise in flow (especially during freezing) may also contribute to the higher degree of heterogeneity in size and partition of lipid droplets and ice crystals within the final products.

5. Functional properties of protein and polysaccharide additives

Fig. 1 also shows the effect of the additives of proteins or polysaccharides on the phase state of a food system. A protein or a polysaccharide added to a food system can be incompatible with its

macromolecular components. This means that an additive may be dissolved
in the bulk system either completely or partially. As a result an added
protein may perform its functional properties either within the bulk of a
system or within the volume of one of system phases. The phase
separation induced by protein additives may be accompanied by the water
redistribution between food system phases. Thus, the functionality of a
protein additive will depend upon its solubility in a given food system,
the composition of co-existing phases and the phase volume ratio. The
phase diagram for the system gives all this information. [2, 7]

5.1 Co-solubility of biopolymers

Now, we discuss the problem of protein solubility. The branches of
binodals (in Fig. 1) do not coincide with the phase diagram axes. This
means that biopolymers are limitedly co-soluble and we are dealing with
the phenomenon of limited thermodynamic compatibility. The presence of
an incompatible polysaccharide in a protein solution does not have any
substantial effect on the equilibrium between the native and denatured
form of proteins. Presumably, this means that the volume of protein
molecules does not become larger on denaturation. However, the presence
of an incompatible polysaccharide can intensify the aggregation of
denatured molecules and the formation of a three dimensional gel
network. The addition of an incompatible polysaccharide to a solution of
a protein gives rise to an increase in thermodynamic activity of both
biopolymers which behave as if they were in a more concentrated
solution. It gives rise to some effects common to many functional
properties of proteins.

5.2. Stabilizing agents for emulsions

Fig. 4 shows the effect of use of mixtures of incompatible biopolymers
as emulsion stabilizers. It shows protein adsorption on n-decane drops
dispersed in aqueous solutions of legumin (curve 1) or in mixed solutions
of legumin with 1% dextran (curve 2). The concentration of dextran was
kept constant whilst the concentration of legumin was increased. When
the protein content becomes high enough the mixed aqueous solution
undergoes phase separation because of incompatibility of these biopolymers.

The presence of a polysaccharide reduces the protein concentration
required for protein multilayer adsorption at the interface. The
comparison of the phase diagram (curve 3) with the protein adsorption
isotherm (curve 2) for the same legumin-dextran-water system shows that
the protein concentration required for multilayer formation is located in
the vicinity and below the binodal curve. This seems to be mainly due to
an increase in the thermodynamic activity of protein.

Fig. 4 shows that, presumably, there is a basic difference in protein
behaviour as an emulsion stabilizer between single-phase (in the
composition region A-B) and two-phase mixed solutions of biopolymers (in
the composition region B-C). In the composition region (A-B) lying under
the binodal curve, protein adsorption gives rise to the formation of a
relatively thin monolayer coverage of dispersed particles. The formation
of protein multilayers (B-C) seems to result from phase separation in the
aqueous continuous phase. This latter gives rise to microencapsulation
of dispersed particles of lipids. Presumably, multilayer formation can

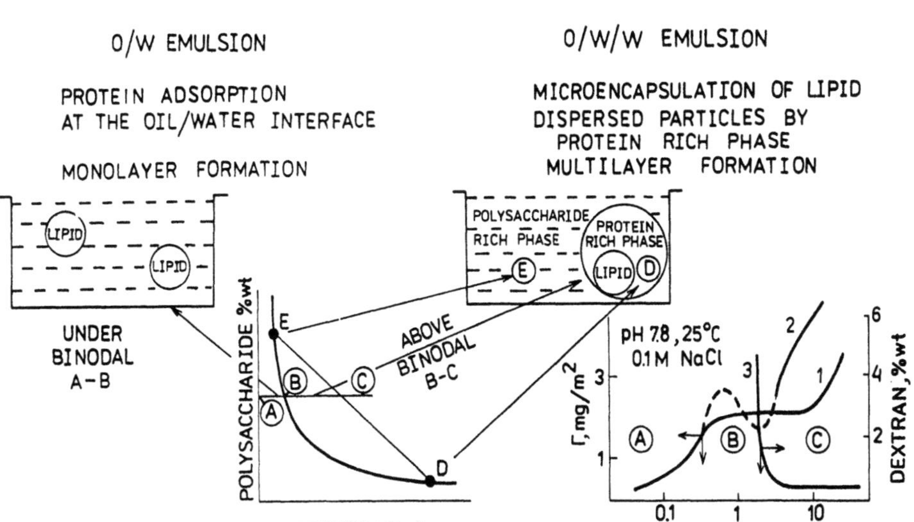

Fig. 4. Protein behaviour as emulsion stabilizer :The effect of thermodynamic incompatibility.

be regarded as a transition from the single O/W emulsion to the double (O/W/W) emulsion of decane in the protein-rich phase of a W/W emulsion. Presumably, a transition from mono- to multilayer adsorption seems to be similar to a transition from lyophobic to lyophilic dispersed particles. [2, 7]

5. 3. Depletion flocculation

The comparison of the protein adsorption isotherms (curves 1 and 2 in Fig. 4) shows the other interesting result. The addition of an incompatible biopolymer to the protein solution contributes to the depletion flocculation of the dispersed particles covered with the protein. The phenomenon of depletion flocculation is likely to be similar to the phenomenon of limited thermodynamic incompatibility of biopolymers in solutions.

However, there is a basic difference between the two phenomena. Unlike biopolymer solutions, colloidal dispersions are thermodynamically unstable in principle because of an inherent excess surface free energy. Therefore, e. g. in the case of flocculation, the effect of a polysaccharide is more strongly pronounced than on demixing of its mixed solutions with proteins. [2] The phase separation threshold for protein - polysaccharides solution mixtures usually exceeds 4%. But polysaccharides as flocculants are generally used at a concentration lower than 1%.

5. 4. Gelforming Agents

Now we will consider the effect of the co-solubility of macromolecular system components on gelforming system properties. For instance, Fig. 1 shows that if point A_1 is the critical concentration for protein gelation and point B_1 represents the critical concentration for polysaccharide gelation, protein-polysaccharide mixtures lying between the binodal and the rectangle $O-A_1-C_1-B_1$ can form mixed gels. However, mixed solutions with lower total concentration of biopolymers give rise mixed gels as well. It occurs in the composition region lying inside the rectangle within the shaded area. This is due to an increase in thermodynamic activity of incompatible biopolymers in their mixed solutions. [1,6,7]

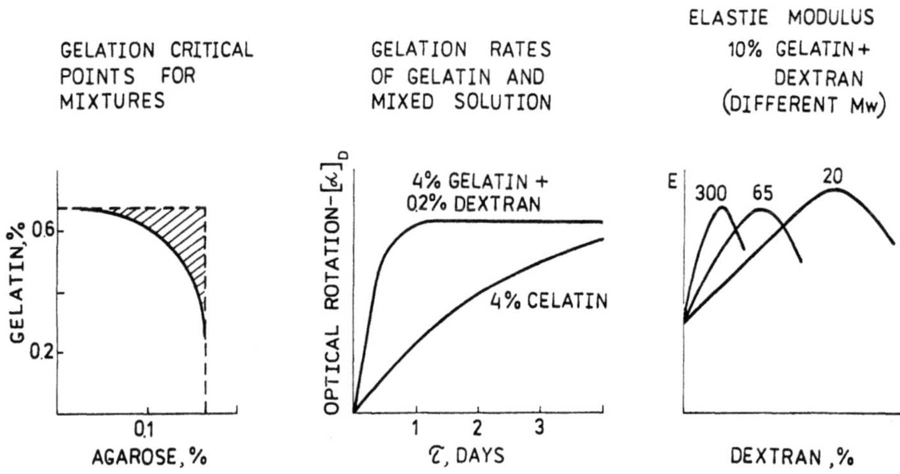

Fig. 5. Some functional properties of biopolymer mixtures.

Fig. 5 shows the effect of agarose in lowering the critical concentrations for gelation of gelatin-agarose mixed solutions and the effect of dextran in increasing gelation rate of gelatin-dextran mixed solutions and the elastic moduli of these gels. The higher concentration of polysaccharide (depended on its molecular weight) leads to a system phase separation producing gels in which the dispersed particles of polysaccharide solution serve as a filler. The modulus of elasticity of gelatin gels filled with liquid dispersed particles decreases as the volume fraction of filler increases. In the composition range corresponding to single phase mixed solutions of thermodynamically incompatible biopolymers, as the total concentration of biopolymers is increased, the gelation rate, and mechanical properties of a gel, rise as well. For two phase mixed solutions the higher the total concentration of biopolymers, or the higher the volume fraction of dispersed particles, the lower the modulus of elasticity of gels. This possibility to change significantly the mechanical properties of gels without substantial alterations in composition permits to study relationships between the organoleptical and physical properties of solid foods.

Small quantities of dextran when added to gelatin solutions can strongly increase the rate of gelation and the elastic moduli of gelatin gels. Presumably, it cannot be only due to an increase in its thermodynamic activity. Some important aspects of the mechanism of the gel formation of gelain, polysaccharides and their mixtures may be related to thermodynamic properties of macromolecular aggregates arising from a worsening of the thermodynamic quality of the solvent and to the depletion flocculation effect. Presumably, depletion flocculation may be responsible for the end-to-end type of aggregation of macromolecular associates of a colloidal size and the increased rate of the formation of gel networks. It can be assumed, that junction zones in biopolymer solutions can only be formed if the bulk concentration of biopolymer molecules exceeds a certain critical value. Such formation of junction zones results from interaction of randomly distributed macromolecules, the centres of mass of which do not coincide. As a result of a local increase in the concentration of polymeric chains a junction zone of a certain longitude within a fibrous aggregate can form. Accordingly, zones with a lower concentration of biopolymer chains arise on both ends of the junction zones. The shape of such an aggregate will be similar to a dumbell. Non-ordered terminal parts of these aggregates are notable for a lower concentration of biopolymer chains having higher flexibility and higher thermodynamic activity. The depletion flocculation of aggregates probably occurs via their collision and the interpenetration of their end zones. This may give rise to an increased concentration within the volume of end zones superimposed on each other and results in the formation of junction zones of a second generation between aggregates and, finally, in the formation of a three dimensional gel network. Generally, all macromolecular additives which are incompatible with gelforming components of a system can affect its gelation rate and rheological properties.

Biopolymer incompatibility affects both the bulk and the surface properties of gels. The relative surface activity and competitive adsorption of biopolymers at the air/water interface may be evaluated by measurement of the contact angle of water drops on a gelatin gel surface formed in contact with air. Here, gelatin can be used as a reference. It is also of importance that gelatin gels are also notable for extreme hydrophobicity. The contact angle of water drops on the surface of a 12-20% gelatin gel lies in the range from 120° to 130°. As the concentration of a protein added to the gelatin solution is increased and exceeds a certain critical value, this protein starts to expel the gelatin from the air/water interface. Accordingly, above certain concentrations of added protein the contact angle of water on the surface of the gelatin gel decreases, and depends on protein content. Normally, less hydrophilic proteins (or other biopolymers) are concentrated at the non-polar phase/water interface. Presumably, we are dealing here with thermodynamic incompatibility of biopolymers within the interfacial layer.

6. Interfacial layer

The presence of interfacial or depletion layer around dispersed particles is the next interesting feature of W/W emulsions which is illustrated in Fig. 6. Both types of incompatible biopolymers forming the phases are mutually depleted from the vicinity of the interfacial surface. Within this layer the concentration of the incompatible biopolymers is reduced.

Fig. 6 also shows possible results of adding lipids to W/W emulsions. An interesting structure formation process may be presumably initiated by the addition of lipids to a W/W emulsion. W/W emulsions can be additionally stabilized by particles of lipids located partly at the W/W interface and partly within one of the aqueous phases. It depends on the relative wettability of lipids by the two aqueous phases. The adsorption of dispersed particles of lipids and flavouring components within the interfacial layer may result in concentrating the dispersed particles of lipids, their flocculation and coalescence within this layer. This may lead to the formation of a three dimensional layer, i.e., a new continuous phase of lipids. This may induce a honeycomb-like type of the

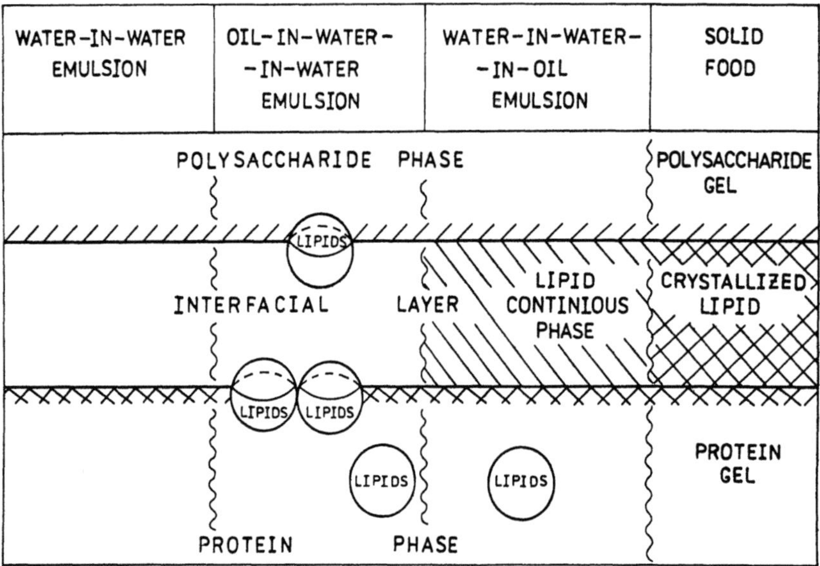

Fig. 6. Schematic representation of the structure formation of low-fat foods.

structure of lipid continuous phase of W/W/O emulsion. This may occur at sufficiently high concentrations or the volume fraction of dispersed phase in W/W emulsions where individual movement of particles is restricted. Presumably, easy deformability and coalescence of aqueous dispersed particles in W/W emulsions may also contribute to the formation of a lipid continuous phase. Of great significance is the use of crystallizable lipids and the choice of "phase forming" macromolecular components which are capable of forming weak thermoreversible gels in both aqueous phases. The formation of a lipid continuous phase in a bulk W/W emulsion with a very low (less than 25-30%) lipid content may be regarded as a conversion of O/W/W emulsion into W/W/O solid emulsion with the honeycomb-like structure of the lipid phase. This new type of a heterogeneity seems to be an important factor providing the fat-like mouthfeel and specific texture usually typical of fats and oils.

7. Spinneretless Spinning and Thermoplastic Extrusion

One of the features of W/W emulsions is very low interfacial tension of about 0.02-0.0001 dyn/cm. Common solvent, water, and marked co-solubility of biopolymers seem to be main contributory factors for low interfacial tension of W/W emulsions and easy deformability of their dispersed particles in flow. These specific features are of great practical importance. This is clarified in Fig. 7. It is the general

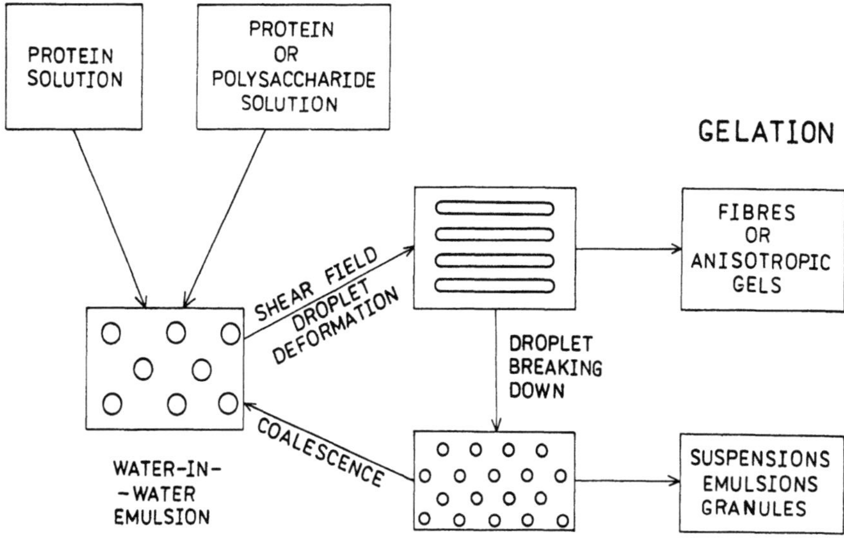

Fig. 7. General scheme of spinneretless spinning.

scheme of spinneretless spinning or fibre shaping process which is responsible for fibrous and lamellar structure of many foods. Liquid dispersed particles of W/W emulsions can be deformed in a shear field. Fibre-like liquid particles are non-stable. They break down into smaller spherical particles. These latter can be deformed and break down to smaller droplets again. They also may coalesce to larger droplets. The dynamic equilibrium: droplet deformation, break down, coalescence may establish itself in a flowing emulsion. Its controlling opens the way of producing various materials. The shape of liquid filaments can be fixed by converting one or both liquid phases into the solid state. Spinneretless spinning implies formation of an anisotropic structure of heterophase liquid systems in flow and fixing of this structure by converting one or both phases of the system into the solid (gelled) state.[1-4]

Three types of products may be obtained by spinneretless spinning, namely fibres, or gels filled with oriented liquid filaments or gel-like fibres, or granules. For instance, one of promising applications of the process is the producing of small protein granules. By heating and mixing a two-phase system containing a globular protein (ovalbumin in particular) in dispersed system phase tiny spherical gel-like particles can be obtained. Owing to low interfacial tension typical of W/W emulsions dispersed particles can be so small that they may be perceived

in mouth as fluid rather than individual gel-like particles. Such
product may be used as a partial replacement for fats in many reduced-fat
foods.

It should be also noted that generally the fact of polymer
incompatibility in a solution suggests their incompatibility in a melt as
well. Therefore, it has been assumed that in the thermoplastic extrusion
process we are dealing with heterophase mixtures of melted water-
plasticized biopolymers. It has been also shown that the thermoplastic
extrusion process is performed by the mechanism of spinneretless
spinning. [1-4]

8. Conclusion

In conclusion it should be stressed that in many cases it is
impossible to consider the functional properties of proteins without
taking into account the role of polysaccharide components of a food
system and inversely. My concern was to show by some examples the
importance of the thermodynamic approach to a better understanding of a
wide range of the functional properties of food proteins and the
processes of structure formation in foods.

9. References

1. V. Tolstoguzov, *Food Hydrocolloids*, 2: 339 (1988).
2. V. Tolstoguzov, *Food Hydrocolloids*, 4: 429 (1991).
3. V. Tolstoguzov, In "Developments in Meat Science" ed. R. A. Lawrie,
   Vol. 5, Chapter 4, p. 159, Elsevier Applied Science, London(1991).
4. V. Tolstoguzov, *Food Technology International*, 71 (1991).
5. V. Tolstoguzov, *Food Hydrocolloids*, 2: 195 (1988).
6. V. Tolstoguzov, In "Gums and Stabilisers for the Food Industry"
   G. O. Phillips, P. A. Willams and D. J. Wedlock, ed., Vol. 5, p. 157, IRL Press,
   Oxford(1989).
7. V. Tolstoguzov, In "Gums and Stabilisers for the Food Industry",
   G. O. Phillips, P. A. Willams and D. J. Wedlock, ed., Vol. 6, p. 241, IRL Press,
   Oxford(1992).

THE EFFECTS OF COOLING ON THE PHYSICOCHEMICAL PROPERTIES OF CASEIN MICELLE

Ryoya Niki[1] and Yoh Sano[2]

[1]Hokkaido University, Sapporo, Hokkaido 060, Japan
[2]National Food Research Institute, Tsukuba City
  Ibaraki 305, Japan

ABSTRACT

The size and the surface properties of casein micelles during the cooling were studied to elucidate the effects of cooling on the casein micelles with dynamic light scattering, electrophoretic light scattering and refractive index measurements. The refractive index of casein micelle is higher than that of soluble casein removed-casein. The casein micelles take more unfoldes structure because of the weak hydrophbic interaction at low temperature, and the diameter of casein micelles increased with the decrease of temperature . The zeta-potential of casein micelles decreased to zero with decreasing temperature because of the release of soluble caseins having the negative charge. The stability model of casein micelles is discussed with these data.

INTRODUCTION

Milk is cooled after milking to prevent the growth of bacteria and to keep the quality of milk during the storage. However,cooling milk causes the several alterations in the properties of milk components[1]. Especially, cooling milk results in the important changes of the chemical and physical properties of casein micelles.

In milk, caseins mostly exist in the form of casein micelles,but also exists in the dissociated form in some extent. The distribution of two forms of caseins is dependent on the temperature. When milk is held at low temperature, some caseins release from casein micelles into milk serum and when the cooled milk is warmed again, the soluble casein reversibly incorporated into casein micelles, which means that there is a temperature-dependent reversible equilibrium between two forms of caseins in milk.[2] In this connection,

Food Hydrocolloids: Structures, Properties, and Functions
Edited by K. Nishinari and E. Doi, Plenum Press, New York, 1994

casein micelles as the soluble caseins. However, the effects of cooling on the casein micelle itself have not been studied thoroughly.

In this paper, we investigated the changes in the size and the surface properties of casein micelles during the cooling using the dynamic light scattering, electrophoretic light scattering, refractive index measurements *et al.* to elucidate the effects of cooling on the casein micelles.

MATERIALS

Casein micelle solution

Casein micelles were fractionated as the pellet from fresh skim milk by ultracentrifugation at 70000Xg and 20°C for 70 min. The pellets of casein micelle fractionated were dissolved in artificial milk serum[7] to prepare the solution of casein micelles. Each casein micelle sample was freshly prepared and was used within two days. The concentration of protein in solution was estimated by the Folin method and the final concentration of protein was adjusted to 3%(w/v).

SCR-casein micelle (Soluble Casein Removed-casein micelle) solution

SCR-casein micelle was prepared as follows. Casein micelle solution which was held at 4°C for 3 day was ultra-centrifuged at 70000Xg and 4°C for 70 min. SCR-casein micelles was obtained as pellets. These pellets were dissolved in the artificial milk serum to prepare the SCR-casein micelle.

RESULTS AND DISCUSSION

Refractive index of casein micelles

The refractive index of the casein micelle was estimated by measuring the turbidity of aqueous dispersion of casein micelle in various refractive indexes of the solvents. Samples of casein micelles were prepared for turbidimetric measurements by diluting with aqueous buffer solution containing different concentrations of glycerol. All measurements were made at room temperature (approximately 20°C). The optical densities of the suspension of casein micelles were read on a UV-Vis spectrophotometer (UV-300,Shimadzu Co. Ltd., Kyoto) .

According to the Rayleigh-Gans-Debye theory,[8] the refractive indices of casein micelles can be measured as the intercept on the abscissa of the plot of $n_0\sqrt{\tau/c}$ against the refractive index of the solvent n_0, where $\sqrt{\tau/c}$ is the specific turbidity per unit concentration measured at 436 nm. As is shown in Fig. 1, good straight lines were obtained between them.

The refractive index of casein micelle is higher than that of SCR-casein micelle, which indicates that SCR-casein micelle has the more leaky structure than intact casein micelle because of the release of β-casein at low temperature treatment.

Quasi-Elastic Light Scattering Measurements

Particle sizes of casein micelle were determined by the quasi-elastic light scattering method which was made with a Light Scattering Analyzer ELS-800 (Ohtsuka Co. Ltd., ,Tokyo,Japan) at each temperature ranging from 35°C to 10°C, the temperature being kept constant by circulating water. The light source was a He-Ne laser of 632.8 nm and the scattering angle was 90°. The autocorrelater was connected to the light scattering analyzer

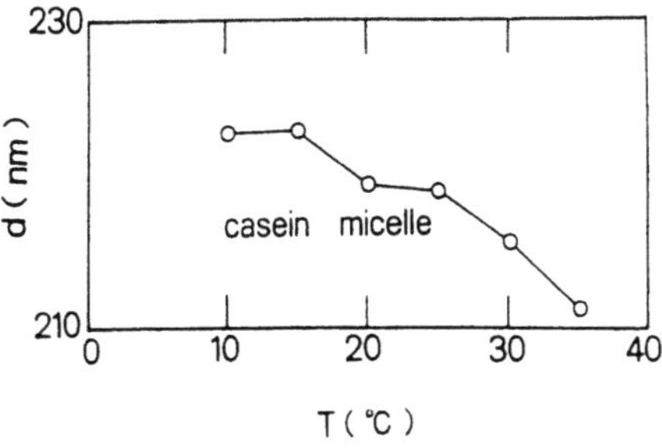

Figure 1. $n_o\sqrt{\tau/c}$ plotted against refractive index of the solvent no for casein micelles diluted

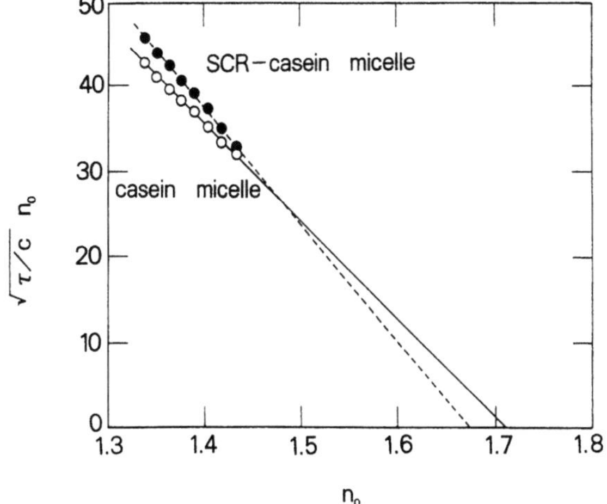

Figure 2. Particle diameter d(nm) of casein micelles at each temperature in buffer/glycerol mixtures

and the autocorrelation function of the photomultiplier photocurrent was automatically analyzed by standard computer fitting techniques to obtain the translational diffusion coefficient of casein micelles. Each casein micelle solutions were usually prepared by diluting a small volume of stock casein micelle solution of about 3%(w/v) into the artificial milk serum solution which was kept in optically clean double-distilled water. The observed single-clipped photocount correlation function could be described by a single exponential curve at each temperature. The translational diffusion coefficients were obtained by the first cumulant of the electric field time correlation function at time=0 and the diameter of casein micelle, d(nm), was calculated according to the Einstein-Stokes equation. The results are shown in Fig.2, where the diameter of casein micelles increased with the decrease of temperature.

The hydrophobic interaction is highly temperature dependent, and it is minimal below 5°C and maximal at higher temperature. When milk or casein micelle solutions are kept at lower temperatures, β-casein and some amounts of α_s- and κ-casein are released from casein micelles. Furthermore it may be assumed that the hydrophobic bonds remaining in micelles also are influenced by lowering temperature. Casein micelles may take more unfolded structure at lower temperatures than at higher temperature.

Electrophoretic Light Scattering Method

The study of electrophoretic behavior of casein micelle affords the important and unique information about surface properties of the micelles. Casein micellar systems were also studied by the moving boundary method. This approach, however, requires the formation of a boundary between two solutions with different concentrations. Therefore, the moving boundary method is not suitable, in principle, for the casein micellar system in a physiological buffer solution. Electrophoretic light scattering method may be suitable for the investigation of casein micelles.

Electrophoretic light scattering was carried out with commercially available Model ELS-800 (Ohtsuka Electronic Co.,Ltd.,Tokyo,Japan). A He-Ne laser was used as a light source. The photo-multiplier was mounted on a computer-controlled goniometer by which the scattering angle could be varied over a range of 5-30°. An electrophoretic cell with a square cross section was used. The dimensions of the cell were 17mm in length (in the direction of applied electric field), 10mm in width (in that of the incident beam), and 2mm in thickness (in that vertical to both of them). The mounting of the electrophoretic cell was designed so that it can be moved in the vertical direction and the scattering volume in the cell can be set at a mechanical precision of 0.01mm. The temperature of the cell was controlled by the circulation of thermostated water through the mounting block of the cell.

The cell holder is provided with a stage, the height of which is adjustable with an automated micrometer so that the laser beam path can be moved from the top to the bottom of the cell. Electroosmosis leading to flow with a parabolic velocity profile cannot be avoided in the case of a closed electrophoresis cell. The true mobility can be calculated by determining the apparent mobilities at various cell positions and the data points were fitted by the least-square method to obtain the velocities at the stationary layer.

The zeta-potential of a spherical particle of diameter d(nm) which exhibits an electro-phoretic mobility U in an electrolyte solution of relative permittivity ε_r and viscosity η is given by

$$\zeta = 3\eta U/2\varepsilon_r \, \varepsilon_0 \, f(\kappa d)$$

where ε_0 is the permittivity of vacuum, and $f(\kappa d)$ is a function of κ, being the Debye-Huckel parameter and U is expressed in units of $m \cdot s^{-1} \cdot V^{-1} \cdot cm$. Results of zeta-potential calculated using this equation is shown in Fig.3. The zeta-potential of casein micelle decreased with decreasing temperature, that is, zeta-potential is coming near zero. The decrease in zeta-potential at lower temperatures may be related to the release of soluble caseins, because soluble casein components have negative charges at pH 6.6 of milk.

Stability model of casein micelles

The stability of casein micelles is controlled by the potential energy function. The potential energy function has two components: an attractive potential due to London-van der Waals forces between the particles and a repulsive potential which arises because of an electric double layer around the particles. The function of the repulsive potential depends on the electric potential which surrounds a particle and arises from a charge on the surface of the particle that induces an imbalance of positive and negative ions in the solution immediately surrounding the particles. The function of the repulsive potential is usually calculated by the DLVO theory of colloid stability. On the other hand, the attractive potential energy arises from the dispersion forces between atoms. The dependence of the attractive potential on the interparticle distance is obtained from a pairwise integration of all the London-van der Waals

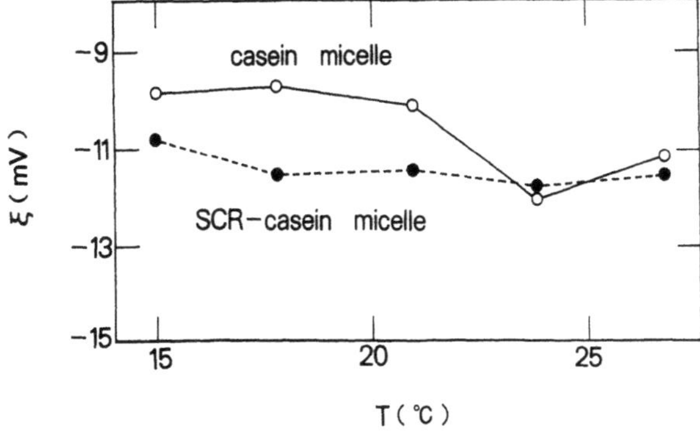

Figure 3. Zeta-potential (mV) of casein micelles at each temperature.

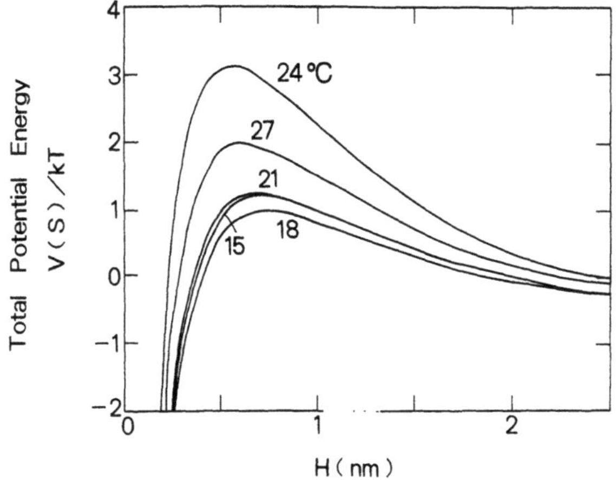

Figure 4. Total potential energy curves for casein micelle at differnt temperature

forces between the component atoms in each particle. The combination of the two potential energy functions gives the total potential energy function for two equal-sized spheres such as casein micelles as a function of the particle distance, H(nm). The Hamaker constant for casein micelles is assumed to be of the order of 10-21 Joule. The results of total potential functions of casein micelle is shown in Fig. 4. The height of the peak showing the extent of energy barrier decreases with decreasing temperature. This means that the stability of casein micelles trends to decrease on cooling. However, the repulsive hydration potential term may be important in this system. We are now studying in more details.

REFERENCES

1. C.V. Morr and R.L. Richter, Chemistry of processing, *in* :"Fundamentals of Dairy Chemistry," N.P.Wong, Van NostrandReinhold Co., New York (1988).
2. W.K. Downey and R.F. Murphy, The temperature-dependent dissociation of β-casein from bovine casein micelles and complexes, *J.Dairy Res.* 37:361 (1970).

3. L.K. Creamer, G.P. Berry, and O.E. Mills, A study of the dissociation of b-casein from the bovine casein micelle at low temperature, *N.Z. J. Dairy Sci. Technol.* 12:58 (1977) .

4. R. Niki, H.J. Lee, and S.Arima, Einfluß der kühlung afu die caseinmicellen der kuhmilch, *Milchwissenschaft.* 33:473 (1978).

5. R. Niki, T. Kimura, and S. Arima, Eigenchaften des bei niedriger temparatur löslichen caseins *Milchwissenschaft.* 35:202 (1980).

6. R. Niki, T. Kimura, and S. Arima, Einfluss der kühlung auf die fraktionierte caseinmicelle, *J. Fac.Agr.Hokkaido Univ.*, 61:436 (1984).

7. R. Jenness and J. Koop, Preparation and properties of a salt solution which simulates milk ultrafiltrate, *Neth. Milk Dairy J.* 16:153 (1962).

8. Y. Sano and M. Nakagaki, Wave length dependence of the turbidity on spheroidal particles caluculated in the Stevenson-Heller approximation, *J. Phys. Chem.* 87:1614 (1983).

GELATINE - RELATING STRUCTURE AND CHEMISTRY TO FUNCTIONALITY

Christopher B. Hudson

Group Executive, Research and Scientific Affairs
Goodman Fielder Wattie Limited
230 Victoria Road
Gladesville, NSW 2111
Australia

INTRODUCTION

Gelatine is an important functional ingredient in many food products, and it is widely used as a culinary ingredient in prepared foods.

Gelatine, a naturally derived hydrocolloid, is a protein which possesses a number of unique functional properties. It is obtained from collagen rich animal tissues, such as skin, tendon and bone, by a controlled hydrolysis of the collagen enabling extraction into hot water. The processes of manufacture for gelatine have been evolving for over 150 years and have taken advantage of many advances in food technology and food engineering.

The sources of collagen generally used for gelatine manufacture are skins of cattle and pigs, and cattle bone.

Gelatine exhibits a very wide range of functional properties which result in its employment in many food categories. There are many traditional uses of gelatine which still have strong market demand today, such as in gelatine desserts, in ice cream and yoghurt, and in confectionery jubes, jellies and marshmallow. There have been several important new product development areas for gelatine, examples being low fat products, including spreads and yoghurts, microparticulation of proteins, microencapsulation of food ingredients, fining of wines and fruit-containing beverages, and hydrolysed gelatines of different specifications, many of which are non-gelling proteins. Cold water soluble "instant" gelatines are receiving increasing interest.

In all of these applications the chemistry and functionality of gelatine is a key factor. Current research and development within GFW Gelatine International, the international gelatine subsidiary of Goodman Fielder Wattie Limited, is aimed at obtaining a better understanding on how the structure and chemistry of gelatine influences functionality, and hence gelatine's role and performance in product applications.

Additionally process development research is geared to take advantage of a better understanding on gelatine structure and chemistry, so as to optimise production of those gelatine types which exhibit the best performance in particular applications.

CHEMISTRY OF GELATINE

The name gelatine really refers to a family of water-soluble protein molecules, of similar structural type, which are related in structure to the collagen protein from which gelatine is derived.

Gelatine is characterised by a unique composition and sequence of amino acids. Characteristic features of gelatine are the high content of the amino acids glycine, proline and hydroxyproline, as is apparent in Figure 1.

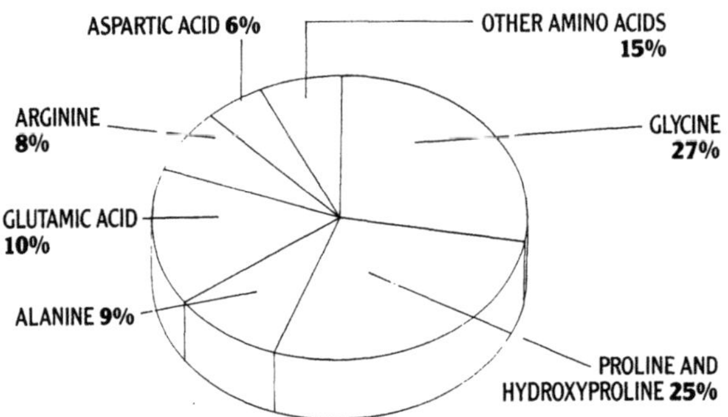

Fig. 1 Gelatine amino-acid composition

The character of gelatine is strongly influenced by the ordered presence of glycine, as each third amino acid, and by the abundance of imino acids proline and hydroxyproline.

Sequences of the amino acids glycine-x-proline and glycine-x-hydroxyproline, where x is an amino acid residue, are often observed. These sequences, as are illustrated in Figure 2, are responsible for the triple helical structure in collagen, and for the ability of gelatine to form thermoreversible gels, where regions of helical character form in the gelatine protein chains, immobilizing water.

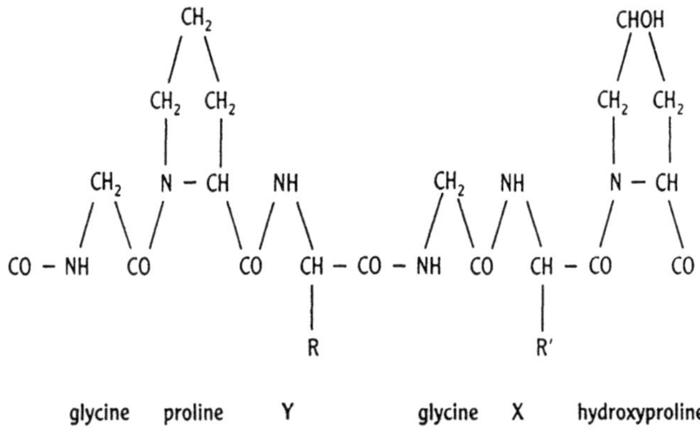

Fig. 2 Gelatine structural chain

Gelatine molecules are quite large, with molecular weight ranging from a few thousand up to several hundred thousand.

In the case of collagen, the precursor of gelatine, very ordered triple helical structures, called tropocollagen units, are linked into fibrils, which in turn build up into a continuous network of fibres, forming the important collagen structure in connective tissues. This collagen structure has a very important bearing upon the processes of extraction of gelatine from collagen, and on the properties of gelatine itself.

Intramolecular and intermolecular cross-links are present within the collagen structure, as is illustrated in Figure 3, and these cross-links have a very great influence on the molecular weight profile of gelatine and upon its functional properties.

Fig. 4 Possible paths of collagen conversion to gelatine

Cross-links are formed in the protein chains of collagen by condensation reactions between the side chain groups of the amino acids lysine, hydroxylysine and histidine. Some of these cross-links are labile and subject to hydrolysis when gelatine is extracted from collagen, but others are very stable and will form a permanent bridge between the helical chains in the tropocollagen units. The cross-links are responsible for the presence in gelatine of higher molecular weight proteins, above 100 kD, since these molecules will contain sections from more than one collagen alpha chain joined by one or more cross-links, as is illustrated in Figure 4.

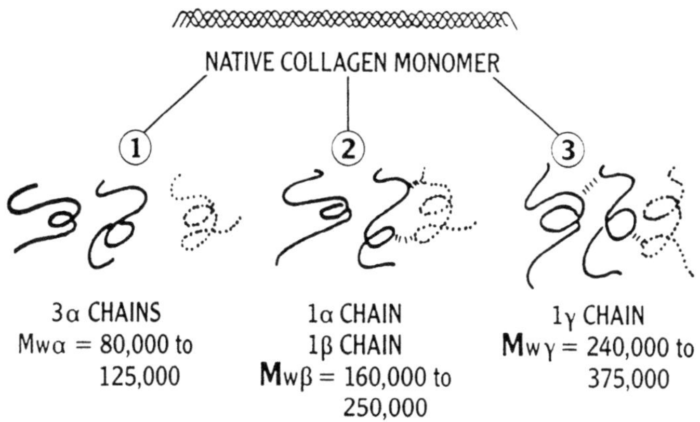

Fig. 3 Types of collagen cross-linking

The chemistry of the cross-links present in the gelatine has become better understood through recent research, including collaborative research between GFW Gelatine International and the Meat Research Laboratory of CSIRO Division of Food Processing, in Australia.

The molecular weight profile for gelatine is quite complex and will be essentially different for each gelatine type and grade. A considerable amount of research is being conducted on molecular weight profiles for different gelatines as a means of gaining a better understanding of the composition and performance of gelatine. For example, there is now evidence that molecular weight profile has a significant bearing on the physical properties of gelatine, such as gel strength which is measured commercially as the "Bloom," and also on viscosity.

Techniques such as gel permeation chromatography and gel electrophoresis have enabled the separation of gelatines into specific molecular weight fractions.

A typical molecular weight distribution for soluble collagen is shown in Figure 5, where very distinct bands at approximately 100 kD for the alpha chain, 200 kD for the beta chain, and 300 kD for the gamma chain are apparent.

For typical commercial gelatines the molecular weight distribution is very complex, as illustrated in Figure 6; however it is apparent that for higher grade gelatines, with higher Bloom and viscosity, there is a shift in average molecular weight to a higher value as compared to lower grades.

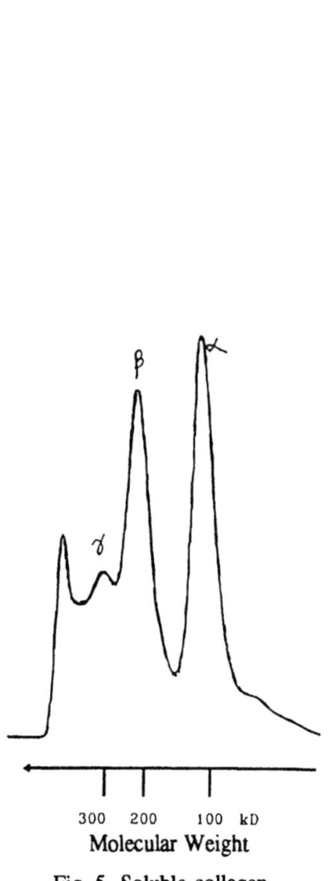

Fig. 5 Soluble collagen

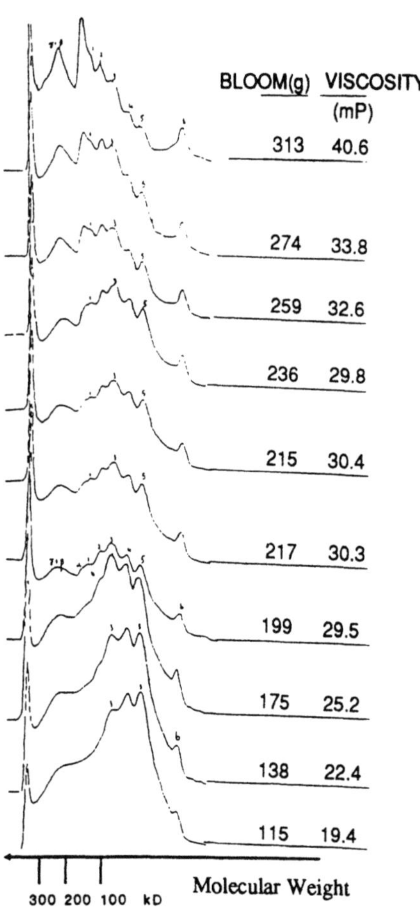

Fig. 6 Commercial gelatines molecular weight profile

Gelatine gels are thermoreversible and have melting point typically in the range 25-32°C.

Above 40°C gelatines are assumed to exist as random coils and, upon cooling, some reformation of the collagen fold is thought to occur, either as double or triple helices, giving rise to junction zones. These junction zones of helical confirguration, which form between chains, immobilise water by hydrogen bonding to form a gel.

The interaction of gelatine with other substances in a food system is greatly influenced by the isoionic point (pI) which controls the charge density and net charge on the protein as a function of the pH of the food system in which the gelatine is employed. In addition the conformational shape of the protein molecules is influenced by the pI in relation to pH. These relationships are illustrated in Figure 7.

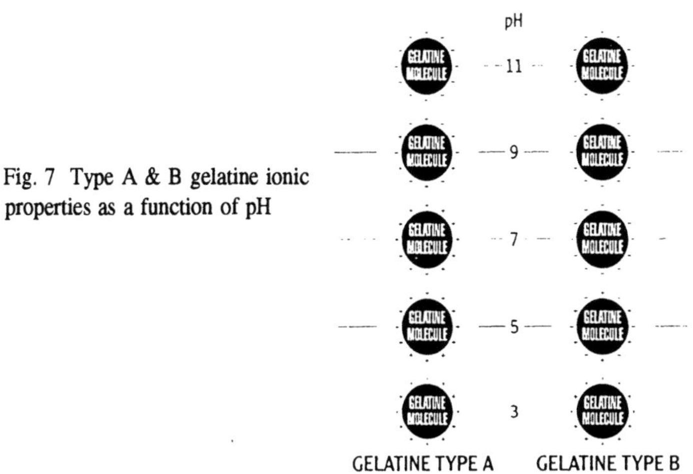

Fig. 7 Type A & B gelatine ionic properties as a function of pH

The isoionic point of gelatine is determined by the type of process of collagen raw material conditioning and subsequent gelatine extraction. The age of the animals from which the collagen tissue has been sourced is very important in determining the type of gelatine process required. For example, with skins from young animals, such as calf and pig, an acid conditioning stage may suffice and this will lead to a gelatine bearing an isoionic point, pI at 8.0-8.5 being quite close to the pI of collagen itself at 9.0-9.5. Gelatines with high pI are termed Type A. Tissues from older animals will require an alkali conditioning frequently followed by an acid conditioning which leads to a Type B gelatine with pI 4.8-5.0; this is typical of cattle skin. The principle differences between collagen tissues of older animals compared to young ones are the degree and nature of the cross-links in the protein structure. Collagens with a higher level of cross-links will require a greater degree of hydrolysis to achieve acceptable extractability of gelatine.

GELATINE APPLICATIONS AND FUNCTIONALITY

The amphoteric nature of gelatine, its ability to form thermoreversible gels, its contribution to viscosity, and its protective colloidal and surface active properties all impart functional properties important to food systems. These include:

- Gelation
- Water binding
- Emulsification
- Adhesion
- Film formation
- Crystallisation control
- Thickening & stabilisation
- Whipping and foam generation
- Beverage fining
- Glaze formation

Gelatine gels, having a melting point close to physiological temperature, are unique in regard to melting and mouthfeel characteristics. The high degree of water solubility of gelatine produces gels of excellent clarity with no tendency to undergo syneresis. These properties are extremely valuable for the formulation of gelatine desserts and many varieties of confectionery jellies and jubes.

The protective colloid functionality of gelatine is employed to control crystallisation, for example, in water systems the control of ice crystal size is important in ice cream. The control of sugar crystal size is important in fondants and confections.

Foaming and whipping properties of gelatine are exhibited in applications such as cream thickening and marshmallow production.

It is however important to recognize that in many product applications more than one function of gelatine comes into play. A good example is marshmallow where foam formation, rate of set and film formation all are key roles. In ice cream the role of gelatine is also complex and several attributes determine the manner in which gelatine influences finished product quality, as the consumer sees it. Looking at one of the more recent product applications, low fat spreads, gelatine plays a role in the formation and stabilisation of a water in oil emulsion where, once again, several attributes make a contribution.

Current research has an objective to determine in more depth the relationship between the chemistry of gelatine and functional performance. It is seen that molecular weight distribution and charge density for gelatines can be related to specific properties. An important aim is to be able to determine factors in the chemistry of gelatines which will allow better specification of the proteins for individual applications than has been possible in the past with traditional modes of gelatine product specification.

Research at GFW Gelatine International laboratories is being directed at separation of gelatine into molecular weight fractions to determine the relationship of specific areas of molecular size to important functional properties such as gel strength and viscosity. A technique has been developed to carry out gel strength (Bloom) measurements on very small quantities of gelatine. In addition the use of membrane separation technologies to fractionate gelatine on the basis of molecular weight is under current investigation.

GELATINE PERFORMANCE IN FOOD APPLICATIONS

The ways in which gelatine is used greatly influences its value and effectiveness in food product applications.

When gelatine solutions are prepared it is important to recognise the need to obtain effective dispersion in water or in the food system involved, so as to prevent formations of lumps. Adequate hydration then needs to occur so that a complete solution is obtained. Factors which influence gelatine performance in this regard are particle size of the gelatine, concentration required, the time available, gelatine viscosity, and mixing equipment employed.

Once in solution, gelatine can suffer a loss in gel strength and viscosity when exposed to elevated temperatures for any prolonged period of time. Loss of both gel strength and viscosity is a result of hydrolysis of the gelatine molecules and consequent reduction in average molecular weight. The rate of hydrolysis is a function of temperature and pH of solution. Figure 8, as an example, clearly illustrates the potential loss of gel strength as a function of pH, temperature and time.

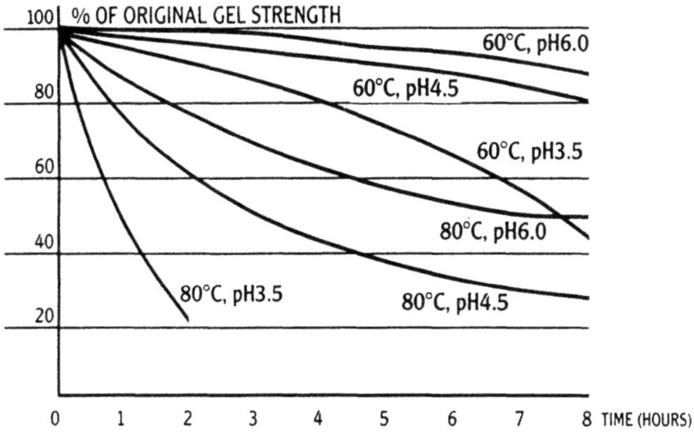

Fig. 8 Rate of loss of gel strength

To avoid unnecessary hydrolysis, the addition of acid to a product formulation should be delayed until the latest possible time in the process.

Important physical properties of gelatine such as gel strength and viscosity are also influenced by the environment in which the gelatine is employed in a food system. Because of the amphoteric properties of gelatine, its structure in an aqueous food system is very much influenced by the pH of that system.

For example, as is illustrated in Figure 9, the viscosity of gelatine in solution is significantly influenced by pH, and viscosity displays a minimum at the isoionic point for each gelatine.

The influence of electrolytes upon the behaviour of gelatine in aqueous solution is also observed in Figure 9.

Likewise the gel strength of gelatine is also influenced by the pH of the food system in which gelatine is employed, as is illustrated in Figure 10.

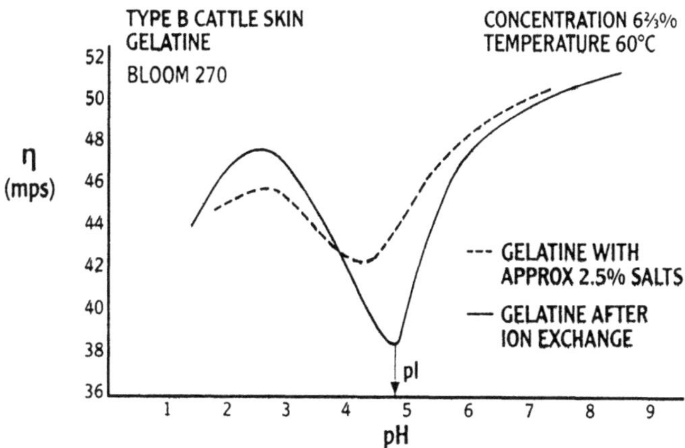

Fig. 9 Gelatine viscosity as a function of pH

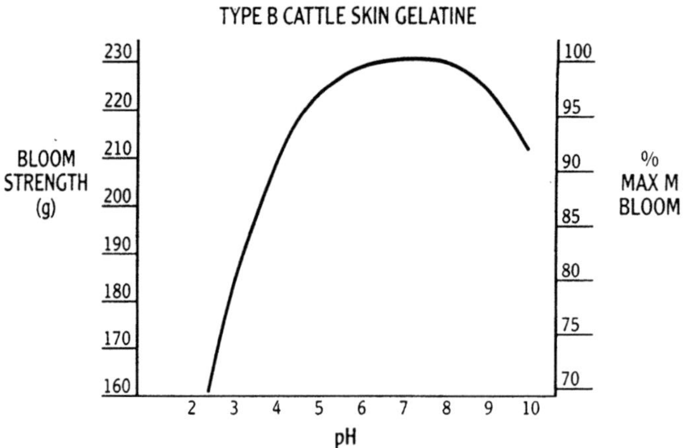

Fig. 10 Relationship of bloom strength to pH

CONCLUSION

Research and development on the molecular structure and chemical behaviour of gelatines will continue to build a better understanding on the various important functional properties of gelatine in regard to its performance in food applications. Such research will also enable the manufacture of gelatines of optimum grade for requirements in each particular food application, and will allow optimisation of the gelatine grade in regard to the conditions under which it is employed.

RELATIONSHIP BETWEEN RHEOLOGICAL PROPERTIES AND CONFORMATIONAL STATES OF 7S GLOBULIN FROM SOYBEANS AT ACIDIC pH

Takao Nagano,[1] Hiroyuki Mori,[1] Katsuyoshi Nishinari[2]

[1]Central Research Institute, Fuji Oil Co., Ltd.,
 4-3, Kinunodai, Yawara, Tsukuba-gun 300-24, Japan
[2]Department of Food and Nutrition, Osaka City University
 Sumiyoshi, Osaka 558, Japan

ABSTRACT

We propose a simple method for isolating 7S and 11S globulins in high purity, and study the effects of pH on the rheological properties of the obtained 7S gel. Using this method, the heating differential scanning calorimetry (DSC) curve of each fraction shows only one endothermic peak. The storage modulus G' of 7S globulin gel rapidly increased with decreasing pH below 5.5 - 6. Fourier transform infrared (FT-IR) measurements suggested that as pH fell below 5.5 - 6, 7S globulin tended to denature with heat, and the number of exposed β-strands increased. Thus, the number of intermolecular bonds which were available for the formation of gel networks increased, and a rigid gel formed at acidic pH.

INTRODUCTION

Soybean proteins are used extensively in processed foods and traditional Japanese foods because of their high nutritional value and functional properties, not the least of which is the ability to form gels. Soybean proteins consist of two major components, 7S and 11S globulins, with different structures and gelation properties.[1] Individual investigation of 7S and 11S is necessary to quantify the gel properties of soybean proteins. We studied 7S for two reasons: 1) the literature contains little information regarding heat-induced gels of 7S, as compared with those of 11S, and 2) 7S forms a gel at a lower temperature than 11S, which may make it an important functional ingredient in meat products.

To make optimal use of soybean proteins as functional ingredients in foods, more information is needed regarding the effect of environmental conditions (temperature, pH, ionic strength, etc.) on the rheological properties of soybean proteins. We previously reported the effects of temperature on the rheological properties of 7S gels.[2] In this paper we examined the effects of pH on the rheological properties of 7S gels. In addition, the relationship between gel properties and protein conformation was investigated by Fourier transform infrared (FT-IR).

MATERIALS AND METHODS

The variety of soybeans used in this study was Enrei. Soybean seeds were defatted with n-hexane. Sodium bisulfite (60-69% SO_2) was purchased from Kishida Chemical Co.,Ltd. The storage modulus G' was determined using a CSL-500 (Carri-Med). The sample was subjected to a shear oscillation of 1Hz and a 5% strain. Infrared spectra were recorded on a FT-300 (Horiba, Ltd.) equipped with DTGS detector. The total number of scans was 100 with a resolution 4 cm^{-1}. All experiments were run using a horizontal attenuated total reflectance (ATR) crystal (i.e., ZnSe, Spectra-Tech, INC.).

RESULTS AND DISCUSSION

Isolation of 7S and 11S globulin from soybeans

Soy protein isolate showed two endothermic peaks in heating curves derived from differential scanning calorimetry (DSC), which are believed to represent the components of 7S and 11S. The presence of two endothermic peaks makes it difficult to clarify the properties of a heat-induced gel. Therefore, We required 7S and 11S which show only one endothermic peak during heating DSC. The isolation of 7S and 11S from soybeans is widely carried out by the method of Thanh and Shibasaki.[3] According to Damodaran's report,[4] the 11S fraction exhibits only one endothermic peak, while the 7S fraction exhibits two peaks in DSC curves because of contamination by 11S in the 7S fraction.

Therefore, an improved method of isolation had to be developed. Suchkov et al.[5] reported a novel tequnique using broad beans and peas which was based on the different solubilities of 7S and 11S in various NaCl concentrations near the isoelectric point. However, we were unable to apply the same technique to soybeans. After various experiments on the different solubilities of 7S and 11S near the isoelectric point, we determined that pH 5 and 0.25 M NaCl provides the best condition for isolating 7S at 4 ℃, and that the addition of a reducing agent has an important effect on the isolation of 11S and 7S. According to the above findings, we proposed the isolation protocol for 7S and 11S shown in Fig. 1. Fig. 2 shows that only one endothermic peak is present in the heating DSC curve of the 7S fraction.

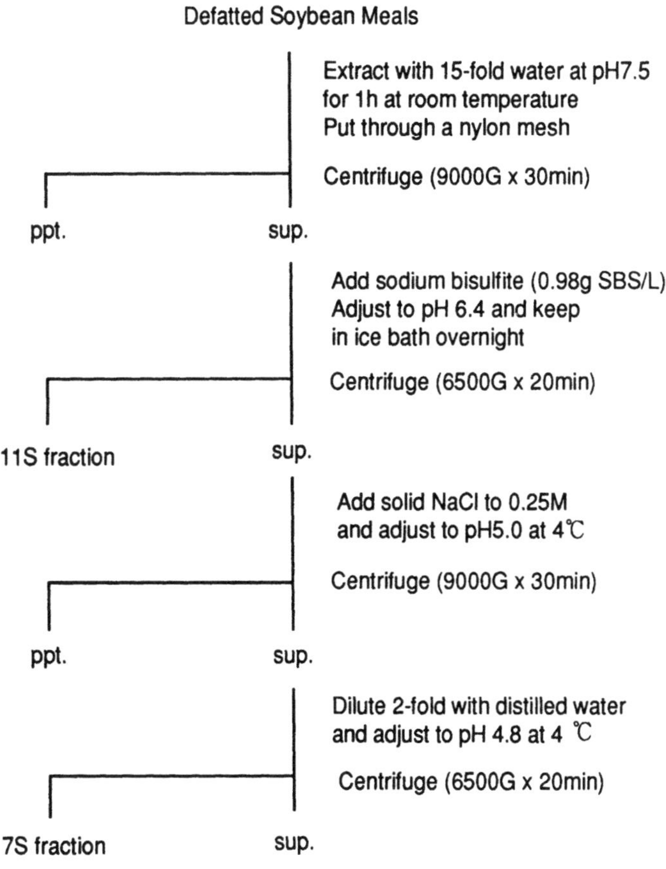

Fig. 1 Schematic diagram for the isolation of 7S and 11S globulins.

The effects of pH on the rheological properties of 7S globulin gels

The gelation curves of 7S were studied at different pHs. Since 7S is insoluble at acidic pH without salts, 7S was dispersed in 2.5 % NaCl solution. The storage modulus G' of 7S globulin gels which have been heated at 80 ℃ for 1800 sec. is shown in Fig. 3 as a function of pH. The storage modulus G' of 7S gel increased rapidly with decreasing pH below 5.5 - 6, and reached a maximum near pH 3.5. These results suggest that electrostatic charges have important effects on the rheological characteristics of 7S gels.

FT-IR measurement is useful for determining the secondary structure of protein molecules. FT-IR measurement was performed in this study to explain why G' increased rapidly below 5.5 - 6. Second derivative FT-IR spectra of the sol and gel at different pHs are shown in Fig.4. The gel was formed by heating at 80℃ for

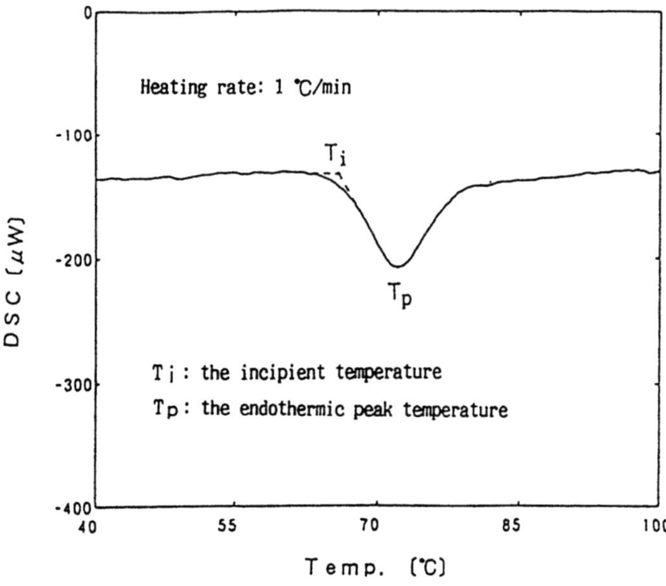

Fig. 2 Heating DSC curve of 10% 7S globulin in 35mM k-phosphate buffer, pH 7.6.

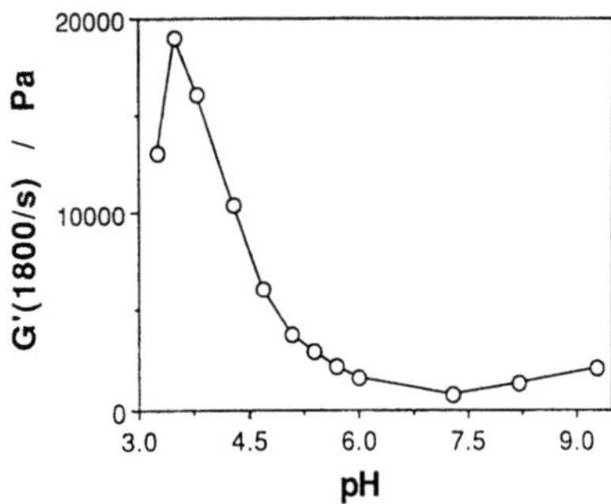

Fig. 3 The storage moduli of 15% 7S globulin gels which have been heated at 80 ℃ for
1800sec. as a function of pH.

1800 sec. in an oil bath and cooled with flowing water. With sol, a relatively strong band at 1631 cm⁻¹ and a band at 1653 cm⁻¹ were observed, while the Amide I region showed no differences at different pHs. The bands at 1631 cm⁻¹ and 1653 cm⁻¹ are associated with β-strand and α-helix, respectively.[6] According to Susi and Byler,[7] a relatively strong band near 1630 cm⁻¹ and a weak band near 1675 cm⁻¹ suggests an antiparallel β-sheet structure. Therefore, 7S is believed to have an antiparallel β-sheet structure.

The three-dimensional structure of phaseolin has recently been determined by X-ray crystallography.[8] Phaseolin is a 7S globulin of the common bean. Since the amino acid sequence of its subunit is very homologous to that of soybean 7S,[9] these two proteins are believed to have very similar structures. The crystal structure of phaseolin consists of β-strands and α-helices, primarily antiparallel

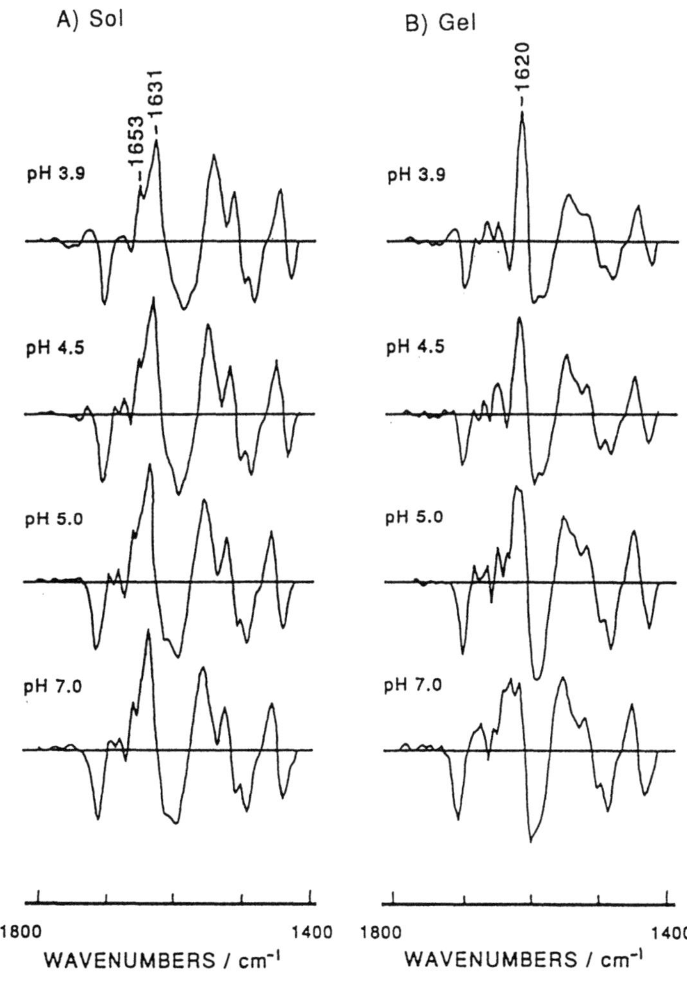

Fig. 4 Second derivative FT-IR spectra of 15% 7S globulin in sols and gels at different pHs. 7S globulin was dispersed in 2.5% NaCl solution.

β-sheet structures. The results from FT-IR spectra of 7S are in good agreement with this reported structure of phaseolin.

On the other hand, the second derivative spectra of gels were different at different pHs. At lower pH, the band at 1620 cm[-1], which is believed to indicate the formation of intermolecular structures, was strong. Casal et al.[10] reported that a band near 1624 cm[-1] was believed to indicate an exposed β-strand. The development of a band near 1620 cm[-1] has also been observed in the gelation of other proteins.[11]

The secondary structure of 7S in sol dose not change at different pHs. On the other hand, the secondary structure of 7S in gel clearly changes with pH (Fig. 4). The changes of protein structure caused by heat treatment depend significantly on pH. These results suggest that, as pH decreases, develops a molten globule state: i.e., the tertiary structure fluctuates while the secondary structure is maintained. Molten globule states have been observed in other proteins at acidic or alkaline pH and at moderate or high salt concentrations.[12]

ACKNOWLEDGMENT

We are indebted to Dr. Y. Morita for numerous discussion and advice.

REFERENCES

1. K. Saio and T. Watanabe, Difference in Functional Properties of 7S and 11S Soybean Proteins., *J. Texture. Stud.*, **9** :135 (1978).

2. T. Nagano, M. Hirotsuka, H. Mori, K. Kohyama and K. Nishinari, Dynamic Viscoelastic Study on the Gelation of 7S Globulin from Soybeans., *J. Agric. Food Chem.*, **40** :941 (1992).

3. V.H. Thanh and K. Shibasaki, Major Proteins of Soybean Seeds., *J. Agric. Food Chem.*, **24** :1117 (1976).

4. S.Damodaran, Refolding of Thermally Unfolded Soy Proteins during the Cooling Regime of the Gelation Process: Effect on Gelation., *J. Agric. Food Chem.*, **36** :262 (1988).

5. V.V. Suchkov, I.A. Popello, V.Y. Grinberg and V.B. Tolstogusov, Isolation and Purification of 7S and 11S Globulins from Broad Beans and Peas., *J. Agric. Food Chem.*, **38** :92 (1990).

6. A. Dong, P. Huang and W.S. Caughey, Protein Secondary Structures in Water from Second Derivative Amide I ., *Biochemistry*, **29** :3303 (1990).

7. H. Susi and D.M.Byler, Fourier Transform Infrared Study of Proteins with Parallel β-chains., *Arch. Biochem. Biophys.*, **258** :465 (1987).

8. M.C. Lawrence, E. Suzuki, J.N. Varghese, P.C. Davis, A.V. Donkelaar, P.A. Tulloch and P.M. Colman, The Three-dimensional Structure of the Seed Storage Protein Phaseolin.,*EMBO J.*, **9** :9 (1990).

9. J.J. Doyle, M.A. Schuler, W.D. Godette, V. Zenger and R.N. Beachy, The Glycosilated Seed Storage Proteins of *Glycine max* and *Phaseolus vulgaris.*, *J. Biol. Chem.*, **216** :9228 (1986).

10. H. L. Casal, U. Kohler and H.H. Mantsch, Structural and Conformational Changes of β-lactoglobulin B., *Biochim. Biophys. Acta*, **957** :11 (1988).

11. A.H. Clark, D.H.P. Saunderson and A. Suggett, Infrared and Laser-Raman Spectroscopic Studies of Thermally-induced Globular Protein Gels., *J. Peptide Protein Res.*, **17** :353 (1981).

12. P.S. Kim and R.L. Baldwin, Intermediates in the Folding Reactions of Small Proteins., *Annu. Rev. Biochem.*, **59** :631 (1990).

ANALYSIS AND PRACTICAL EXPLOITATION OF THE MOLECULAR BASES FOR THE FORMATION AND STABILISATION OF PROTEIN FOAMS

Maria Velissariou and Andrew Lyddiatt*

Biochemical Recovery Group
School of Chemical Engineering
University of Birmingham
Birmingham, UK. B15 2TT

ABSTRACT

The variation in quality of foams generated from homogeneous and binary mixtures of standard proteins (bovine serum albumin, lysozyme and cytochrome $\underline{c}$) has been studied with reference to protein concentration, system pH and the ionic strength of feedstocks. The composition of foams generated from selected protein mixtures in batch and continuous foaming systems is described in terms of protein enrichment. Conditions for increasing enrichment in protein recoveries from extracts of waste brewer's yeast are described, and the foaming quality of recovered products discussed.

INTRODUCTION

An understanding of the molecular characters which predispose the partition of proteins and other solutes to the gas-liquid interfaces of foams is essential to the practical exploitation (or combating) of foaming phenomena. Such practices might focus upon specific solutes (or multiple solute interactions under defined conditions) identified as essential to the formation or elimination of stable foams in food and drink manufacture. Alternatively, it might take the form of the identification of circumstances whereby foaming processes can be used for the selective fractionation of solute mixtures in bioseparations.

The relationship of solute behaviour between the related processes of foam stabilisation/destabilisation and foam fractionation has not been commonly addressed.

Work in our laboratory has established an empirical molecular basis for the enhancement of the stabilisation of foams generated from beers and from defined solutions of mixtures of protein and carbohydrate solutes[1-3]. Extension of these studies to the examination of protein behaviour in analytical and preparative systems[4] has attempted to

---

* To whom correspondence should be addressed

Food Hydrocolloids: Structures, Properties, and Functions
Edited by K. Nishinari and E. Doi, Plenum Press, New York, 1994

draw correlations between solute occupancy of stable foams and the establishment of foam fractionation processes. Of particular interest is the proposal that foaming processes *de facto* may be exploited to fractionate surfactant or foam positive solutes from complex mixtures[5]. Such materials, particularly when generated from food and drink processing waste[5], may be of benefit to generic processes requiring natural surfactant behaviour[6].

MATERIALS AND METHODS

Bovine serum albumin (BSA; A-7906), lysozyme (L-2879) and cytochrome $\underline{c}$ (C-3006) were purchased from Sigma Chemical Company, Poole, UK. Yeast extracts were prepared from waste cake kindly supplied by Bass Cape Hill Brewery, Birmingham. Yeast cake was suspended (40% wet w/v) in 20 mM potassium phosphate buffer (pH 8.0 unless otherwise stated) and milled in a 0.6 l KDL Dynomill operated at 3,200 rpm with a flow rate of 15 l/hr temperature controlled by chilled water (8°C). Extracts of milled cells, containing greater than 95% broken cells (as judged by direct microscopy), were clarified by batch centrifugation for 90 minutes at 30,000 x g.

Foam quality of homogeneous or complex solutions of proteins was quantified by determination of head retention values (HRV) using a modified apparatus of Rudin as described by Velissariou[4]. Foam fractionation of protein mixtures was undertaken in a 1.6 l foam tower operated batch wise or continuously as described by Velissariou[4] to yield bottom, foam and foam precipitate fractions. Protein concentrations were determined by spectrophotometry using published extinction coefficients in direct assays, or following size fractionation by high performance liquid chromatography (HPLC)[4]. Alternatively, the Coomassie blue method (described in Velissariou[4]) was applied with BSA as standard (0-30 µg/ml). The protein content of samples was visualised by sodium dodecylsulphate polyacrylamide gel electrophoresis (SDS-PAGE) using a Pharmacia Phast system and procedures recommended by the manufacturer[7]. Ribonucleic acid (RNA) concentrations were determined by the standard orcinol method described by Velissariou[4] using Torrula yeast RNA (Sigma Chemical Co., Poole, UK) as standard (0-120 µg/ml).

RESULTS AND DISCUSSION

HRV For Standard Protein Solutions

HRV determinations for BSA in 50 mM citrate buffer containing 4% (v/v) ethanol at pH 4.8 (close to the pI for BSA) demonstrated an increase with protein concentration in the range 1-6 mg/ml (Fig. 1). Lysozyme (1 mg/ml; pI 10.5) yielded negligible HRV in similar homogenous solution at 1 mg/ml (data not shown[3]), and little enhancement when mixed with BSA (Fig. 1). However, in 50 mM citrate buffer containing 4% (v/v) ethanol at pH 6.0, addition of 1 mg/ml lysozyme or cytochrome $\underline{c}$ (pI 10.0) incrementally increased HRV for BSA in the concentration range 1-6 mg/ml. This range was equivalent to protein molar ratios (BSA:additive) of approximately 0.2 to 1.0. It is assumed that protein association through opposite charges on BSA (-tive) and lysozyme or cytochrome $\underline{c}$ (+tive) at pH 6.0 synergistically enhanced the foam stability of such binary systems (Fig. 2). This enhancement could be quenched at moderate ionic strength (0.4 M NaCl) but re-established at higher values or in acidic conditions (pH 3.6) at low pH, presumably through increased hydrophobic interactions (data not shown[4]).

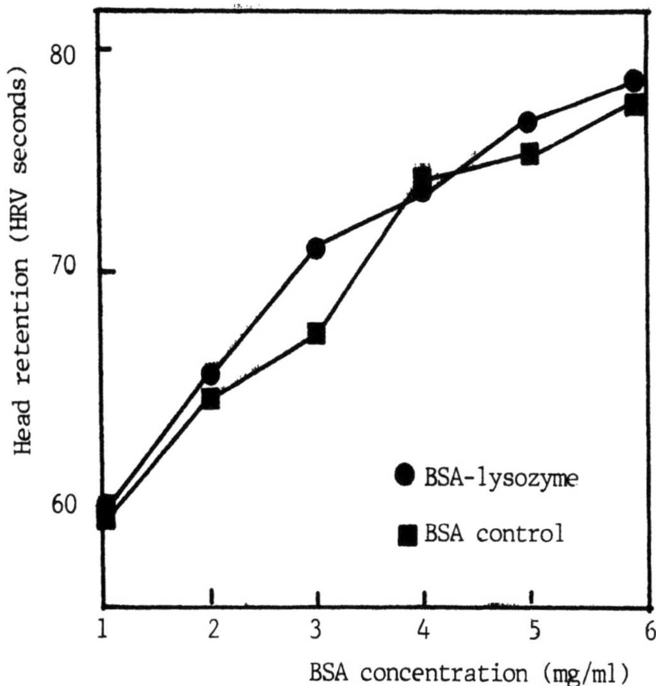

Figure 1. Head retention values (HRV) for BSA solutions at pH 4.8 in the presence or absense of lysozyme (1mg/ml).

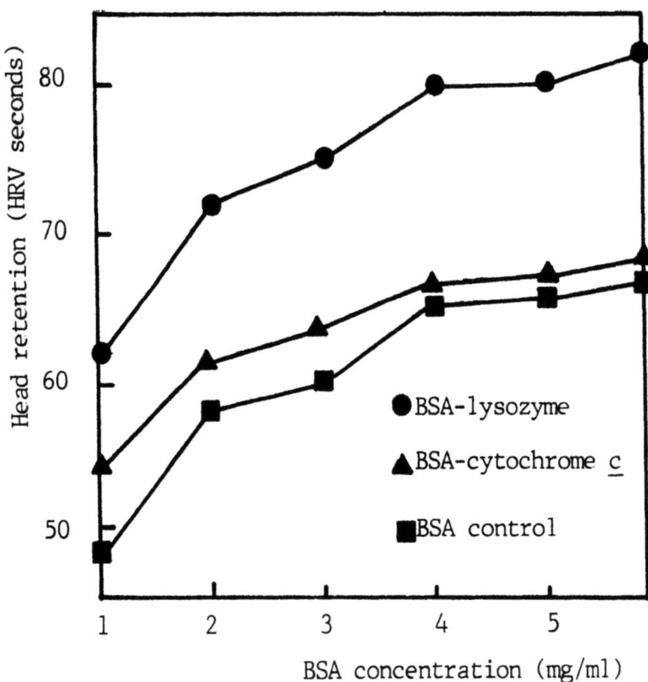

Figure 2. Head retention values (HRV) for BSA solutions at pH 6.8 in the presence or absense of lysozyme or cytochrome $\underline{c}$ (1mg/ml).

Foam Fractionation of Standardard Proteins

Correlations were sought between enhancement of foam quality, and the occupancy of foam fractions by protein solutes. Protein mixtures of BSA and lysozyme (molar ratios 0.2 to 2.0) were prepared in 20 mM sodium phosphate at pH 8.0 and foamed batchwise in a 1.6 l tower at 20°C for 30 minutes by sparging nitrogen at 80 ml/min[4]. Under conditions where proteins in solution carried opposite electrostatic charges, it was shown that BSA enrichment (concentrations in collected foam relative to the initial feedstock) declined with added lysozyme (increasing molar ratio), and increasing BSA concentration (Fig. 3). Under such conditions, lysozyme enrichment remained low and constant (approximately 1.2; data not shown[4]). Thus under conditions of increasing protein concentration where HRV (or foam stability) increased (compare Figs. 1 and 2), the BSA occupancy of foam decreased. However, the volume, liquid content and bubble size increased with protein concentration and BSA:lysozyme molar ratio (data not shown[4]).

Foam Fractionation of Yeast Extracts

Broadly similar results were obtained with continuous foaming of yeast extracts for the fractionation of bulk protein from RNA (see Fig. 4), where protein enrichment was greatest at the lowest solute concentrations in feedstocks (ie. 0.26 mg/ml protein and 0.1 mg/ml RNA). Such conditions yielded a fractionation index of greater than 7 for the separation of protein and RNA. However, estimates of protein enrichment (e_t) include contributions from the soluble fraction (e_s) in collected foam plus precipitated solids which could be recovered by centrifugation. It is of considerable interest that precipitated material recovered from yeast protein foams could be resolubilised in 2M NaOH and assayed for protein content following dilution[4]. The protein fraction was readily maintained in solution both during exhaustive dialysis against 2.5 mM potassium phosphate pH 7.6 using a membrane sac having 10 k Dalton molecular weight rejection characteristics, and following lyophilisation[4].

Desalted and freeze-dried samples were prepared from yeast foaming experiments at pH values pH 7 to 8 with feedstocks having average protein and RNA concentrations of 0.26 mg/ml and 0.53 mg/ml respectively. Samples dissolved in 40 mM potassium phosphate buffer containing 4% (v/v) ethanol at pH 7.6 were subjected to head retention value (HRV) analyses in a miniaturised Rudin apparatus. The results shown in Table 1 indicate that the protein material recovered as a precipitate in foaming experiments exhibited exceptional foaming characteristics when compared to feedstock, bottoms and soluble foam fractions. Silver-stained SDS-PAGE Phast analyses[7] of these fractions indicated that protein precipitates also exhibited characteristic banding patterns of polypeptides with molecular masses in the range 15 to 30 k Daltons (data not shown [4]). Although the foaming process, subsequent alkaline solubilisation, and lyophilisation may 'condition' the precipitate fraction and enhance HRV, it is perhaps more probable that the foaming process itself can be interpreted as specifically selecting for such foam-positive material. Interesting parallels can be drawn with beer foam manufactured in bulk continuous processes (but without noticeable precipitation) from beer which, following exhaustive dialysis, lyophilisation and normalisation of solute concentration, outperformed the mother beer in HRV analyses[1,2].

CONCLUSION

We conclude that experiments with simple protein mixtures indicate that the enhancement of foam stability by such solutes can be interpreted in terms of simple stoichiometric interactions based upon electrostatic and hydrophobic forces. In contrast, a

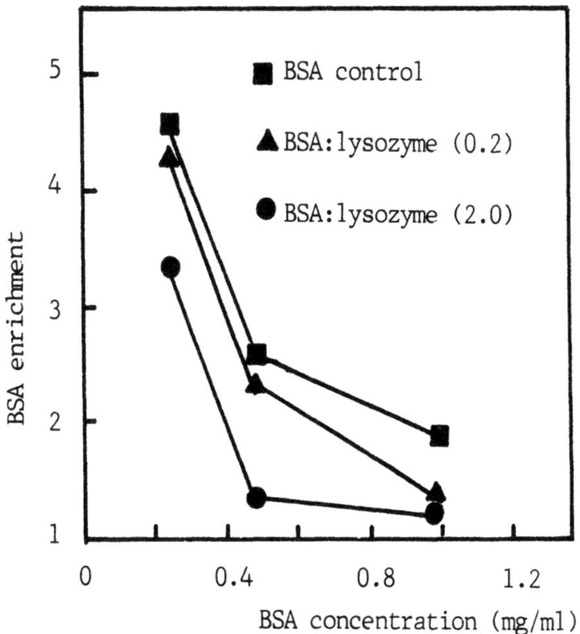

Figure 3. Variation of BSE enrichment with feedstock concentration in batch foam production in the presence or absence of lysozyme. BSA:lysozyme molar ratios were fixed at 0.2 and 2.0.

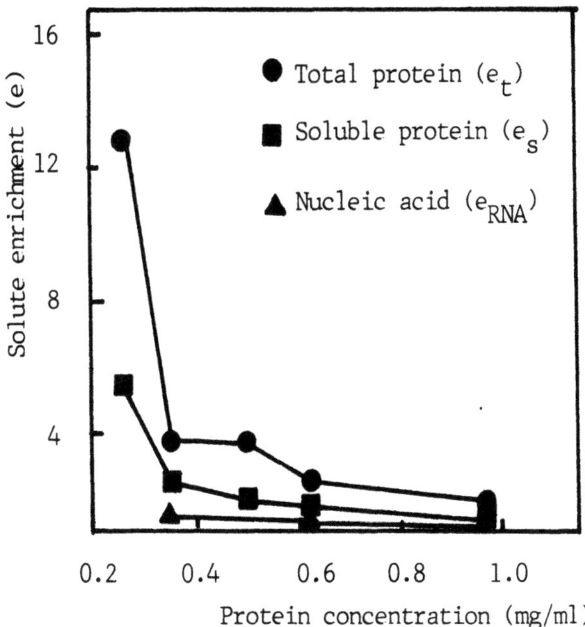

Figure 4. Effect of feedstock concentration upon foam composition in continuous foam production from yeast extracts containing protein and RNA in the mass ratio 10:1.

Table 1. Foam stability of feedstock and products* recovered from continuous foam processing of clarified extracts of wet-milled, waste brewer's yeast.

System pH	Head retention values (HRV secs)			
	Feedstock	Bottoms	Foam	Foam solids
7.0	202	177	70	450
7.6	210	200	122	480
8.0	180	136	65	447

* Samples were solubilised, dialysed and lyophilised before dissolution at 1 mg/ml in 40 mM potassium phosphate containing 4% (v/v) ethanol at pH 7.6.

simple interpretation of foam stabilisation in terms of foam occupancy by protein solutes is less clear cut. However, there is evidence that batch or continuous foam fractionation of synthetic binary mixtures of proteins in complex feedstocks derived from brewery waste may achieve significant protein enrichment at low feedstock concentrations. The foaming qualities of protein derived from the latter process may invite applications in food and drinks manufacture.

ACKNOWLEDGEMENT

Maria Velissariou and Andrew Lyddiatt gratefully acknowledge the support of the SERC Rolling Programme in the Centre for Biochemical Engineering, University of Birmingham.

REFERENCES

1. T.J. Leeson, M. Velissariou and A. Lyddiatt, Biochemical and physical analysis of beers - roles for macromolecular species in foam stabilisation at dispense, in: "Food Polymers, Gels and Colloids", E. Dickinson, ed., Royal Society of Chemistry, Cambridge, pps. 477-481 (1990).
2. T.J. Leeson, "Beer Foam Stability: Biochemical and Physical Studies Directed Toward Improved Manufacturing Processes". PhD Thesis, University of Birmingham (1988).
3. M. Velissariou and A. Lyddiatt, Foaming proprties of well defined protein systems, in: "Food Polymers, Gels and Colloids", E. Dickinson, ed., Royal Society of Chemistry, Cambridge, pps. 477-481 (1990).
4. M. Velissariou, "Foam Fractionation of Biopolymers; Studies of Protein Behaviour in Analystical and Preparative Systems". PhD Thesis, University of Birmingham (1992).
5. A. Lyddiatt, unpublished experiments (1988).
6. A. Lyddiatt, Use of biotechnological processes for the utilisation of spent yeast, in: "Biotechnology in the Food Industry; New Applications and Novel Approaches", B. Moseley, ed., IBC Technical Publications, pps. 60-67 (1990).
7. "Pharmacia Technical Manual for Phast Electrophoresis", Pharmacia Ltd., Uppsala, Sweden (1987).

PHYSICAL MODIFICATION OF PROTEINS

H.M. Rawel [1] and G. Muschiolik [2]

[1] German Institute for Human Nutrition
 D-1505 Bergholz-Rehbrücke
[2] WIP Research Group
 Functionality and Application of Proteins
 University of Potsdam
 D-1505 Bergholz-Rehbrücke

ABSTRACT. Protein isolates from faba beans, soya beans and egg albumin were mechanolysed using the mill VIBRATOM depending on the factors. The native and modified proteins were investigated to characterize the chemical and physico-chemical properties. Following chemical tests were carried out: Estimation of free amino groups using TNBS reagent, estimation of tryptophan content, available lysin and of cystein/cystin content. The investigated physico-chemical and functional properties include surface hydrophobicity, heterogenity and molecular weight distribution (using gel electrophoresis, LPLC and HPLC-techniques), foaming, waterbinding, emulsifying and solubility.

The results show that there are some conformation changes leading to aggregation of protein molecules at the beginning of the mechanolysis while in latter stages a splitting of molecules can also occur. Both these processes during a short mechanolysis improve the waterbinding, while under conditions of longer mechanolysis a better foaming and emulsifying can be observed.

INTRODUCTION

Preliminary experiments on mechanolysis of proteins have been documented by Kroll et al.[1-2]

Two principle types of changes are normally observed in mechanolysed products, which can be classified as being of physical or chemical nature. The former includes changes of the particle size, increase of surface area, structural deterioration and the latter leads to denaturation processes resulting in changes of the chemical composition due to hydrolysis-, oxydation-, photolysis- and thermal processes. Both physical and chemical changes effect the functional properties like solubility, waterbinding, emulgation and foaming.

The aim of this paper is to visualize such changes in egg and selected plant proteins as well as to show possibilities of characterisation of mechanolysed protein products.

Food Hydrocolloids: Structures, Properties, and Functions
Edited by K. Nishinari and E. Doi, Plenum Press, New York, 1994

MATERIAL AND METHODS

Faba bean protein isolate (FBPI) was prepared from flour by extraction at pH 7.0, precipitating the extracted protein at pH 4.2, neutralising it to pH 7.0 followed by spray drying. Acid precipitated soy protein isolate (SPI) and egg albumin (EIA) from Henningson represented further starting materials.

The physical modification of the dry products was conducted using a swing mill (VIBRATOM 0,6, Siebtechnik GmbH) depending on time factors. Mechanolysed and control proteins were chara-cterised regarding their N-content according to KJELDAHL-method. Estimation of free amino groups (using a modified TNBS-method[3]), of tryptophan,[4] of available lysine,[5-6] and of cystein/cystin content (amperometric)[7] was carried out for the investigated samples.

Physico-chemical and functional properties characterised were: surface hydrophobicty in native condition using cis-parinar acid,[8-9] Nitrogen solubility profile by suspending and stirring the protein samples at pH 7.0 and determining the soluble N-content after centrifugation at 5800g for 10 min. The emulsifying activity (EA, 2.5 % w/v protein solution, 50 ml sunflower oil, 47.4 % water, centrifugation at 1000 g for 10 min), emulsion stability (ES, heated EA emulsion, 80 °C 15 min, centrifugation as for EA), water absorption[10] and foaming propeties were also investigated. Molecular weight distribution was determined using SDS-PAGE.[11]

Table 1 Summary of investigated physico-chemical and functional properties

No.	EIA				SPI			
	control	0,5 h mec.	5 h mec.	30 h mec.	control	0,5 h mec.	5 h mec.	30 h mec.
1	0,73	0,13	0,16	0,22	0,81	0,81	0,74	0,70
2	6,20	5,90	6,20	5,80	5,80	5,60	5,90	5,30
3	100,00	46,70	3,60	2,40	100,00	69,30	59,40	64,30
4	--	--	--	--	100,00	--	69,00	71,00
5	38,30	53,70	39,10	23,40	13,70	6,50	6,90	0,00
6	34,50	30,60	34,50	32,20	0,00	0,00	0,00	0,00
7	100,00	57,70	20,80	16,10	90,50	80,20	61,30	55,50
8	69,50	63,10	59,30	62,20	89,30	99,20	99,00	71,20
9	89,40	78,20	74,30	68,40	83,20	87,40	98,80	66,10
10	0,10	0,40	0,60	0,40	1,20	2,00	6,40	2,50

No.	FBPI				No.	Property
	control	0,5 h mec.	5 h mec.	30 h mec.		
1	0,61	0,62	0,51	0,88	1.	Free amino groups [umol/mg Prot.]
2	6,40	6,40	6,50	6,40	2.	Available Lys [g /100 g Prot.]
3	100,00	41,00	41,00	5,10	3.	Tryptophan [%]
4	100,00	--	91,00	61,00	4.	Hydrophoby Index [%]
5	9,90	2,10	4,00	7,00	5.	SS-groups [umol /g Prot.]
6	3,50	4,60	0,00	0,00	6.	SH-groups [umol /g Prot.]
7	83,70	78,60	63,50	53,70	7.	Solubility Index [%]
8	66,50	67,30	70,70	77,10	8.	Emulsion activity [%]
9	63,60	69,50	73,60	75,80	9.	Emulsion stability [%]
10	1,00	1,40	2,60	3,80	10.	Waterbinding [g /g sample]

RESULTS AND DISCUSSION

Table 1 shows some of the properties of the three investigated protein samples. We observe a general fall in solubility with only a slight increase after 30 hours of mechanolysis which is accompanied with a similiar trend for the amount of free amino groups. A small

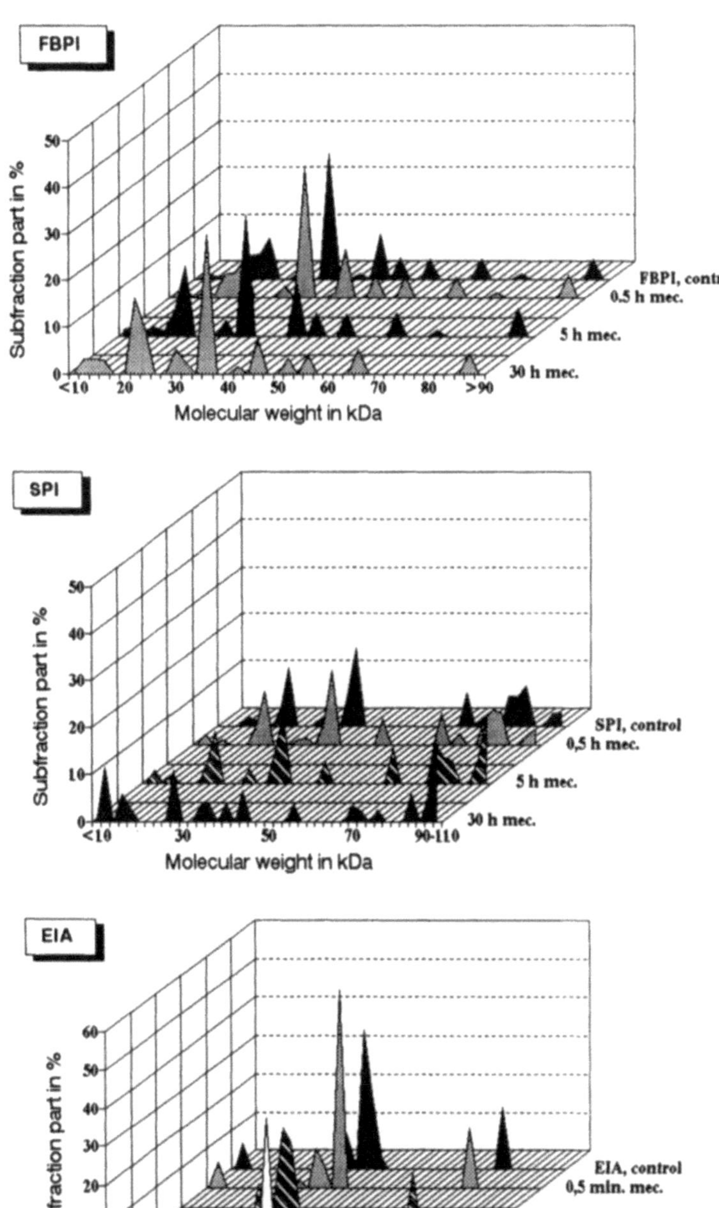

Fig.1 SDS-PAGE of the mechanolysed and control proteins

change in available lysine and a decrease in the surface hydrophobicity (for plant proteins) as well as in the content of tryptophan can be noted. We further register a fall in the amount of buried disulfide bonds with a parallel increase in the sulfhydryl groups. Most of this changes are largely due to thermal denaturation during mechanolysis and a MAILLARD-reaction with other components of the protein isolates is also likely to take place.

The functional properties of the investigated plant proteins seem to be positively influenced by mechanolysis esp. for SPI which shows a very high waterbinding capacity and an improvement of emulsifying behaviour after 5 hours of treatment. Preliminary investigations on foaming properties of mechanolysed proteins also predict a general improvement which is dependent on conditons of mechanolysis (results will be published later).

The results of SDS-PAGE as represented in Figure 1 shows that both aggregation and breakdown processes take place resulting in a decrease of the solubility at different stages of treatment. The fact that under electrophoresis conditions most of the investigated proteins can be made soluble (in comparison to solubility profile and results of LPLC/HPLC) points the importance of disulfide and secondary binding forces in aggregation processes occuring during mechanolysis. We further register that esp. for FBPI the basic beta chains with molecular weights around 20 kDa are reactiver than the acidic alpha chains. This confirms the assumption that a sort of speading or opening of the molecule due to grinding occurs making possible such reactions to take place. This could explain not only the loss of tryptophan but also the decrease in lys, arg and his as shown by results of amino acid analysis. Comparable changes in acidic amino rests (glu and asp) can be observed only for SPI.

Further investigations concentrate on the optimisation of mechanolysis to improve selected functional properties with shorter treatment conditions.

ACKNOWLEDGEMENTS

We would like to thank and appreciate the support of FRG Ministry of Economy for sponsoring this research project. We further thank coworkers K. Lengfeld and S. Großmann for their skillful technical assistance.

REFERENCES

1. J.Kroll and B.Gassmann, Manufacture of egg white substitute material based on mechanolysis and ultrafiltration, Nahrung 30:93 (1986).
2. J.Kroll et al., Beeinflussung funktioneller Eigenschaften von Proteinen durch gekoppelte mechanolytische und chemische Modifizierung, Nahrung 28:389 (1984).
3. J. Adler-Nissen, Determination of the hydrolysis of food protein hydrolysates by Trinitrobenzenesulfonic Acid, J. Agric. Food Chem., 27 (1976).
4. J.Chrastil, Spectrometric Determination of Tryptophan and Tyrosin based on new color Reactions, Analtical Biochem., 158:443 (1986).
5. K.J.Carpenter, The estimation of available lysine in animal-protein foods, Biochem J., 77:604 (1960).
6. V.H.Booth, Problems of the determination of FDNB-available lysine J. Sci. Food Agric., 22:658 (1971).
7. H.Schmandke and M.Schultz, Zur Untersuchung und Beeinflussung funktioneller Eigenschaften von Proteinrohstoffen und Proteinfasern, Nahrung 21:775 (1977).
8. A.Kato, S.Nakai, Hydrophobicity determined by a fluoresence probe method and its correlation with surface properties of proteins, Biochim. Biophys. Acta 624:13 (1980).
9. L.P.Voutsinas, S.Nakai and V.R.Harwalkwar, Relationships between protein hydrophobicity and thermal functional properties of food proteins, Can. Inst. Food Sci. Technol. J. 16:185 (1983).
10. L.Heinevetter and J.Kroll, Zur Bestimmung der Wasserbindung pulverförmiger, quellfähiger Substanzen mittels einer Kapillarsaugmethode, Nahrung 26:K17 (1982).
11. V.K.Laemmli, Cleavage of structural proteins during the assembly of the head of Bacteriophage T4, Nature 227:680 (1970).

RELATIONSHIP BETWEEN THE FUNCTIONAL PROPERTIES AND MICROSTRUCTURE OF FOOD PROTEIN GEL: EFFECTS OF FATTY ACID SALTS

Naoko Yuno-Ohta,[1] Hiroyuki Ohta,[2] Hideo Maeda,[1]
Mitsuko Okada,[1] and Kiyozo Hasegawa[3]

[1]Faculty of Health and Living Science, Naruto University of Education
 Naruto City, Takashima, Tokushima 772 Japan
[2]Department of Biological Sciences, Faculty of Bioscience and Biotechnology
 Tokyo Institute of Technology
 Nagatsuta, Yokohama 227, Japan
[3]Faculty of Home Economics, Mukogawa Women's University
 Nishinomiya, Hyogo 663, Japan

ABSTRACT

The effects of fatty acid salts (FAS) with various carbon chain length on heat-induced gel of sesame 13S globulin were investigated. The softness, water-holding ability and transparency of the gel markedly increased in the presence of sodium caprate ($NaC_{10:0}$) or sodium laurate ($NaC_{12:0}$) at 150 or 125 molar ratio to a molecule of the 13S globulin, respectively, but not sodium caprylate ($NaC_{8:0}$). Similar effects were also observed in sodium dodecyl sulfate (SDS), an ionic fatty acid derivative (FAD), whereas no effect was observed in sucrose monolaurate, a nonionic FAD. Moreover, such phenomena in FAS-protein gel was also observed on rice globulin. These results suggested that the addition of suitable amount of FAS or FAD with more than ten carbon atoms could improve water-holding ability and transparency of the various heat-induced protein gels.

INTRODUCTION

Although sesame and rice seed proteins have not been as greatly exploited as proteins from other seeds such as soybean, they are important food resources rich in essential amino acids. We investigated the functional properties of those proteins,[1-3] in order to improve the utilization to levels comparable with soybean globulin, a major food protein of plant origin. We have shown that sesame 13S globulin, a predominant protein in sesame seed, formed a very hard gel which showed very low water-holding ability in contrast to soybean protein gel.[3] On the other hand, rice globulin showed intermediate hardness and water-holding ability between sesame and soybean globulins.[3]

Food Hydrocolloids: Structures, Properties, and Functions
Edited by K. Nishinari and E. Doi, Plenum Press, New York, 1994

We have reported that the softness, water-holding ability and transparency in heat-induced gel of sesame 13S globulin markedly increased in the presence of proper amount of sodium oleate ($NaC_{18:1}$) or sodium linoleate ($NaC_{18:2}$) without chemical and enzymatic modifications.[4] Furthermore, scanning electron microscopy have shown that a very fine network structure was formed in the transparent gels.[4] In contrast, such positive effects were not observed by the addition of sodium caprylate ($NaC_{8:0}$). Although these results expect the utilization of the FASs for various food proteins to improve the gel properties, it is still uncertain how transparency and water-holding ability were improved by the addition of the FASs.

In this report, to clarify the factors which are necessary for FAS to improve the gel properties, the effects of carbon chain length and hydrophilic regions on the hardness, water-holding ability and transparency of the FAS-protein gels were investigated. In addition, the effects of FAS on rice globulin gel were also examined.

MATERIALS AND METHODS

Preparation of Sesame and Rice Globulins

Sesame 13S globulin was purified from sesame seeds (white seed) according to the procedure in our previous paper.[4] Rice globulin fraction was prepared from rice embryos (Nipponbare, kindly given from Gekkeikan Ltd. Kyoto, Japan) according to the procedure of Morita et al.[5] without final purification step by column chromatography.

Preparation of the Gels

The gelation of sesame globulin with and without FAS was performed using 150 μL of the 15% protein solution in Borax buffer (pH 9.2, I=0.24) as described in our previous paper.[4]
The gelation of rice globulin was achieved according to the same procedure as the sesame globulin, except for the usage of 80 mM Borax buffer (pH 7.4, I=0.23) and the protein concentration (10%) for gelation.

Measurement of the Gel Properties

The hardness of the gels was measured using a rheometer (Fudoh NRM-2010J-CW) under the measurement conditions described in our previous paper.[4] As a criterion of the water-holding ability of gels, the syneresis of the gels was measured. After the gels were centrifuged, the syneresis of the gels was measured as the amount of separated water from the gels according to the method in our previous paper.[4]

RESULTS

Effects of FASs on the properties of the sesame 13S Globulin Gel

The effects of three kinds of FAS, $NaC_{8:0}$, $NaC_{10:0}$ and $NaC_{12:0}$, on the hardness of sesame 13S globulin gel were investigated (Fig.1A). Fig. 1A shows changes in gel hardness with increase in molar ratio of FAS to a 13S globulin molecule. In the presence of $NaC_{10:0}$ ($NaC_{12:0}$), the gel hardness increased below 100 (75) molar ratio, and then decreased to much less than that of native gel above 150 (125) molar ratio. On the other hand, a slight increase was observed by the addition of $NaC_{8:0}$.

The effects of these FASs on the syneresis of the gels were also investigated (Fig.1B). In $NaC_{10:0}$ and $NaC_{12:0}$, the syneresis remarkably decreased with increases in the amounts of FASs, and furthermore, at 150 molar ratio, the gel syneresis was not observed. In contrast, the gel syneresis did not decrease in $NaC_{8:0}$, and it increased to about 1.2 times higher than that without FAS at 175 molar ratio.

The appearance of the gels containing FAS having various carbon chain length were compared (Fig. 2). At 100 molar ratio of FASs, the gradual increase in transparency of the gels was observed with increase in carbon chain length (Fig. 2A). At 150 molar ratio, both $NaC_{10:0}$ and $NaC_{12:0}$ formed transparent gels but not $NaC_{8:0}$ (Fig. 2B).

These results indicated that the addition of the FASs having above ten carbon atoms could improve the hardness, water-holding ability and transparency to the sesame gels. These changes in the gel properties by three FASs described above also showed that the FAS with a longer carbon chain was more effective to form the soft, high water-holding and transparent gel.

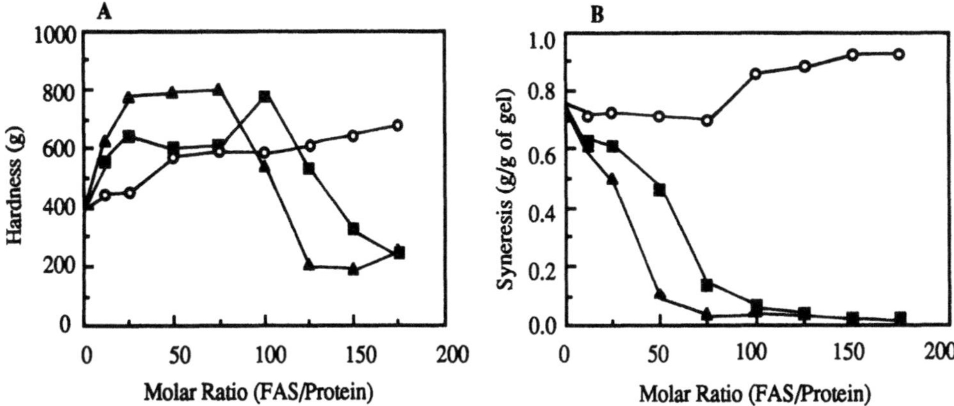

Fig. 1. Effects of FASs on hardness (A) and syneresis (B) of sesame 13S globulin gels. The preparation of the gels was described in MATERIALS AND METHODS. $\bigcirc$,$NaC_{8:0}$; $\blacksquare$, $NaC_{10:0}$; $\blacktriangle$, $NaC_{12:0}$.

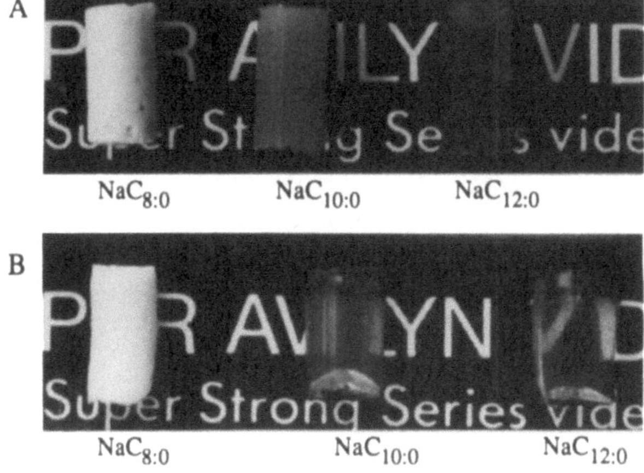

Fig. 2. Effects of FASs on transparency of sesame 13S globulin gels. The molar ratios of FASs to a molecule of 13S globulin were 100 (A) and 150 (B), respectively.

Effects of FADs on Sesame Globulin Gel

We investigated the effects of FADs with ionic or nonionic hydrophilic group having carbon chain length as the same as $NaC_{12:0}$ on the physicochemical properties of the gels. No change was observed in the gel properties by the addition of sucrose monolaurate. On the other hand, in SDS, an ionic FAD, the hardness increased below 50 molar ratio, and then decreased above 125 molar ratio, similarly to $NaC_{12:0}$. The syneresis also remarkably decreased above 25 molar ratio . Moreover, only in SDS, the transparent gel was formed at 100 molar ratio.

These results suggested that the changes in gel characteristics were influenced by the electrostatic properties of the hydrophilic region of FAD added .

Effects of FAS on Hardness and Syneresis of Rice Globulin Gels

We have previously investigated the gelation properties of rice globulin in comparison with those of sesame and soybean globulins and showed that the hardness and syneresis of rice globulin gels were intermediate among these globulins.[3] The effects of the FASs on the gel properties of rice globulin were also examined. The soft and transparent gel with high water-holding ability was also formed in rice globulin using $NaC_{10:0}$ or $NaC_{12:0}$ but not $NaC_{8:0}$ under the similar conditions to sesame globulins.

DISCUSSION

Our results indicated that transparent and hydrophilic gels from sesame and rice globulins were formed by a very simple method using FASs. From these observations, we propose a mechanism on the formation of the transparent and hydrophilic gels as follows. When FAS having more than ten carbon atoms is added to these globulins at a proper molar ratio, adequate interaction between the carbon chain of FAS and the hydrophobic region of the protein moiety will occur, and consequently, the hydrophilicity of the protein will increase by introduction of negative charges of carboxyl groups in fatty acids into hydrophobic regions of these globulins. As a result of such interaction, aggregation of denatured protein produced by hydrophobic interaction among proteins was suppressed, and then a more hydrophilic gel will be formed.

REFERENCE

1. N. Yuno, T. Matoba, and K. Hasegawa, Emulsifying properties of native and citraconylated sesame 13S globulin, Agric. Biol. Chem., 52: 685 (1988).
2. N. Yuno-Ohta, Y. Segawa, S. Fujii, S., Y. Fujiwara, M. Kuchiba, T. Matoba, and K. Hasegawa, Intermolecular forces involved in the gelation and gel stability of sesame 13S globulin, Agric. Biol. Chem.,52: 1747 (1988)
3. N. Yuno-Ohta, H. Maeda, M. Okada, and H. Ohta, Gelation characteristics of rice globulin, in preparation.
4. N. Yuno-Ohta, H. Maeda, M. Okada, and K. Hasegawa, Formation of transparent gels of sesame 13S globulin: Effects of fatty acid salts, J. Food. Sci., 56:86 (1992).
5. Y. Morita, and C. Yoshida, Studies on γ globulin of rice embryo. Part I. Isolation and Purification of γ globulin from rice embryo. Agric. Biol. Chem., 32: 664, (1968).

GELATION MECHANISM OF 11S GLOBULINS OF SOYBEANS BY Ca^{+2}-INDEPENDENT TRANSGLUTAMINASE

I.J. Kang, Y. Chanyongvorakul, Y. Matsumura,
H. Sakamoto[1], M. Motoki[1] and T. Mori

Research Institute for Food Science, Kyoto
Univ.,Gokasho, Uji 611, Japan
[1] Food Research & Development Laboratories, Ajinomoto
Co.,INC., 1-1 Suzuki-cho, Kawasaki-shi, Japan

ABSTRACT

In this study, Ca^{+2}-independent transglutaminase, isolated from culture filtrate of *Streptoverticillium* sp., was
used to enzymatically crosslink 11S globulin of soybean. In
the case of native glycinin, 70% of its acidic subunits
participated in the reaction, while most of the basic subunits did not. For pre-heated glycinin, both the acidic and
basic subunits were susceptible to the reaction. In the
presence of N-Ethylmaleimide(NEM), pre-heated glycinin formed
a better crosslinked network structure and had the higher
values of storage and loss modulus as compared to the case of
native glycinin. In the absence of NEM, only loss modulus
values of pre-heated glycinin were significantly higher than
those of native one.

INTRODUCTION

There have been many reports on the modification of
proteins, particularly, crosslinking of food proteins with
transglutaminase (TG). However, most of TG used in the studies for crosslinking food proteins were derived from mammalian organs which involved high cost for purification. The
mammal TG is Ca^{2+}-dependent enzyme. The presence of Ca^{2+} in
protein solution sometimes affects physical state of proteins, for instance, solubility. Ando et al.,[1] have extracted
a unique Ca^{2+}-independent TG from *Streptoverticillium* sp.
This purified enzyme could polymerize many kinds of proteins
including soybean globulins, α_{s1}-casein and albumin . Therefore, to understand the mechanism of protein polymerization
by this microbial TG is important both from scientific and
technological viewpoints. In the present study, main compo-

nent of soybean protein, 11S globulin, was used as the substrate for TG reaction to investigate the influence of structural change of the substrate protein on crosslinking reactions catalyzed by TG. The structural change of 11S globulin was induced by pre-heating treatment.

MATERIALS AND METHODS

Microbial transglutaminase was prepared from the culture filtrate of *Streptoverticillium* S-8112 according to the method of Ando et al.,[1]. The 11S globulin was prepared from acetone powder according to method of Thanh and Shibasaki,[2].

RESULT AND DISCUSSION

The number of lysine and glutamine residues in the protein substrate relates directly to enzyme activity. The higher the effective lysine and glutamine contents, the more crosslinks will be formed. When glycinin solutions with and without NEM (blocking reagent of SH group) were pre-heated at 100 °C, surface lysine and glutamine contents increased with increasing pre-heating times (data not shown). Furthermore, amount of intermolecular ε-(γ -glutamyl)lysine crosslinks increased with pre-heating times for both cases . The number of crosslinks of native showed 2.69 and pre-heated for 120sec with and without NEM showed 7.3 and 6.16 μmole/g dry weight, respectively. Enhancement of enzyme reactivity by pre-heating was also confirmed by electrophoresis experiment. All samples, i.e., native glycinin and pre-heated glycinin with and without NEM at various pre-heating periods, were diluted to 0.4% and subjected to transglutaminase reaction for 2 hr at 37°C. Fig.1a shows the electrophoretic patterns of all samples and Fig.1b shows the densitometric determination. The polymer fractions (lane a-d) increased whereas the subunits contents decreased when pre-heating periods were increased in the presence and absence of NEM. Ikura et.al.,[3] reported that basic subunits did not participate in enzymatic polymerization at all. However, it seems that not only acidic subunits but also basic subunits were decreased by TG reaction following the pre-heating treatment (Fig. 1a and 1b). This result suggests that basic subunits became susceptible to TG as a result of the exposure of basic subunits to the molecular surface. The decrease of acidic subunit was more dramatic and more than 90% of subunits were polymerized by TG in the pre-heated sample for 120 sec.
Viscoelastic properties of glycinin gels formed by microbial TG have been studied. Fig 2a shows the rapid increase of storage modulus(G') and loss modulus(G") during TG reaction. This indicates that a viscoelastic gel matrix was formed quickly by intermolecular crosslinks catalyzed by TG. The pre-heating treatment had an effect on the rheological behavior of enzymatic glycinin gel. Increase in the pre-heating periods, resulted in slow reaction rate and decrease in maximum G', particularly in the glycinin pre-heated for 60 sec. However, loss modulus, G" value of pre-heated glycinin

increased more rapidly with incubation time and showed maximum at 60 sec.

The effect of NEM on viscoelastic properties is shown in
Fig. 2b. The G' and G" values of glycinin modified by NEM
were two third of those of native glycinin when subjected to
TG reaction for 2 hr. Such a decrease in G' and G" can be
explained by conformational changes of glycinin caused by
modification of sulfhydryl groups with NEM, which affects the
susceptibility to enzyme reaction. However, pre-heating
treatment in the presence of NEM increased not only G" but
also G' values of glycinin gels especially at 60 sec which
gave higher values than those of native. The result was

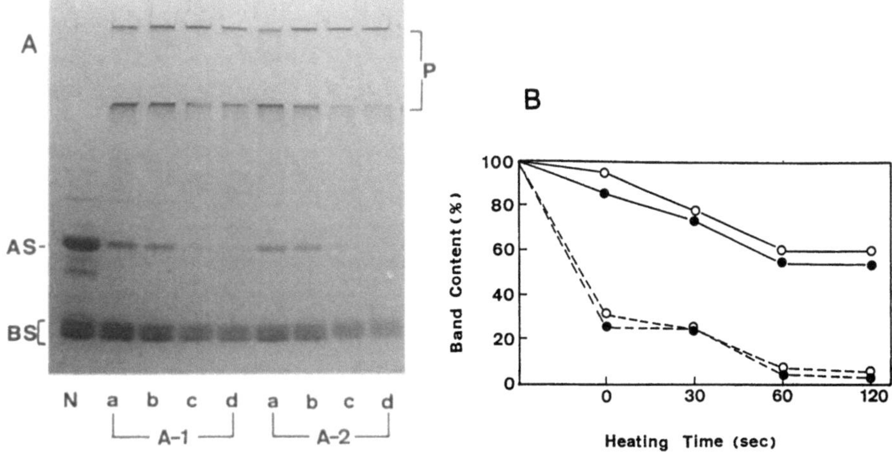

Fig. 1. SDS. PAGE pattern of the samples reacted with TG. One part of TG
was added to fifty parts glycinin by weight. Before TG reaction, samples
were heated at 100°C for 0, 30, 60 and 120 sec in the absence (A-1, O)
and presence (A-2, ●) of NEM, respectively, and corresponding gels were
labeled a, b, c, and d. N indicates native glycinin (no enzyme addition); p, polymer products; AS, acidic subunits; BS, basic subunits.
(B) Changes in the content of AS (---) and BS (——) subunits shown
in A (1 and 2) were analyzed quantitatively with densitometer tracings
based on the amount obtained from the lane loaded with native glycinin.

contrary to the case of in the absence of NEM (Fig. 2a). This
contrasting result between the pre-heated samples with and
without NEM could be explained clearly by three-dimensional
microstructure of the gels. In the case of native glycinin,
the gel has a well crosslinked network with some clumps of
aggregated proteins. In the absence of NEM, glycinin gel pre-
heated for 120 sec has a poor network structure showing
neither fibrillar nor sheetlike structures . Such a poorness
may be responsible for decrease in storage modulus. However,
glycinin gel pre-heated for 120 sec in the presence of NEM
produced an extremely well crosslinked network(data not

shown). These results indicate that pre-heating treatment in the presence or absence of NEM affected differently the structure of glycinin gels formed by TG reaction.

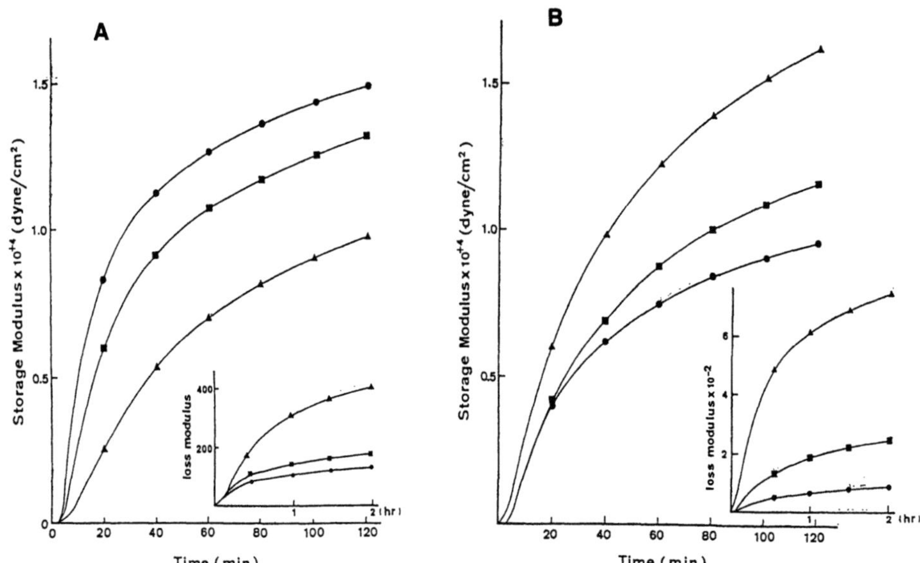

Fig. 2. Changes in storage modulus (G′) and loss modulus (G") of native and heated glycinin during TG reaction. Glycinin(6%) was heated at 100 °C in various times, and then reacted with TG at 37°C.
(2A) native (●); heated for 30 sec (■) and heated for 60 sec (▲) in the absence of NEM. (2B) native + NEM (●);   heated + NEM for 30 sec (■) and heated + NEM for 60 sec (▲). The sample were subjected to 3Hz sinusoidal shear oscillations with an amplitude (displacement ±100 μm).

REFERENCES

1. H.Ando, M.Adachi, K.Umeda, A.Matsuura, M.Nonaka, R.Uchio, H.Tanaka and M.Motoki,Purification and characteristics of a novel transglutaminase derived from microorganisms, *Agric. Biol. Chem.*53:2613(1989).
2. V.H.Thanh and K.Shibasaki,Major proteins of soybean seeds. A straight forward fractionation and their characterization, *J. Agric. Food Chem.*24:1117 (1976).
3. K.Ikura,T.Kometani,M.Yoshikawa,R.Sasaki and H.Chiba, Crosslinking of soybean 7S and 11S protein by transglutaminase, *Agric. Biol. Chem.*44:2979(1980).

MOISTURE SORPTION CHARACTERISTICS OF STEER HIDE
COLLAGEN FIBERS TREATED WITH CALCIUM HYDROXIDE

Keito Boki and Naohito Kawasaki

Kinki University
Higashi-Osaka, Osaka 577

ABSTRACT

Hydration properties of steer hide collagen fibers treated with calcium hydroxide were investigated based on the adsorption-desorption moisture isotherm at 11 to 30°C. An increase in the amount sorbed on alkali-treated collagen was explained due to the increase in the number of carboxyl- and amino-groups. In the range of water activity of 0 to 0.85, a hysteresis loop was observed. The results of heat of sorption suggest that sorbed moistures were composed of the moistures bound on the active sites and those fixed in the inter triple helix and inter microfibrillar phase. The Freundlich equation, which represents a monolayer model, fitted the adsorption isotherm of untreated collagen fiber. The Henderson and Chung and Phost equations which represent an experimental model proposed for food sorption fitted the adsorption isotherm of alkali-treated collagen fiber.

INTRODUCTION

Collagen functional properties such as swelling, wettability, water absorption, water holding capacity, etc., are directly related to the manner in which the collagen interacts with water.[1] Collagen hydration studies have been made by many investigators.[2-4] Collagen fibers are prepared commercially by the method of alkali or acid treatment. The interchain forces composed of hydrogen bonds, van der Waals interaction, electrostatic interaction and hydrophobic interaction in the collagen molecule are weakened by treating collagen fiber with calcium hydroxide.[5] Since interchain force bonds are gradually and irreversibly broken by an alkali-treatment, it results in an increased number of polar groups and in a transformed structure of the collagen fold.

In this paper, hydration properties of steer hide collagen fiber treated with calcium hydroxide were investigated on the basis of adsorption-desorption isotherms, thermodynamic properties and applicability of the isotherm equation.

Food Hydrocolloids: Structures, Properties, and Functions
Edited by K. Nishinari and E. Doi, Plenum Press, New York, 1994

MATERIALS AND METHODS

Materials

Steer hide was washed with water and soaked for 0(C-I), 10(C-II), 20(C-III) and 30(C-IV) days in twice the weight of 1% calcium hydroxide solution. The treated collagen was decalcificated with nine-fold weight of 1% ammonium chloride solution and washed with water. The decalcificated collagen was degreased and dehydrated with three-fold volumes of acetone and the collagen fibers were dried at room temperature.

Isoelectric Point and Gelatization Degree

The isoelectric point of the collagen fiber was measured by the Japanese Industrial Standard method.[6] Gelatization degree was measured as follows: Collagen fiber and warm water (approx. 65°C) were put into a conical beaker and the beaker was incubated for 2hrs at 65°C. The slurry was filtered through a filter paper (No. 5A) and was diluted to a 1% solution. The dilution and the biuret-reagent were put in a test tube and mixed vigorously. The mixture was kept at room temperature for 30 min. The absorbance of the mixture was determined at 540 nm. The gelatization degree was calculated using a calibration curve.

Pore Volume and Moisture Isotherm

The pore volume of the collagen fiber was calculated from the N_2 adsorption isotherm reported by Boki et al.[7] The adsorption-desorption moisture isotherm on collagen fiber was measured by the method of Boki et al.[8]

RESULTS AND DISCUSSION

The isoelectric point values decreased from 6.9 to 5.2 with extended treatment period, indicating an increase in the number of carboxyl groups due to the longer process. Conversely, gelatization degree increased from 9.7% to 34.1% with extended treatment period.

The amounts sorbed of the first circle adsorption over the whole water activity (Aw) range were larger in the order, C-I, C-II and C-III or C-IV, as shown in Fig. 1. The amounts sorbed in the range of Aw larger than 0.90 were proportional to the pore volume of the collagen fibers.[7] The large difference in amount sorbed was observed between C-I and C-II, C-III and C-IV. The increase in amount sorbed seemed to be due to the increase in the number of carboxyl groups by hydrolysis of asparagine and glutamine[3] and the increase in the number of carboxyl and amino groups by hydrolysis of the peptide-linkage. The amounts sorbed of the second and third circle adsorption of C-II, C-III and C-IV were a little larger than those of the first circle ones. The micropores with radii less than 20Å were produced by treating collagen fibers with calcium hydroxide.[7] The moisture sealed in the smallest pores during the first adsorption could not be removed through vacuum evaporation, that is, the first desorption. In the range of an Aw less than approx. 0.85, the desorption branches of the isotherms, the hysteresis loops, were observed and in the range of an Aw larger than 0.85, the moisture sorbed seemed to be reversible because of no desorption branches. The larger hysteresis loop of C-1 suggests the moisture sorbed was strongly held in the collagen fold. On the other hand, the decrease in magnitude of the hysteresis loop of alkali-treated collagen fibers indicates the decrease in moisture holding capacity.

The differential heats of sorption, calculated using the Clausius-Clapeyron equation and the entropy of sorption, calculated using the equation given by Hill et al.[9] are shown in

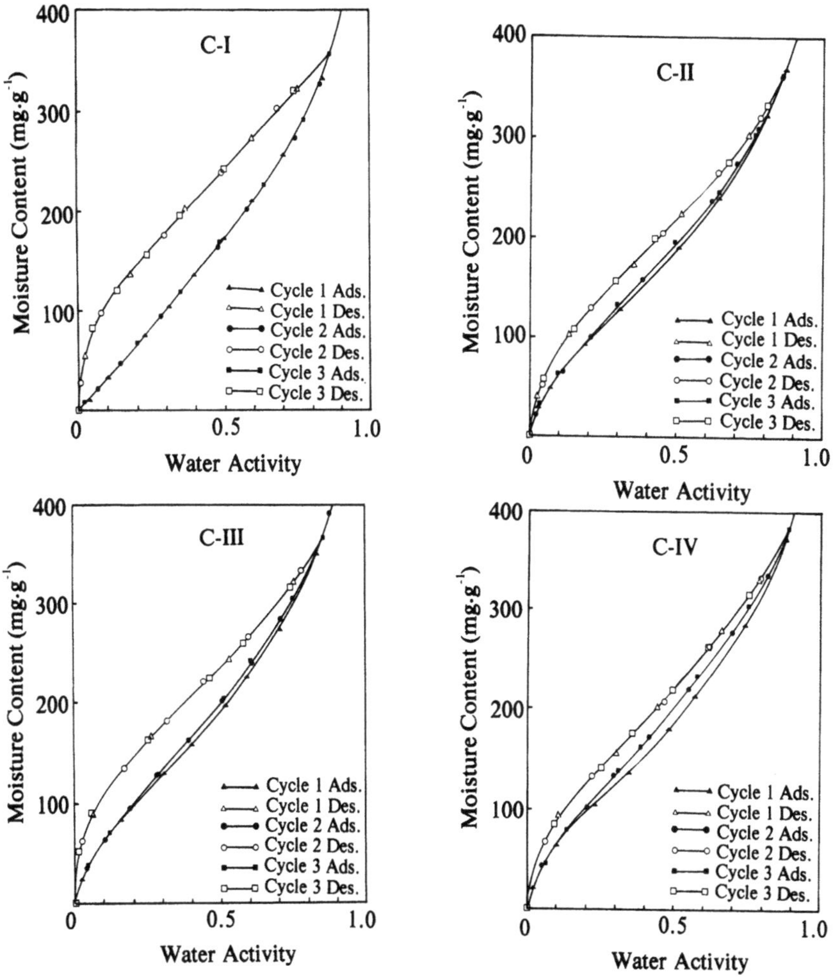

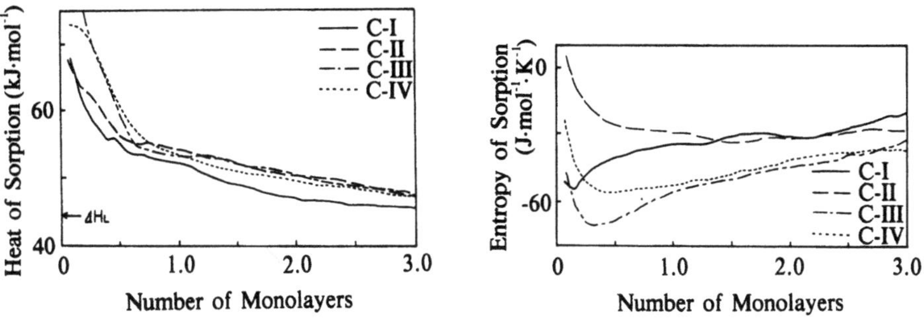

Fig. 1 Adsorption-desorption moisture isotherms on collagen fibers C-I, C-II, C-III and C-IV at 20°C.

Fig. 2. The heat of sorption decreased rapidly with an increase in the number of monolayers up to approx. 0.75 and gradually decreased to a condensation heat of water with increasing amount sorbed. The relation between the values of heat of sorption and the amount sorbed agreed with that reported by Pineri et al.[10] The results indicate that moistures sorbed are divided into two types: the moistures bound strongly on the active sites and those fixed in the inter triple helix and inter microfibrillar phase.[10] The facts that

Fig. 2 Differential heat of sorption and entropy of sorption.

heats of sorption of C-II, C-III and C-IV were larger than that of C-I indicate that stronger sorption sites were produced by the alkali-treatment. Entropy of sorbed moisture sharply decreased with increasing number of monolayers, passed through minima, and then gradually increased with increasing amount sorbed. The lowest entropy of C-III seemed to result from the sorption of moisture on the active sites in the smallest micropores.[7]

The parameter constant and applicable Aw range for a good fit, calculated by regression using the linear form of the eight isotherm equations:[11] Langmuir, Freundlich, BET, Harkins and Jura, Halsey, Smith, Henderson, and Chung and Pfost equations. To the adsorption data of C-II, C-III and C-IV, the Freundlich,[12] Henderson[13] and Chung and Pfost[14] equations fitted over the wide range of Aw (0.015—0.850). However, average residue, standard deviation and standard error (10.5—63.3) for the Freundlich equation were larger than those (0.6—7.3) for the Henderson and Chung and Pfost equations. Thus Henderson and Chung and Pfost equations best represented the experimental data for the treated collagen fibers. On the other hand, the Freundlich equation, which represents a monolayer model,[12] best fitted the adsorption data of the untreated collagen fiber C-I. The Henderson and Chung and Pfost equations are empirical models proposed for food sorption.[13,14] The difference in sorption models of C-I and C-II, C-III and C-IV can be well explained by assuming that both carboxyl- and amino-groups and porous structure susceptible to moisture were produced by the alkali-treatment.

REFERENCES

1. D.H. Chou and C.V. Morr, Protein-water interactions and functional properties, *J. Am. Oil Chem. Soc.* 56:53(1979).
2. E.P. Katz and S.T. Li, The intermolecular space of reconstituted collagen fibrils, *J. Mol. Biol.* 73:351 (1973).
3. K. Iwamoto and Y. Fujii, Adsorption of water vapor by collagen and gelatin films, in "Photographic Binders," H. Irie, et al., ed. Research Group of Photographic Binders in Japan, (1976).
4. N. Sasaki, S. Kazuma and K. Hikichi, Structural and molecular motion aspects of collagen hydration, *Rep. Prog. Polym. Phys. Jpn.* 23:721 (1980).
5. K. Shirai, "Hikaku kagaku," Japan Leather Technical Society, ed., (1985).
6. Japanese Industrial Standard, K-6503, 8. Isoelectric point, (1977).
7. K. Boki, N. Kawasaki, K. Minami and H. Takahashi, Structural analysis of collagen fibers by nitrogen adsorption method, *J. Colloid Interface Sci.* accepted (1992).
8. K. Boki, S. Ohno and S. Shinoda, Determination of sweet potato starch mixed in kudzu starch by moisture sorption method, *J. Food Sci.* 54:487 (1989).
9. T.L. Hill, P.H. Emmett and L.G. Joyner, Calculation of thermodynamic functions of adsorbed molecules from adsorption isotherm measurements: nitrogen on graphon, *J. Am. Chem. Soc.* 73:5102 (1951).
10. M.H. Pineri, M. Escoubes and G. Roche, Water-collagen interactions: calorimetric and mechanical experiments, *Biopolymers*, 17:2799 (1978).
11. K. Boki and S. Ohno, Equilibrium isotherm equations to represent moisture sorption on starch, *J. Food Sci.* 56:1106 (1991).
12. H. Freundlich, "Colloid and Capillary Chemistry," Methuen, London, (1926).
13. S.M. Henderson, A basic concept of equilibrium moisture, *Agri. Eng.* 33:29 (1952).
14. D.S. Chung and H.B. Pfost, Adsorption and desorption of water vapor by cereal grains and their products, *Transactions of the ASAE* 10:552 (1967).

CALCIUM - INDEPENDENT TRANSGLUTAMINASE DERIVED FROM A MICROORGANISM : ITS CHARACTERISTICS AND CAPABILITY IN PROTEIN CROSSLINKING AND GEL FORMATION

Hiroko Sakamoto, Masahiko Nonaka, and Masao Motoki

Food Research & Development Laboratories
Ajinomoto Co., Inc.
1-1 Suzuki-cho, Kawasaki-ku, Kawasaki-shi, 210 Japan

ABSTRACT

Characteristics of a novel transglutaminase which was derived from a microorganism thought to belong to *Streptoverticillium* sp. were observed. The molecular weight of the purified enzyme was found to be about 40,000 on SDS-polyacrylamide gel electrophoresis, the isoelectric point 8.9 and the optimum pH of the reaction 6 ~ 7. The enzyme requires no calcium ions for its activity.

Several food proteins, such as milk, soybean, poultry and fish proteins were polymerized and gelatinized by the microbial transglutaminase. Reverse phase-HPLC analysis after proteolytic digestion of the polymerized proteins showed existence of the ε-(γ -Glu)Lys bond. It was suggested that the ε-(γ -Glu)Lys bonds which were formed by a catalytic reaction of the transglutaminase were concerned with the improvement of the texture of foods.

INTRODUCTION

Transglutaminase (R-glutaminyl-peptide: amine γ-glutamyl-transferase; EC 2. 3. 2. 13) is one of the enzymes which are capable of introducing covalent cross-links between proteins. It is a Ca^{2+} dependent enzyme that catalyzes acyl-transfer reaction, in which γ-carboxyamide groups of peptide-bound glutaminyl residues are the acyl donors. When amino groups of peptide-bound lysine residues are the acyl acceptor, ε-(γ -Glu)Lys bonds are generated.

Transglutaminases are widely distributed in various organisms and their various biological functions and enzymatic properties have been investigated. In recent years,

Food Hydrocolloids: Structures, Properties, and Functions
Edited by K. Nishinari and E. Doi, Plenum Press, New York, 1994

Ikura et al.(1980), Motoki and Nio(1983) and Kurth and Rogers (1984) reported polymerization and crosslinking of some food proteins which were catalyzed by mammalian transglutaminases. In spite of those reports, it was thought to be difficult to use such mammalian transglutaminase for food production because of its high supply cost. Thus, we had screened for microorganisms that produced transglutaminase for the purpose of mass production. A microorganism was screened and a transglutaminase was purified from the culture filtrate of the microorganism thought to belong to *Streptoverticillium* sp. of actinomycetes (Ando et al., 1989). Described in this report are the characteristics of the present enzyme, and capability in protein crosslinking and gel formation.

MATERIALS AND METHODS

Microbial transglutaminase (11.3 units/mg) was prepared from a culture filtrate of *Streptoverticillium* sp. (Ando et al., 1989).

SDS-polyacrylamide gel electrophoresis(SDS-PAGE) was carried out on a slab gel. The isoelectric points were obtained by isoelectric focusing using Pharmalyte pH 3 ~ 10 (Pharmacia Co.). The enzyme activity was measured by a colorimetric hydroxamate procedure with N-carbobenzoxy-L-glutaminyl-glycine (CBZ-glutaminylglycine) (Folk and Cole, 1966).

Crude α_{s1}-casein, crude soybean 7S and 11S globulin, rabbit myosin and actin, bovine serum albumin, human serum albumin and conalbumin were prepared or obtained for protein substrates (Nonaka et al., 1989). Polymerization of the proteins was carried out without the calcium ion. α_{s1}-Casein and soybean globulin solutions (10 mg/ml) were prepared in a 0.1 M Tris-HCl buffer (pH 7.6). When the albumins were protein substrates, protein solutions were prepared in a 0.1 M Tris-HCl buffer (pH 7.6) containing 10 mM dithiothreitol. Rabbit myosin and actin solutions were diluted to 10 mg/ml with a buffer of 0.5 M KCl and 20 mM Tris-maleate (pH 7.5). Except for myosin and actin, each protein solution (1.0ml) was mixed with 0.2 units of microbial transglutaminase and incubated for 60 min at 55°C. Following the addition of the present enzyme (1.0 unit), the myosin and actin solutions (1.0 ml) were incubated for 60 min at 10°C. After incubation, the reaction mixtures were immediately analyzed by SDS-PAGE.

Protein solutions (50 mg/ml) were prepared in a 0.1 M Tris-HCl buffer (pH 7.6). Aliquots of 2.0 ml from the protein solutions and 2 units of microbial transglutaminase were mixed and put into a small test tube, with subsequent incubation at 55°C for 60 min. Protein gel formation was confirmed by standing the test tubes on their heads.

Polymerized α_{s1}-casein was digested by the sequential addition of proteolytic enzymes and analysis of ϵ-(γ-Glu)Lys bond was performed by the method of Griffin et al. (1982).

RESULTS AND DISCUSSION

The molecular weight of microbial transglutaminase (called MTGase hereafter) was about 40,000, as judged from the results of SDS-PAGE and gel chromatography on a Sephadex G-75 column (data not shown), and it was found to be a monomeric enzyme consisting of a single polypeptide chain. This was about 1/2 the molecular weight of the transglutaminase (called GTGase hereafter) derived from the liver of guinea pigs. The isoelectric point, pI, obtained on electrofocusing of the Pharmalyte was about 8.9, which was completely different from the pI of GTGase, 4.5.

When hydroxylamine and CBZ-glutaminylglycine were used as substrates, the optimum pH was about 6 ~ 7 with the reaction time of 10 min at 37°C, the optimum temperature when the reaction time was 10 min at pH 6.0 was about 50°C, and the stable pH range on treatment for 10 min at 37°C was 5 ~ 9. As to thermal stability, 100% activity remained at 40°C on treatment for 10 min at pH 7.0, and 74% activity at 50°C.

The enzyme being investigated acts in both the presence (1 mM, 5 mM) and the absence of calcium ions. And the activity was not inhibited on adding EDTA. Thus this enzyme clearly differs from GTGase, which is defined as calcium-dependent enzyme.

SDS-PAGE of α_{s1}-casein and soybean globulins treated with MTGase shows that the monomeric fraction of the intact protein diminished or disappeared, and a novel polymer fraction, which could not enter into the separation gels, was formed (Fig.1). Since enough 2-mercaptoethanol was contained in the sample solution for SDS-PAGE, the polymer detected could not have been formed by disulfide bonds. On the SDS-PAGE of rabbit myosin treated with MTGase, polymerization of rabbit myosin was observed. Although rabbit actin was incubated with the same enzyme under the same condition, it revealed no change. Bovine serum albumin, human serum albumin and conalbumin were polymerized by MTGase when dithiothreitol was added to the reaction mixture (data not shown).

We have reported the gelation of a protein solution by GTGase (Nio et al., 1985). The protein solutions of α_{s1}-casein and soybean globulins with higher concentrations turned into self-supporting gels with the addition of MTGase as well as was shown in the case for GTGase. The protein solution was not gelatinized unless transglutaminase was added.

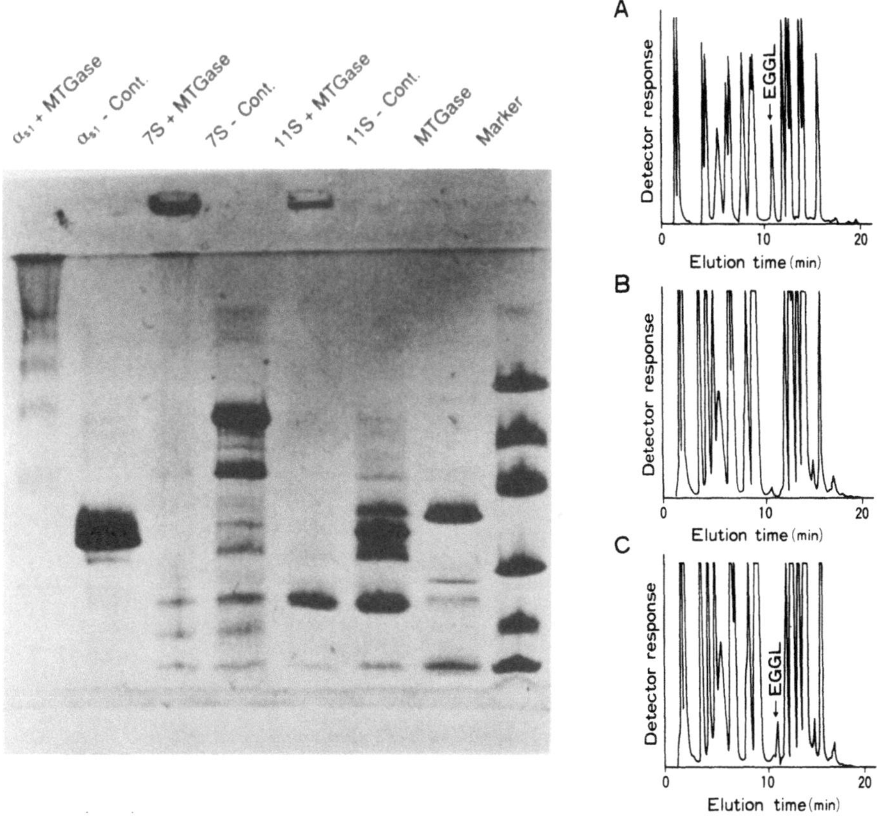

Fig.1 . (left side) SDS-PAGE of proteins polymerized by microbial transglutaminase (MTGase). α_{s1}, 7S and 11S represent α_{s1}-casein, soybean 7S and 11S globulins, respectively. Each protein solution was incubated with inactivated MTGase at 55°C for 60 min (– Cont), and with MTGase (+MTGase).

Fig.2. (right side) HPLC analysis of the ε-(γ -Glu)Lys (EGGL) bond in the proteolytic digest of polymerized α_{s1}-casein. (A) mixture of 20 amino acids and EGGL; (B) proteolytic digest of α_{s1}-casein incubated with inactivated MTGase; (C) proteolytic digest of α_{s1}-casein polymerized by MTGase.

The peak of synthetic ε-N-γ-glutamyllysine on HPLC (RP-HPLC) was detected at 10.8 min. The peak of the ε-(γ-Glu)Lys bond was recognized on HPLC chromatogram of the sample solution derived from polymerized α_{s1}-casein, but not of control (Fig.2). Thus, the ε-(γ-Glu)Lys bond in the polymer was formed by the transacylation between γ-carboxyamides of glutamine residues and ε-amino groups of lysine residues in α_{s1}-casein catalyzed by MTGase. This fact confirmed that MTGase is one of the transglutaminases.

We also found solutions of carp myosin were gelatinized and polymerized in the presence of MTGase. The ε-(γ-Glu)Lys bond was derived from polymerized carp myosin.

Our results may indicate feasibility of application for food processing. It was observed that several suspensions and emulsions containing industrially used sodium caseinate or skim milk were gelatinized by incubating with MTGase following heat-treatment while the samples with no enzyme were not. The characteristics of prepared gels were largely affected by the enzyme concentration. Further study of physical properties of commercial food materials (soy protein products, fish paste products etc.) treated by MTGase is now in progress.

ACKNOWLEDGMENTS

The screening and purification of MTGase, and the study on characteristics of MTGase were performed through collaboration with Amano Pharmaceutical Co., Ltd. (Nagoya, Japan).

REFERENCES

Ando, H., Adachi, M., Umeda, K., Matsuura, A., Nonaka. M., Uchio, R., Tanaka, H., and Motoki, M., 1989, Purification and characteristics of a novel transglutaminase derived from microorganisms, *Agric. Biol. Chem.* 53:2613

Folk, J.E., and Cole, P.W. , 1966, Mechanism of action of guinea pig liver transglutaminase. I. purification and properties of the enzyme : identification of a functional cysteine essential for activity, *J. Biol. Chem.* 241:5518

Griffin, M., Wilson, J., and Lorand, L., 1982, High-pressure liquid chromatographic procedure for the determination of ε-(γ-glutamyl)lysine in proteins, *Anal. Biochem.* 124:406

Ikura, K., Kometani, T., Yoshikawa, M., Sasaki, R., and Chiba, H., 1980, Crosslinking of casein components by transglutaminase, *Agric. Biol. Chem.* 44:1567

Kurth, L., and Rogers, P. J., 1984, Transglutaminase catalyzed cross-linking of myosin to soya protein, casein and gluten, *J. Food Sci.* 49:573

Motoki, M., and Nio, N. , 1983, Crosslinking between different good proteins by transglutaminase, *J. Food Sci.* 48:561

Nio, N., Motoki, M., and Takinami, K., 1985, Gelation of casein and soybean globulins by transglutaminase, *Agric. Biol. Chem.* 49:2283

Nonaka, M., Tanaka, H., Okiyama, A., Motoki, M., Ando, H., Umeda, K., and Matsuura, A., 1989, Polymerization of several proteins by Ca^{2+}- independent transglutaminase derived from microorganisms, *Agri. Biol. Chem.* 53:2619

EMULSION STABILITY

Eric Dickinson

Procter Department of Food Science
University of Leeds
Leeds LS2 9JT, England

ABSTRACT   The properties of adsorbed protein layers have a crucial influence on the stability of food emulsions. Creaming and flocculation of oil-in-water emulsions are also affected by the nature and strength of biopolymer–biopolymer and biopolymer–surfactant interactions in the aqueous phase and at the interface. Overall stability is determined by a delicate balance of structural, thermodynamic and rheological considerations. In systems containing protein-coated droplets, the addition of small quantities of calcium ions or polysaccharides can have a profound effect on stability behaviour.

INTRODUCTION

Many foods that contain oil or fat exist as oil-in-water emulsions. They are typically stabilized by an adsorbed layer of protein at the oil–water interface which forms a protective steric barrier around the oil droplets. Additional stabilization may be provided by soluble polysaccharide acting as a thickening or stucturing agent in the aqueous continuous phase. This article summarizes some of the important physico-chemical factors affecting stability of food colloids.[1] Concepts are illustrated by reference to recent research in the field.

Emulsion stability is a kinetic concept. Good stability implies that there is no discernible change in the size distribution or the spatial arrangement of droplets over the experimental time-scale. Depending on the context, the observational time-scale can be anywhere from a few seconds to a few years ($\sim 10^8$ s), which implies that stability is also a relative concept. Loss of stability may involve a number of processes which take place simultaneously or consecutively depending on the circumstances.

Creaming arises from the action of gravity on droplets of lower density than that of the continuous phase. This produces a vertical concentration gradient of droplets within the sample but no change in overall droplet-size distribution. Severe creaming leads to a distinct cream layer at the top and a rather clear serum layer at the bottom. *Flocculation* involves aggregation of droplets without rupture of the protective stabilizing layer at the oil–water interface. The strength of the interdroplet attractive forces determines whether flocculation is weak (reversible) or strong (not easily reversed). The end result of the creaming process is, in essence, the formation of a macroscopic flocculated state (the cream layer).

Food Hydrocolloids: Structures, Properties, and Functions
Edited by K. Nishinari and E. Doi, Plenum Press, New York, 1994

This article is mainly concerned with the processes of creaming and flocculation. There are, however, several other instability processes which occur in food emulsions. Coalescence involves the irreversible coming together of creamed or flocculated liquid droplets to produce larger spherical droplets. Partial coalescence occurs when semi-solid fat globules clump into non-spherical aggregates with rupture of the stabilizing layer. Ostwald ripening is the growth of larger droplets at the expense of smaller ones by diffusion of soluble dispersed phase material through the continuous phase. Phase inversion is the change in morphology from an oil-in-water emulsion to a water-in-oil emulsion. The fundamental science of these processes is to be found in standard textbooks,[2-5] although it should be noted that the theories have been mostly developed to describe dilute systems stabilized by simple small-molecule surfactants or synthetic polymers. Our particular concern here is with the role of complex biopolymers (proteins and polysaccharides) in the stabilization of concentrated emulsions.

STABILIZATION BY ADSORBED PROTEIN

Protein-coated droplets in a stable oil-in-water emulsion are kept apart by a combination of charge stabilization and steric stabilization.[3] In charge stabilization the short-range van der Waals attraction between droplets is overcome by the repulsive interaction between overlapping electric double-layers, whereas in steric stabilization the repulsion arises from overlap of "hairy" polymer layers. Electrostatic (charge) stabilization is favoured by dense monolayers of highly charged globular proteins at low ionic strength. Polymeric (steric) stabilization is favoured by thick adsorbed layers of flexible proteins under good solvent conditions. The use of existing theories to predict the stability of systems of protein-coated droplets is severely limited by our current knowledge of the structure of adsorbed protein layers.

The structure of a layer of disordered protein (such as β-casein or α_s-casein) is quite different from that of a globular protein like β-lactoglobulin or bovine serum albumin (BSA). Recent neutron reflectivity measurements of β-casein adsorbed at the oil–water or the air–water interfaces have shown[6,7] a high-density inner layer of thickness ~2 nm and a low-density outer layer of thickness 5–6 nm extending into the aqueous phase. In contrast, a spread monolayer of BSA is reported[8] to have a dense inner layer of thickness ~1 nm and a very low-density outer layer of thickness only ~2 nm. Dynamic light-scattering data[9] for protein-coated polystyrene latices are consistent with a hydrodynamic layer thickness of 10–15 nm for adsorbed β-casein as compared with just ~2 nm for adsorbed β-lactoglobulin. We can represent adsorbed β-casein in terms of a train–loop–tail model, where the train segments lie in direct contact with the surface and the dangling loops and tails contribute to the steric stabilization. It has been suggested[10] that the large hydrodynamic thickness of the β-casein layer is due to a long tail composed of 40–50 hydrophilic residues at the N-terminus of the protein. In contrast, adsorbed monolayers of β-lactoglobulin or BSA have no large regions dangling into the aqueous phase; they are therefore best regarded as close-packed two-dimensional assemblies of interacting deformable particles.[7]

To be an effective emulsifier, the protein adsorbing during emulsification must protect the newly formed droplets against spontaneous flocculation and coalescence. The proportion of total protein becoming permanently associated with the droplets depends on a number of factors: the emulsification conditions, the oil/protein ratio, the structure of the protein, and its interactions with other species in the emulsion (including inorganic ions). Figure 1 shows the effect of calcium ion concentration on the fraction of protein adsorbed in emulsions made with pure egg-yolk protein phosvitin under controlled homogenization conditions.[11] Phosvitin is a highly charged protein ($\sim 4 \times 10^4$ daltons) which gives excellent electrostatic

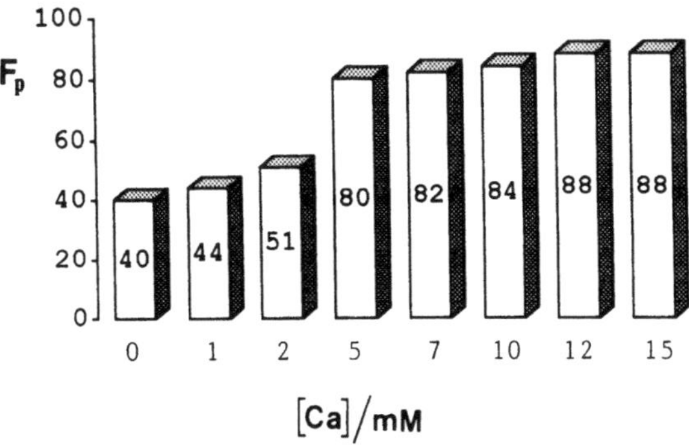

Fig. 1 Fraction of protein adsorbed in emulsions (0.5 wt% phosvitin, 20 wt% n-tetradecane, pH 7) as a function of calcium ion concentration in aqueous phase prior to homogenization. Values of F_p represent the percentage of the total protein associated with the cream layer after centrifugation.

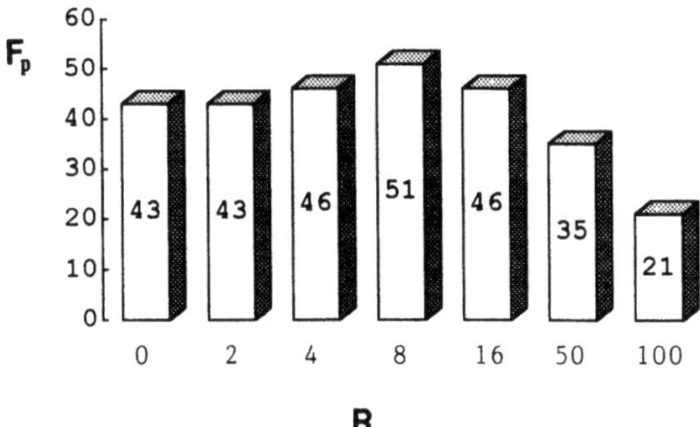

Fig. 2 Fraction of protein adsorbed in emulsions (0.4 wt% β-lactoglobulin, 20 wt% soya oil, pH 7) as a function of the amount of phosphatidylethanolamine present prior to homogenization. The percentage of the total protein associated with the cream layer after centrifugation, F_p, is shown for various values of the lecithin/protein molar ratio R.

stabilization of emulsion droplets under neutral pH conditions.[12] Its very high degree of phosphorylation leads to strong calcium ion binding;[13] this lowers ionic repulsion between adsorbed phosvitin molecules, producing a greater protein surface concentration, and eventually flocculation (see below). Many food emulsions contain surface-active lipids in addition to surface-active protein. Figure 2 shows the effect of soybean phosphatidylethanolamine on the fraction of protein associated with oil droplets in fine emulsions made with β-lactoglobulin under controlled homogenization conditions.[14] Large concentrations of lecithin partially displace milk protein from the droplet surface[15] (as is found also with other oil-soluble surfactants[16,17]). The presence of small amounts of oil-soluble surfactant may give a rather larger protein load due to more efficient packing of protein molecules in the more mobile mixed protein/surfactant film. The data in Figures 1 and 2 refer to emulsions where the protein concentration is low enough and the homogenization conditions severe enough to produce monolayer protein coverage. In food emulsions containing slightly higher coverages of commercial proteinaceous emulsifiers, the influence of emulsifying conditions on the protein load is reasonably well characterized in empirical terms,[18] although the mechanisms involved are not yet properly understood at a fundamental level.

The dynamics of the adsorbed protein layer depends on the packing density and on the nature of the interfacial interactions. For BSA adsorbed at a solid surface, the long-time self-diffusion coefficient measured by fluorescence recovery after photobleaching (FRAP) has been found[19] to decrease by a factor of 10 as the surface packing fraction increases from 10% to 70%. At surface concentrations below saturation coverage, the mechanism of surface diffusion has been described[20] as resembling molecular motion in an archipelago of impermeable particles. For β-lactoglobulin adsorbing at the oil–water interface, the FRAP technique has indicated[21] that the saturated protein monolayer is effectively immobile. This is consistent with the high surface shear viscosity[22,23] of adsorbed films of β-lactoglobulin and other globular proteins. In contrast, adsorbed β-casein has a low surface shear viscosity, reflecting the substantial mobility of casein molecules in the stabilizing layer and their rapid exchange with other casein molecules in the bulk aqueous phase.[24,25] Phosvitin also forms a mobile layer; this is presumably because extreme close packing is prevented by the unscreened ionic repulsion between the highly charged adsorbed molecules. Addition of calcium ions has been shown[11] to produce a large increase in the surface shear viscosity of adsorbed β-casein or phosvitin films due to screening of ionic interactions and interfacial aggregation by calcium bridging.

FLOCCULATION

Flocculation takes place when the interaction free energy between a pair of protein-coated droplets becomes appreciably negative at some separation.[1] Several ways in which this may happen are listed below.

- **Lowering of the net charge on the protein** This produces a reduction in the surface charge density associated with the adsorbed layer and therefore a possible loss of any charge stabilization. Typically, such destabilization takes place on approaching the isoelectric pH or in the presence of strongly binding counter-ions (e.g. Ca^{2+}).[26,27]

- **Addition of electrolyte** This screens out the double-layer repulsion. The critical salt concentration required to induce aggregation is much lower for divalent counter-ions. Large emulsion droplets exhibit secondary minimum flocculation[1] even at rather low salt concentrations due to the predominant influence of van der Waals interactions.

- **Lowering of the solvent quality of the aqueous medium** In a poor solvent protein molecules tend to form aggregates in bulk solution; loops and tails on adsorbed molecules tend not to extend very far into the dispersion medium; and the entropic stabilizing "steric repulsion" is converted into an enthalpic *de*stabilizing "steric attraction". Changes in solvent quality arising from ethanol addition, for example, can significantly affect the properties of casein films.[28,29]

- **Emulsification with too little protein** When insufficient emulsifier is available during emulsification to cover all the newly created interface, adsorbed protein molecules or aggregates thereof (e.g. casein micelles) are shared between pairs of droplets. This type of bridging flocculation is exemplified by the clustering of fat globules during homogenization of cream.[30]

- **Competitive adsorption during emulsification** In an emulsion made from mixed emulsifiers of differing surface activity, competitive adsorption may lead to strong bridging flocculation.[31] This phenomenon may occur in systems containing protein + protein,[32] protein + polysaccharide[33] and protein + water-soluble surfactant.[34]

- **Addition of polysaccharide** Flocculation is induced by low levels of ionic or neutral polysaccharides.[35] The type and degree of flocculation depends on the nature and strength of the protein–polysaccharide interaction at the emulsion droplet surface. Attraction leads to bridging flocculation, and repulsion to depletion flocculation.[1]

It can be inferred from this catalogue that emulsion flocculation is a diverse and ubiquitous phenomenon. Let us now consider some specific cases in a little more detail.

Figure 3 illustrates the destabilizing effect of some food polymers on charge stabilized colloidal particles. The data are taken from the light-scattering investigation by Lips and coworkers[36] of aggregation of polystyrene latex particles (diameter 88 nm) by biopolymers in 0.067 M NaCl at pH 6. Results are expressed in terms of the stability ratio W; this is the ratio of the fast diffusion-limited rate measured in excess electrolyte ($\sim$0.5 M NaCl) to the rate measured under more stable conditions. A value of $\log_{10}W \approx 1$ corresponds to a highly unstable system, whereas a value of $\log_{10}W > 4$ corresponds to a stable system. Without added biopolymer, the interparticle pair potential of the model colloid has an electrostatic energy barrier of height $\sim$13 kT and range $\sim$4 nm. We see from Figure 3 that flocculation is induced by addition of dextran, casein, gum arabic, xanthan or guar. The mechanism appears not to be the same, however, for each of these five cases. Behaviour typical of bridging flocculation is seen for sodium caseinate, dextran and gum arabic [and also for gelatin (not

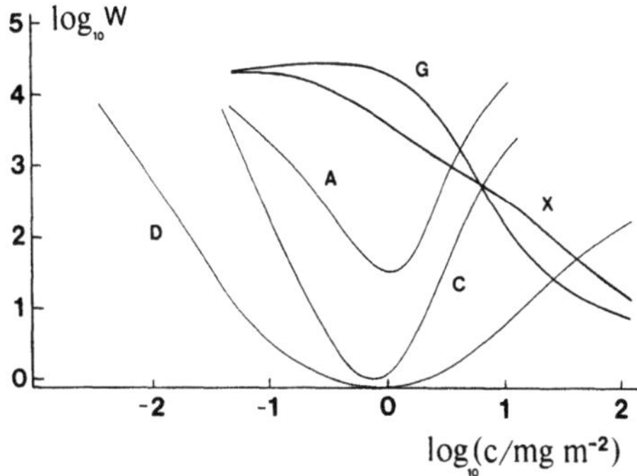

Fig. 3 Stability of polystyrene latex particles with respect to aggregation in the presence of various biopolymers (pH 5.9, ionic strength 0.067 mol dm^{-3}, 28 °C) as measured by light scattering (36). The logarithm of the stability ratio W is plotted against the logarithm of the biopolymer concentration c expressed as total mass added per unit area of available particle surface: (A) gum arabic, (C) sodium caseinate, (D) dextran (5×10^5 daltons), (G) guar and (X) xanthan.

shown)]. Bridging flocculation is maximized at a biopolymer concentration corresponding to approximately half the monolayer coverage (0.5–1 mg m^{-2}). The strongly adsorbing casein displays a narrow window of instability, in contrast to the wide window of instability shown by the weakly adsorbing dextran. The low flocculating power of gum arabic is attributed[36] to its compact copolymer character. The non-adsorbing guar and xanthan show instability at higher concentrations, but there is no sign of any restabilization. The latter behaviour is typical of depletion flocculation in which particles are driven together by an osmotic pressure gradient caused by the exclusion (depletion) of biopolymer molecules from the region of continuous phase between them.[37]

Calcium ions have a crucial influence on the state of flocculation of casein-coated emulsion droplets.[26,38] This is also the case for emulsions made with phosvitin.[13] Figure 4(a) shows droplet-size distributions determined by light scattering (Malvern Mastersizer) for emulsions with calcium ions (5, 7 or 12 mM) added prior to homogenization. In 5 mM $CaCl_2$ the mean volume–surface diameter ($d_{32} = 1.00 \pm 0.05$ μm) is the same as in the calcium-free system. But higher $CaCl_2$ concentrations give wider droplet-size distributions indicative of bridging flocculation. (The presence of flocs is confirmed by microscopy.[11,13])

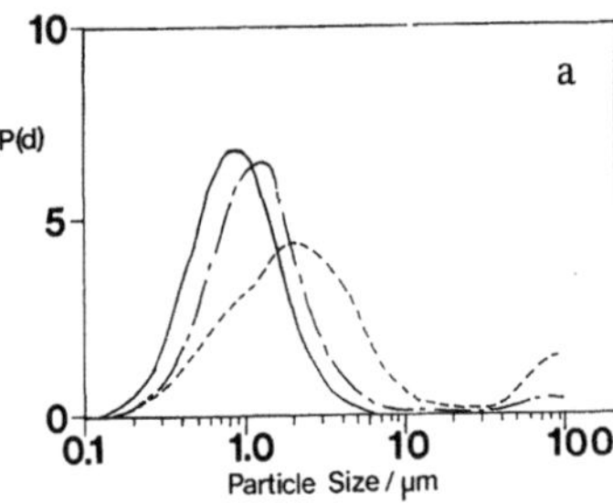
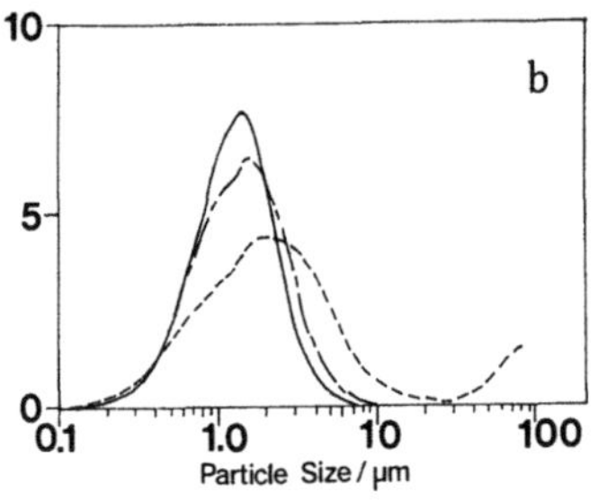

Fig. 4 Droplet-size distributions of emulsions (0.5 wt% phosvitin, 20 wt% n-tetradecane, pH 7) measured by the Malvern Mastersizer. Curves denote relative probability $P(d)$ of finding droplet (or floc) of size d. (a) Fresh emulsions made with calcium ions added before homogenization: ——, 5 mM Ca^{2+}; — — —, 7 mM Ca^{2+}; – – –, 12 mM Ca^{2+}. (b) Diluted emulsions containing 12 mM Ca^{2+} added before homogenization: – – –, immediately after dilution; — – – —, after 45 minutes; ——, after 90 minutes.

The flocculation can be reversed by diluting the emulsion in calcium-free buffer solution (pH 7) as indicated in Figure 4(b). After circulating for a period of 90 minutes in the Mastersizer sample cell, where the droplets are subjected to moderate shear flow, the diluted emulsion exhibits a distribution that is essentially identical to that of the freshly made calcium-free emulsion. A similar kind of reversible calcium-induced flocculation is obtained also with β-casein emulsions.[13,26]

CREAMING

In the early stages of creaming there is merely a vertical concentration gradient of oil droplets. Later on in the process, a distinct boundary may appear between a dense upper cream layer and a lower depleted serum layer (with maybe a third middle region with appearance similar to the original emulsion). Kinetics of creaming may be monitored from the changing thicknesses of cream or serum layers as estimated by eye, or from oil volume fraction profiles determined ultrasonically.[1]

Three key factors affecting creaming behaviour are droplet-size distribution, the rheology of the continuous phase, and the state of flocculation of the droplets.[39] In food emulsions these factors are in turn influenced by interactions amongst the adsorbed proteins, the small-molecule surfactants and the polysaccharides. Biopolymer–surfactant interactions may be important in the adsorbed layer itself as well as in the bulk aqueous medium.

The effect of droplet size on creaming is illustrated in Figure 5 for emulsions made with the same (small) constant amount of sodium caseinate (0.1 wt%) together with various concentrations c_s of water-soluble surfactant $C_{12}E_8$ (octaoxyethylene glycol n-dodecyl ether). Expressed as a percentage of total sample height (20 cm), the serum layer thickness after 14 days storage at 25 °C decreases from ~50% in the absence of $C_{12}E_8$ to ~15% for $c_s = 1.0$ wt%. The reason for the improved stability is primarily the reduction in the median droplet size d_{med} during emulsification. On the basis of recent competitive adsorption data[16] for emulsions made with β-casein + $C_{12}E_8$, we can estimate that the protein is completely displaced from the interface for $c_s \geq 0.1$ wt%.

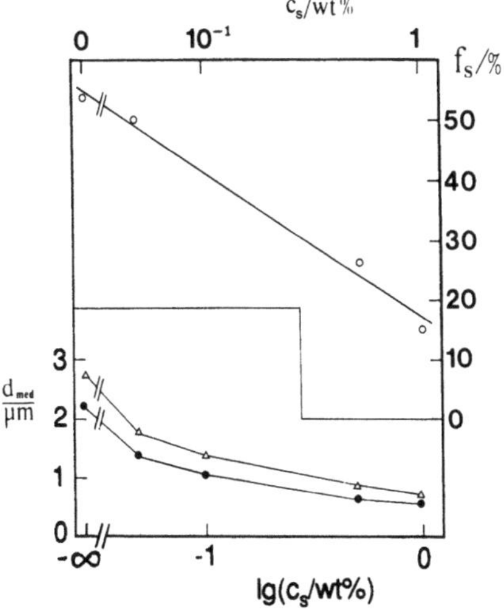

Fig. 5 Influence of nonionic surfactant $C_{12}E_8$ on creaming and average droplet size in emulsions (0.1 wt% caseinate, 20 wt% n-tetradecane, pH 7, 25 °C). Upper plot shows serum fraction f_s estimated by eye after 14 days storage *versus* logarithm of surfactant concentration c_s. Lower plot shows Coulter counter median droplet diameter d_{med} *versus* logarithm of c_s: •, fresh emulsion; △, after 14 days.

The presence of high-molecular-weight polysaccharide may have a positive or negative influence on creaming stability depending on biopolymer concentration and the nature of the protein–polysaccharide interaction. Where the interaction is not net attractive, creaming is typically enhanced at low added polymer concentrations due to depletion flocculation, but it is inhibited at higher concentrations due to thickening and gelation behaviour.[41,42] Figure 6 shows the effect of succinoglycan, an anionic microbial polysaccharide, on the developing oil volume fraction profile $\phi(h)$ for a casein-stabilized emulsion at neutral pH. Creaming profiles were determined ultrasonically[43] in samples stored at 20 °C. Data are presented for emulsions containing (a) 0.0 wt%, (b) 0.025 wt% and (c) 0.04 wt% succinoglycan. At zero time (curve A in each plot) each emulsion has a uniform volume fraction $\phi(h) \approx 0.2$ at each height h. This emulsion exhibits reasonably good stability in the absence of the polysaccharide: after 9 days storage the composition in the middle three-quarters of the sample still remains very close to $\phi = 0.2$. In the presence of 0.025 wt% succinoglycan, however, the creaming is much faster: after 2 days storage there develops a sharp boundary separating a dense cream layer ($\phi \approx 0.5$) at the top from a clear serum layer ($\phi \approx 0$) at the bottom. But further increase in the concentration of added polysaccharide to 0.04 wt% leads to slower creaming than with 0.025 wt%, although it is still faster than for the original emulsion. Complete inhibition of creaming over this time-scale is achieved at succinoglycan concentrations of ~0.2 wt% or above. Similar behaviour is also found with xanthan,[42] although here the added biopolymer content required for stabilization is slightly higher. The criterion for elimination of creaming over a storage period of several weeks is that the emulsion should have a limiting low-stress shear viscosity that is 10^5–10^6 times larger than the approximately Newtonian shear viscosity of the equivalent emulsion containing no polysaccharide.[7]

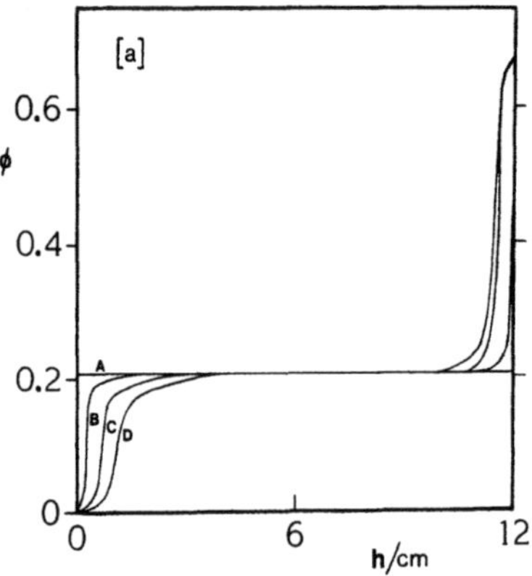

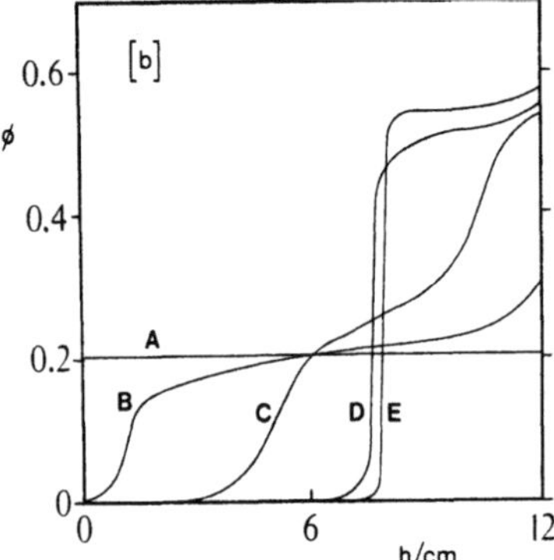

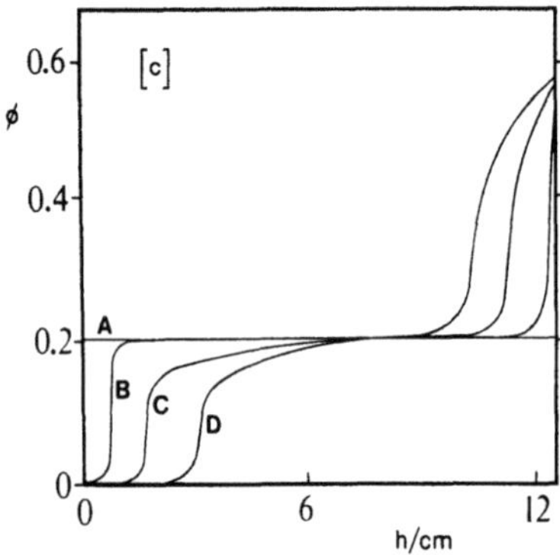

Fig. 6 Influence of succinoglycan on creaming of emulsions (0.75 wt% caseinate, 15 wt% n-tetradecane, pH 7, 20 °C) as measured by ultrasound velocity scanning. The oil volume fraction ϕ is plotted as a function of height h. (a) Profiles for emulsion with no polysaccharide present after (A) 0 days, (B) 3 days, (C) 7 days and (D) 9 days. (b) Profiles for emulsion containing 0.025 wt% polysaccharide after (A) 0 days, (B) 0.3 days, (C) 0.9 days, (D) 2 days and (E) 6 days. (c) Profiles for emulsion containing 0.04 wt% polysaccharide after (A) 0 days, (B) 1 day, (C) 5 days and (D) 10 days.

Complexation between protein and polysaccharide at the emulsion droplet surface can improve steric stabilization. Droplet sizes may be smaller if the polysaccharide is present during homogenization, and so the rate of creaming may be reduced so long as there is no bridging flocculation. Figure 7 shows the effect of complexation of BSA with the nonionic polysaccharide dextran (5×10^5 daltons) on kinetics of serum separation at pH 7.[44] The covalent complex was made by dry-heating a mixture of BSA + dextran (1:3 by weight) for 3 weeks at 60 °C as described by Kato *et al.*[45] The emulsion prepared with the complex is very much more stable with respect to coalescence and creaming than that prepared with BSA alone at the same protein concentration.[44] The emulsion prepared with a simple BSA + dextran mixture exhibits faster serum separation than the polysaccharide-free emulsion; this is due to depletion flocculation (or possibly weak bridging flocculation) of BSA-coated droplets by uncomplexed dextran.[46]

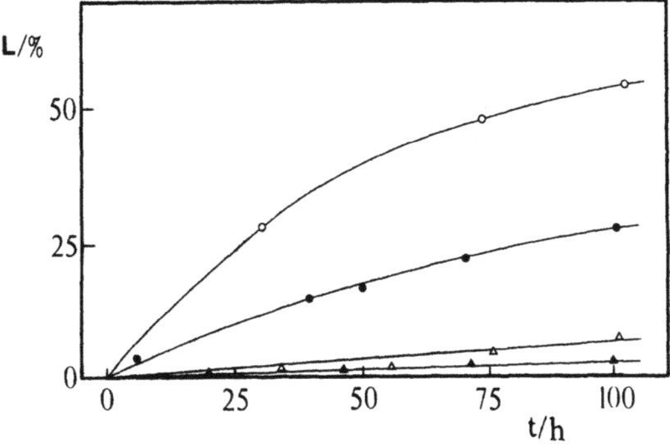

Fig. 7 Influence of protein–polysaccharide complexation on creaming of emulsions (0.5 wt% BSA, 1.5 wt% polysaccharide, 10 wt% n-hexadecane, pH 7). Serum layer thickness L is plotted against storage time t: •, no polysaccharide present; o, mixture of BSA + dextran (5×10^5 daltons); Δ, BSA–dextran conjugate made by dry heating; ▲, mixture of BSA + dextran sulphate (5×10^5 daltons).

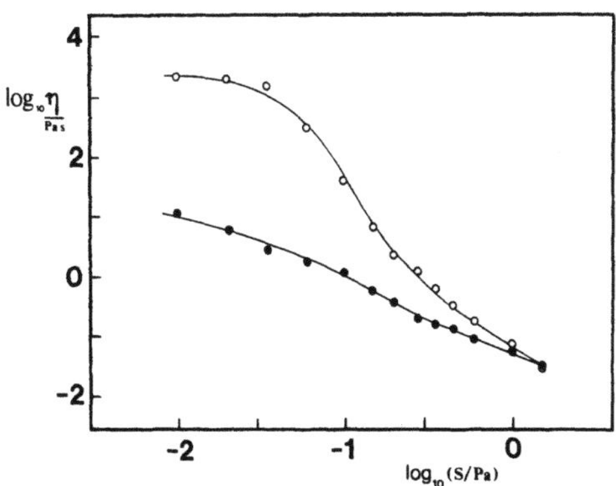

Fig. 8 Rheology of emulsions (1 wt% SDS or Tween 20, 0.08 wt% rhamsan, 30 wt% mineral oil, pH 7, ionic strength 0.05 mol dm^{-3}) measured in creep experiments with constant-stress Bohlin rheometer using a cell of concentric cylinder geometry. Logarithm of the apparent shear viscosity η is plotted against logarithm of the applied shear stress S: •, SDS; o, Tween 20.

Non-covalent protein–polysaccharide complexation can also provide effective emulsion stabilization.[35] Figure 7 also shows serum separation data for an emulsion made with a mixture of BSA + dextran sulphate (1:3 by weight). The great improvement in stability arising from the presence of the polysaccharide during emulsification is attributable to the formation of a thicker, stronger steric-stabilizing layer around the droplets. Even though both biopolymers carry a net negative charge at pH 7, a soluble ionic complex can be formed *via* local electrostatic interaction between the highly charged anionic polymer and positively charged patches on the globular protein.[47] Surface shear viscosity measurements give independent evidence for an interfacial complex between BSA and dextran sulphate.[44]

Where a system contains a mixture of polysaccharide (hydrocolloid) and small-molecule surfactant (emulsifier), the nature of the surfactant–polysaccharide interaction will influence emulsion stability, structure and rheology. Figure 8 compares the rheological behaviour of fine emulsions of similar droplet size made with (i) the nonionic surfactant Tween 20 [polyoxyethylene(20) sorbitan monolaurate, 1.0 wt%] and (ii) the anionic surfactant SDS (sodium dodecyl sulphate, 1.0 wt%). Each emulsion contains 0.08 wt% of the anionic microbial polysaccharide rhamsan.[48] Creep measurements were made on a Bohlin constant-stress rheometer.[49] At moderate shear stresses (>1 Pa) the apparent viscosity is roughly the same for emulsions (i) and (ii). But at very low stresses ($\sim 10^{-2}$ Pa) the apparent viscosity of the Tween 20 system is more than 2 orders of magnitude larger than that of the SDS emulsion. (In fact, the latter emulsion gives viscosity data in the low-stress regime which are fairly close to the values obtained with mixed *solutions* of 1 wt% surfactant (SDS or Tween 20) + 0.1 wt% rhamsan without any oil droplets present.) The much higher low-stress viscosity of the Tween 20 emulsion correlates with its much better creaming stability: ultrasound velocity scanning shows[49] no significant change in volume fraction profile $\phi(h)$ after 50 days storage of the Tween 20 emulsion (1 wt% surfactant, 0.08 wt% rhamsan), whereas the corresponding SDS emulsion was found to exhibit extensive serum separation after just a few hours. It seems that the limiting low-stress (low shear-rate) viscosity is a good guide to the relative creaming stability of these emulsions.

CONCLUSIONS

This brief review has focussed on factors affecting the flocculation and creaming of oil-in-water emulsions. An underlying theme is the way in which stability is sensitive to the strength and nature of molecular interactions. The study of model systems gives useful insight into the behaviour of emulsions containing small-molecule surfactants and high-molecular-weight polysaccharides as well as adsorbed proteins. Basic understanding of such systems provides a sound foundation for the intelligent formulation of real food colloids.

REFERENCES

1. E. Dickinson, "An Introduction to Food Colloids," Oxford University Press, Oxford (1982).
2. P. Becher, "Emulsions: Theory and Practice," 2nd edn, Reinhold, New York (1965).
3. E. Dickinson and G. Stainsby, "Colloids in Food," Applied Science, London, chap. 4 (1982).
4. Th. F. Tadros and B. Vincent, Emulsion stability, *in* "Encyclopedia of Emulsion Technology," P. Becher, ed., Marcel Dekker, New York, vol. 1 (1983).
5. R. J. Hunter, "Foundations of Colloid Science," Oxford University Press, Oxford, vol. 2, chap. 16 (1989).
6. E. Dickinson, D. S. Horne, J. S. Phipps and R. M. Richardson, A neutron reflectivity study of the adsorption of β-casein at fluid interfaces, *Langmuir*, in press.
7. E. Dickinson, Structure and composition of adsorbed protein layers and the relationship to emulsion stability, *J. Chem. Soc. Faraday Trans.* 88:2973 (1992).
8. A. Eaglesham, T. M. Herrington and J. Penfold, A neutron reflectivity study of a spread monolayer of bovine serum albumin, *Colloids Surf.* 65:9 (1992).
9. D. G. Dalgleish and J. Leaver, Possible conformations of milk proteins adsorbed on oil/water interfaces, *J. Colloid Interface Sci.* 141:288 (1991).
10. A. R. Mackie, J. Mingins and A. N. North, Characterization of adsorbed layers of a disordered coil protein on polystyrene latex, *J. Chem. Soc. Faraday Trans.* 87:3043 (1991).
11. J. A. Hunt, Surface and emulsifying behaviour of egg-yolk protein, Ph. D. thesis, University of Leeds (1992).

12. E. Dickinson, J. A. Hunt and D. G. Dalgleish, Competitive adsorption of phosvitin with milk proteins in oil-in-water emulsions, *Food Hydrocolloids* 4:403 (1991).

13. E. Dickinson, J. A. Hunt and D. S. Horne, Calcium induced flocculation of emulsions containing adsorbed β-casein or phosvitin, *Food Hydrocolloids* 6:000 (1992).

14. E. Dickinson and G. Iveson, unpublished results.

15. J.-L. Courthaudon, E. Dickinson and W. W. Christie, Competitive adsorption of lecithin and β-casein in oil-in-water emulsions, *J. Agric. Food Chem.* 39:1365 (1991).

16. J.-L. Courthaudon, E. Dickinson and D. G. Dalgleish, Competitive adsorption of β-casein and nonionic surfactants in oil-in-water emulsions, *J. Colloid Interface Sci.* 145:390 (1991).

17. E. Dickinson and S. Tanai, Protein displacement from the emulsion droplet surface by oil-soluble and water-soluble surfactants, *J. Agric. Food Chem.* 40:179 (1992).

18. E. Tornberg, A. Olsson and K. Persson, The structural and interfacial properties of food proteins in relation to their function in emulsions, *in* "Food Emulsions," 2nd edn, K. Larsson and S. E. Friberg, eds, Marcel Dekker, New York (1990).

19. R. D. Tilton, A. P. Gast and C. R. Robertson, Surface diffusion of interacting proteins, *Biophys. J.* 58:1321 (1990).

20. M. J. Saxton, Lateral diffusion in an archipelago, *Biophys. J.* 52:989 (1987).

21. D. C. Clark and P. J. Wilde, Surfactant-induced surface diffusion of protein as a determinant of disperse phase stability, *in* "Gums and Stabilisers for the Food Industry," G. O. Phillips, D. J. Wedlock and P. A. Williams, eds, IRL Press, Oxford, vol. 6 (1992).

22. J. Castle, E. Dickinson, B. S. Murray and G. Stainsby, Mixed protein films adsorbed at the oil–water interface, *ACS Symp. Ser.* 343:118 (1987).

23. E. Dickinson, S. E. Rolfe and D. G. Dalgleish, Surface shear viscosity as a probe of protein–protein interactions in mixed protein films adsorbed at the oil–water interface, *Int. J. Biol. Macromol.* 12:189 (1990).

24. E. Dickinson, S. E. Rolfe and D. G. Dalgleish, Competitive adsorption of α_{s1}-casein + β-casein in oil-in-water emulsions, *Food Hydrocolloids* 2:397 (1988).

25. J. R. Hunter, R. G. Carbonell and P. K. Kilpatrick, Coadsorption and exchange of lysozyme/β-casein mixtures at the air–water interface, *J. Colloid Interface Sci.* 143:37 (1991).

26. E. Dickinson, R. H. Whyman and D. G. Dalgleish, Colloidal properties of model oil-in-water food emulsions stabilized separately by α_{s1}-casein, β-casein and κ-casein, *in* "Food Emulsions and Foams," E. Dickinson, ed., Royal Society of Chemistry, London (1987).

27. E. Dickinson, S. K. Narhan and G. Stainsby, Factors affecting the properties of cohesive creams formed from cream liqueurs, *J. Sci. Food Agric.* 48:225 (1989).

28. S. Bullin, E. Dickinson, S. J. Impey, S. K. Narhan and G. Stainsby, Stability aspects of casein-containing emulsions, *in* "Gums and Stabilisers for the Food Industry," G. O. Phillips, D. J. Wedlock and P. A. Williams, eds, IRL Press, Oxford, vol. 4 (1988).

29. E. Dickinson and C. M. Woskett, Effect of alcohol on adsorption of casein at the oil–water interface, *Food Hydrocolloids* 2:187 (1988).

30. H. Mulder and P. Walstra, "The Milk Fat Globule," Pudoc, Wageningen (1974).

31. E. Dickinson and V. B. Galazka, Bridging flocculation induced by competitive adsorption: implications for emulsion stability, *J. Chem. Soc. Faraday Trans.* 87:963 (1991).

32. E. Dickinson, F. O. Flint and J. A. Hunt, Bridging flocculation in binary protein-stabilized emulsions, *Food Hydrocolloids* 3:389 (1989).

33. E. Dickinson and V. B. Galazka, Bridging flocculation in emulsions made with a mixture of protein and polysaccharide, *in* "Food Polymers, Gels and Colloids," E. Dickinson, ed., Royal Society of Chemistry, Cambridge (1991).

34. E. Dickinson, M. I. Goller and D. J. Wedlock, Bridging flocculation in emulsions stabilized by gelatin and a low-molecular-weight surfactant, *in* "Gums and Stabilisers for the Food Industry," G. O. Phillips, D. J. Wedlock and P. A. Williams, eds, IRL Press, Oxford, vol. 6 (1992).

35. E. Dickinson and S. R. Euston, Stability of food emulsions containing both protein and polysaccharide, *in* "Food Polymers, Gels and Colloids," E. Dickinson, ed., Royal Society of Chemistry, Cambridge (1991).

36. A. Lips, I. J. Campbell and E. G. Pelan, Aggregation mechanisms in food colloids and the role of biopolymers, *in* "Food Polymers, Gels and Colloids," E. Dickinson, ed., Royal Society of Chemistry, Cambridge (1991).

37. E. Dickinson, The role of hydrocolloids in stabilizing particulate dispersions and emulsions, *in* "Gums and Stabilisers for the Food Industry," G. O. Phillips, D. J. Wedlock and P. A. Williams, eds, IRL Press, Oxford, vol. 4 (1988).

38. E. Dickinson, Food colloids—an overview, *Colloids Surf.* 42:191 (1989).

39. E. Dickinson, The structure and stability of emulsions, *in* "Food Structure: Its Creation and Evaluation," J. M. Blanshard and J. R. Mitchell, eds, Butterworths, London (1988).

40. C. M. Woskett, Competitive adsorption and protein–surfactant interactions in food emulsions, Ph. D. thesis, University of Leeds (1989).

41. Y. Cao, E. Dickinson and D. J. Wedlock, Creaming and flocculation in emulsions containing polysaccharide, *Food Hydrocolloids* 4:185 (1990).

42. Y. Cao, E. Dickinson and D. J. Wedlock, Influence of polysaccharides on the creaming of casein-stabilized emulsions, *Food Hydrocolloids* 5:443 (1991).

43. Y. Cao, The instability of emulsions containing polysaccharide, Ph. D. thesis, University of Leeds (1991).

44. E. Dickinson and V. B. Galazka, Emulsion stabilization by protein–polysaccharide complexes, *in* "Gums and Stabilisers for the Food Industry," G. O. Phillips, D. J. Wedlock and P. A. Williams, eds, IRL Press, Oxford, vol. 6 (1992).

45. A. Kato, Y. Sasaki, R. Furuta and K. Kobayashi, Functional protein–polysaccharide conjugate prepared by controlled dry-heating of ovalbumin–dextran mixtures, *Agric. Biol. Chem.* 54:107 (1990).

46. E. Dickinson and M. G. Semenova, Emulsifying behaviour of protein in the presence of polysaccharide under conditions of thermodynamic incompatibility, *J. Chem. Soc. Faraday Trans.* 88:849 (1992).

47. V. B. Tolstoguzov, Functional properties of protein–polysaccharide mixtures, *in* "Functional Properties of Food Macromolecules," J. R. Mitchell and D. A. Ledward, eds, Elsevier Applied Science, London (1986).

48. G. Robinson, C. E. Manning and E. R. Morris, Conformation and physical properties of the bacterial polysaccharides gellan, welan and rhamsan, *in* "Food Polymers, Gels and Colloids," E. Dickinson, ed., Royal Society of Chemistry, Cambridge (1991).

49. E. Dickinson, M. I. Goller and D. J. Wedlock, unpublished results.

PROTEINS AND SUGARS AFFECTING THE ZETA POTENTIAL
AND STABILITY OF DISPERSED VESICULAR GLOBULES
IN W/O/W EMULSIONS

Sachio Matsumoto

Laboratory of Food Chemistry
College of Agriculture
The University of Osaka Prefecture
Gakuen-cho, Sakai-shi, Osaka 593, Japan

ABSTRACT

W/O/W emulsions have numerous applications, as dispersed vesicular globules in emulsions in this type can encapsulate a variety of water-soluble ingredients in the inner aqueous phase, which is separated from the aqueous suspending fluid by a thin layer of the oil phase. Zeta potential of the negatively-charged vesicular globules at nuetral pH was found to be influenced systematically by immobilizing proteins (bovine serum alubumin, chymotrypsin and lysozyme) in the inner aqueous phase, or by adding sugars (ribose, xylose, fructose, glucose, sucrose and maltose) in the two aqueous phases. Due to the surface activity of proteins, their adsorbed molecules on the inside surface of oil layer in the vesicular globules may play a role in the surface potential of oil layer according to the magnitude of their isoelectric points. In fact, a rapid coagulation occured among the globules containing 1% lysozyme (pI:10.5-11) in the inner aqueous phase, which showed the lowest zeta potential in a series of the samples to be tested. The zeta potential was also reduced by increasing amount of sugars in the two aqueous phases, while the stability of the globules against coagulation was improved by addition of sugars. This may be brought about by expanding location of the slipping plane during electrophoretic movement with increasing concentration of sugars due to the formation of a viscous hydration layer on the surface of the globules. An attempt was made to consider the effect of proteins on the total interactions between the vesicular globules.

INTRODUCTION

The term 'W/O/W emulsion' describes a series of complex two-phase systems, i.e.,the two aqueous phases being separated by the oil phase. Microscopic examination indicates clearly that the dispersed globules in emulsions in this type can be characterised by a vesicular structure with single or multiple inner aqueous phase separated from the aqueous suspending fluid by a thin layer of the oil phase components. [1,2]

The morphological feature of the globules may appear irrespective of the kind of oils, while a series of hydrophobic emulsifiers play an important role in the formation of W/O/W emulsions. In a generalized phase diagram of the ternary mixture of water-oil-emulsifier,[3] the region of W/O/W emulsion formation can be specified by the location near the D-phase, so that the two separated steps of emulsification[4,5] or the phase inversion technique[6,7] may be useful for providing a variety of W/O/W emulsions.

W/O/W emulsions in a stable form could find a variety of potential applications in the formulation and processing of foods. For example, the inner aqueous phase will be able to immobilize certain water-soluble ingredients, so that the emulsions may be used as vehicles for nutrient administration for special dietary purposes.[8] Low calorie creams could also be developed using the W/O/W-type dispersion, as it is possible to prepare thick emulsion systems with a small fraction of oil phase.[9] In the past few years, patents on the applications of W/O/W emulsions in food manufacturing have certainly come from several laboratories.

On the other hand, there are so many types of foods and food materials in the world that food scientists are left with more and more complex behaviours to deal with.[10] The author has been concerned with developing a generalized model for understanding the textural features of a variety of foods, and has concentrated on the macroemulsion state as one of the models of non-equilibrium multicomponent systems. Even if it is only a temporary step in the research activities of food science, it seems to be important to obtain further insights into factors affecting the dispersion state of a variety of macroemulsions so as to consider the physicochemical properties of foods consisted of poly-phases with multicomponents.

This article deals with some of the dynamic characteristics detected in the dispersed vesicular globules of W/O/W emulsions.

DYNAMIC ASPECTS OF THE DISPERSED VESICULAR GLOBULES
(BACKGROUNDS OF THE STUDIES)

Thinning of Oil Layer on the Vesicular Globules

The oil layer on the surface of the inner aqueous phase in the various W/O/W emulsions is extremely thin irrespective of the amount of the oil phase. This is brought about by a thinning process in the oil layer immediately after the W/O/W emulsion formation, as identical with the liquid films in egg white foam or soap bubble. Consequently, the inner aqueous phase is being surounded by the thin layer of the oil phase components in a steady state, while the surplus of the oil phase is condensed and located heterogeneously in the oil layer. An examination[2] on a large sized aqueous droplet (about 0.06 μl) in water prepared with the oil phase components suggests that the Marangoni effect plays a role in the induction period of thinning due to the ununiformmity of adsorption of the emulsifier molecules on the inside and outside surfaces of the freshly prepared layer of the oil phase. This induces a rapid migration of oil phase components in the oil layer so as to attain an equilibrium of the adsorption of emulsifier used. The migration of oil phase components may also cause an appearance of the Plateau region and then Laplace pressure in the oil layer. The thinning process appears all over the oil layer, so that more than 90% of the surface of the aqueous droplet are covered by an extremely thin layer

of the oil phase within 90sec after the droplet preparation.   Although the
thinning process described above can be observed throughout the examination
with a model system, it is postulated that the same phenomenon may take
place in the oil layer on the surface of the inner aqueous phase of W/O/W
emulsons.

A steady state of the oil layer after the thining dynamics may be
brought about by a repulsion between the hydrophobic residues of emulsifier
molecules adsorbed on the two surfaces of oil layer due to a decrease of
entropy in the oil layer system, since the micro-Brownian motion of the
hydrocarbon chains of the adsorbed emulsifiers on the two inner surfaces
of oil layer will be restricted by the steric hindrancee of each molecule,
when the two surfaces of the oil layer come closer to each other.

Water Permiability of Oil Layer

The viscosity of W/O/W emulsions is influenced by the osmotic pressure
gradient between the inner aqueous phase and the aqueous suspending fluid.[11]
Although the changes in viscosity occur on either sides of increase or
decrease according to the drift of the osmotic pressure, the rate of changes
in viscosity in an initial stage of the measurement is proportional to
the degree of the osmotic pressure gradients between the two aqueous phases.
Therefore, this phenomenon can be analysed by means of the swelling or
shrinkage of the vesicular globules due to the migration of water molecules
permeating through the oil layer[11].

The author and his colleagues[12] examined the swelling of the vesicular
globules in a series of W/O/W emulsions containing an equi-molar glucose
in the two aqueous phases.   The measurement was commenced by dilution of
the emulsion more than 200 times with pure water.   A drop of the emulsion
was placed on a glass slide with a small depression immediately after the
dilution, and then mounted in an optical microscope with a photographic
attachment so as to record periodically the changes in volume of the
vesicular globules at a fixed microscopic field.   This dilution resulted
an increase of volume of the vesicular globules, because a concentration
gradient between the two aqueous phases of the W/O/W emulsion was induced
by the dilution of the aqueous susupending fluid.   The water permeation
coefficient (P_0) of the oil layer was calculated by the following relation
applied to the lipid bilayer sysytems.[13]

$$P_0 = (dv/dt) / \alpha \, (g_2 c_2 - g_1 c_1) \cdot \bar{V} \qquad\qquad (1)$$

where dv/dt is the increasing rate of volume of the globules, α is the
surface area of the globule, g and c are the osmotic coefficient and
molar concentration of glucose, the suscripts 1 and 2 refer to the outside
and inside of the globules, and $\bar{V}$ is the partial molar volume of water.

The values for the water permeation coefficient of the oil layer in
a series of W/O/W emulsions calculated were covered a range from 2×10^{-4}
cm/sec to 8×10^{-4} cm/sec.   It is worth noting that the values obtained
are roughly identical with those of the lipid bilayer systems, for which
values ranging from 10^{-3} to 10^{-4} cm/sec have been reported from several
laboratories.[13]

The author and his colleague[14] have recently re-examined the water
permeation coefficient of the oil layer under the concentration gradients
of a series of sugars between the two aqueous phases of W/O/W emulsions
using the same technique described above.   The results obtained show that
the coefficient increases from 5.97×10^{-4} to 8.89×10^{-4} cm/sec with reducing number of e-OH groups in a sugar molecule, as summarized in Table 1.

Table 1. Water permeation coefficient (P_o) of oil layer under a concentration gradient (0.2M) of a series of sugars between the two aqueous phases of W/O/W emulsions.[14]

Sugar	P_o (10^{-4} cm/sec)	SD^a (10^{-4} cm/sec)	Number of e-OH[15]
Ribose	8.89	0.84	2.1
Fructose	8.18	0.38	3.0
Xylose	7.65	0.47	3.5
Glucose	6.59	0.59	4.6
Sucrose	6.69	0.30	6.3
Maltose	5.97	0.55	7.2

[a] Standard deviation within more than six data.

Enzymatic activity in the Vesicular Globules

An attempt was made to immobilize invertase[16] or β-amylase[17] in the dispersed vesicular globules of W/O/W emulsions so as to examine their enzymatic activities in the inner aqueous phase. In the case of invertase immobilized, an increasing amount of reducing sugars in the aqueous suspending fluid was observed steadily during a span of incubation time (over 10days) with the sample emulsions, when a difinite amount of the substrate sucrose was introduced into the aqueous susupending fluid. This suggested that the sucrose molecules are being hydrolyzed in the inner aqueous phase due to the migration of sucrose to the inner aqueous phase from the aqueous suspending fluid permeating through the oil layer. The molar flux of sucrose across a unit area of the oil layer ranging from about 2 to 6 pM/sec could be estimated from increasing amount of the reducing sugars.

The recent studies[14] on the sugar permeability of oil layer in W/O/W emulsions without enzyme have indicated that the values for the molar flux of a variety of sugars are much slower than those of the systems with enzymes described above. Such the discrepancy was thought to be brought about by the surface activity of protein molecules, i.e., the molecules adsorbed tightly on the inside surface of the oil layer. Therefore, it also seemed that several active sites of the adsorbed invertase may exist on the outside surface of the oil layer facing to the aqueous suspending fluid.

In order to observe the enzymatic activity on the outside surface of the vesicular globules, it was tried to provide the W/O/W emulsions[17] whose inner aqueous phase contained 10μg/ml of β-amylase, while the aqueous suspending fluid dissolved a definite amount of the substrate amylose which was not possible to permeate the oil layer because of giant molecule. The results obtained suggest that the activity of β-amylase appears in the aqueous suspending fluid, since an increasing pattern of the certain amount of maltose could be identified steadily in the suspending fluid during a span of incubation time (about 3days) with the W/O/W emulsions.

A series of the studies described above may elucidate that a part of protein molecules are possibly integrated in forming the thin oil layer of the dispersed vesicular globules in the W/O/W emulsion preparation.

ZETA POTENTIAL AND STABILITY OF THE DISPERSED VESICULAR GLOBULES CONTAINING PROTEINS IN THE INNER AQUEOUS PHASE

The author has found that the dispersion state of the dispersed vesicular globules in W/O/W emulsions is much influenced by a kind of proteins immobilized in the inner aqueous phase at neutral pH. Despite the fact that the globules were already being charged negatively in the aqueous suspending fluid due to differences in the affinity of the oil and the aqueous phases for ions, the above finding may result from changes in the charged potential of the globules according to the surface activity of proteins. Protein molecules adsorbed tightly onto the inside surface of the oil layer, as has been described for the case of enzymes. It was, therefore, tried to obtain informations on the correlation between the zeta potential of the vesicular globules and the isoelelctric point of proteins immobilized in the globules of W/O/W emulsons.

The emulsion samples to be tested were divided into liquid paraffin and olive oil systems, respectively. The former was stabilized by Span 80 (sorbitan monooleate) and Tween 80 (polyoxyehtylene sorbitan monooleate), while the latter was prepared by use of TGCR (tetraglyceryl condensed ricinolate) and DGMO (decaglyceryl monooleate). Each system was also subdivided into four samples according to a kind of proteins immobilized in the inner aqueous phase, such as the control (without protein), BSA (bovine serum albumin, pI:4.4-4.9), chymotrypsin (pI:8.1-8.6), and lysozyme (pI:10.5-11.0) samples, respectively.

An assembly[18] with a rectangular cell was provided for evaluating the electrophoretic mobility of the dispersed vesicular globules at neutral pH at room temperature. Each sample of the W/O/W emulsions was diluted about 100times with an aqueous solution of 1 mmol/1 KCl immediately before the measurement. The field strengths within a range from 3.3 to 10.0 V/cm were used to ensure that the electrophoretic mobility is independent of the actual applied voltage. The zeta potential (ζ) on the slipping plane of the diffused portion in the electrical double layer around the vesicular globules was calculated from the Smoluchowski equation for the electro-phoresis on non-conducting spheres, as follows:

$$\mu = \varepsilon_0 \, D \, \zeta / \, \eta \qquad (2)$$

where μ is .the electrophoretic mobility, ε_0 is the permittivity of free space, and D, and η are the relative permittivity, and viscosity of the aqueous suspending fluid.

Table 2. Electrophoretic data of the dispersed vesicular globules in W/O/W emulsons at neutral pH at room temperature.

Inner aq. phase	μ (μm·sec^{-1}/ V·cm^{-1})	ζ (mV)	Deviation (mV)
— Liquid paraffin system —			
Without protein	-2.50	-32.1	±0.9
1% BSA	-2.14	-27.4	±0.5
1% Chymotrypsin	-1.67	-21.4	±0.4
1% Lysozyme	-1.06	-13.7	—
— Olive oil system —			
Without protein	-5.56	-71.3	2.8[a]
1% BSA	-4.30	-55.1	3.1[a]
1% Chymotrypsin	-3.53	-45.3	2.2[a]
1% Lysozyme	-2.27	-29.0	2.0[a]

[a] Standard deviation.

Table 2 summarizes the values for the electrophoretic mobility and
zeta potential of the vesicular glbules of the samples to be tested. The
effect of proteins on the zeta potential of the globules appears in relation
to the different magnitudes for the isoelectric point of the proteins immobilized in the inner aqueous phase, although the values obtained with
the liquid paraffin system are smaller than those with the olive oil system.
This may be explained by the relatively longer distance to the slipping
plane from the surface of the globules in the liquid paraffin system, since
the globules become to be covered by a deep hydration layer due to the
existence of polyoxyethylene chains in the adsorbed molecules of Tween 80.[19]

However, Table 2 indicates that the zeta potential of the globules
can be ranked with the magnitudes of isoelectric point of the proteins irrespective of the kind of emulsifiers used, while a rapid coagulation occurs
irreversively among the vesicular globules in the lysozyme samples showing
the lowest zeta potential of the globules in a series of the samples to be
tested. As has been described hereinbefore, the surface activity of protein
molecules results so as to form an adsorbed layer on the inside surface of
the oil layer in the dispersed vesicular globules, so that the adsorbed
protein molecules play an important role in the surface potential of the
dispersed vesicular globules according to the value of pH in the two aqueous
phases.

An attempt was made to estimate the total potential energy (V_T) of
interactions between the two vesicular globules as a function of the separation (H) of the slipping plane in view of the DLVO theory, as follows:

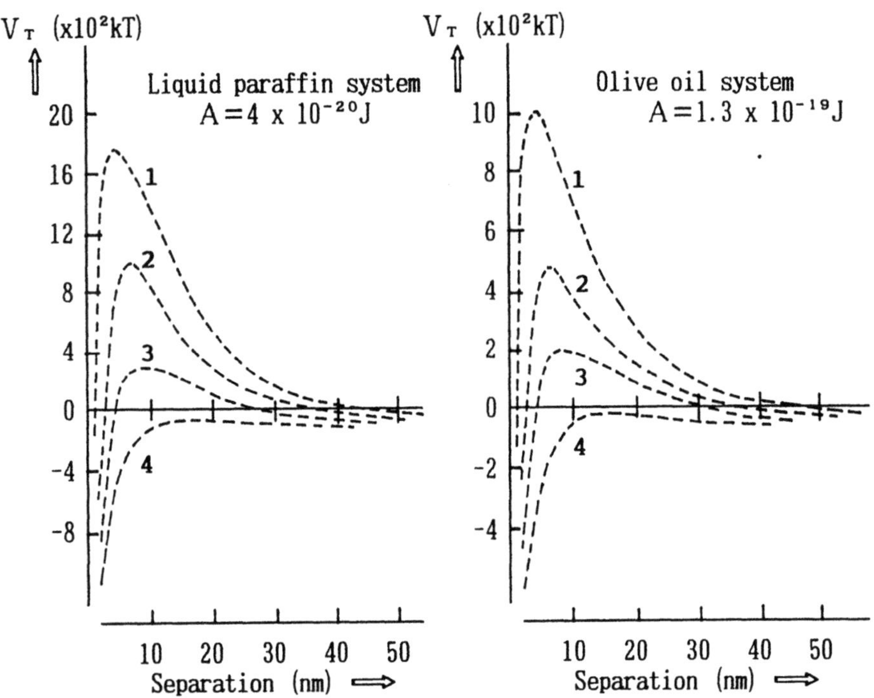

Fig. 1 Total potential energy (V_T) of interactions between the two
vesicular globules containing proteins in the inner aqueous phase
at neutral pH at room temperature. Curve 1 ; without protein, 2 ;
1% BSA, 3 ; 1% chymotrypsin, 4 ; 1% lysozyme, respectively.

$$V_T = 2\pi\varepsilon_0 D r \zeta^2 \ln[1 + \exp(-\kappa H)] - r A/12H \qquad (3)$$

where r is the radius of the globules, A is the Hamaker constant, and κ is the Debye-Hückel parameter given by

$$\kappa = (2 n e^2 z^2/\varepsilon_0 D k T)^{1/2}. \qquad (4)$$

Hence, n, e, and z are the concentration, elementary charge, and valency of counter ions, and kT is the thermal kinetic energy.

Fig. 1 compares the total potential energy of interactions between the two dispersed vesicular globules with each sample in the W/O/W emulsion systems. It was assumed that there is no energy barrier against the separation between the globules containing lysozyme because of the instability in the dispersion state in each system, as described hereinbefore. Consequently, the value of the Hamaker constant A in eq. 3 could be estimated as being 4×10^{-20}J for the liquid paraffin system and 1.3×10^{-19}J for the olive oil system.

CHANGES IN LOCATION OF SLIPPING PLANE ON THE VESICULAR GLOBULES IN THE PRESENCE OF SUGARS IN THE TWO AQUEOUS PHASES

Since the zeta potential of the dispersed colloidal particles in aqueous media is generally identical with an electric potential on the slipping plane located over the surface of the particles, the zeta potential seems to be more or less influenced by the condition of hydration on the surface of the particles at a constant surface potential. Table 1 in this article suggests that the water molecules around the dispersed vesicular globules tend to the increase of the viscous property in the presence of sugars in the two aqueous phases with increasing number of equatorial-OH (e-OH) groups in a sugar molecule relating to the formation of tridymite structure with water molecules.[15]

An attempt has been made to obtain informations on the correlation between the zeta potential of the vesicular globules and the number of

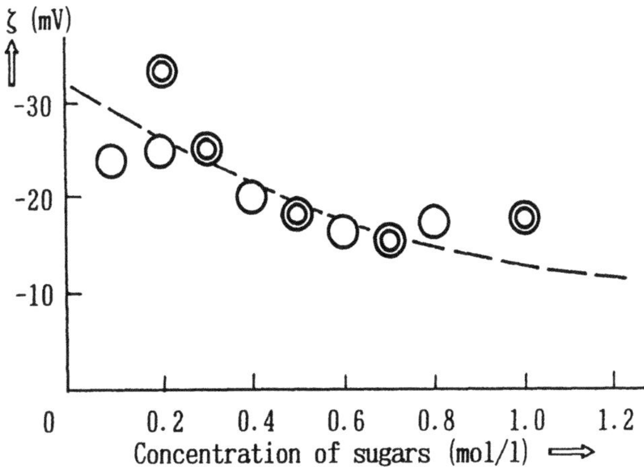

Fig. 2   Plots of zeta potential of the dispersed vesicular globules vs. concentration of sugars in the two aqueous phases at neutral pH. ○ : For the case of ribose (e-OH=2.1), and ◎ : maltose (e-OH=7.2).

equatorial-OH groups of sugars (ribose, fructose, xylose, glucose, sucrose, and maltose) dissolved in the two aqueous phases with the various concentrations using the liquid paraffin system (see Table 2). The same assembly and techniques described hereinbefore were employed for measuring the zeta potential of the vesicular globules in W/O/W emulsions at neutral pH at room temperature, while each emulsion was diluted with an aqueous mixed solution of 1 mmol/1 KCl and the same sugar and concentration used for preparing the emulsion.

Fig. 2 shows the effect of sugar concentrations on the zeta potential of the vesicular globules in comparing the minimal number of equatorial-OH groups (ribose) to the maximal number of those (maltose) within the measurements in this study. The same data could be obtained as shown in Fig. 2 for the other sugars to be tested with the different numbers of equatorial -OH groups. It is obvious from the results obtained that the zeta potential certainly decreases with increasing concentrations of sugars, although the equatorial-OH groups in a sugar molecule do not play a role in the above results within the experimental error against to the author's expectation. The thing is, however, that the existence of sugars in the aqueous phases may bring a drift of location of the slipping plane over the surface of each vesicular globule.

The distribution of electric potential around the surface of the globules along the vertical axis (x axis) from the slipping plane will be assumed to obey the Boltzmann's probability, as follows:

$$\zeta(x) = \zeta \exp(-\kappa x) \quad \text{or} \quad x = [\ln\zeta - \ln\zeta(x)]/\kappa \qquad (5)$$

where $\zeta(x)$ is the potential at the distance of x from the slipping plane, and κ is the Debye-Hückel parameter (see eq. 4) calculated as 0.104 nm^{-1} for all cases of the measurements in this study. The value for the zeta potential ζ of the dispersed vesicular globules without sugar (control) in the aqueous phases was already evaluated as -32.1 mV for the liquid paraffin system (see Table 2), so that it is possible to estimate the drift of slipping plane on the x axis from the decreasing values for the zeta potential in the presence of sugars.

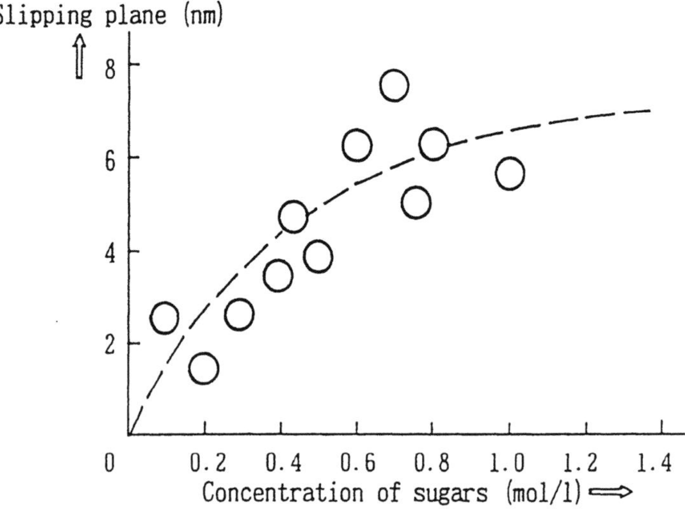

Fig. 3 Correlation between the drift of slipping plane over the surface of the vesicular globules and the concentration of a series of sugars employed in this study.

Fig.3 illustrates the drift of slipping plane calculated from the mean values of zeta potential at each concentration of all sugars used.   The expanding location of slipping plane during electrophoretic movement of the globules due to the increasing concentration of sugars seems to be brought about by a viscous hydration layer surrounding the globules.   Each plot in Fig.3 can be summarized by an expression on the basis of the least squares, as follows:

$$S = 7.5[1 - \exp(-c/0.49)] \tag{6}$$

where S is the drift of slipping plane from the control (without sugar), and c is the molar concentration of each sugar.   The above relation suggests that the existence of sugars over 1 mol/l in the aqueous phases of W/O/W emulsions results to extend the slipping plane up to about 7.5 nm outside from the standard position.   It should be mentioned that the changes in zeta potential caused by the presence of sugars do not apparently affect to the interactions between the dispersed vesicular globules contrary to the general concept (see eq.3).   The fact that the W/O/W emulsions to be tested were all stable even in the case of the lower zeta potentials.

REFERENCES

1. S.Matsumoto, Formation and stability of water-in-oil-in-water emulsions, in:"Macro- and Microemulsions," D.O.Shah,ed., ACS Symposium Series 272, Am.Chem.Soc., Washington,D.C. (1985).
2. S.Matsumoto, W/O/W-type multiple emulsions, in:"Nonionic Surfactants, Physical Chemistry," M.J.Schick,ed., Marcel Dekker, New York (1987).
3. K.Shinoda and S.Friberg,"Emulsions and Solubilization," John Wiley and Sons, New York (1986).
4. S.Matsumoto, Y.Kita and D.Yonezawa, An attempt at preparing water-in-oil-in-water multiple emulsions, J.Colloid Interface Sci. 57:353 (1976).
5. S.Matsumoto, Y.Ueda, Y.Kita and D.Yonezawa, Preparation of water-in-oil-in-water multiple phase emulsions in an edible form, Agric.Biol.Chem. 42:739(1977).
6. S.Matsumoto, Development of W/O/W-type dispersion during phase inversion concentrated W/O emulsions, J.Colloid Interface Sci. 94:362(1983).
7. S.Matsumoto, Y.Koh and A.Michiura, Preparation of W/O/W emulsions in an edible form on the basis of phase inversion technique, J.Dispersion Sci.Technol. 6:507(1985).
8. S.Matsumoto and P.Sherman, A preliminary study of W/O/W emulsions with a view to possible food applications, J.Texture Studies 12:243(1981).
9. Y.Koh and S.Matsumoto, Preparation of butter cream in the state of W/O/W-type dispersion (in Japanese), Nippon Nougeikagaku Kaishi 62:1345 (1988).
10. S.Matsumoto, Preparation and state of macroemulsions stabilized by small molecule surfactants, Food Hydrocolloids 3:249(1989).
11. S.Matsumoto and M.Kohda, The visocosity of W/O/W emulsions; an attempt at the water permeation coefficient of the oil layer from the viscosity changes in diluted systems on aging under osmotic pressure gradients, J.Colloid Interface Sci. 73:13(1980).
12. S.Matsumoto, T.Inoue, M.Kohda and K.Ikura, Water permeability of oil layer in W/O/W emulsions under osmotic pressure gradients, J.Colloid Interface Sci. 77:555(1980).

13. A.Cass and A.Finkelstein, Water permeability of thin lipid membranes, J.General Physiology 50:1765(1967).

14. K.Ueda and S.Matsumoto, Effect of sugars on the physicochemical properties of W/O/W emulsions, J.Colloid Interface Sci. 147:333(1991).

15. H.Uedaira, M.Ishikura, S.Tsuda and H.Uedaira, Hydration of oligosaccharides, Bull.Chem.Soc.Japan 63:3376(1990).

16. S.Matsumoto and W.W.Kang, An attempt at measuring saccharose flux permeating through the oil layer in W/O/W emulsions on the basis of the hydrolytic activity of invertase, Agric.Biol.Chem. 52:2689(1989).

17. S.Matsumoto, K.Kataoka and K.Ueda, An aspect of proteins in dispersed globules of W/O/W emulsions, 7th Intern.Confern.on Surface.Colloid Sci. Compiegne, France (1991).

18. K.Ueda and S.Matsumoto, Effect of proteins on the zeta potential of dispersed vesicular globules in W/O/W emulsions, Bull.Chem.Soc.Japan 64:3163(1991).

19. J.V.Boyd, C.Parkinson and P.Sherman, Factors affecting emulsion stability and the HLB concept, J.Colloid Interface Sci. 41:359(1972).

FILLER EFFECTS OF OIL DROPLETS ON PHYSICAL PROPERTIES OF EMULSION GELS

Y.Matsumura[1], H.Sakamoto[2], M.Motoki[2]
and T.Mori[1]

[1] Research Institute for Food Science
Kyoto University, Gokasyo, Uji, 611 Japan
[2] Food Research & Development Laboratories
Ajinomoto, Co.,INC. 1-1 Suzuki-cho, Kawasakiku
Kawasaki-shi, 210 Japan

ABSTRACTS

Filler effects of oil droplets on viscoelastic properties of emulsion gels were investigated by small deformation
mechanical measurements. Emulsions were made with soya oil
and soy 11S globulin(15 wt % oil, 1.5 - 7 % protein concentrations). Emulsions and soy 11S globulin solutions were
gelated by Ca++-independent transglutaminase from microorganism. The shear storage modulus(G') and loss modulus(G")
of proteinous gels and emulsion gels depended on the protein
concentration. The concentrate-dependence of modulus was
approximately by an exponential function of the form :
$G \propto C^n$, in which n=4 and n=2 was found in the case of the
proteinous gel and the emulsion gel, respectively. Shear
moduli of emulsion gels were much higher than those of
proteinous gels. The gel made from fine emulsion (containing smaller size oil droplets) exhibited higher G' and G"
than the gel from coarse emulsion. Addition of Tween 20 was
found to reduce G' and G" of emulsion gel even if the protein matrix is not significantly displaced from the oil-
water interface by the surfactant.

INTRODUCTION

Emulsion type gel(a complex gel filled with emulsified
fat in a matrix of protein network)has been one of the most
popular colloidal foods, for example, Chinese-Japanese
traditional soybean curd 'Tofu' and western sausage. The
state of oil droplets in addition to protein network structure should affect physical properties of emulsion gel.[1] In
this study, emulsion stabilized by soy protein '11S globulin' was gelated by transglutaminase[TG] and the filler
effect of oil droplets on physical properties of gels was
investigated.

Food Hydrocolloids: Structures, Properties, and Functions
Edited by K. Nishinari and E. Doi, Plenum Press, New York, 1994

MATERIALS AND METHODS

Soy 11S globulin was prepared from defatted flake by the procedure of Thanh and Shibasaki.[2] TG was purified from the culture filtrate of Streptoverticillium S-8112 according to the method of Ando et al.[3] Oil-in-water emulsion was made from 11S globulin solution(various protein concentrations) and soybean oil. The ratio of aqueous phase and oil was 85:15 by weight. Dynamic shear storage modulus [G'] and loss modulus[G"] of emulsion gels or 11S globulin gels(no oil) were measured by a Rheolograph-sol(Toyo seiki seisakusyo, Tokyo, Japan) during TG reaction for 120 min.[4]

RESULTS AND DISCUSSIONS

G' and G'' of 11S globulin gels and emulsion gels depended on the protein concentrations. The concentration dependence of modulus was approximated by an exponential function of the following form like other polymer gels.[5]
$$G \propto C^n$$
In the case of 11S globulin gels, n is 3.6 and 3.8 for G' and G", respectively. In most cases of biopolymer gels, the modulus is proportional to approximately C^4 at relatively low concentrations. Kohyama et al.[6] also obtained n=3.4 for 11S globulin gel formed by heating followed by the addition of δ-glukonolactone. In emulsion gels, G' and G" was proportional to $C^{2.0}$ and $C^{1.9}$, respectively. Within a high concentration range(10 – 20 %), biopolymer gels usually shows n=2 .[7] The reason why emulsion gels behaved like high concentration biopolymer gels could be explained by assumption that 11S globulin molecules were 'concentrated' by filling of oil droplets.

In order to show clearly the effect of the filling of oil droplets, dynamic moduli of the same concentrations(5%) of proteinous gel and emulsion gel were compared in Table 1. There was no significant difference in the amount of ε-(γ-glutamyl)lysine isopeptide bonds between proteinous and

Table 1. Differences in viscoelastic parameters between proteinous gels and emulsion gels.

	protein[1]	emulsion[1]
G'(10^3Nm^{-2})	0.24	1.65
G"(10Nm^{-2})	0.30	3.75
tan δ (10^{-2})	1.25	2.27
Glu-Lys(μmole/g)	9.30	9.05

[1]Both gels contain 5% 11S globulin and emulsion gels contain 15% soybean oil.

emulsion gel, revealing that the presence of oil did not affect the TG reaction. Both moduli of the emulsion gel exhibited much higher values than those of the proteinous gel. G' and G" increased 7 times and 12 times, respectively. Such large increases in shear moduli could be mainly ascribed to the filler effect of oil droplets, since emulsion gel had the same amount of ε-(γ-glutamyl)lysine isopeptide as proteinous gel. It may be general tendency that small deformation moduli of gels increased by filling particulates.[8]

The value of tan δ of the emulsion gel was significantly higher than that of the proteinous gel, showing high ratio of lost energy to stored energy in emulsion gel network. For other protein concentrations, emulsion gels exhibited higher moduli and tan δ than proteinous gels similarly.

The effect of droplet size of oil on viscoelastic properties of emulsion gels was investigated. Coarse(mean oil droplet diameter : 4 μm), medium(2 μm), fine(1 μm) emulsions(5% 11S globulin, 15% soybean oil) were gelated by TG for 2 hr and small deformation shear moduli were measured. Results of small deformation mechanical measurements are shown in Table 2 with the amount of ε-(γ-glutamyl)lysine isopeptide bond of each gel. There was no significant difference in the amount of ε-(γ-glutamyl)lysine isopeptide bond among three emulsion gels. Oil droplet size affected viscoelastic properties of emulsion gels. G' and G" of 'fine' emulsion gel is twice and third times higher than those of 'coarse' one, respectively. 'Medium' emulsion gel shows medium G' and G" value between fine and coarse ones. These results clearly show that small size oil droplet is more effective to increase shear moduli than large size one. This finding is of importance for controlling texture of emulsion gels by physical techniques. Normally, texture control of emulsion type and other composite gels is attempted by the change of filling volume of particulates. High viscoelasticity requires high content of filler particulates. Considering the problem of excessive calory

Table 2. Effect of oil-droplet size on viscoelastic parameters of emulsion gels [1].

	Large	Medium	Small
Mean droplet diameter(μm)	4.07	2.13	0.95
G'(10^3Nm^{-2})	0.7	1.0	1.5
G"(10Nm^{-2})	1.2	2.0	3.7
tan δ (10^{-2})	1.7	2.0	2.5
Glu-Lys(μmole/g)	10.0	9.0	9.3

[1]All emulsion gels contain 5% 11S globulin and 15% soybean oil.

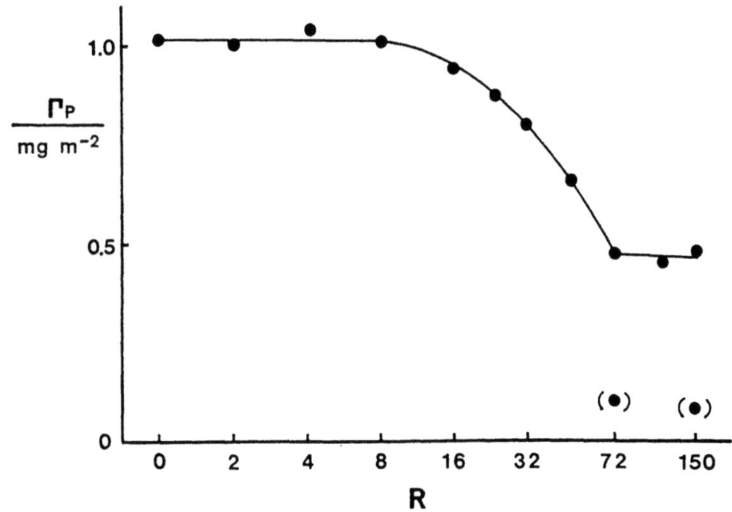

Fig.1. Competitive displacement by Tween 20 of 11S globulin from the oil-water interface in soya oil-in-water emulsions containing 5% 11S globulin and 15% soybean oil. Protein surface concentration Γ is plotted against Tween 20 - to -protein molar ratio R. (●); Tween 20 was completely displaced from the oil-water interface by 11S globulin (see text).

intake, however, high contents of oil is not desirable. Our finding suggests that we could change texture of emulsion gels by controlling physically oil droplet size as well as by increasing oil volume.

The effect of the interaction or adsorption of the protein network to the oil droplet surface was also investigated. Competitive displacement of 11S globulin from the oil-water interface in emulsions (5% 11S globulin, 15% soybean oil) was attempted by Tween 20 according to the methods of Courthaudon et al.[9], prior to the gelation by TG. The results are shown in Fig. 1. The protein surface concentration of 11S globulin was about 1 mg/m^2 in the absence of Tween 20. The increase of Tween 20 addition up to R=8 did not cause any displacement of proteins from the surface. The protein surface concentration started to decline at R>16, but the completely displacement could not be achieved even by addition of Tween 20 at R=150 and about 50% 11S globulin was remained at the interface. In order to obtain the emulsion in which 11S globulin is completely desorbed from the oil droplet surface, emulsions were made from Tween 20 solution and soya oil by using same method, and later the emulsions were mixed with 11S globulin solution. The ratio of Tween 20 to the protein (R) in these emulsions is 72 and 150, respectively, and data of them were represented in parenthesis in Fig. 1. Pre-existing of Tween 20 could not reject the 11S globulin adsorption onto the oil surface perfectly, nevertheless only 0.1mg/m^2 coverage was achieved. Emulsions with various protein surface concentrations, thus obtained, were gelated by TG and viscoelastic parameters were measured (Table 3). The amount of $\varepsilon-(\gamma-$glutamyl)lysine isopeptide bond was nearly constant except

Table 3. Effect of Tween 20 addition on viscoelastic parameters of emulsion gels.

Tween/11S(mol. ratio)[1]	0	8	24	72	72[2]
Γ (mg/m^{-2})[1]	1.02	1.02	0.88	0.51	0.11
G'(10^{3}Nm^{-2})	1.51	1.3	0.78	0.75	0.70
G"(10Nm^{-2})	3.7	3.8	1.5	1.2	1.5
tan δ (10^{-2})	2.5	2.9	1.9	1.6	2.1
Glu-Lys(μmole/g)	9.3	9.4	9.2	9.0	8.2

[1]Results in Fig.1.
[2]Tween 20 was competitively displaced from the oil-water interface by 11S globulin.

little decrease in the case of pre-existing of Tween 20 (72**). This means that Tween 20 addition has no or little effect on TG reaction, and all changes of viscoelastic parameters in Table 3 could be attributed to the change in the interactive force between the oil surface and 11S globulin matrix. Tween 20 addition at R=8 (no change of Γ) showed similar viscoelastic parameters as those of the emulsion gel in the absence of Tween 20. The displacement of only 10% 11S globulin by Tween 20 at R=24, however, caused fairy decreases of G', G" and change of tan δ. G' and G" decreased to half and to one-third, respectively. Further displacement of 11S globulin from oil droplet surfaces did not cause significant changes of G' and G". Even in the case that 90 % of 11S globulin was desorbed from the interface(R=72**), viscoelastic parameters were similar to those of emulsion gel at R=24. It is noteworthy that only 10% displacement of 11S globulin from the oil droplet surface attained minimum level of G' and G", and further displacement of protein has no effect on the viscoelastic parameters of emulsion gels. This means that completely displacement of proteins from the oil droplet surface is not necessary for changing overall physical properties of gels. In other words, addition of relatively a small amount of Tween 20 with low percentage displacement of proteins from oil droplet surfaces is adequate to change the physical properties of emulsion gels. This finding should be important for the designing of real food emulsion systems containing both of proteins and surfactants.

REFERENCES

1. K.R.Langley, M.L.Green, and B.E.Brooker, Mechanical properties and structure of model composite foods, *in*:"Food Polymers, Gels and Colloids", E.Dickinson, ed., Royal Society of Chemistry, London, (1991).
2. V.H.Thanh and K.Shibasaki, Major proteins of soybean seeds. A straight forward fractionation and their characterization, *J. Agric. Food Chem.*, 24: 1117: (1976).

3. H.Ando, M.Adachi, K.Umeda, A.Matsuura, M,Nonaka, R.Uchio, H.Tanaka, and M.Motoki, Purification and characteristics of a novel transglutaminase derived from microorganisms, *Agric. Biol. Chem.*, 53: 2613 (1989).
4. M.Yoshida, K.Kohyama and K.Nishinari,Gelation properties of soy milk and soybean 11S globulin from Japanese -grown soybeans, *Biosci. Biotech. Biochem.*, 56: 735 (1992).
5. N.Hirai, The gel elasticity of high polymers, *Bull. Inst. Chem. Res. Kyoto Univ.*, 33: 21 (1955).
6. K.Kohyama, M.Yoshida and K.Nishinari, Rheological study on gelation of soybean 11S protein by glucono-δ-lactone, *J. Agric. Food Chem.*, 40: 740 (1992).
7. A.H.Clark, S.B.Ross-Murphy, K.Nishinari and M.Watase, Shear modulus-concentration relationships for Biopolymer gels. Comparison of independent and cooperative crosslink descriptions, *in*: "Physical networks. Polymer and gels" W.Burchard and S.B.Ross-Murphy, ed., Elsevier Applied Science, London, (1990).
8. R.K.Richardson, G.Robinson, S.B.Ross-Murphy, and S.Todd, Mechanical spectroscopy of filled gelatin gels, *Polymer Bulletin*, 4: 541 (1981).
9. J-L,Courthaudon, E.Dickinson and D.G.Dalgleish, Competitive adsorption of β-casein and nonionic surfactans in oil-in-water emulsions, *J. colloid Interface Sci.*, 145: 390 (1991).
10. J-L.Courthaudon, E.Dickinson, Y.Matsumura, and D.C.Clark, Competitive adsorption of β-lactoglobulin + Tween 20 at the oil-water interface, *Colloids Surfaces*, 56: 293 (1991).

NMR STUDY ON LYSOPHOSPHATIDYLCHOLINE-PROTEIN INTERACTIONS AND THEIR FUNCTIONAL PROPERTIES

Yoshinori Mine,[1] Kazuhiro Chiba,[2] and Masahiro Tada[2]

[1]Research Institute of Q. P. Corporation, Fuchu-shi, Tokyo 183, Japan
[2]Laboratory of Bio-organic Chemistry,
Tokyo University of Agriculture and
Technology, Fuchu-shi, Tokyo 183, Japan

ABSTRACT

We analyzed the emulsion state of phospholipid-protein complexes from their dynamic structure. The interfacial adsorption of the complexes was evaluated by the mean droplet size of emulsions and NMR. The interfacial adsorptivity of ovalbumin (OA) was increased with the interaction of lysophosphatidylcholine (LPC) to form smaller droplets, and the formation of a fine emulsion was further promoted by the addition of linoleic acid in the mixture. The motional property of Ser-P^{68} of OA was much restricted at the interface when the complex formed a fine emulsion. The LPC changed the conformation of OA to increase the α-helix and to reduce the β-sheet content by interaction with the protein, and the heat stability of the protein was enhanced. The results of ^{31}P and ^{13}C NMR spectrum indicated that LPC is bound to OA through hydrophobic interaction and lead to conformational change of protein structure. These results suggested that LPC and linoleic acid mutually changed the structure of protein to increase the interfacial adsorption and heat stability. The region of Ser-P^{68} in OA was tightly adsorbed at the interface when the complexes formed fine emulsion.

INTRODUCTION

The ability to form a stable emulsion is one of the most important functional properties of proteins in food systems. Proteins display a versatility in surface-active behaviors due to their amphiphilic nature and their varying physicochemical properties. In the last decade, both their backbone flexibility and proper hydrophobic-hydrophilic balance of the protein molecule have assigned primary factor in its emulsifying properties.[1] Casein from cow's milk is an example as a superior emulsifier of a food protein which possesses both high flexibility and amphiphilicity. However, many food proteins are globular in nature and are often limited in both molecular flexibility as well as hydrophobic groups. For example, ovalbumin (OA) and lysozyme from hen egg white have a compact globular structure which

Food Hydrocolloids: Structures, Properties, and Functions
Edited by K. Nishinari and E. Doi, Plenum Press, New York, 1994

exhibit poor emulsifying properties. OA is known as a major protein of egg white and is critical ingredient to functional properties of egg white. In order to improve the emulsifying property of proteins, various ways of conventional chemical and enzymatic modification of proteins have been attempted. Lysophosphatidylcholine (LPC), although, it exists as a minor component in phospholipids (PLs) is an interesting emulsifier with higher water solubility and emulsifying properties. We found that LPC and free fatty acids(FFAs) promote the interfacial adsorptivity of OA and the LPC/FFA/OA complexes showed heat stability.[2]

The interaction between the PLs and proteins has been an active research topic for a number of years and the molecular processes which occur during emulsification in food system have been the subject of several recent reviews and books. However, the protein and PLs conformation at an oil-in-water interface and its role in stabilizing emulsion are not clearly understood. There was much difficulty in the evaluation of the interfacial adsorptivity of proteins, PLs, and their complexes by conventional physicochemical techniques. We have recently reported the interfacial adsorptivity of egg phosphatidylcholine (PC), LPC and OA by means of nuclear magnetic resonance (NMR) which provides an valuable information about their interfacial adsorptivity.[3-7] In this work, we describe the interfacial adsorptivity of LPC/FFA/OA complexes and their heat stability, including a conformational analysis of the interaction between LPC and OA.

MATERIALS AND METHODS

Materials. Ovalbumin (OA) was prepared by the method of Kekwick et al.[8] PC and LPC, 98% pure, were obtained from egg PC (Q.P. Corp., PC-98N, Tokyo) and LPC (Q.P., LPC-1). Triolein (Wako Pure Chemicals, Osaka) and FFAs (Wako) were purified by silica gel column chromatography (hexane-diethylether, 97:3).

Preparation of Complexes and Emulsification. Unilamellar PC vesicies (SUV-PC) , LPC micelles and LPC/FFA vesicles were prepared by microfluidizer (Model M-110F, Microfluidics Corp., MA; 8000 psi at 4 °C) and sonicating (Branson sonifier Model 250, out put 40w) for 5 min at 60 °C, respectively. The SUV-PC, LPC micelles and LPC/FFA vesicles were mixed with OA solution to yield various molar ratios and sonicated for 10 min at 25 °C. The buffer used was 20 mM HCl. Emulsification was achieved by adding triolein dropwise to the complex solution in 20 mM HCl with agitation by a disperser (physcotron NS-10 fitted with a generator shaft) at 12 000 rpm for 2 min. The size distribution of the emulsion droplet was measured by a laser light scattering photometer (submicron particle sizer, Pacific Scientific Nicomp Model 370-HPL, MD). The data were expressed as mean volume droplet size.

Measurements of NMR spectra. ^{13}C and ^{31}P NMR spectra were recorded on a Varian VXR-4000s spectrometer at 20 °C. The ^{31}P NMR spectra were acquired with 32 K data points, a 40 000 Hz spectral window, 45° pulse (25 μsec) and 2.0 sec pulse delay. Proton was fully decoupled. Chemical shifts were referenced to 85% H_3PO_4 at 0 ppm. T_2^* relaxation times for choline, glycerol and acyl chain carbons of LPC were calculated approximately from the line widths of the carbon signals ($v_{1/2}$), $T_2^*=1/(\pi \times v_{1/2})$.

Circular Dichroism measurement. Circular dichroism (CD) measurement was made with a JASCO J-720 spectropolarimeter using a 1 mm cell at 25 °C. The relative proportions of secondary structure were calculated according to the modified CD curve-fitting method.[9]

RESULTS AND DISCUSSION

Fig. 1 shows the changes of the mean droplet size of emulsion composed of triolein and aqueous dispersion of PL/protein complexes. The mean droplet size of emulsion composed of alone aqueous dispersion of OA and triolein was about 5.8 μm and the emulsion was unstable. As the molar ratio of PLs/protein increased, the mean droplets diameter of the emulsion became smaller. The LPC/OA complex formed smaller droplet compared to PC/OA complex. Moreover, the LPC/linoleic acid/OA complex formed smaller droplets. These results indicated that the interfacial adsorptivity of OA was promoted by the interaction with PC or LPC and the formation of fine emulsion was further promoted by linoleic acid. OA contains two phosphoryl residues at the serine 68 and 344 residues, and these two phosphate residues are well resolved by their ^{31}P NMR spectra.[10] Choline and phosphate residues constitute the head group of PC and LPC. The phosphorus signals of ^{31}P NMR are influenced by the motional properties of phosphate moieties in molecules and the phosphorus signals of LPC and OA were well correlated with their interfacial adsorptivity.[3,5,6] Thus, phosphate residues of OA and PLs can be used as a sensitive probe for studying the protein -PLs interactions and their interfacial adsorptivity. Fig. 2 shows ^{31}P NMR spectra of LPC/linoleic acid/OA complex in aqueous dispersion (a) and emulsion (b). The molar ratio of LPC/OA was 10. The complex gave three separated phosphorus signals and these signals

(μm)

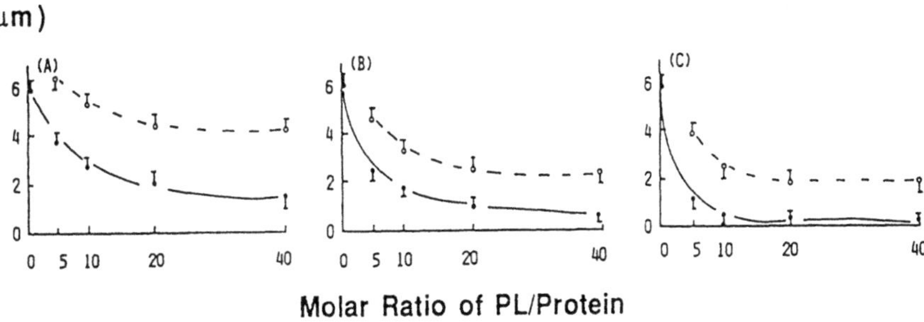

Fig. 1. Changes of the mean droplet size of emulsion composed of PL and protein (—) or just PL (---) as the emulsifier. HCl (20mM) and triolein were used as aqueous and oil phase, respectively. The ratio of oil/protein (w/w) was 3. (A) PC (o), PC/OA, (●); (B) LPC (o), LPC/OA (●); (C) LPC/linoleic acid (o), LPC/linoleic acid/OA (●). (Reprinted with permission from ref. 2. Copyright 1992 American Chemical Society)

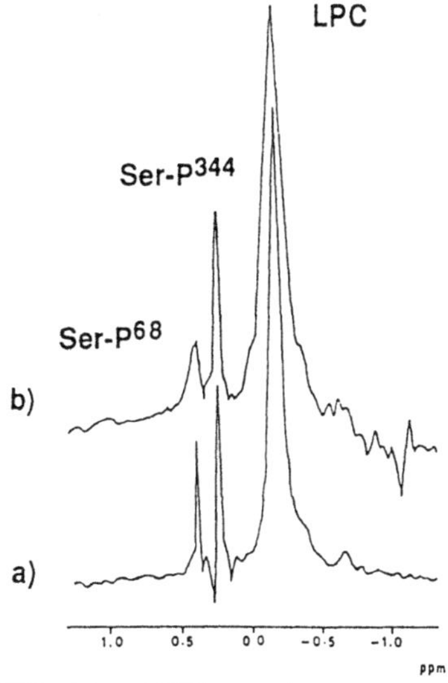

Fig. 2. ^{31}P NMR spectra of LPC/linoleic acid/OA complex in an aqueous dispersion (a) and emulsion (b). The emulsions were prepared with triolein and the complex dispersed in 20mM HCl. The ratio of oil/protein (w/w) was 3. A total of 12 000 acquisitions were made, and additional line broadening was 0.1 Hz. Chemical shifts were referenced to 85% H_3PO_4 at 0 ppm. (Reprinted with permission from ref. 2. Copyright 1992 American Chemical Society)

were assigned to phosphoserine 68 (Ser-P[68]) and phosphoserine 344 (Ser-P[344]) of OA and phosphate residue of LPC as shown in Fig. 2.[2)] The line widths of ^{31}P NMR spectra of Ser-P[68] and Ser-P[344] of OA and LPC micelles were 7, 5, 12 Hz, respectively. In aqueous dispersion of LPC/linoleic acid/OA complex, the phosphorus signal of LPC scarcely changed with the interaction of OA. The results indicated that the head group motion of LPC was scarcely restricted in the interaction of the protein. On the other hand, ^{31}P NMR spectra of the phosphorus signals of OA in the complex at the interface showed markedly different spectra. The line widths of Ser-P[68] of OA were broadened about 3-fold compared to that in aqueous dispersion and the peak height was reduced when the complex formed a fine emulsion. The peak broadening of the phosphorus signals in ^{31}P NMR spectra was observed with increasing magnitude of the negative phosphorus chemical anisotropy. Thus, peak broadening of the ^{31}P NMR spectra was interpreted as indicating the motion of the phosphoserine and phosphate moieties in OA and LPC being restricted by adsorption at the interface. As mentioned above, linoleic acid promoted the interfacial adsorptivity of the LPC/OA complex. Fig. 3 shows the changes of the mean droplet size under the influence of emulsifying temperature and the kind of, or unsaturation degree of FFA which coexisted with LPC. The mean droplet size of emulsion was little influenced by saturated FFAs. On the other hand, the mean droplet size decreased with decreasing degree of unsaturation of FFA . Smaller droplets were obtained by heating during emulsification. The peak broadening of Ser-P[68] in OA was not observed when saturated FFA (palmitic acid or stearic acid) was used as a component of the LPC/FFA/OA complex. From these results, it is suggested that the peak broadening of Ser-P[68] in OA is only shown when the complexes formed fine emulsions. The free energy of binding of PLs to protein consists of two main contributions a) from electrostatic interactions b) from hydrophobic interactions. NMR studies can provide direct information about a) the binding site b) changes in protein structure and c) the equilibrium and exchange rates between bound and free PL molecules. In addition, NMR relaxation time measurement constitute a very sensitive probe for the study of the molecular environment, and T2* values are also known to be sensitive to intermolecular interactions. The line widths of ^{31}P NMR and T2* values for choline C1 - C3, glycerol C4 - C6, and CH3 group of acyl chain of LPC micelles, LPC/linoleic acid liposome and LPC/linoleic acid/OA complex are shown in Table 1. The binding molar ratio of LPC/OA

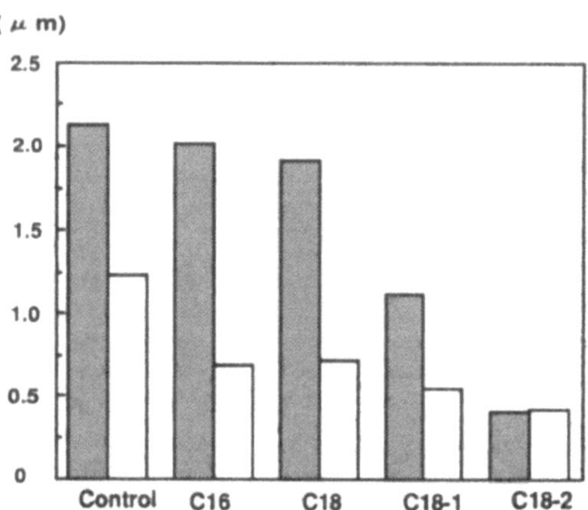

Fig. 3. Mean droplet size of the emulsions composed of complex and triolein. The complex was composed of LPC/FFA/OA (10:10:1, mol/mol) in 20 mM HCl. The ratio of oil/protein (w/w) was 3. The emulsions were prepared at 25(dotted bars) or 80°C (open bars): control; LPC/OA (10:1, mol/mol), C16, palmitic acid; C18, stearic acid; C18-1, oleic acid; C18-2, linoleic acid. (**Reprinted with permission from ref. 2. Copyright 1992 American Chemical Society**)

Table 1. Line widths of ^{31}P NMR and the T_2* relaxation times for choline, glycerol and acyl chain carbons of LPC.

| | ^{31}P NMR (Hz) | ^{13}C NMR (ms) | | | | | | |
| | | C1 | C2 | C3 | C4 | C5 | C6 | CH$_3$ |
			choline			glycerol		acyl
LPC	20	26	19	28	23	21	32	48
LPC/C$_{18\text{-}2}$	33	28	21	13	19	15	19	36
LPC/C$_{18\text{-}2}$/OA	27	25	18	17	17	12	16	31

was 20. In aqueous dispersion of LPC micelles, the choline, glycerol carbons and acyl chain carbons were clearly resolved and showed the sharp signals. Aqueous dispersion of equimolar mixture of LPC and long-chain FFA form bilayer.[7] In the bilayers composed of LPC and linoleic acid, the line width of ^{31}P NMR was broader and T2* values for choline C-3, glycerol C4 - C6 and CH3 group of acyl chain were low in comparison with those of LPC in globular micelles. In the case of LPC/linoleic acid/OA complex, T2* relaxation times for glycerol C4 - C6 and acyl chain carbons were more reduced by interacting with OA, resulting from restriction of molecular motion. The T2* value for choline C-3 was decreased, but that of choline C1 - C2 scarcely changed. The line width of ^{31}P NMR of LPC in the complexes became sharper compared to LPC/linoleic acid liposome by interacting with the protein. These results shows that acyl chains, glycerol and a part of choline carbons of LPC interact with the protein, while the motion of head group remains unaffected.

In general, the emulsion composed of globular proteins are sensitive to heat treatment. The destruction of the emulsion by heat treatment is known to be caused by the denaturation of the constitutive components at the interface. Fig. 4 shows the effect of heat treatment on emulsion composed of OA and PLs/OA complexes. The molar ratio of PLs/OA was 20. The emulsion composed of just OA was affected by the heat treatment by heating 80°C for 20 min. On the other hand, the emulsion of LPC/linoleic acid/OA complex showed heat stability and mean droplet size of the emulsion was hardly changed by heating. The difference of heat sensitivity of these emulsion may be closely correlated with conformational change of OA

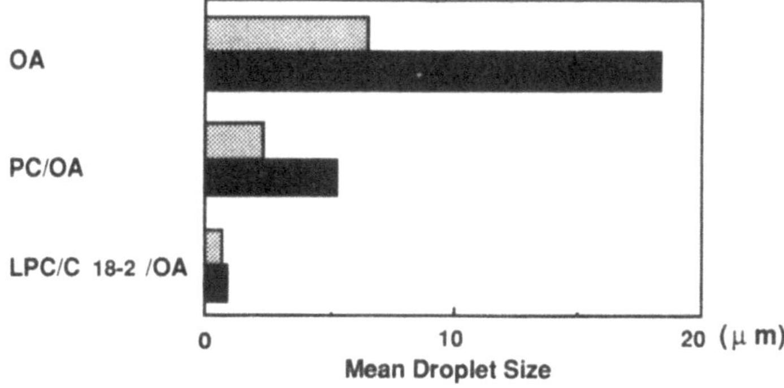

Fig. 4. Effect of the heat treatment on the emulsions composed of OA. PC/OA complex and LPC/linoleic acid/OA complex. The oil/protein (w/w) was 6. The emulsions were untreated (dotted bars) or heated in a water bath at 80°C for 20 min (solid bars). The complex was composed of PL/OA (20:1, mol/mol) in 20mM HCl.

Table 2. Secondary structures of native and heated OA and LPC/$C_{18\text{-}2}$/OA complex derived from the CD Spectra.

treatment		% secondary structure parameter			
		α-helix	β-sheet	β-turns	random coil
OA	native	30.9	34.9	7.8	26.5
	heated at 80 °C[a]	19.2	44.0	4.5	32.3
LPC/$C_{18\text{-}2}$[b]/OA	native	44.4	14.0	11.2	30.4
	heated at 80 °C	36.1	22.3	6.3	35.3

a. Heated in a water bath at 80 °C for 20 min. b. Linoleic acid.

caused by the heat treatment. Table 2 shows the secondary structure of native and heated OA and LPC/linoleic acid/OA complexes in 20 mM HCl solution derived from the CD analysis. In the absence of LPC and linoleic acid, the α-helix and β-sheet contents of OA were 30.9% and 34.9%, respectively. The α-helix content of OA was decreased from 30.9% to 19.2% by heating at 80°C for 20 min. The β-structure content of the protein was increased by heat treatment. On the other hand, the α-helix content of OA was increased from 30.9% to 44.4% by coupling with LPC and linoleic acid and the secondary structure of LPC/linoleic acid/OA complexes were not influenced by heat treatment compared to OA. These results suggest that the heat stability of the emulsion composed of the complexes may be closely correlated with its protein structure and dynamic state of the complexes.

REFERENCES

1. S. Nakai, and E. Li-Chan, Hydrophobic Interactions in Food Systems, CRC Press, Boca Raton, Florida (1988).
2. Y. Mine, H. Kobayashi, K. Chiba, and M. Tada, ^{31}P NMR Study on the Interfacial Adsorptivity of Ovalbumin Promoted by Lysophosphatidylcholine and Free Fatty Acids, *J. Agric. Food Chem.* 40:1111 (1992).
3. K. Chiba, and M. Tada, Study of the Emulsion Stability and Headgroup Motion of Phosphatidylcholine and Lysophosphatidylcholine by ^{13}C and ^{31}P NMR, *Agric. Biol. Chem.* 53:995 (1989).
4. K. Chiba, and M. Tada, Effect of the Acyl Chain Length of Phosphatidylcholines on their Dynamic States and Emulsion Stability, *J. Agric. Food Chem.* 38:1177 (1990).
5. K. Chiba, and M. Tada, Interfacial Adsorptivity of Lysophosphatidylcholine Measuring Its Interaction with Triacylglycerols and Free Fatty Acids, *Agric. Biol. Chem.* 54:2913 (1990).
6. Y. Mine, K. Chiba, and M. Tada, Effect of a Limited Proteolysis of Ovalbumin on Interfacial Adsorptivity Studied by [31]P Nuclear Magnetic Resonance, *J. Agric. Food Chem.* 40:22 (1992).
7. Y. Mine, K. Chiba, and M. Tada, Effect of Oxidation of Free Fatty Acids on the Interfacial Adsorptivity of Lysophosphatidylcholine/Free Fatty Acid/Ovalbumin Complexes, *Biosci. Biotech. Biochem.* 56:1814 (1992).
8. R.A. Kekwich, and R.K. Cannan, The Hydrogen Ion Dissociation Curve of the Crystalline Albumin in the Hen's Egg, *Biochem. J.* 30:227 (1936).
9. J.T. Yang, and C-S.C. Wu, Calculation of Protein Conformation from Circular Dichroism, *Methods in Enzymology* 130:208 (1986).
10. H.J. Vogel, and W.A. Bridger, Phosphorus-31 Nuclear Magnetic Resonance Studies of the Two Phosphoserine Residues of Hen Egg White Ovalbumin, *Biochemistry* 21:5825 (1982).

EMULSIFYING PROPERTIES OF BOVINE SERUM ALBUMIN
AND ITS ENZYMATIC HYDROLYZATE

Masayoshi Saito

National Food Research Institute, Ministry of Agriculture,
Forestry and Fisheries, Tsukuba 305, Japan

ABSTRACT

Bovine serum albumin (BSA) was hydrolyzed with protease and the
emulsifying properties of the hydrolyzate were studied. When the
hydrolysis was limited with trypsin, the emulsifying activity index (EAI)
increased 40% above its original level. To determine which peptides
contribute to the improved emulsifying properties of the hydrolyzate,
peptides adsorbed onto a emulsified oil globule surface were extracted and
analyzed. SDS-polyacrylamide gel electrophoresis showed that a peptide
with a molecular weight of approximately 24 KD was preferentially adsorbed
onto the oil surface. Amino acid composition and sequence analysis showed
that this peptide corresponded to the third domain of BSA, residues 377-582.
Hydrophobic chromatography showed that the 24 KD peptide had higher
hydrophobicity than that of BSA. Our results suggest that the third domain
of BSA, which has a hydrophobic region for fatty acid binding, may play an
important role in emulsification.

INTRODUCTION

Interest in the utilization of blood proteins from slaughtered animals is
increasing in Japan. In order to use wasted protein resources, such as
plasma proteins from animal blood in food processing, it is necessary to
understand their functional properties. Plasma protein has already been
shown to have good emulsifying properties. However, some problems were
discovered while using plasma protein, such as decreases in emulsifying
activity during heat treatment due to the formation of protein polymers.[1,2]
There have been previous studies on improving the emulsifying properties of

proteins by enzymatic hydrolysis.[3,4,5] In one such study, we hydrolyzed plasma protein with proteases and found that the emulsifying activity could be improved under several conditions.[3] Shimizu et.al.[4] reported that a certain peptide from the peptic hydrolyzate of α s1-casein has good emulsifying properties. Some synergistic effects between peptides in the emulsification have also been studied,[5] but the ideal peptide structure needed in emulsification is still not known.

Bovine serum albumin (BSA) is a main component of bovine plasma proteins and its primary structure has already been determined.[6] Because BSA acts as a fatty acid binding protein in plasma, the hydrolyzate of BSA may provide a good model for the analysis of peptide emulsification. In this study, BSA was hydrolyzed with proteases and the peptide fragments having good emulsifying and adsorption properties were investigated. From these results, we estimated what types of peptides are effective in emulsification.

MATERIALS AND METHODS

Materials

Bovine serum albumin (A-4378), trypsin (T-8253), soybean trypsin inhibitor (T-9128), pepsin (P-6887), and tricaprylin (T-9126) were purchased from Sigma Chemical Co. (St. Louis).

Enzymatic hydrolysis of BSA

i) Tryptic hydrolysis. Tryptic hydrolysis was carried out under two conditions (a or b). After incubating the solution for 10, 20, 30, or 60 min, a trypsin inhibitor solution was added to stop hydrolysis. Then the mixture was cooled in an ice bath, dialyzed against deionized water with a Spectra/por 6 dialysis membrane (molecular weight cut-off 2,000) (Spectrum, Los Angeles), and freeze-dried.

ii) Peptic hydrolysis. Peptic hydrolysis was carried out under two conditions (a or b). After incubating the solution for 10, 20, 30, or 60 min, the pH of each solution was adjusted to 7.0 with 2.0 M tris solution to stop hydrolysis. Then the mixture was cooled in an ice bath, dialyzed against deionized water, and freeze-dried.

Measurement of emulsifying activity

The emulsifying activity index (EAI) was measured according to the turbidimetric procedure of Pearce and Kinsella.[7] Two milliliters of protein solution (1% w/v) in a 5.0 mM phosphate buffer (pH 7.0) and 0.5 g of tricaprylin were homogenized at 30 °C for 3 min, using a Polytron PT-7 (Kinematica, Switzerland) at its maximum speed. An aliquot of the emulsion was diluted with a 0.1% sodium dodecyl sulfate (SDS) solution and the turbidity was measured at 500 nm.

Sodium dodecyl sulfate-polyacrylamide gel electrophoresis (SDS-PAGE)

A Phastsystem (Pharmacia LKB Biotechnology, Uppsala) with an 8-25% acrylamide gel was used for SDS-PAGE. After electrophoresis, the gels were stained with Coomassie Blue.

Extraction of adsorbed peptide from oil globule surface

Twenty milliliters of hydrolyzate solution (1% w/v) in a 5.0 mM phosphate buffer (pH 7.0) and 5.0 g of tricaprylin were homogenized at 30 °C for 8 min, using a Polytron PT-20 at its maximum speed. The emulsion was immediately diluted with 20 ml of water and centrifuged. The aqueous portion was removed by aspiration. Twenty milliliters of water were added to the cream portion and this was gently mixed. This mixture was then centrifuged and the aqueous portion was removed. A washed cream was obtained by repeating the above procedure twice. To a given volume of the washed cream, 1/3 volume of 6 % SDS, 10 % glycerol solution was added. This mixture was heated at 100 °C for 3 min and then centrifuged. The aqueous portion was analyzed by SDS-PAGE.

Gel filtration and ion-exchange chromatography

Gel filtration chromatography was carried out at 4 °C with a Sephadex G-100 column (2.6 X 90.0 cm) (Pharmacia LKB Biotechnology, Uppsala) by eluting the sample with a 0.2 M phosphate buffer (pH 7.2) containing 0.05% NaN_3 at a flow rate of 30 ml/hr. Ion-exchange chromatography was carried out at 4 °C with a DEAE Toyopearl 650 M column (2.6 X 30.0 cm) (Tosoh, Tokyo) equilibrated with a 0.01 M phosphate buffer (pH 7.1) containing 0.05% NaN_3, with a linear NaCl gradient from 0.0 M to 0.5 M at a flow rate of 30 ml/hr. Absorbance of the eluted fractions was measured at 280 nm. The fractions were dialyzed against deionized water and were freeze-dried.

High performance liquid chromatography (HPLC)

An LC-6AD liquid chromatograph (Shimadzu, Kyoto), equipped with a Vydac C_{18} column (0.46 X 25.0 cm) (Cypress International, California) and a UV-6A ultraviolet spectrophotometer set at a wavelength of 230 nm was used. The hydrolyzate was eluted at 40 °C with a linear gradient from 0.1% (v/v) trifluoroacetic acid to 95% (v/v) acetonitril at a flow rate of 1.0 ml/min.

Determination of amino acid composition and sequence

The samples were hydrolyzed in 6.0 M HCl at 110 °C for 22 hr in vacuo. An amino acid analysis of each hydrolysate was carried out on an automatic amino acid analyzer (model 835, Hitachi, Tokyo). An amino acid sequence analysis was carried out using an automatic protein sequencer (model 477A, Applied Biosystems, California) with an amino acid analyzer (model 120A, Applied Biosystems, California).

Determination of hydrophobicity

Hydrophobic chromatography was carried out to determine hydrophobicity. A column of Phenyl-Sepharose CL-4B (1.3 x 25.0 cm) (Pharmacia LKB Biotechnology, Uppsala) equilibrated with a 0.01 M phosphate buffer (pH 6.8) containing 1.0 M ammonium sulfate was used. The samples were eluted at 25 °C with a linear ammonium sulfate gradient from 1.0 M to 0.0 M at a flow rate of 60 ml/hr. The absorbance of the eluted fractions was measured at 280 nm.

RESULTS AND DISCUSSION

Change in emulsifying activity during hydrolysis

Figure 1 shows the changes in EAI during protease hydrolysis. In tryptic hydrolysis under condition (a), EAI decreased during the first 10 min and then slightly increased. However, EAI increased by 40% with tryptic hydrolysis under condition (b) after 60 min of hydrolysis. No increase in EAI was observed in the peptic hydrolysis samples. Tryptic hydrolysis (condition b) was then carried out for 6 hr. The EAI reached maximum value after 2 hr of hydrolysis, and began to decrease thereafter. This suggests that peptides having high EAI values were produced during the first 2 hr and that the further hydrolysis of these peptides caused the decrease in EAI.

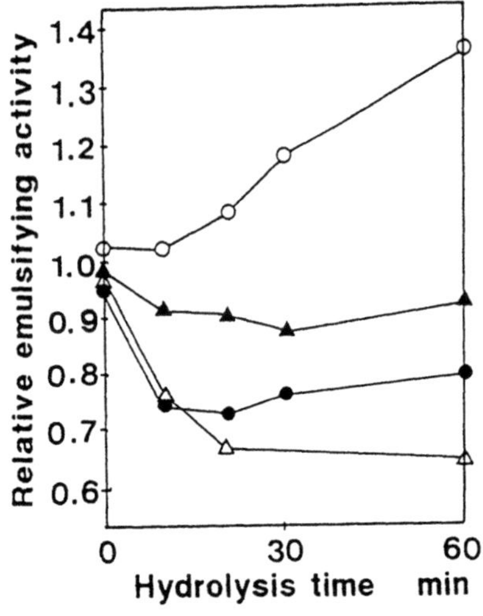

Fig. 1 Changes in emulsifying activity during protease hydrolysis.

●: treated with trypsin (condition a) (2.5 % BSA in 0.04 M tris-HCl buffer pH 8.15, trypsin:BSA=1:1000, 30 °C)
○: treated with trypsin (condition b) (2.5 % BSA in 0.04 M tris-HCl buffer pH 8.15, trypsin:BSA=1:100, 37 °C)
▲: treated with pepsin (condition a) (2.5 % BSA in 0.12 M ammonium formate buffer pH 3.7, pepsin:BSA=1:1000, 37 °C) △: treated with pepsin (condition b) (1.25 % BSA in HCl pH 2.2, pepsin:BSA =1:25, 37 °C).

Figure 2 shows the SDS-PAGE patterns of the tryptic hydrolyzates (condition b). In the first 2 hr, a peptide with a M.W. of approximately 24 KD was produced (arrowhead). Several fragments with smaller molecular weights were also produced. The increase in EAI may be related to the formation of these peptide fragments.

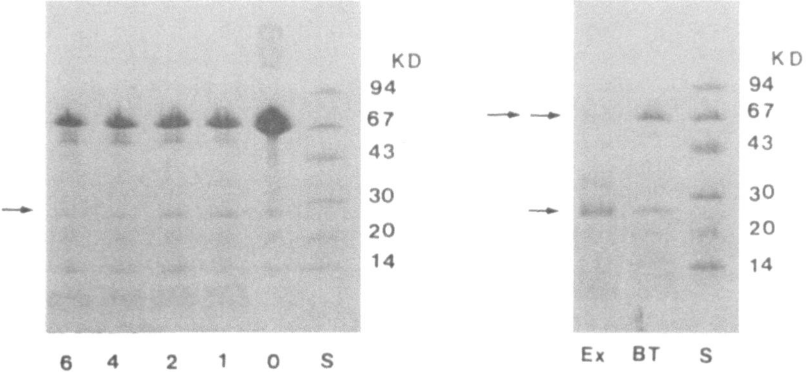

Fig. 2 SDS-PAGE patterns of tryptic hydrolyzate.

S:molecular weight marker,
0,1,2,4,6: tryptic hydrolysis for
0,1,2,4, and 6 hr, respectively.

Fig. 3 SDS-PAGE patterns of the tryptic hydrolyzate (BT) and peptides extracted from the oil glouble surface (Ex).

S: molecular weight marker, Tryptic hydrolysis was carried out for 2 hr.

Composition of material adsorbed onto the oil globule surface

Since proteins and peptides with good emulsifying properties are believed to be preferentially adsorbed onto an emulsified oil globule surface, and also lower the surface tension, peptides adsorbed onto the emulsified oil globule surface were extracted and analyzed.

Figure 3 shows the SDS-PAGE patterns of the peptides extracted from the oil globule surface compared to that of the tryptic hydrolyzate used for emulsification. A peptide with a M.W. of approximately 24 KD was observed to be preferentially adsorbed onto the oil globule surface (arrowhead), whereas the adsorption of the intact BSA (double arrowhead) was much less. These results suggest that this 24 KD peptide plays an important role in emulsification of the hydrolyzate.

Separation and purification of hydrolyzate

Separation and purification of the 24 KD peptide from tryptic hydrolyzate were performed. The BSA was hydrolyzed with trypsin to obtain the hydrolyzate with maximum EAI. The hydrolyzate was first separated into five fractions (F1 to F5) by gel filtration chromatography. The SDS-PAGE patterns showed that F3 contained the 24 KD peptide. The F3 fraction was further separated into two fractions (F3a and F3b) by ion-exchange chromatography. The SDS-PAGE patterns showed that the 24 KD peptide was concentrated in F3a. The yield of F3 was 11 % and that of F3a was 3.2 %.

Identification of peptides

Fraction 3a was subjected to an amino acid composition analysis. Peters and Feldhoff[8] reported that the trypsin selectively cleaved the peptide bond between Lys^{376} and His^{377} of BSA. Since the fragment contains the residues 377-582, and has a molecular weight of approximately 24 KD, amino

acid composition of the 24 KD peptide and that of the residues 377-582 of BSA
were compared. The compositions of these two amino acids were found to be
similar.

Fraction 3a was further purified to a single peak using HPLC. This peak
was then subjected to the sequence analysis. The sequence of ten amino
acids from the N-terminal was found to be exactly the same as that of
residues 377-386 of BSA (Table 1). These results strongly suggest that the
24 KD peptide contains the residues 387-582 of BSA.

Table 1 Amino acid sequences of the 24 KD peptide and residues 377-386
of BSA.

	1	2	3	4	5	6	7	8	9	10
The 24 KD peptide*	His-Leu-Val-Asp-Glu-Pro-Gln-Asn-Leu-Ile-
BSA (377-386)	His-Leu-Val-Asp-Glu-Pro-Gln-Asn-Leu-Ile

* Sequence of ten amino acids from the N-terminal.

Hydrophobicity of hydrolyzate

The tryptic hydrolyzate was separated into four fractions (Fa to Fd) by
hydrophobic chromatography, and each fraction was analyzed by SDS- PAGE
(Fig. 4). The SDS-PAGE patterns showed that the Fc fraction was BSA rich and
that the Fd fraction was the 24 KD peptide rich. The retention time of BSA
was 45 min, whereas that of the 24 KD peptide was 69 min. The retention times
of the other peptides were shorter than that of BSA, ranging from 16 to 26
minutes. These results indicate that the hydrophobicity of the 24 KD peptide
was the highest among the tryptic hydrolyzate.

Bovine serum albumin is composed of three domains and the pattern of the
24 KD peptide (residues 387-582 of BSA) almost matches the third domain.[6]
The third domain of BSA is known to include the major fatty acid binding
site.[9] The present study demonstrates that the hydrophobicity of this
region is higher than that of whole BSA or other peptides. The preferential
adsorption characteristics of the 24 KD peptide is possible because of the
presence of the hydrophobic region for fatty acid binding. The removal of
the first and the second domain may facilitate the binding of the third
domain to the hydrophobic surface. However, the interaction of the 24 KD
peptide and other smaller peptides during emulsification might not be
negligible. Further investigation into the role of smaller peptides in
emulsification and the mechanism of increase in emulsifying properties is now
in progress.

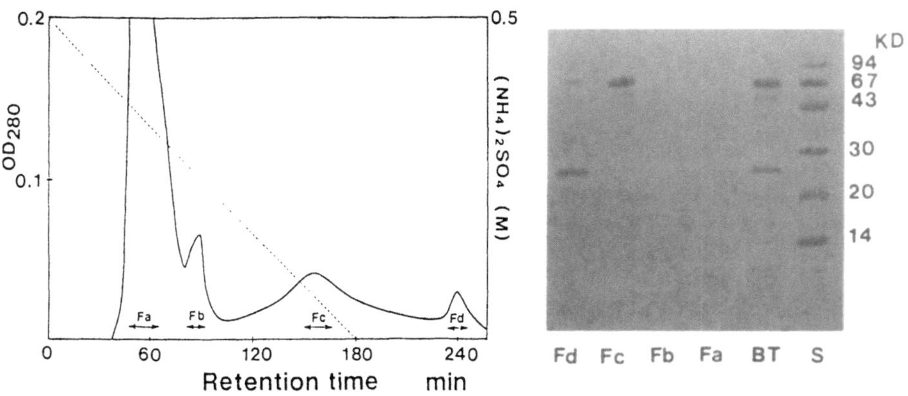

Fig. 4 Hydrophobic chromatogram of the tryptic hydrolyzates and
SDS-PAGE patterns of each fraction (Fa~Fd).

REFERENCES

1. M. Saito and H. Taira, Agric. Biol. Chem., **51**: 2787 (1987).

2. M. Saito, N. Ichikawa, and H. Taira, Agric. Biol. Chem., **52**: 2831 (1988).

3. M. Saito, H. Taira, M. Shimizu, and K. Yamauchi, J. Jap. Soc. Food Sci. Tech., **38**: 920 (1991).

4. M. Shimizu, S. W. Lee, S. Kaminogawa, and K. Yamauchi, J. Food Sci., **51**: 1248 (1986).

5. S. W. Lee, M. Shimizu, S. Kaminogawa, and K. Yamauchi, Agric. Biol. Chem., **51**: 1535 (1987).

6. J. R. Brown, Fed. Proc., **35**: 2141 (1976).

7. K. N. Pearce and J. E. Kinsella, J. Agric. Food Chem., **26**: 716 (1978).

8. T. Peters, Jr. and R. C. Feldhoff, Biochemistry, **14**: 3384 (1975).

9. R. G. Reec, R. C. Feldhoff, O. L. Clute, and T. Peters, Jr., Biochemistry, **14**: 4578 (1975).

ENHANCED EMULSIFYING ABILITY OF FOOD SURFACTANTS BY ADDITION OF LYSOPHOSPHOLIPIDS

Satoshi Fujita, Atsushi Suzuki* and Emi Yahisa*

S. Fujita Consultant Office, 3-7-1, Tsukusigaoka,
Kashiwa-shi, Chiba 277; *Nagoyaseiraku K.K., 310,
Nakasunatyo, Tenpaku-ku, Nagoya, 468 Aichi Japan

ABSTRACT: Soy lysophospholipids (SLP) and soy lysophosphatidylcholine (LPC) were added to food surfactants for the emulsification of vegetable oil in soy sauce and acidic fruit juice, and of simulated coffee whitener. The addition of soy LPC or SLP to nonionic food surfactants (sucrose fatty acid esters, polyglycerine fatty acid esters and monoglycerides) improved stability of emulsions by which they were prepared. The emulsions became tolerant to salty or acidic aqueous phase. The quality and stability ofemulsions were improved when SLP was added to those food surfactants in the preparation of the emulsions with milk proteins. The preferred amount of SLP or LPC to the food surfactant was found to be more than 20-40 Wt% (depending to the purpose) for the stableemulsions having long shelf life.

INTRODUCTION

Lysophospholipids are unique natural amphoteric surfactants, and they are after meal hydrolysis products of bile and food lecithins in the small intestine. Recently soy and egg lecithins have been improved by addition of phospholipase A-2, getting lysophospholipid(LP), leading to wider applications as the more useful surfactant and oil in water (O/W) emulsifier[1-4]. The authors studied the synergistic effects of soy LP (SLP) and lysophosphatidylcholine (LPC) with other surfactants (emulsifiers), such as glycerine mono fatty acid ester (MG), fatty acid (FA), sucrose fatty acid ester (SE), polyglycerine fatty acid ester (PGE) and sorbitan fatty acid ester (SOE). In the case of the mixtures of natural surfactants, SLP/MG/FA, SLP/FA, SLP/MG, many of them showed significant surface activity, such as very low surface tension and high wettability, when their MG and FA components consisted of polyunsaturated FA and/or medium chain FA[5]. It was shown that these lipid mixtures might be useful surfactants when proper compositions were selected for the specific purposes. Surface activity, water solubility and acid/salt tolerance of aqueous solution of food surfactants, such as MG,

Food Hydrocolloids: Structures, Properties, and Functions
Edited by K. Nishinari and E. Doi, Plenum Press, New York, 1994

SE, PGE and SOE, were also enhanced by blending with SLP or soy LPC[6],[7]. The present investigation relates to O/W food emulsions stabilized by these food surfactants blended with LPs.

MATERIALS AND METHODS

Materials. Soy LPC, from Nihon Shyouji K.K., and egg LP from QP.Corp. were used. SLP were prepared by the modified method of Nakai[8]. Glycerine mono stearate, Emulsy MS (MGS), from Riken Vitamin K.K.; sucrose mono- distearate, Ryouto Ester S-1170 (SE11) and S-970 (SE9), from Mitsubishikasei Shokuhin K.K.; hexaglycerine sesquistearate, SY Grister MS-500 (PGE-10), from Sakamoto Yakuhin K.K.; Span 60 (SOE), from Kao K.K.. Analytical data and standard values of these materials are shown in Table 1.

Emulsion preparation and its evaluation. (a) Corn oil/soy sauce emulsions were formed with 50g of corn salad oil, 45g of soy sauce (NaCl=15%) and 5g of 10% (w/w) aqueous solution of mixed surfactant, the emulsions were prepared by using high speed mixer (AM-8, Nihonseiki K.K.), mixed at 13,000rpm at 60 °C for 5 min, under clean condition. Emulsion samples in 30 ml graduated glass tubes were stored in an incubator in which the temperature varied between 20°C and 40°C for 12hr. (b) Corn oil/grape fruit juice emulsions were formed by the same way as descrived in (a), using aseptic grape fruit juice (Kagome K.K., pH=3.3) instead of soy sauce. (c) Hardened vegetable oil/skim milk emulsion (simulated coffee whitener) were formed with 25g of hydrogenated canola oil (MP=35°C), 4.5g of skim milk powder in 60.5g of 0.2% aqueous solution of hexametaphosphate, and 10g of 7% aqueous solution of LPs or mixed surfactant (SLP/SE11, SLP/PGE10 and SLP/(MGS/SE11=1:1)). The pre-emulsions were prepared by using AM-8 Homogenizer (14,000rpm) at 60°C for 5 min, and further emulsified by ultrasonic homogenizer (US-300, Nihonseiki K.K.), with ϕ20 generator, output adjust 8 (300μA), at 50-60°C for 2 min. After incubating the emulsions at 4°C for 16hrs, their creaming layers were observed. Then, the sampls were warmed to room temperature, their mean droplet size and surface area were determined by particle size distribution analyzer (CAPA-700, Horiba-seisakusyo). The emulsion stability was measured by centrifugal method: the samples (ca. 25°C) were centrifuged by H-103 RLH centrifuge , Kokusan-Enshinki K.K., with rotor F at 10,000rpm (10400 x G) for 5 min, after standing at 45°C for 45 min, they were observed for their conditions.

Table 1. Analytical data of lysophospholipids and food surfactants used (%,Wt.)

Surfactant	Abbreviation (HLB value)	%lysolipid /monoester	14:0	16:0	18:0	18:1	18:2	18:3
Soy lysophosphatidylcholine	soy LPC	>99	–	24	7	11	51	5
Egg lysophospholipid	egg LP	>95	–	66	24	6	1	–
Soy lysophospholipid	SLP	85	–	29	6	10	47	6
Glycerin mono stearate	MGS	97	1	30	64	–	–	–
Sucrose mono- distearate	SE11(11)*	≒55	2	27	68	–	–	–
Sucrose mono- distearate	SE9 (9)*	≒50	2	28	67	–	–	–
Hexaglycerine sesquistearate	PGE10(10)*		2	27	69	–	–	–
Sorbitan monostearate	SOE (4.7)*		not determined					

* Standard value, not determined.

RESULTS AND DISCUSSION

Fig. 1 shows the stability of corn oil/soy sauce emulsions (a), and of corn oil/grape fruit juice emulsions (b), after 1 and 2 month. Creaming layer, thin emulsion water layer and separated oil phase are shown in each figures. Emulsions with soy LPC showed fine stability. With MGS or SOE only (Fig. 1-1), the emulsification was impossiblein both cases of (a) and (b), and it became possible and stabilized by blending 25-30% soy LPC. In the case (b), addition of SLP above 30% to MGS or SOE stabilized the emulsions (data not shown). With SE9 or PGE10 only (Fig. 1-2, 1-3), the emulsification was possible in both cases, although they were unstable. Adding 10% of soy LPC to these preparations, the emulsions were stabilized, and it is preferred that the addition of LPC was more than 30% in soy sauce emulsion. Seimiya[9] reported that the addition of phosphatidylcholine inhibited Na^+ and Ca^{++} binding with phosphatidylethanolamine due to the hydrophobic hydration of $-N(CH_3)_3$ radical of choline. LPC, its head group having strong hydrating tendency[10,11], shows high level of salt and acid tolerance[3,4]. The results obtained could be due to the reason that the nonionic surfactant molecules incorporated with LPC molecules form the mixed hydrated surface layer which prevent coagulation of droplets under acid and/or salt conditions. LPC was preferred for emulsions having salt, because of its betaine like zwitterionic property. SLP was preferred for acidic emulsions because of its rich content of anionic lipids. Fig. 2 shows the conditions of simulated coffee whitener emulsions (c) after incubation at 4°C for 16hr. The emulsion with 100% SLP or egg LP showed good stability, and with 100% PGE10 was half demulsified. Fig. 3 shows the mean droplet size(μm) and the surface area (m^2/gr fat) of the emulsions. Fig. 4 shows the centrifuged emulsions. The results from these data are as follows: (A) SLP and egg LP were also excellent emulsifiers for emulsions containing skim milk. (B) The addition of SLP to food nonionic surfactants improved the quality and stability of the emultions. (C) When the ratio of SLP/SE or SLP/PGE exceeded 4:6, the mean oil droplet size and its surface area were not so changed. However, the data of centrifugal demulsification showed that the emulsion stability was proportional to SLP content. Barratt[12] and Korver[13] reported complex formation between LPC and caseins or whey proteins. Korver suggested that the addition of 8 mol of LPC/mol β-lactoglobulin unfold the proteins, after that, dozens of LPC moleculs bound to the proteins, and compounds that more effectively promote the unfolding of protein should accelerate the rate of complex formation more. Jones[14] reported protein unfolding by anionic and nonionic surfactant, the former complex is formed by ionic binding and the latter by hydrophobic binding. Dickinson[15-17] described about competitive adsorption between proteins and surfactants in food emulsions, and showed that it is important to adjust protein/surfactant ratio for preparing stable food emulsions[16]. Recently Mine[18] reported the complex formations between LPC and ovalbumin, and the O/W interfacial adsorptibility of the protein increased with LPC addition, making finer emulsion droplet. It is well known, when an emulsion of similar composition to the (c) emulsions is prepared with skim milk only or surfactants without skim milk, the emulsion stability is very poor. In the present study, we showed that SLP and LPC are useful surfactants as they bind proteins on the surface of O/W food emulsions, improve their stability, also useful when they are added to nonionic food surfactants especially.

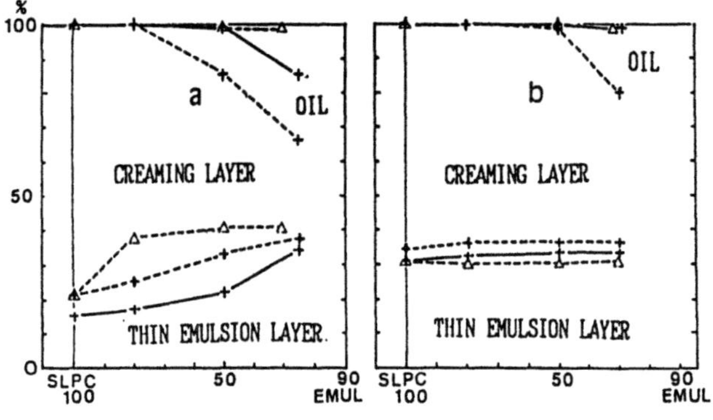

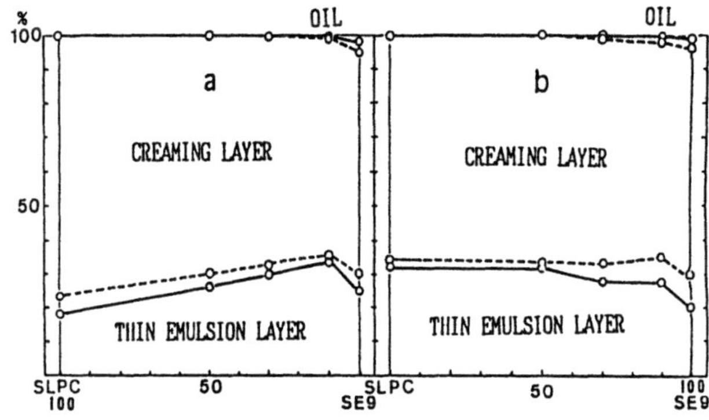

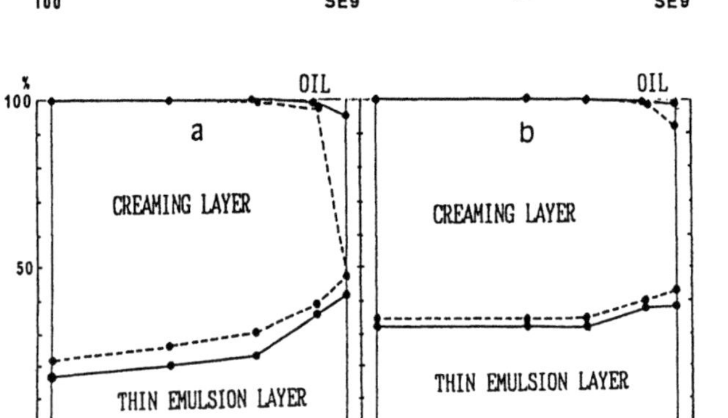

Fig. 1 Stability of the emulsions formed with soy LPC/food surfactant mixtures after 1 and 2 months. (a)Corn oil/ soy sauce emulsions. (b)Corn oil/grape fruit juice emulsions. Upper: with MGS or SOE Middle: with SE9 Lower: with PGE10.
———— : 1 month
-------- : 2 months
+: MGS, △: SOE. ○:SE9 ●: PGE10.

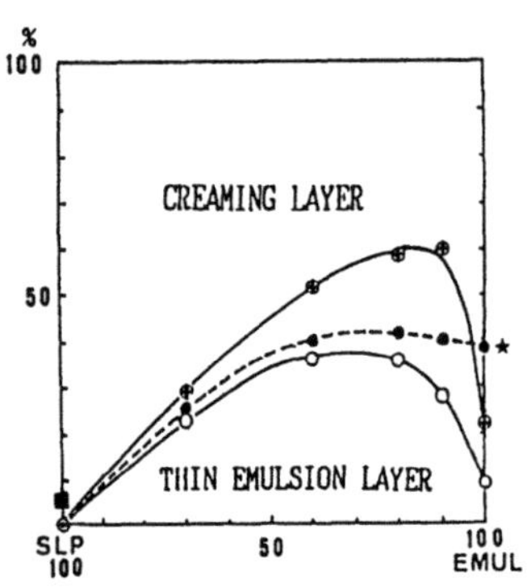

Fig. 2 Creaming of the simulated coffee whitener emulsions prepared with SLP/food surfactant mixtures. after incubating at 4℃ for 16 h.
○: SE11. ●: PGE10.
⊕: SE11/MGS. ■:Egg LP.

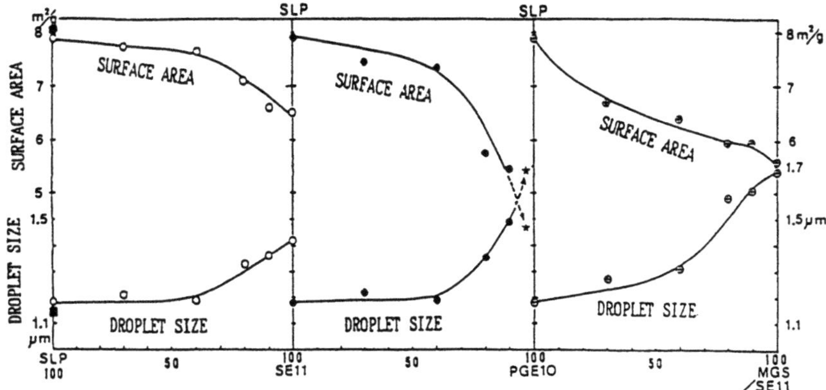

Fig. 3 Mean droplet size and surface area of the simulated coffee whitener emulsions prepared with SLP/food surfactant mixtures after incubating at 4 for 16h.   ○: SE11, ●: PGE10, ⊕: SE11/MGS, ■:Egg LP. ★: demulsified.

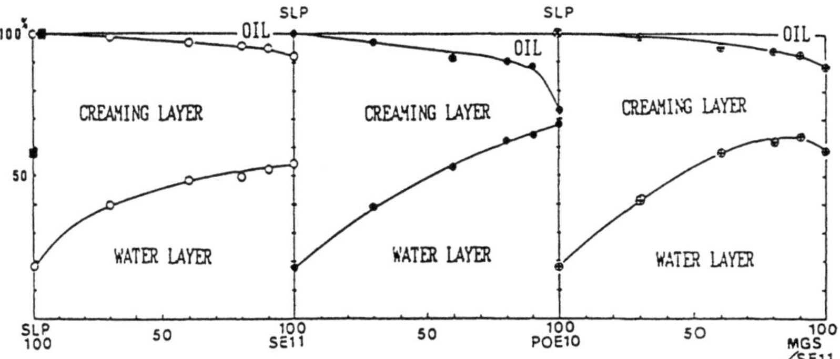

Fig. 4 Centrifugal demulsification of simulated coffee whitener emulsions with SLP/food surfactant mixtures. ○: SE11, ●: PGE10, ⊕: SE11/MGS, ■:Egg LP.

REFERENCES

1. R. Hell, H. Menz, and T. Wieske, Brit. Patent. 1, 215, 868(1970).
2. S. Fujita, H. Edo, US patent. 4, 794, 015(1988).
3. S. Fujita, and K. Suzuki, Nippon Nogeikagaku Kaishi. 64:1355(1990).
4. S. Fujita, and K. Suzuki, Nippon Nogeikagaku Kaishi. 64:1361(1990).
5. S. Fujita, and K. Suzuki, J. Am. Oil Chem. Soc.. 67:1008(1990).
6. S. Fujita, and K. Suzuki, J. Japan Oil Chem. Soc.. 40:1105(1991).
7. S. Fujita, and K. Suzuki, J. Jap. Soc. Food Sci. Tecnol.. 39:151(1992).
8. E. Nakai, K. Suzuki, S. Sato, and M. Kato, Japan Pat. laid Open. 89/16595(1989).
9. T. Seimiya, H. Miyasaka, T. Kato, T. Sirakawa, and M. Iwahasi, Chem. Phys. Lipids. 43: 161(1987).
10. K. Chiba, M. Tada, Nippon Nogeikagaku Kaishi. 62:859(1988).
11. K. Chiba, M. Tada, Agric. Biol. Chem.. 53:995(1989).
12. M. D. Barratt, and L. Rayner, Biochim. Biophys. Acta. 255:974(1972).
13. O. Korver, and H. Meder, J. Dairy Research. 41:9(1974).
14. M. N. Jones, and A. Brass, in:"Food Polymers, Gels, and Colloids, "E. Dickinson, ed., p. 65, Royal Soc. Chem., Cambridge(1991).
15. E. Dickinson, and C. M. Woskett, in:"Food Colloids, "R. D. Bee, ed., p. 74, Royal Soc. Chem., Cambridge(1991).
16. E. Dickinson, S. K. Narhan, and G. Stainsby, in:"Food Colloids, "R. D. Bee, ed., p. 372, Royal Soc. Chem., Cambridge(1991).
17. E. Dickinson, in:"Microemulsions and Emulsions in Food, " M. El-Nokaly Ed., p. 114, Am. Chem. Soc., Washington, DC(1991).
18. Y. Mine, H. Kobayashi, K. Chiba, and M. Tada, J. Agric. Food Chem. 40:1111(1992).

MIXED POLYSACCHARIDE GELS FORMED BETWEEN XANTHAN GUM AND GLUCOMANNAN

P.A. Williams, P. Annable, G.O. Phillips and K. Nishinari [1]

Polymer and Colloid Chemistry Group, The North East Wales Institute
Connah's Quay, Deeside, Clwyd CH5 4BR UK
[1] Dept. of Food and Nutrition, Osaka City University, 3-3-138
Sumiyoshi-Ku, Osaka, 558, Japan

INTRODUCTION

The ability of polysaccharides to enhance viscosity and to form gels has led to their widespread use in food products. Often synergistic combinations of polysaccharides are employed in order to manipulate rheological characteristics and, in addition, to provide effective savings in production costs. Common examples of polysaccharide blends used include kappa-carrageenan-galactomannan, xanthan-galactomannan and alginate-pectin. The origin of the synergy in these polysaccharide mixtures has been the subject of extensive research studies over the last twenty years or so and still gives rise to considerable controversy[1,2]. Cairns et al [3] have identified four different types of gel structure that might arise in binary mixtures of polysaccharides and these are illustrated in figure 1. In structure a) only one of the polysaccharides contributes to the gel network and the second polymer is simply entrapped within the matrix. In structure b) both polysaccharides form independent penetrating networks. Structure c) arises if polymer demixing occurs prior to gelation giving a phase separated network and structure d) is termed a coupled network and arises when sections of the two polysaccharides associate.

This paper is concerned with studies on gels formed by mixing xanthan gum with galactomannans and glucomannans. Although individually none of these polysaccharides form gels at neutral pH and room temperature, on mixing very strong gels can be produced.

Food Hydrocolloids: Structures, Properties, and Functions
Edited by K. Nishinari and E. Doi, Plenum Press, New York, 1994

STRUCTURE AND CONFORMATION OF XANTHAN GUM

Xanthan gum is a bacterial polysaccharide and has a β–D-(1,4)-linked celullosic main chain with trisaccharide side-chains, consisting of α-D-mannose, β-D-glucuronic acid and β-D-mannose on every other glucose residue. The mannose residues linking the side-chain to the backbone may be acetylated and the terminal mannose residues may be pyruvated to degrees dependent on the <u>Xanthomonas campestris</u> substrain. There is no fully accepted model for the conformation of the xanthan molecules in solution and considerable controversy surrounds whether they exist as single or dimeric strands. Studies using a number of techniques including optical rotation, circular dichroism, differential scanning calorimetry, light scattering and nmr [4-8] have indicated that the molecules undergo a conformational transition in solution as a consequence of changes in ionic strength, pH and temperature. The precise nature of the ordered and disordered structures is still not completely resolved. Thus it has been argued that in the single stranded model, the ordered structure consists of a single helix which transforms to a coil on heating and that the temperature of this transition is a function of ionic strength or pH whilst in the double stranded model the ordered form is a double helix which converts to a dimerised expanded coil partially retaining double helical sequences. It has also been suggested that the ordered structure may be stabilised by association of the side-chains with the backbone and recent ^{13}C nmr data has indicated that considerable side-chain mobility is established prior to the increase in backbone mobility as the molecules undergo the order to disorder transition [8].

INTERACTION OF XANTHAN GUM WITH GALACTO- AND GLUCOMANNANS

Rocks first reported that mixing solutions of xanthan and locust bean gum resulted in the formation of thermoreversible gels [9]. Gelation was not evident in xanthan-guar gum mixtures and he attributed this to the increased degree of galactose substitution along the mannan chain for guar gum which prevented binding to xanthan molecules. Rees and co-workers[10-12] undertook an extensive study of xanthan-galactomannan interactions using optical rotation and rheological techniques and confirmed that interaction decreased as the proportion of galactose increased in keeping with Rocks' observations. Their results led them to propose that gelation involved binding of smooth regions on the galactomannan backbone with the ordered xanthan helix as shown in figure 2a. McCleary modified this model slightly since he found that the galactomannan did not necessarily require long unsubstituted sections along its backbone but rather sections where the galactose residues are substituted only on one side of the chain (figure 2b)[13].

Tako and Nakamura [14-16] studied the rheological behaviour of mixed systems and since deacetylation of the xanthan gave rise to stronger gels they concluded that interaction occurred through insertion of the xanthan side-chains into unsubstituted segments along the

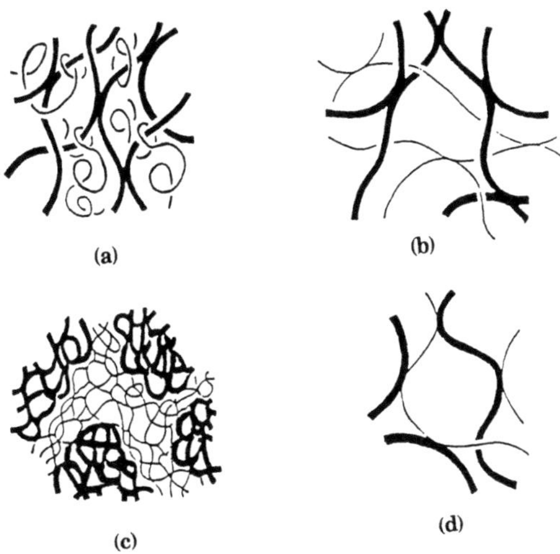

Figure 1. Various types of network structures for binary polysaccharide gels. a) a single polymer network with the second polymer entrapped b) interpenetrating networks c) phase separated networks and d) coupled networks. From Cairns et al [3] with permission.

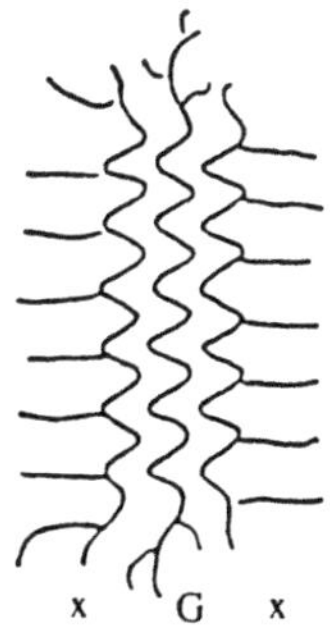

Figure 2. Schematic representations of the molecular origin of xanthan interaction with galactomannan a) according to Dea et al [11] b) according to McCleary [13].

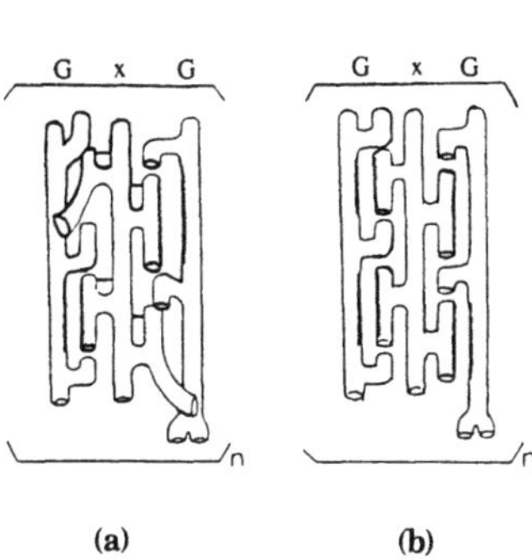

Figure 3. Schematic representation of the possible mode of interaction between xanthan gum and locust bean gum. a) interaction is incomplete due to association of the side chains with the xanthan backbone b) increased interaction after deacetylation. From Tako and Nakamura [15].

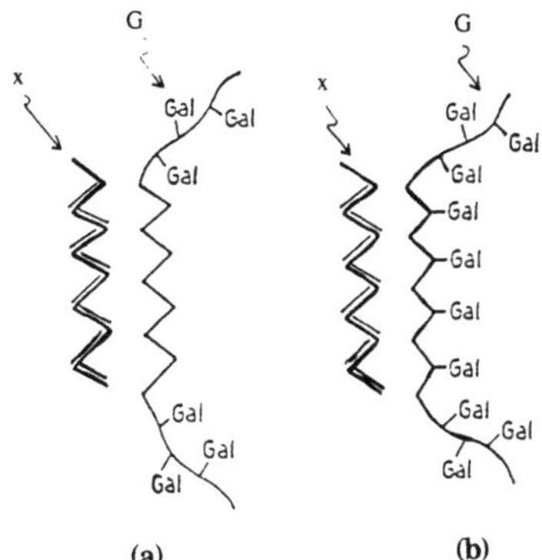

Figure 4. Schematic representation showing the xanthan-galactomannan sandwich structure. From Cairns et al [3].

galactomannan backbone. A schematic illustration of their model is given in figure 3.

Direct evidence for an interaction between xanthan and galacto- or glucomannan was presented by Cairns et al [17-19] using X-ray diffraction. However, these workers found that gelation only occurred when mixtures were heated above the order->disorder xanthan transition and argued, therefore, that interaction occurred between disordered xanthan chains and the galactomannan in contrast to the model of Rees and co-workers. They suggested that since optical rotation studies by Dea et al [11] showed substantial recovery of optical rotation on gelation and that if the binding involved the disordered xanthan chains, only a small proportion of the xanthan molecule could be involved with the rest adopting an ordered conformation. Since their X-ray diffraction data indicated an axial advance per chemical repeat distance of 0.52nm they suggested a possible binding scheme such as that illustrated in figure 4 where a galactomannan chain is "sandwiched" between two xanthan chains.

Cheetham and co-workers [20-24] carried out extensive studies on xanthan-galactomannan systems using gel permeation chromatography and optical rotation and undertook a series of simple mixing experiments. They concluded that it was necessary for the xanthan molecules to be at least partially disordered and that interaction occurs between the disordered sequences and the galactomannan chains as depicted in the "sandwich" model. They pointed out that the xanthan side-chains would extend vertically from the backbone thus enabling hydrogen bonding to occur between the xanthan backbone and the galactomannan. They also pointed out that substantial intermolecular hydrophobic association may occur through overlapping of C-H groups. Their model is depicted in figure 5.

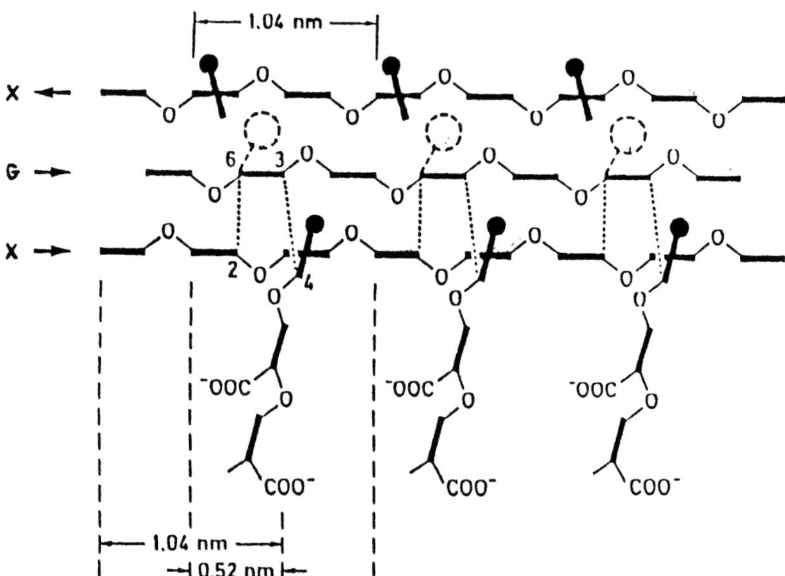

Figure 5. Schematic representation for the xanthan (X) and galactomannan (G) mixed junction zone. = galactose side group. From Cheetham and Mashimba [23] with permission.

Recently Lopes et al [25] studied the rheological properties of native xanthan-guar gum mixtures. They concluded that intermolecular association could occur when the xanthan molecules were present in both the ordered and disordered form but that interaction was stronger in the latter as indicated by the higher critical aggregation temperature.

It is clear, therefore, that although there is a general concensus that gelation is a consequence of molecular association of xanthan gum with galacto- and glucomannans i.e a coupled network, the nature of the interaction is not fully understood. Our studies over recent years have been directed towards addressing this problem. We have used a number of techniques including rheology, differential scanning calorimetry (dsc) and electron spin resonance spectroscopy (esr) and we now report a summary of our findings for the xanthan-glucomannan system.

Rheological Investigations

Small deformation oscillation measurements [26] were carried out on xanthan-glucomannan mixtures as a function of temperature. Samples were cooled (approx. $1^{\circ}C/min$) to the desired temperature and equilibrated for 5 minutes prior to 3 consecutive measurements. Results showed that gelation occurred at about $63^{\circ}C$ on cooling in the absence of electrolyte irrespective of the polysaccharide mixing ratio in agreement with reports by Dea et al [11]. The gelation temperature was reduced in the presence of electrolytes and was found to be sensitive to the nature of the cations. Thus in the presence of NaCl and $CaCl_2$ the onset of gelation was found to be $50^{\circ}C$ and $40^{\circ}C$ respectively. This is illustrated in figure 6. Gel strength measurements showed that the gels formed were stronger in the absence of electrolyte and the optimum xanthan-glucomannan mixing ratio was ~1:1 compared to ~2:1 in the presence of 0.04 mol dm^{-3} NaCl [27] (figure 7). The results, therefore, indicated that the interaction may be different in the two solvent systems.

Differential Scanning Calorimetry

Studies were undertaken using dsc, with a scan rate of $0.2^{\circ}C/min$, and the heating and cooling curves obtained for the xanthan gum and xanthan-glucomannan mixtures are given in figure 8[26]. The peak observed for the xanthan alone with a midpoint transition temperature, T_m, at $51^{\circ}C$ was attributed to the xanathan conformational transition and there was no hysteresis observed on cooling and heating. For the mixtures an exothermic peak occurred on cooling at T_m~$57^{\circ}C$ (i.e. several degrees above the temperature for the xanthan conformational change) with the start of the transition occurring at 62-$66^{\circ}C$, consistent with the gelation temperature observed rheologically. The endothermic peak, which was evident on heating, had a T_m about $2^{\circ}C$ higher than the corresponding peak in the cooling curve indicating that some molecular aggregation had occurred. The enthalpy associated with the process was dependent on the xanthan-glucomannan mixing ratio and this is shown in figure 9. The maximum enthalpy was obtained for ~1:1 mixtures. As

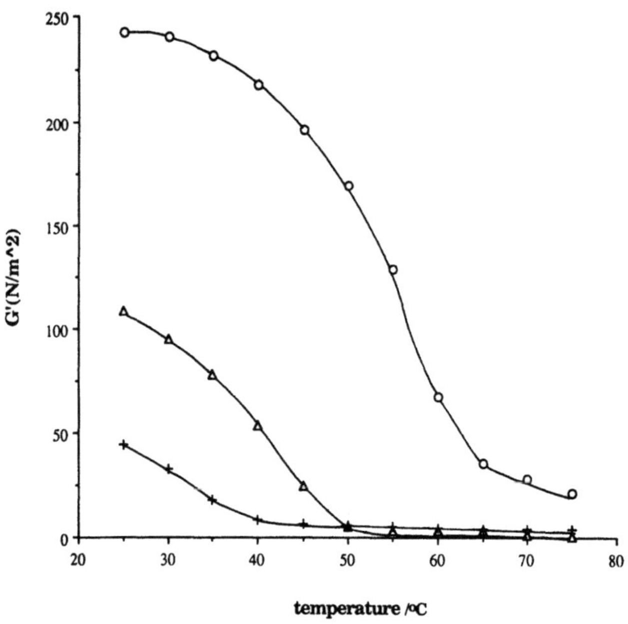

Figure 6. Storage modulus on cooling (approx. 1°C/min) 1:1 mixtures of xanthan and glucomannan (total polysaccharide concentration 1.2%); O, in the absence of electrolyte; Δ, in the presence of 0.04 mol dm^{-3} NaCl; +, in the presence of 0.02 mol dm^{-3} CaCl$_2$.

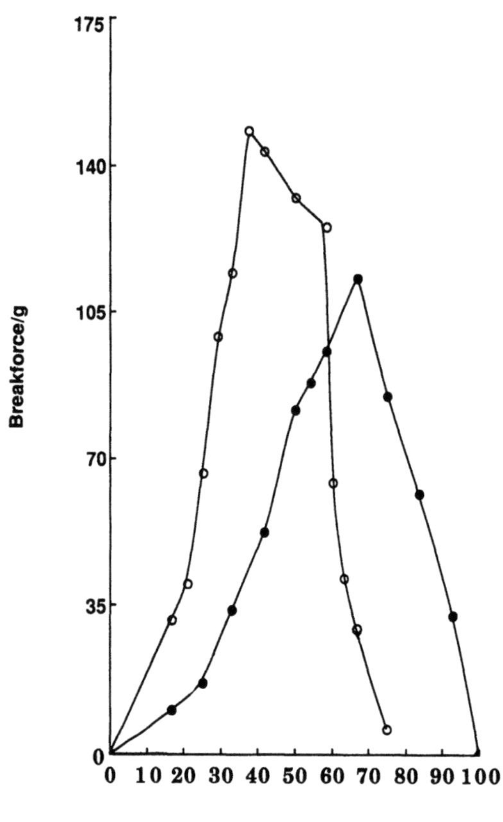

Figure 7. Breakforce as a function of xanthan/glucomannan ratio; O, in the absence of electrolyte; ●, in the presence of 0.04 mol dm^{-3} NaCl.

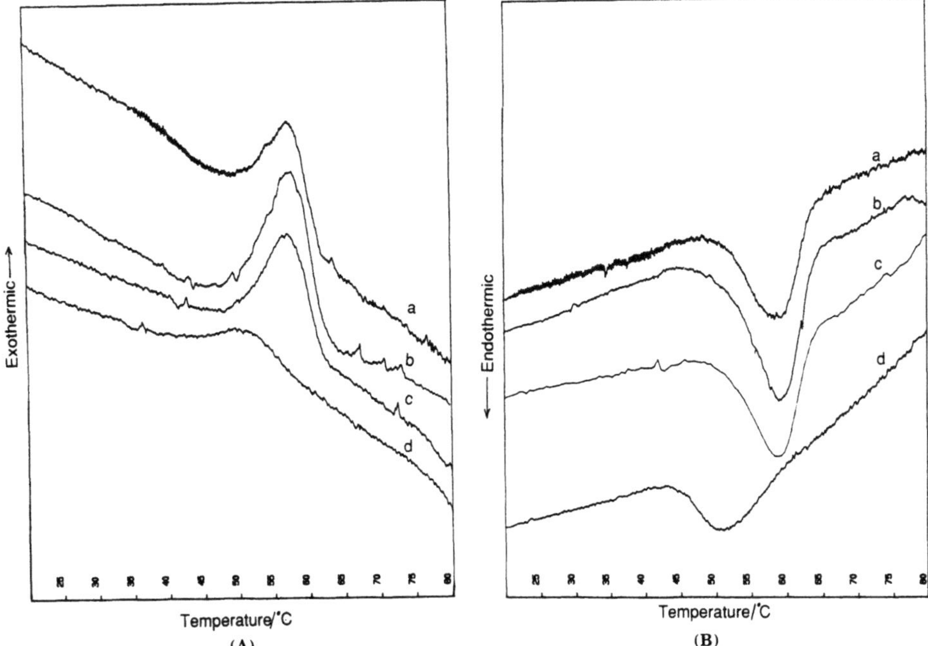

Figure 8. (A) Dsc cooling curves (scan rate = 0.2°C/min) for a) 1:3, b) 1:1, c) 3:1 and d)1:0 xanthan-glucomannan mixtures at 1.2% total polymer concentration; (B) Dsc heating curves for mixtures as above. From Williams et al [26] with permission.

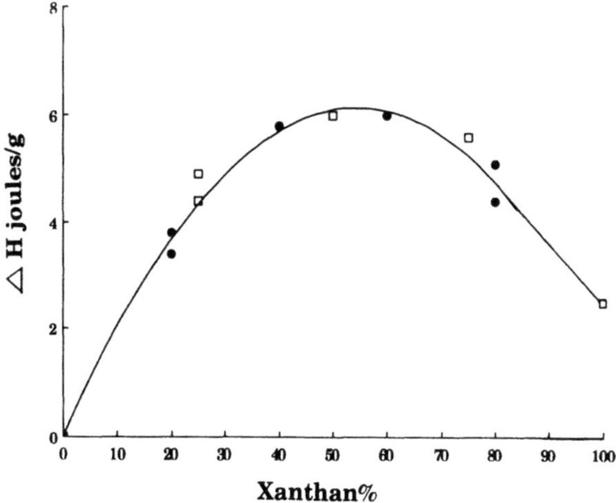

Figure 9. Enthalpy of gelation of xanthan-glucomannan mixtures at various ratios; □, 1.2% and ●, 0.5% total polymer. From Williams et al [26] with permission.

reported above this ratio was also found to produce the strongest gels. It could be argued that since gelation occurs above the <u>expected</u> temperature for the xanthan conformational change, the glucomannan interacts with disordered xanthan molecules in accordance with the findings of Cairns et al., and Cheetham and co-workers. However it might also be argued that the xanthan ordering process shifts to slightly higher temperatures in the presence of glucomannan and that interaction only occurs following formation of the ordered structure. A shift in the temperature of the conformational change to higher temperatures in the presence of galactomannan was reported by Dea et al.,[11] from optical rotation studies.

For dsc studies in the presence 0.04 mol dm^{-3} NaCl, xanthan alone showed an exothermic peak at T_m~82.5°C (figure 10) and this was attributed to the conformational change which is known to shift to higher temperatures in the presence of electrolyte. For a 1:1 mixture the cooling curve showed two peaks, one at T_m~82.5°C corresponding to the xanthan conformational transition and a second at T_m~42°C which corresponds to the gelation temperature. It is evident therefore, that in this system gelation occurs as a consequence of glucomannan interaction with ordered xanthan molecules, contrary to the findings of Cairns et al. It is possible however, that binding occurs between disordered sequences within the ordered structure as reported by Cheetham and co-workers. As noted earlier, Cairns et al. reported that gelation only occurred if solutions were heated to above the temperature for the xanthan conformational transition. We have found this not to be the case and this is illustrated in figure 11 which shows the storage modulus (G') - v - frequency sweeps for xanthan-glucomannan mixtures in the presence of 0.04 mol dm^{-3} NaCl after subjecting to different heat treatments. The solutions were prepared by mixing the xanthan and glucomannan solutions together at 25°C, heating to various temperatures below the xanthan order->disorder transition temperature (82.5°C) and equilibrating at each temperature for 5 minutes before cooling to 25°C (approx. 1°aC/min). Results for G' were accumulated after equilibration at 25°C for 5 minutes. G' is seen to increase as the maximum heating temperature increases and becomes less frequency dependent indicating stronger gels are formed. It is possible that the heating process is simply disrupting xanthan aggregates thereby enabling greater interaction with the glucomannan chains. It is evident, therefore, that the differences in gel strengths and gelation temperatures for the mixed systems in the presence and absence of electrolyte can be attributed to the fact that the glucomannan interacts with xanthan molecules that are different conformations.

Electron Spin Resonance Spectroscopy

The molecular motion of the xanthan and glucomannan molecules was monitored using the nitroxide spin labelling technique which involves covalently attaching a nitroxide moiety randomly along the polymer chains. Incorporation was of the order of about one label per several thousand polymer segments so that the intrinsic properties of the polymer were not affected to any significant extent. The technique relies upon the fact that the

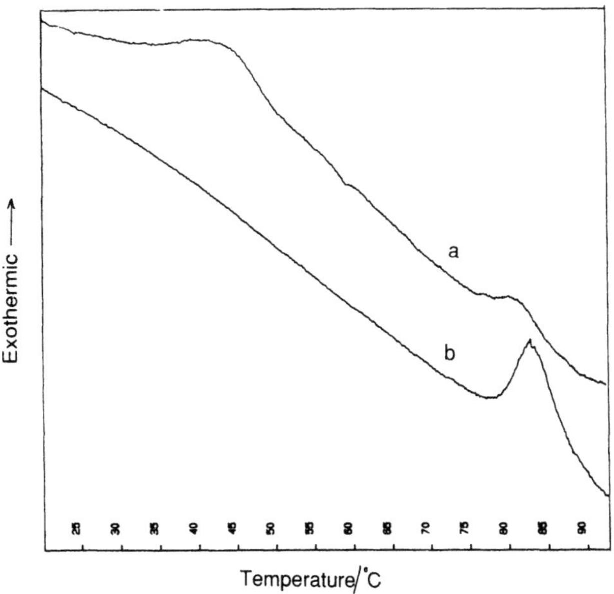

Figure 10. Dsc cooling curve for a) 1:1 xanthan-glucomannan solution b) xanthan alone at a total polysaccharide concentration of 1.2% in 0.04 mol dm^{-3} NaCl. Scan rate = 0.2°C/min. From Williams et al [26] with permission.

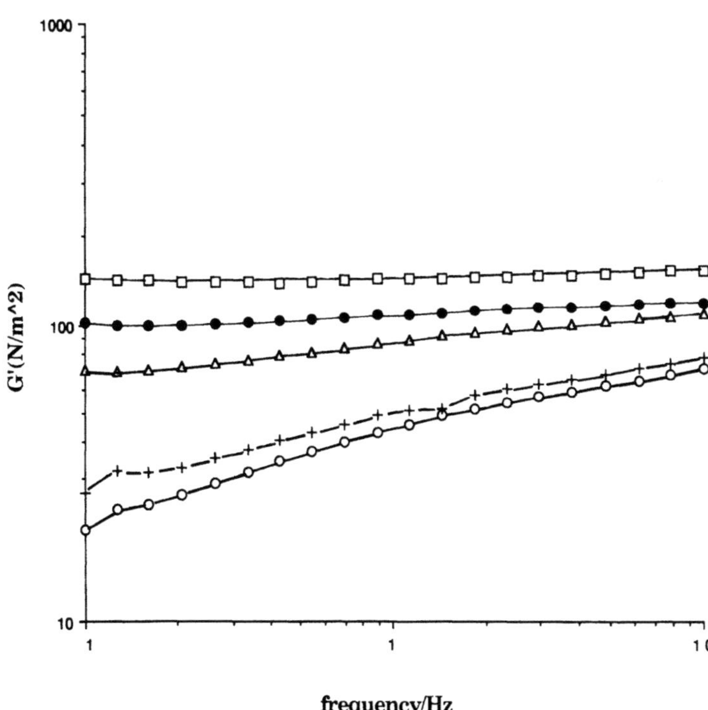

Figure 11. Storage modulus as a function of frequency for xanthan in 0.04 mol dm^{-3} NaCl in the presence of glucomannan measured at 25°C after heating to the following values; O, 25°C; +, 35°C; Δ, 45°C; ●, 55°C; □, 65°C. (Cooling rate, approx. 1°C/min).

shape of the nitroxide esr spectrum changes with the nitroxide mobility. If there is a high degree of mobility then the spectrum is narrowed due to rapid isotropic tumbling whilst if the polymer chains have restricted movement then the spectrum broadens due to anistropic effects. This is illustrated in figure 12.

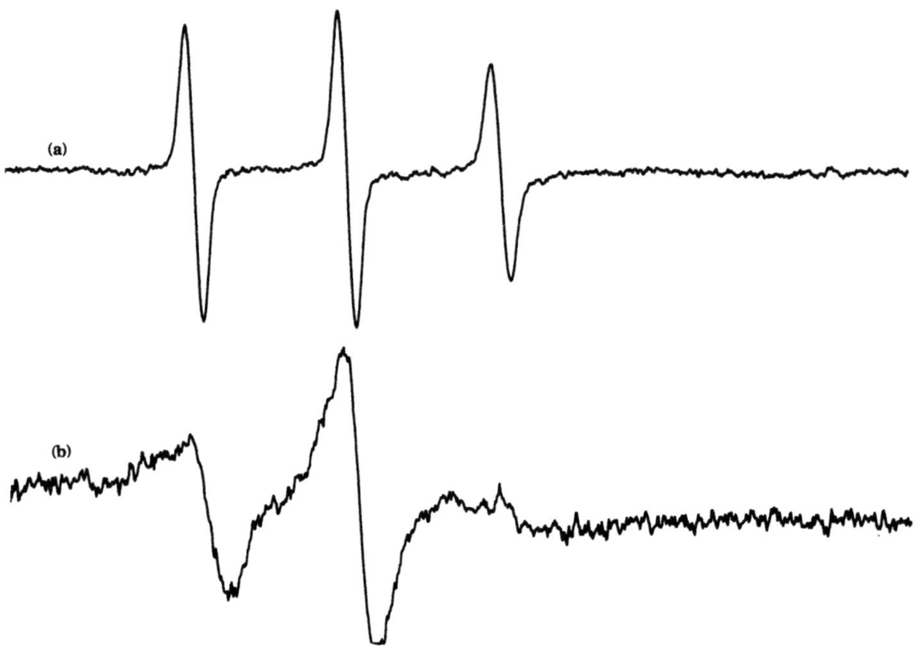

Figure 12. Esr spectra for a nitroxide spin-labelled glucomannan (0.22%) a) mobile, isotropic b) immobile, anisotropic.

The technique is particularly attractive for studying mixed polysaccharide systems since it is possible to monitor the behaviour of each polymer independently. In our studies the glucomannan was spin labelled through the hydroxyl groups of the sugar residues while the xanthan was spin labelled through the carboxyl groups on the side chain. The esr spectra obtained for spin labelled glucomannan alone and in the presence of xanthan gum are shown at various temperatures in figure 13. For glucomannan alone isotropic spectra are obtained at all temperatures while for the mixtures the spectra are essentially isotropic at high temperatures, indicative of a high degree of mobility, but on cooling composite spectra are observed consisting of both isotropic and anisotropic components. The proportions of each were resolved by computer analysis and the % anisotropic component, p, is plotted as a function of temperature in figure 14. p is seen to become evident below ~65°C which corresponds closely to the gelation temperature observed rheologically. The anisotropic component represents ~65% of the signal at low temperature indicating that this proportion of glucomannan chain segments have aggregated.

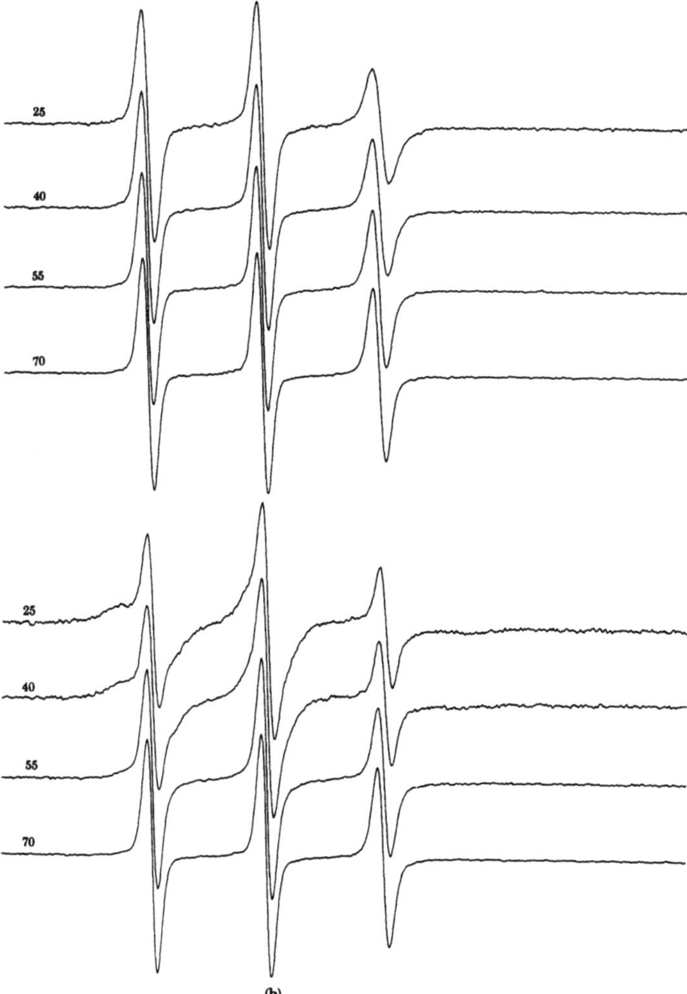

Figure 13. Esr spectra for spin-labelled glucomannan (0.22%) at various temperatures a) in solution alone b) in solution in the presence of xanthan. (Cooling rate, approx. 1°C/min).

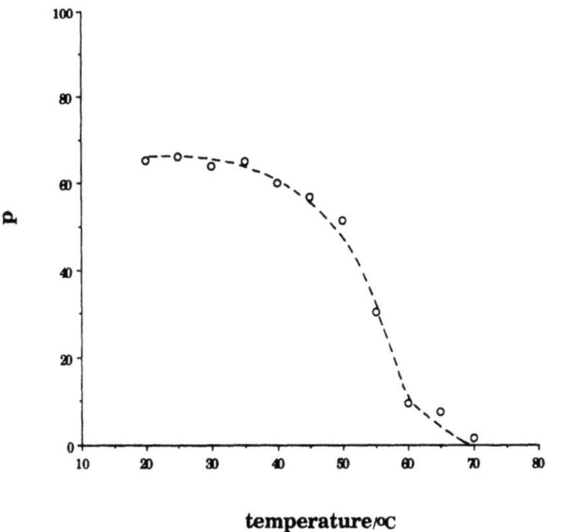

Figure 14. % anisotropic component, *p*, on cooling spin-labelled glucomannan in the presence of xanthan gum (1:1.5, 0.5% total polysaccharide).

The esr spectra obtained for spin labelled xanthan alone in solution are given in figure 15 at various temperatures. At high temperatures the spectra are essentially isotropic but on cooling composite spectra were obtained consisting of both isotropic and anisotropic components with p increasing to ~50% as the temperature decreased below ~65°C (figure 16a). This is close to the temperature for the onset of the conformational change as observed by dsc (figure 8). Since the spin label is attached to the xanthan side-chains the increase in p is an indication of association between the side-chains and the backbone. In the presence of 0.04 mol dm^{-3} NaCl, the anisotropic component becomes evident at higher temperatures (figure 16a) and corresponds to the conformational transition temperature monitored by dsc.

When glucomannan was present the p-v-temperature curves in water (figure 16b) showed that p began to appear at a slightly higher temperature than for xanthan alone indicating that the xanthan conformational change shifted to a slightly higher temperature in the presence of glucomannan. The observations are consistent with the findings of Dea et al[11] for xanthan-galactomannan systems using optical rotation. Since the esr studies using spin-labelled glucomannan indicate that glucomannan chains associate below ~65°C corresponding to the gelation temperature it is evident that xanthan-glucomannan interaction occurs following the ordering of the xanthan chains. Indeed it is probable that the driving force for the process is the increased chain rigidity and accompanying changes in xanthan-water structure.

For xanthan-glucomannan mixtures in the presence of 0.04 mol dm^{-3} NaCl, the esr data shows that the xanthan side chains start to lose mobility at ~82°C (figure 16b) in agreement with the findings from dsc confirming that interaction occurs following xanthan chain ordering. It would appear, therefore, that glucomannan interacts following chain ordering in both water and 0.04 mol dm^{-3} NaCl and casts some doubt that interaction involves disordered xanthan chain segments. Differences in the gelation temperatures and gel strengths may be a consequence of variations in the nature of the xanthan ordered structures in the different solvent systems. The ordered structure in water may adopt a more extended conformation thereby enabling it to form a more extensive three dimensional network thus giving rise to increased gel strength.

References

1. P.A.Williams and G.O.Phillips "Interactions in mixed polysaccharide systems" in "Food Polysaccharides" A.M.Stephen and I.C.M.Dea eds Marcel Dekker (in press).
2. E.R.Morris "Mixed Polymer Gels" in "Food Gels" P.Harris ed Elsevier Applied Science Series (1990) (p291).
3. P.Cairns, M.J.Miles, V.J.Morris and G.J.Brownsey "X-ray fibre diffraction studies on synergistic, binary polysaccharide gels" Carbohydr. Res. 160:411 (1987).
4. E.R.Morris, D.A.Rees, G.Young, M.D.Walkinshaw and A.Darke "Order-disorder transition for a bacterial polysaccharide in solution" J. Mol. Biol. 110:1 (1977).

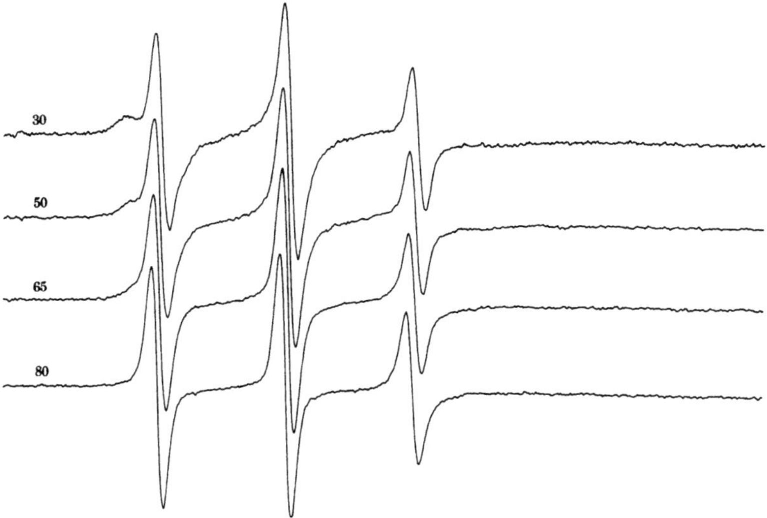

Figure 15. Esr spectra for spin-labelled xanthan gum in solution at various temperatures. (Cooling rate, approx. 1°C/min).

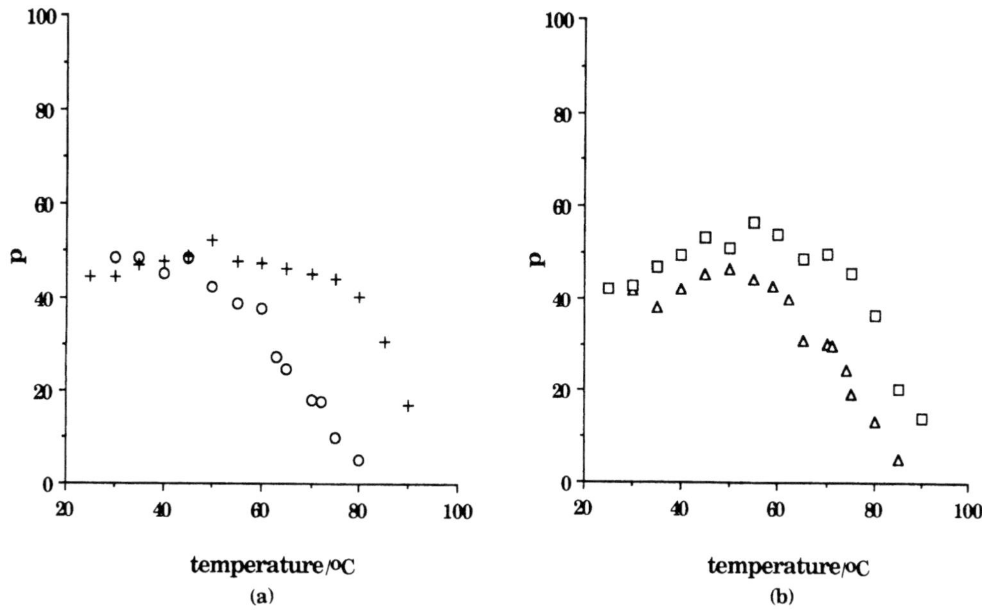

Figure 16. % anisotropic component, p, on cooling spin-labelled xanthan gum a) alone in aqueous solution $\bigcirc$, and in 0.04 mol dm^{-3} NaCl $+$; b) in the presence of glucomannan in aqueous solution $\triangle$, and in 0.04 mol dm^{-3} NaCl $\square$.

5. M.Milas and M.Rinaudo "Properties of xanthan gum in aqueous solutions: Role of the conformational transition" Carbohydr. Res. 158:191 (1986).

6. I.T.Norton, D.M.Goodall, S.A.Frangou, E.R.Morris and D.A.Rees "Mechanism and dynamics of conformational ordering in xanthan polysaccharide" J. Mol. Biol. 175:371 (1984).

7. V.Crescenzi, M.Dentini and L.Pietrelli " Protonation equilibriums of the polysaccharide xanthan in aqueous solution" Period. Biol. 83:125 (1981).

8. A.Gamini, J. de Bleijser and J.C.Leyte "Physico-chemical properties of aqueous solutions of xanthan : An nmr study" Carbohydr. Res. 220:33 (1991).

9. J.K.Rocks "Xanthan gum" Food Technology 25:22 (1971).

10. D.A.Rees "Shapely Polysaccharides" Biochem. J. 126:257 (1972).

11. I.C.M.Dea, E.R.Morris, D.A.Rees, E.J.Welsh, H.A.Barnes and J.Price "Associations of like and unlike polysaccharides: mechanism and specificity in galactomannans, interacting bacterial polysaccharides and related systems" Carbohydr. Res. 57:249 (1977).

12. E.R.Morris, D.A.Rees, G.Robinson and G.Young "Competitive inhibition of interchain interactions in polysaccharide systems" J. Mol. Biol. 138:363 (1980).

13. B.V.McCleary "Enzyme hydrolysis, fine structure and gelling interaction of legume-seed D-galacto-D-mannan" Carbohydr. Res. 71:205 (1979).

14. M.Tako and S.Nakamura " D-Mannose specific interaction between xanthan and D-galacto-D-mannan" Feb. Letts. 204:33 (1986).

15. M.Tako and S. Nakamura "Synergistic interaction between xanthan and guar gum" Carbohydr. Res. 138:207 (1985).

16. M.Tako "Synergistic interaction between deacetylated xanthan and galactomannan" J. Carbohydrate Chem. 10:619 (1991).

17. P.Cairns, M.J.Miles and V.J.Morris "Intermolecular binding of xanthan gum and carob gum" Nature 322:89 (1986).

18. V.J.Morris "Designing polysaccharides for synergistic interaction" in "Gums and Stabilisers for the Food Industry 6" G.O.Phillips, P.A.Williams and D.J.Wedlock eds Oxford University Press Publishers (1992) p161.

19. G.J.Brownsey, P.Cairns, M.J.Miles and V.J.Morris "Evidence for intermolecular binding between xanthan and the glucomannan konjac mannan" Carbohydr. Res. 176:329 (1988).

20. N.W.H.Cheetham, B.V.McCleary, G.Teng, F.Lum and Maryanto "Gel permeation studies on xanthan-galactomannan interactions" Carbohydr. Polymers 6:257 (1986).

21. N.W.H.Cheetham "Xanthan-galactomannan interactions-A gpc study" in "Industrial Polysaccharides" S.S.Stivala, V.Crescenzi and I.C.M.Dea eds Gordon and Breach Publishers (1987) p325.

22. N.W.H.Cheetham and A.Punruckvong "Gel permeation and optical rotation studies on xanthan-galactomannan interactions" Carbohydr. Polymers 10:129 (1989).

23. N.W.H.Cheetham and E.N.M.Mashimba "Conformational aspects of xanthan-galactomannan gelation. Further evidence from optical rotation studies. "Carbohydr. Polymers 14:17 (1991).

24. N.W.H.Cheetham and E.N.M.Mashimba "Conformational aspects of xanthan-galactomannan gelation" Carbohydr. Polymers 9:195 (1988).

25. L.Lopes, C.T.Andrade, M.Milas and M.Rinaudo "Role of conformaton and acetylation of xanthan on xanthan-guar interactions" Carbohydr. Polymers 17:121 (1992).

26. P.A.Williams, S.M.Clegg, D.H.Day, G.O.Phillips and K.Nishinari "Mixed gels formed with konjac mannan and xanthan gum" in "Food Polymers, Gels and Colloids" E.Dickinson ed Royal Society of Chemistry Special Publ. 82 (1991) p339.

27. P.A.Williams, D.H.Day, M.J.Langdon, G.O.Phillips and K.Nishinari "Synergistic interaction of xanthan gum with glucomannans and galactomannans" Food Hydrocolloids 4:489 (1991).

LARGE DEFORMATION RHEOLOGY OF MIXED
GELS OF KONJAC – KAPPA-CARRAGEENAN

Hiroki Iida[1], Takashi Ochi[1], Shiro Ohashi[1], Kaoru Kohyama[2],
Katsuyoshi Nishinari[2]*, Peter A. Williams[3] and Glyn O. Phillips[3]

[1]San-Ei Gen F.F.I., Inc. 1-11, Sanwa-cho 1-chome, Toyonaka
City, Osaka 561, Japan, [2]National Food Research Institute,
Tsukuba, Ibaraki 305, Japan and [3]North East Wales Institute,
Deeside, Clwyd CH5 4BR, UK
*Present address, Dept. Food and Nutrition, Faculty of Science of Living
Osaka City University, Sumiyoshi, Osaka 558, Japan

ABSTRACT

Konjac flour (KF), a Japanese traditional food material, itself can not make a gel unless deacetylated by alkali. KF can modify the texture of kappa-carrageenan, which is gel forming carrageenan (CGN) with monovalent cations and so on. KF and kappa-carrageenan mixed gels are firmer and highly elastic in comparison to locust bean gum and kappa-carrageenan mixed gels. With the temperature controlled tensile test apparatus (from 5°C to 40°C), ring shaped mixed gels of konjac flour and kappa carrageenan were investigated. From the true breaking stress of gel rings, mixed gels showed a maximum peak at the mixing ratio of KF/CGN from 4 to 6. Temperature dependence of breaking strain and breaking stress was discussed.

INTRODUCTION

In recent Japanese food industries, the use of hydrocolloids as food additives, generally as rheological agents, is increased and indispensable. It is well known that kappa-carrageenan, an extract from certain marine red algae, itself makes rigid and brittle gels with certain monovalent cations. However, the combined use of kappa-carrageenan and galactomonnans, such as locust bean gum provides very elastic gelatin-like gels.[1] In fact, many Japanese food manufacturers utilize these synergistic gels for water dessert jelly because of their high viscoelasticity, higher setting temperature and so on.[2] Meanwhile, konjac flour which has partial mannan main chains[3] like galactomannans is popular as the so-called traditional konjac foods in Japan since 1600's.[4] In a similar manner as galactomannans, the combined use of konjac flour and kappa-carrageenan or xanthan gum were found to form thermoreversible and very elastic gels,[5] hence recently it has attracted special interest as a texture modifier. Breaking strength of mixed gels of KF and kappa-carrageenan is about twice than that of mixed gels of locust bean gum and kappa-carrageenan at a certain mixing ratio.[6]

Large deformation testing is commonly used for the evaluation or comparison of gel strength of commercially available gelling agents in food industries. Compressive testing has been used more frequently than tensile testing to determine gel strength. However, for the highly elastic and strong gels like KF and kappa-carrageenan mixed gels, it is hard to determine the exact rupture point by compressive tests. In compressive tests, the stress in the gels is the sum of tensile and shear stress, but shear stress is negligible in tensile tests.[7]

In this research, four fractions of konjac flour and kappa-carrageenan ring shaped mixed gels were investigated at various temperatures using tensile measurements at large deformation.

MATERIALS AND METHODS

Materials

KF and CGN used in the present work were obtained from FMC Corporation Food Ingredients Division (Philadelphia, USA). Ethanol, potassium chloride and hydrochloric acid of pure reagent form (Kishida Chemical Co., Ltd., Osaka) were used without further purification. CGN was purified by a method shown in Fig.1. 50g of CGN was suspended in 5 ℓ of acid ethanol and washed with ethanol until the supernatant was neutralized. The operation was conducted below 10°C. Cation content of CGN and KF is shown in Table 1 and the viscosity of KF at 20°C is shown in Fig.2. Cation content of CGN shown was analysed with Inductively Coupled Argon Plasma Emission Spectrophotometer, ICAP Mark II (Nippon Jarrell Ash, Kyoto) and the viscosity of KF was measured with B type viscometer, Model BL (Tokimec Inc., Tokyo). KF (mannose/glucose ratio: M/G=5/3) used in the present work showed relatively low viscosity in comparison with usual konjac flour (M/G=6/3).

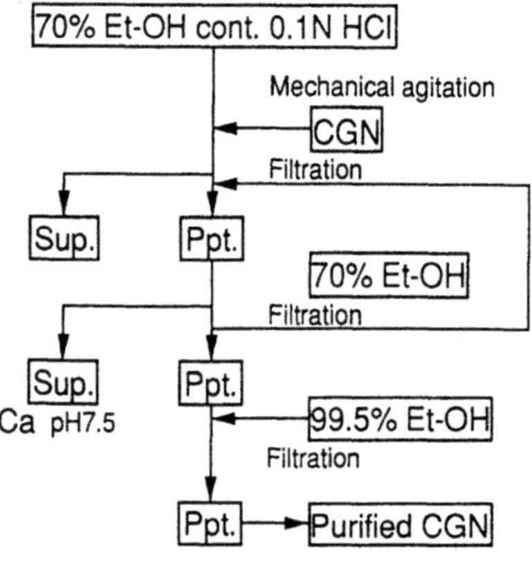

Fig. 1. Purification of CGN

	CGN	KF
Na	21	0.7
K	540	38.3
Ca	160	12.9
Mg	5	4.5
Fe	0.4	0.3

$(\times 10^{-2}\%)$

Table 1. Cation content of purified CGN and KF

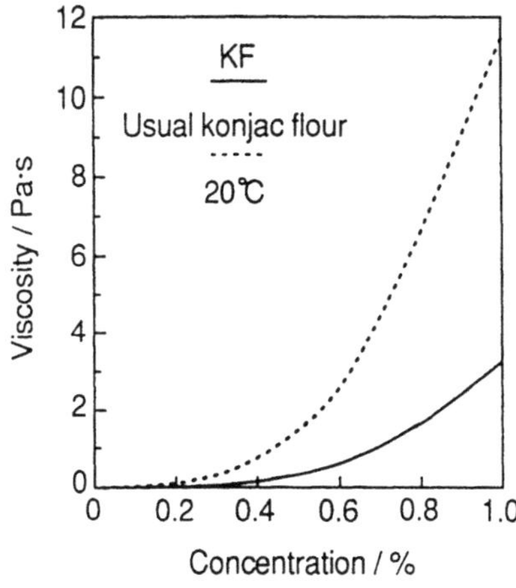

Fig. 2.   Viscosity of KF and usual konjac flour at 30rpm.

Methods

Preparation of Gel Rings — CGN and KF were dissolved together in distilled water at various ratios with KCl (0.08M). Total polysaccharide content was 1.5 % and solutions were heated at 85℃ for 30 minutes with mechanical agitation. After correction of water content lost by evaporation, the solutions were poured into the mold (220 x 220 x 10 mm) which had been heated at 85℃ beforehand in order to make homogeneous gels. The mold was covered with a glass plate and cooled at 25℃ for 7 hours, then cooled at 4℃ for 15 hours and mixed gel slabs were prepared. The gel slabs were cut into ring-shaped gels with double cylindrical cutter. Outer diameter of the gel ring was 30 mm and inner diameter was 20 mm.

Tensile tests — Temperature controlled tensile testing apparatus is illustrated in Fig.3. For the tensile test, the gel ring was hung on two round bars of diameter 7 mm in silicon oil bath with Instron Universal Testing Machine model 4502 (Buckinghamshire, UK). The extention of a gel ring was carried out at 300 mm/min cross-head speed. The temperature of silicon oil bath was adjusted with a cooling and heating unit, which were controlled with a thermostat, and mechanical agitator. Gel rings were soaked in a silicon oil bath which was adjusted to the required temperature for 30 minutes before the tensile test.

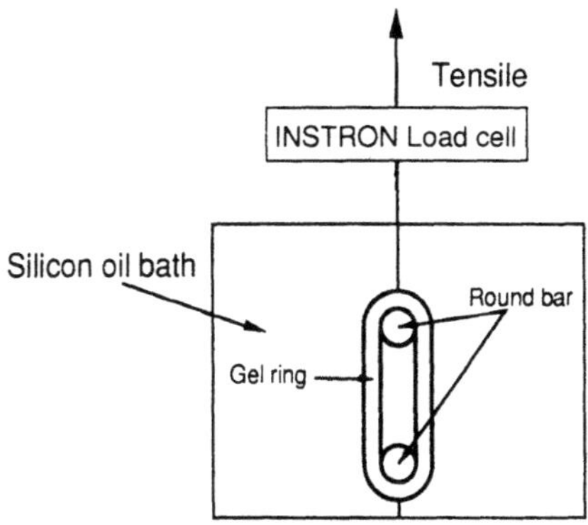

Fig. 3.   Outline of tensile test apparatus.

Breaking strain and true breaking stress of gel rings which are defined in Fig.4[8] were measured. 20 gel rings were prepared for each sample to be used for measurement. The average value and standard deviation were calculated.

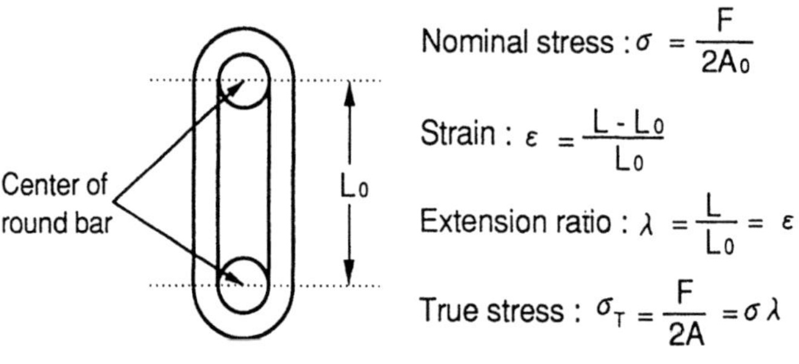

Fig. 4. Definitions
F : Force A_0 : Initial cross-section area
L_0 : Initial length   L : Length

RESULTS

Fig.5 shows a plot of true breaking stress versus KF concentration for CGN and KF mixed gels at a total gum content of 1.5%. This curve shows a maximum in true breaking stress at KF concentration 0.60% (CGN/KF ratio : 6/4). The order of true breaking stress of KF concentration is 0.60>0.75>0.45>0.30%. In this series of tensile tests, gels of 1.5% CGN alone were too brittle and fragile to hang on the round bar of the tensile test apparatus.

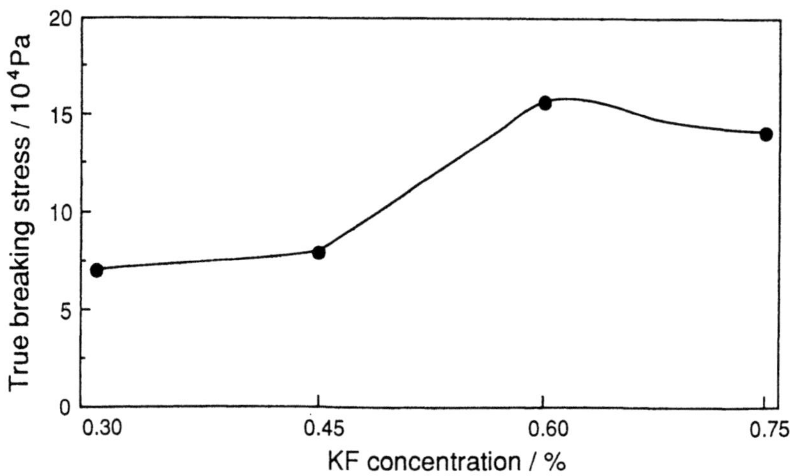

Fig. 5. True breaking stress of CGN and KF mixed gels at 5℃.
Total gum content : 1.5%

Fig.6 shows a plot of true breaking stress versus temperature of mixed gels with various KF concentrations. Mixed gels which contain 0.60% KF showed maximum true breaking stress at the temperature range from 5℃ to 40℃. True breaking stress σ_T for the mixed gels with 0.3% KF decreased with increasing temperature, whilst σ_T for the mixed gels with higher KF content (>0.45%) increased slightly up to a certain temperature and then decreased with increasing temperature.

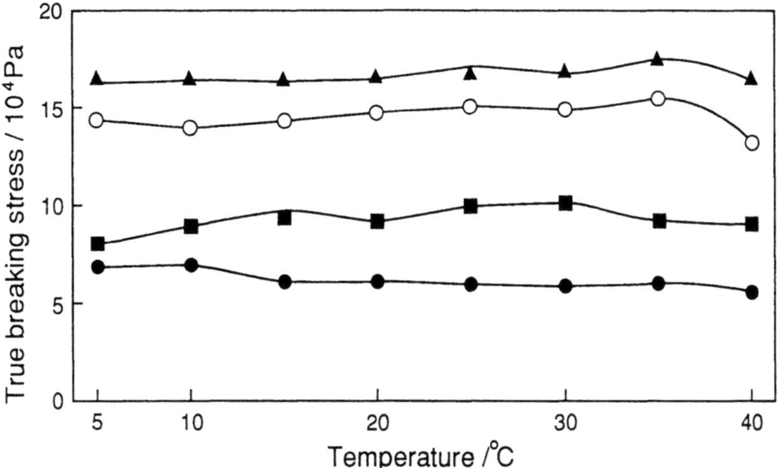

Fig. 6. The effects of temperature on true breaking stress for mixed gels of CGN and KF.
Solid lines are drawn for the guide of eyes.
Total gum content : 1.5%
KF concentration
●: 0.30%,   ■: 0.45%,   ▲: 0.60%,   ○: 0.75%

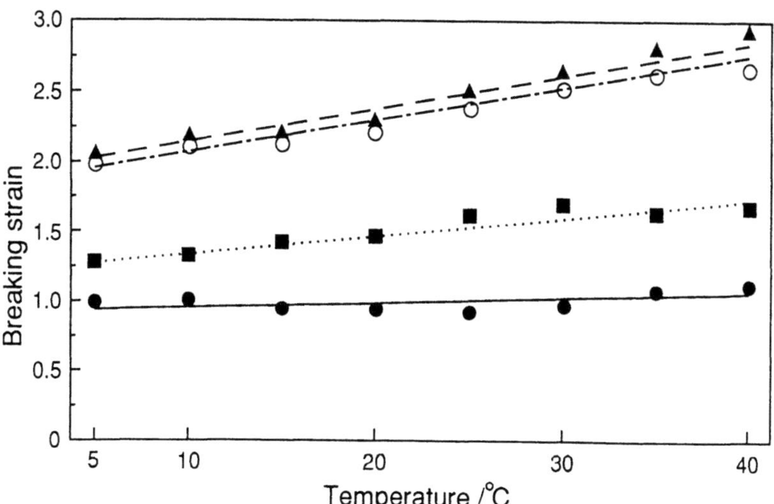

Fig. 7. The effects of temperature on breaking strain for mixed gels of CGN and KF.
Total gum content : 1.5%
KF concentration
● : 0.30%,   ■ : 0.45%,   ▲ : 0.60%,   ○ : 0.75%

Fig.7 shows a plot of breaking strain versus temperature for mixed gels with variuos KF concentrations. The breaking strain of mixed gels of KF and CGN with higher KF content (> 0.45%) increased monotonically with increasing temperature.

DISCUSSION

The temperature dependence of elastic modulus of thermoreversible gels has been explained by a reel-chain model.[9] Although the elastic modulus E defined in the small deformation rheology is not always simply proportional to the breaking stress, the temperature dependence of σ_T can be explained similarly to the temperature dependence of E. According to the reel-chain model, the temperature dependence of σ_T as shown in Fig.6 may be explained in terms of bonding energy ε, the mean end-to-end distance r_m of chains which connect junction zones, and the ceiling number ν, $i.e.$, the upper limit of the number of segments which can be liberated from junction zones just before the transition from gel to sol occurs. According to this treatment, the elastic modulus E increases monotonically for large values of ε, r_m or ν, while E decreases monotonically for small values of these three parameters. Rubber is a typical example of the first family which shows the monotone increasing behavior; carrageenan and gelatin and many other thermoreversible gels belong to the third family which show the monotone decreasing behavior.

The temperature dependence of the upper three curves in Fig.6 (with KF content 0.45%, 0.60% and 0.75%) belongs to the second family where the elastic modulus increases up to a certain temperature T_0 and then decreases with increasing temperature. As is seen in Fig.6, T_0 shifted to higher temperatures with increasing KF content. As was shown by DSC analysis,[10] the konjac mannan interacts with kappa-carrageenan and makes junction zones which is more heat-resistant than junction zones in gels of kappa-carrageenan alone.

REFERENCE

1. N.F. Stanly, Carrageenans, in : *"Food Gels"*, P. Harris, ed., Elsevier Applied Science, London (1990).
2. S. Ohashi, *The Food Industry*, 27, 22, 37(1984).
3. K. Nishinari, P. A. Williams, and G. O. Phillips, *Food Hydrocolloids*, **6** : 199 (1992).
4. M. Yamamoto, Konjac, in : "An Illustrative Guide of the Food Industry of Japan", M. Ikeda, Y. Yuize and T. Yano, eds., Korin Publishing Co., Ltd, Tokyo (1990).
5. I.C.M Dea , A. A. McKinnon, and D. A. Rees, *J. Mol. Biol.*, **68** : 153 (1972).
6. G. J. Shelso, Commercialisation of new synergistic, application of carrageenan in : *"Gums and Stabilisers for the Food Industry 5"*, G.O. Phillips, D.J. Wedlock and P.A. Willams, eds., IRL Press Oxford (1990)
7. M. Stading and A. M. Hermansson, *Food Hydrocolloids*, **5** : 339 (1991).
8. H. McEvoy, S. B. Ross-Murphy, and A. H. Clark, Large deformation and failure properties of biopolymer gels, in : *"Gums and Stabilisers for the Food Industry 2"*, G.O. Phillips, D.J. Wedlock and P.A. Williams, eds., Pergamon Press, Oxford (1984).
9. K. Nishinari, S. Koide, and K. Ogino, *J. Phys. (France)*, **46** : 793 (1985).
10. P. A. Williams, S. M. Clegg, M. J. Langdon, K. Nishinari, and G. O. Phillips, Studies on the synergistic interaction of konjac mannan and locust, in : *"Gums and Stabilisers for the Food Industry 6"*, G.O. Phillips, D.J. Wedlock and P.A. Williams, eds., IRL Press, Oxford (1992).

RHEOLOGICAL STUDY ON A MIXED SYSTEM
OF KONJAC GLUCOMANNAN AND CARRAGEENAN:
EFFECTS OF MOLECULAR WEIGHT OF KONJAC GLUCOMANNAN

Kaoru Kohyama,[1] Hiroki Iida,[2] Takashi Ochi,[2]
Shiro Ohashi,[2] and Katsuyoshi Nishinari[3]

[1]National Food Research Institute
 2-1-2 Kannondai, Tsukuba, Ibaraki 305, Japan
[2]San-Ei Gen F. F. I., Inc.
 1-11 Sanwa-cho, 1-chome, Toyonaka, Osaka 561, Japan
[3]Faculty of Science of Living, Osaka City University
 3-3-138 Sugimoto, Sumiyoshi, Osaka 558, Japan

ABSTRACT

Rheological properties of mixed gels of konjac glucomannan (KGM) with different molecular weights and kappa–carrageenan (CAR) were studied by a dynamic viscoelastic measurement and a tensile testing. Both storage and loss moduli for sol state mixtures containing higher molecular weight of KGM were higher. All systems showed a similar gelling temperature. The breaking stress and the breaking strain for mixed gels became larger with increasing molecular weight of KGM.

INTRODUCTION

Konjac glucomannan (KGM) is a main component of the tuber of *Amorphophallus konjac* K. Koch and makes a thermally stable gel (Konnyaku) by addition of alkaline coagulant. KGM shows an interaction with other hydrocolloids and most of works in the West have been devoted to a binary system such as a mixture of KGM and carrageenan or xanthan gum.[1]

Generally, rheological properties of gels reflect the molecular structure of a gelling agent. Mitchell[2] reported that the elastic modulus of a gel is independent of molecular weight above a certain limiting value. However, the rupture strength continues to rise with increasing molecular weight.

Food Hydrocolloids: Structures, Properties, and Functions
Edited by K. Nishinari and E. Doi, Plenum Press, New York, 1994

We obtained four fractions of KGM with different molecular weights and investigated rheological properties of mixed gels of the KGM with CAR in this work by a dynamic viscoelastic measurement and a tensile testing.

MATERIALS AND METHODS

Materials

Powdered KGM sample and three fractions of KGM with different molecular weights prepared by an enzymatic degradation of a cellulase were kindly supplied by Shimizu Chemical Co. (Hiroshima). A purified kappa–carrageenan (San–Ei Gen F. F. I., Inc., Osaka) was used.

Molecular Weight

KGM samples were dissolved in cadoxen[3] with stirring at room temperature. Intrinsic viscosity was measured at 25 °C using an Ubbelohde viscometer. Molecular weight of each KGM fraction was determined by gel permeation chromatography (GPC) at room temperature.

Dynamic Viscoelasticity

Powders of KGM and CAR were mixed in the ratio of 1:1 and dispersed in deionized water at 85 °C for 30 min. The total gum content was adjusted to 1.5 w/w%. The hot mixture was injected into a cell of a Rheolograph Sol (Toyoseiki Seisakusho, Tokyo). The temperature was lowered from 75 to 20 °C and then heated to 60 °C at the rate of 1.0 °C/min. The storage and loss moduli were recorded every 1 °C.

Tensile Testing

The hot mixture of KGM and CAR was poured into molds with 11 mm depth and then covered with glass plates. They were allowed to stand at 25 °C for 15 h. Then, the gel was cut into rings which have inner diameter of 20 mm and outer diameter of 30 mm. Tensile testing of gels was carried out using a Rheoner RE–33005 (Yamaden Co. Ltd., Tokyo). A gel ring was hanged on two bars of 7 mm diameter arranged perpendicularly. It was initially deformed under its own weight and then vertically elongated with the rate of 5.0 mm/sec.

RESULTS AND DISCUSSION

Chemical Structure and Molecular Weight

The ratios of mannose to glucose for four KGMs were about 2.0 and did not differ by the enzyme treatment. All four fractions of KGM were not different in chemical structure, but they had different molecular weights. Informations about the molecular weight

Sample	$[\eta]^*$(dL/g)	Mw^+	Mw/Mn^+
LM1	1.98	2.56×10^5	6.50
LM2	2.63	4.38	7.72
LM3	3.50	5.96	5.68
ND	3.91	6.89	8.02

*) by Ubbelohde viscometer at 25 °C in cadoxen.
+) by GPC at room temperature in cadoxen.

are shown in Table 1. Intrinsic viscosity and weight average molecular weight (Mw) increased in order of LM1, LM2, LM3, and non–degraded sample (ND). The molecular weight distribution was not narrow.

Dynamic Viscoelasticity

At higher temperatures, all samples tested were sol and the loss modulus was a little bit larger than the storage modulus. The both moduli increased little by little with lowering the temperature until a certain temperature, and then increased steeply at around 30 °C. Below this temperature, samples lost the fluidity and the storage modulus became more than 5 times larger than the loss modulus for all the samples. Gelling temperature (Tgel) was determined as the temperature at which the storage modulus began to increase in the cooling process. When the gel was heated, the both moduli decreased with temperature, however the values were larger than those observed in cooling process at the same temperature. The transition temperature of gel to sol (Tsol) was about 20 °C higher than Tgel for all the samples. Important parameters are shown in Table 2. Both storage and loss moduli for sol state mixtures at 60 °C containing KGM with higher molecular weight were higher. Gel state mixtures of higher molecular weight KGM at 25 °C showed larger values in both moduli except KGM–ND which showed smaller increases in both moduli at

Table 2. Dynamic viscoelastic characteristics for CAR and KGM mixtures.

Sample	Tgel (°C)	Tsol (°C)	G'25*(Pa)	G"25 (Pa)	G'60 (Pa)	G"60*(Pa)
CAR 1.5 %	34	55	4135	680	0.3	0.5
CAR 0.75%	27	43	46	8	0.2	0.0
CAR+LM1	30	50	594	91	0.5	1.2
CAR+LM2	30	49	626	95	0.7	1.3
CAR+LM3	30	50	1010	118	3.8	6.8
CAR+ND	32	53	1311	115	99.6	93.3

*) G'25 and G"60 represent the storage modulus at 25 °C
and the loss modulus at 60 °C respectively.

lower temperatures. All four systems showed similar Tgel and Tsol which were between those of 1.5 % and 0.75 % CAR. Then, differences in molecular weight of KGM showed little effect on transition temperatures.

Tensile Testing

Tensile characteristics for gels are shown in Table 3. CAR gels are firm and brittle as shown by large Young's modulus and small breaking strain. The addition of KGM to CAR made Young's modulus smaller and breaking strain larger. Two systems containing KGM of smaller molecular weights (LM1 and 2) showed almost the same properties in the tensile testing. The breaking stress and the breaking strain for mixed gels of higher molecular weight KGM became larger. Therefore, it is considered that KGM of large molecular weight contributes to make a strong network structure in the mixed system. Breaking stress greatly increased with increasing molecular weight of KGM in CAR–KGM mixed gels.

Table 3. Tensile characteristics for mixed gels of CAR and various KGM. Total gum contents: 1.5 w/w%.

Sample	Breaking stress (kPa)	Breaking strain (cm/cm)	Young's modulus (kPa)
CAR	16.47 $\pm$2.54	0.210 $\pm$0.025	60.64 $\pm$2.04
CAR+LM1	8.57 $\pm$0.97	0.388 $\pm$0.043	15.15 $\pm$1.43
CAR+LM2	9.23 $\pm$0.84	0.369 $\pm$0.030	15.46 $\pm$1.24
CAR+LM3	37.36 $\pm$1.18	0.870 $\pm$0.048	20.76 $\pm$1.09
CAR+ND	74.66 $\pm$4.55	1.958 $\pm$0.103	9.97 $\pm$0.96

mean $\pm$standard deviation from more than 20 times measurements.

ACKNOWLEDGEMENT

We thank H. Tomizawa of Tosoh Corporation Tokyo Research Center for GPC measurement.

REFERENCES

1. K. Nishinari, P. A. Williams, and G. O. Phillips, Review of the physico–chemical characteristics and properties of konjac mannan, *Food Hydrocolloids*, 6:199 (1992).
2. J. R. Mitchell, The rheology of gels, *J. Texture Stud.*, 11:315 (1980).
3. D. Henley, A macromolecular study of cellulose in the solvent cadoxen, *Arkiv Kemi*, 18:327 (1961).

INFLUENCE OF XANTHAN GUM ADDITION ON FROZEN STARCH PASTE PROPERTIES

Cristina Ferrero[a], Miriam Martino[a] and Noemi Zaritzky[a,b]

[a] Centro de Investigación y Desarrollo en Criotecnología de Alimentos (CIDCA) [a,b] Depto. de Ingeniería Química, Facultad de Ingeniería, Universidad Nacional de la Plata, 47 y 116 (1900) La Plata, Argentina

ABSTRACT

The effect of freezing rate, storage temperature (-5 to -20°C) and gum addition (0.3 % w/w) on corn starch and wheat flour pastes (10 % w/w) was studied to analyze various aspects: rheological behavior, exudate production, ice recrystallization by indirect microscopic observation and amylopectin retrogradation by Differential Scanning Calorimeter (DSC). High freezing rates produced a higher paste quality due to smaller ice crystals formation and the absence of starch retrogradation. During frozen storage, starch retrogradation and ice recrystallization contributed to the deterioration of the system (spongy structure, decrease of apparent viscosity and increase of exudate production). The addition of xanthan gum improved system stability. Exudate production decreased and original rheological properties were maintained. No effect of hydrocolloid addition was observed on ice crystal sizes. DSC analysis showed that xanthan gum does not modify amylopectin retrogradation.

INTRODUCTION

Textural changes and syneresis of sauces that arise after thawing precooked frozen foods are major concerns faced by the frozen food industry. These quality problems have been traditionally attributed to starch retrogradation, that is the crystallization of amylose and amylopectin. Other theories have emphasized the importance of ice formation on the stability of these systems[1]. Hydrocolloids besides their stabilizing, gelling and thickening properties are also claimed to act as ice crystal inhibitors but their way of action is not fully understood[2]. The objective of the present work was to analyze the influence of freezing rate, frozen storage and xanthan gum addition on the stability of gelatinized starch pastes in order to relate macroscopic changes with structural parameters.

MATERIALS AND METHODS

Corn starch (Refinerías de Maíz, Argentina) and wheat flour (Molinos Río de la Plata S.A., Argentina) were suspended in distilled water, 10 % w/w wet basis. Half of the samples

Food Hydrocolloids: Structures, Properties, and Functions
Edited by K. Nishinari and E. Doi, Plenum Press, New York, 1994

contained xanthan gum (Saporitti Hnos, Argentina) in a concentration of 0.3%. Gelatinization was performed under controlled conditions in a thermostatic bath at $90 \pm 0.2°C$.

The process was followed with a polarizing microscope (Ortholux II, Leitz, Germany) to verify that all the starch granules had lost their Maltese Crosses. Dry matter content of the gelatinized pastes was determined on each batch. The pastes were placed in small cylindrical holders and cooled in a constant temperature room at $20 \pm 0.5°C$ during 2 hours before being frozen. The samples were processed in freezers at -20 and -80°C to a final temperature of -20°C with different freezing rates. The highest velocity was obtained by dipping the samples in liquid nitrogen. Thermal histories during freezing were recorded with Cu-Constantan thermocouples. Freezing rates were determined according to the IIR[3]. The measured initial freezing point of the starch paste was -0.6°C. Freezing rates ranged from 0.2 cm/h (slow freezing) to values higher than 100 cm/h (ultra-rapid freezing). Frozen samples were stored in cold chambers at -5, -10 and -20°C ($\pm0.5°C$) during three months. Samples were thawed under controlled conditions, at $20 \pm 0.5°C$ for exudate production and retrogradation measurements and at $60 \pm 0.1°C$ for rheological tests.

Water holding capacity was measured in thawed samples by capillary suction of a porous material (filter paper) in contact during 1 min with the sample as described by Ferrero[4]. Triplicate values of the equivalent diameters of the wet area left in the paper were taken as the syneresis index (% exudate). The equivalent diameters, defined as the diameter of a circle that has the same surface area as the problem figure, were measured with an Image Analyzer (Morphomat 30, Zeiss, Germany).

For the rheological measurements a rotational viscometer Haake Rotovisco RV2 (Germany) with a sensor MVIP of concentric cylinders with profiled surfaces and a thermostatic system was used. Flow properties were measured at 60°C with shear rates (D) ranged from 0 to 1024 s^{-1}.

Isothermal freeze fixation technique[5] was applied to measure ice crystal sizes in the frozen samples during storage. Samples in triplicate were processed. Ice crystal sizes were obtained from the micrographs by measuring the holes left in the system. At least 150 crystals were measured with the Image Analyzer to determine equivalent diameter distributions.

Differential Scanning Calorimetry (DSC) measurements were performed with a DuPont 910 DSC system and Series 99 Thermal Analyzer with a temperature control in order to determine the enthalpies (ΔH) associated to starch retrogradation. Storage temperatures were -1, -5, -10 and -20°C. Duplicates of the homogenized thawed samples were placed in aluminum pans and then hermetically sealed. The temperature was raised from 20 to 120°C at a rate of 10°C/min with an instrument sensitivity of 0.1×10^{-3} Joule/s. Dry mass of the samples were also determined.

RESULTS AND DISCUSSION

Freezing rate has an important effect on exudate production in corn starch and wheat flour gels (Fig. 1a). Higher velocities led to lower exudate values. The addition of xanthan gum, in all cases, decreased exudate production. During frozen storage, an increase of exudate production was observed in corn starch pastes without xanthan gum (Fig. 1b,c). Exudate percentages of samples stored at -10 and -20°C had not significant differences ($P \geq 0.05$). The addition of xanthan gum allowed to maintain the initial exudate levels up to 8 weeks of storage. In both pastes, at -5°C, a spongy matrix was observed. This structure did not appear when immersion in liquid nitrogen was the freezing method, nor when xanthan gum was added even at high storage temperatures. Wheat flour gels showed lower exudate values due to a higher water holding capacity related to the presence of proteins.

Effect of freezing rate and xanthan gum on rheological curves of frozen pastes are frozen pastes are shown in Fig. 2. For both pastes non significant differences were observed when comparing the curves of unfrozen (UF) and frozen samples in liquid nitrogen (N); both

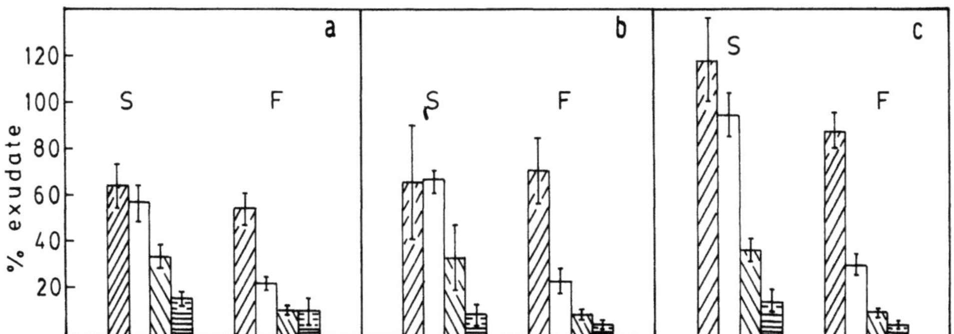

Fig. 1: Effect of freezing rate and storage at -20°C on exudate production of corn starch (S) and wheat flour (F) pastes with and without xanthan gum. (a) just frozen, (b) 1 week and (c) 8 weeks. ☐ frozen at 270 cm/h, without xanthan gum; ▨ frozen at 0.6 cm/h, without xanthan gum; ▤ frozen at 0.6 cm/h, with xanthan gum; ▨ frozen at 0.3 cm/h, with xanthan gum. Bars= std. errors (P<0.05).

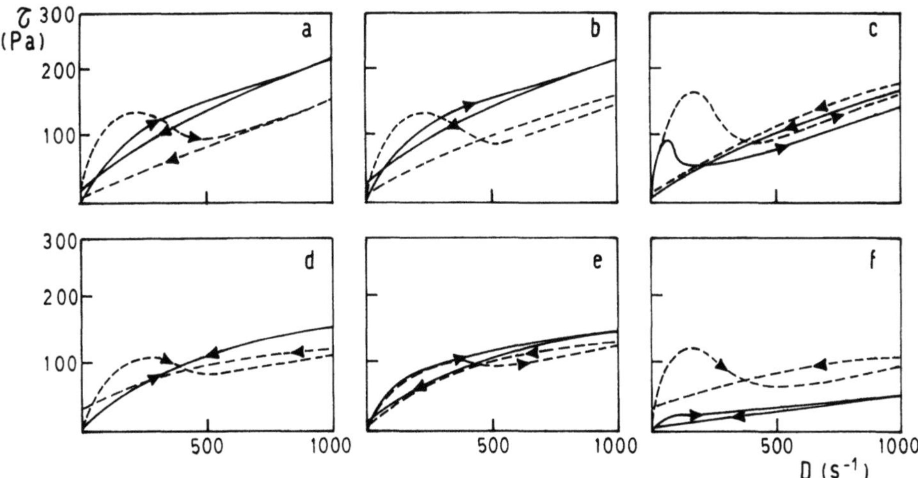

Fig. 2: Influence of freezing rate on shear stress (τ) vs. shear rate (D) curves of corn starch pastes (a,b,c) and wheat flour pastes (d,e,f) without (—) and with (---) xanthan gum. a,d) UF: unfrozen, b,e) N: frozen at 216 cm/h, c,f) S: frozen at 0.3 cm/h.

of them exhibited a smooth texture. Rapid (R) and slow (S) freezing rates produced "grainy" and "spongy" structures respectively, related to amylose retrogradation. Lower freezing rates led to a viscoelastic peak at D<500 s⁻¹. The addition of xanthan gum to both pastes modified the shape of the τ vs. D curves. A notorious viscoelastic behavior was observed in frozen and unfrozen samples for D<500 s⁻¹. Freezing rate had no significant effect on the rheological curves in accordance with the smooth and homogeneous appearance observed in all samples with xanthan gum.

Power law was applied to the results for $D \geq 500$ s⁻¹. Non significant differences ($P \geq 0.05$) were found in the flow behavior indexes (n) corresponding to UF (n= 0.337 ± 0.040), N (n= 0.309 ± 0.013) and R (n= 0.363 ± 0.021) corn starch samples without xanthan gum; however this index increased significantly for the lower freezing rate S (n= 0.754 ± 0.032). A similar behavior was observed in wheat flour pastes. Consistency index (m, Pa.s⁻ⁿ) of UF and N corn starch samples were non significantly different (22.21 ± 0,50 and 23.59 ± 3.02 respectively). Values of m in R and S samples (9.30 ± 1.41 and 0.58 ± 0.13 respectively) differed significantly from the previous ones. In wheat flour pastes only m value for the lower freezing rate (S) was significantly different. Non significant differences were found for both indexes in samples with xanthan gum frozen at different velocities; xanthan gum addition suppressed the deleterious effect of freezing.

During frozen storage, samples frozen in liquid N_2 did not change their behavior. Pastes frozen at lower velocities increased the rheodestruction peak during storage, less notoriously in wheat flour pastes than in corn starch ones. Indexes m and n and apparent viscosities did not exhibit significant modifications ($P \geq 0.05$) during frozen storage. In conclusion, freezing rate was the more important factor in corn starch and wheat flour pastes deterioration while storage time did not show relevant effect. The addition of xanthan gum in both pastes minimized the effect of freezing rate and storage temperatures.

Ice crystal equivalent diameters and the corresponding histograms were obtained from micrographs of corn starch and wheat flour pastes frozen under different conditions. As freezing rate decreased ice crystal size increased, leading to a microstructure deterioration. For example, in corn starch pastes frozen at 2 cm/h the mean equivalent diameter (D) was 37.34 ± 2.50 μm and for 15 cm/h, D was 30.89 ± 1.81 μm. Wheat flour proteins did not show a particular protective effect against freezing since D of corn starch and wheat flour pastes did not differ significantly ($P \geq 0.05$). The addition of xanthan gum did not produce a significant effect on ice crystal size at different freezing rates compared to the pastes without the gum.

Micrographs of the frozen pastes stored at the different tested temperatures showed an enlargement of the ice crystal diameters that is representative of the recrystallization process, as can be observed in Fig. 3 for corn starch paste stored at -10°C. Curves of equivalent ice crystal diameter vs. storage time (Fig. 4) showed that D increased with storage time but the tendency is to reach different limit equivalent diameters (Dl) depending on storage temperature. During recrystallization, as crystal size increases, the driving force declines and stops when a stable crystalline size is attained. The mathematical model[5] used to interpret recrystallization of ice in these systems was:

$$\ln\left(\frac{Dl-Do}{Dl-D}\right)+\frac{(Do-D)}{Dl}=\frac{kt}{Dl^2} \tag{1}$$

where Do is the initial equivalent diameter and k the kinetic constant.

Arrhenius law was applied to obtain the activation energy (Ea) for ice recrystallization of the four tested formulations. Non significant differences ($P \geq 0.05$) were observed, thus the mean Ea value obtained was 54.11 KJ/mol (standard deviation, s= 11.59). We can conclude that xanthan gum addition did not alter ice recrystallization rate, nor its Ea. Commonly, hydrocolloids are recommended as ice crystal growth inhibitors but this study together with some others[2,6] showed that at this concentration the hydrocolloid action should be explained on other basis like high water retention after thawing.

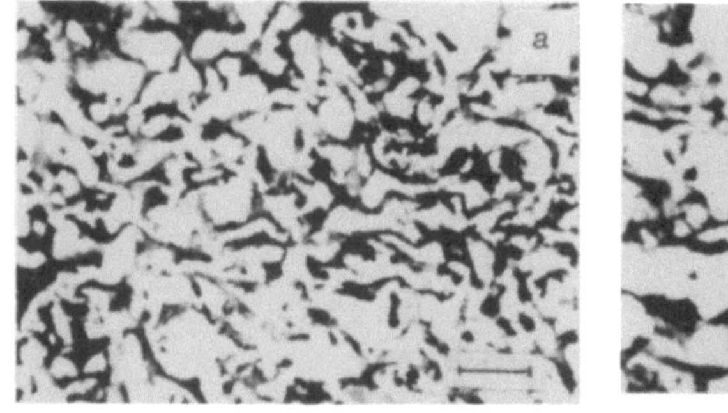
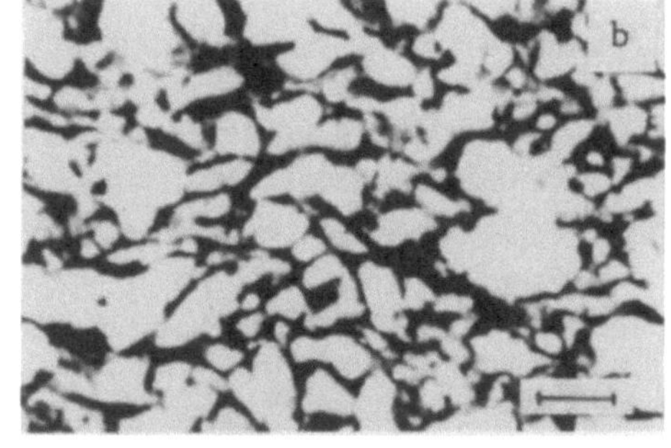

Fig. 3: Effect of frozen storage at -10°C on ice crystal size for corn starch paste frozen at 2 cm/h: (a) 8 days, (b) 60 days. Bar = 100 μm

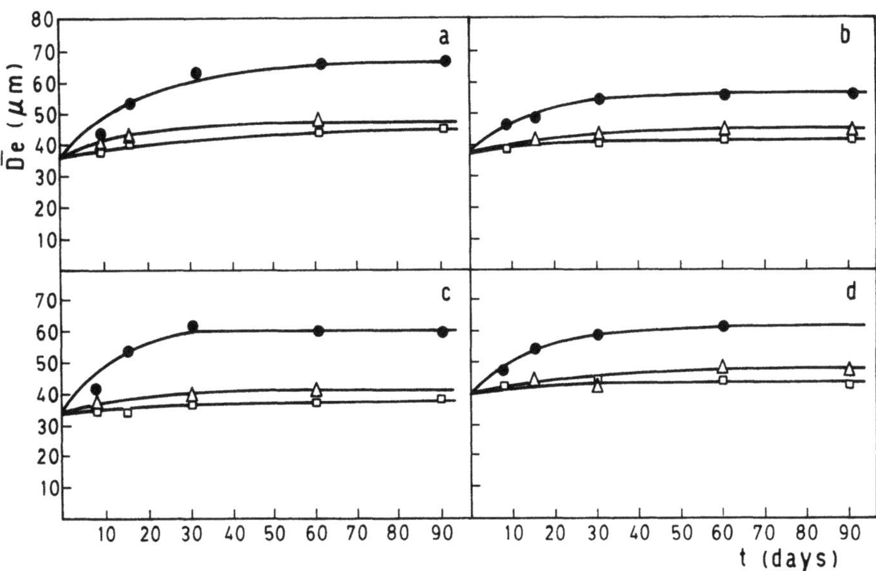

Fig. 4: Mean equivalent ice crystal diameter (De) vs. storage time of corn starch pastes without xanthan gum (a), with xanthan gum (b) and wheat flour paste without xanthan gum (c) and with xanthan gum (d). (●) stored at -5°C, (▲) stored at -10°C, (□)stored at -20°C. Pastes were frozen at 2 cm/h.

Starch retrogradation can be considered as a two step process[7] a short term one related to amylose crystallization (thermally irreversible in the assayed range) and a long term one which involves amylopectin crystallization (thermally reversible). DSC only allows the quantification of the reversible process.

For corn starch pastes, high freezing rates did not produce a detectable amylopectin retrogradation peak. The lower freezing rates, 1 cm/h and 0.3 cm/h, produced peaks corresponding to 2.05 J/g (s= 0.35) and 2.97 J/g (s= 0.20) respectively, but both values did not differ significantly (P≥0.05). Xanthan gum presence did not modify the effect of freezing rate on amylopectin retrogradation. A different result was obtained for flour pastes with and without hydrocolloid; retrogradation peaks were not detected for any assayed freezing rate.

Amylopectin retrogradation kinetics was determined from the enthalpy values (ΔHt) obtained at different storage times. Avrami equation was fitted to the experimental values as follows:

$$\theta = \frac{\Delta H_t - \Delta H_o}{\Delta H_\infty - \Delta H_o} = 1 - e^{(-kt^n)} \qquad (2)$$

where ΔH_∞, k and n were coefficients determined by non linear regression analysis and θ is the retrograded fraction. ΔH_∞ is the maximum enthalpy and ΔH is the enthalpy of the just frozen sample. Obtained values of ΔH_∞ and k for two storage temperatures are shown in Table 1. Amylopectin retrograded fraction (θ) at -1 and -5°C as a function of storage time is shown in Fig. 5. At storage temperatures of -10 and -20°C, DSC retrogradation peaks were not detected during the tested period.

In conclusion xanthan gum addition did not avoid amylopectin retrogradation but it inhibited the development of the characteristic spongy matrix which is related to amylose retrogradation. It had no effect on ice crystal formation. Thus, the low exudate values observed can be attributed to the high water holding capacity of this hydrophilic gum.

Amylose and amylopectin structures may explain differences in retrogradation behavior

Table 1. Retrogradation kinetic parameters during frozen storage

paste composition	Storage temperature			
	−1°C		−5°C	
	$k(days^{-1})$	$\Delta H_m(J/g)$	$k(days^{-1})$	$\Delta H_m(J/g)$
corn starch	0.309(0.064)	8.883(0.330)	0.114(0.033)	9.288(0.836)
corn starch + x. gum	0.369(0.227)	6.730(0.372)	0.216(0.063)	6.132(0.271)
wheat flour	0.289(0.079)	2.897(0.142)	0.130(0.024)	2.575(0.150)
wheat flour + x. gum	0.248(0.023)	3.285(0.071)	0.235(0.034)	2.383(0.084)

standard error between parenthesis (P<0.05); the lowest correlation coefficient of the non linear regression was 0.970; enthalpies were calculated on dry basis.

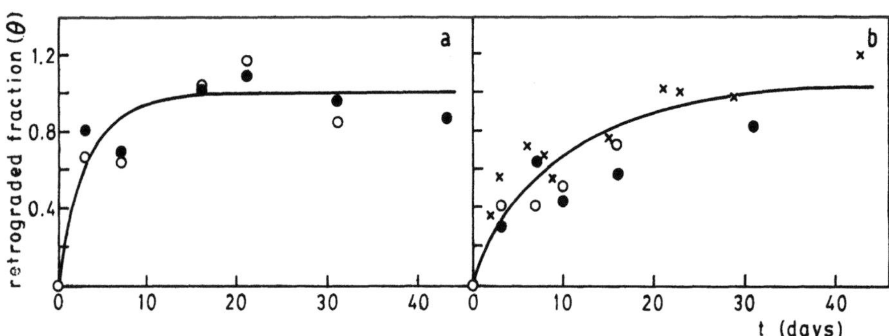

Fig. 5: Retrograded fraction (θ) vs. storage time of corn starch pastes without xanthan gum, stored at -1°C (a) and -5°C (b), frozen at (●) 0.3 cm/h, (o) 1.0 cm/h, (X) 12 cm/h.---- Avrami model

when xanthan gum is present. Amylose chains are linear and longer than amylopectin ones; during gelatinization amylose is released outside the granule and forms an external matrix. Amylose has then a higher exposure to other components of the paste like hydrocolloids. Amylose-hydrocolloid interaction competes with amylose-amylose aggregation decreasing the probability of retrogradation occurrence.

Although high freezing rates and low storage temperatures are highly recommended, the addition of low levels of xanthan gum allows the use of slower freezing rates, maintaining quality characteristics and decreasing equipment investment.

References

1. W.S. Chan and R.T. Toledo, Dynamics of freezing and their effects on the water-holding capacity of a gelatinized starch gel, Journal of Food Science, 4:301-303 (1976).
2. E.R. Budiaman and O. Fennema, Linear rate of water crystallization asinfluenced by temperature of hydrocolloid suspensions, Journal of Dairy Science 70:534-546 (1987).
3. International Institute of Refrigeration. "Recommendations for the Processing and Handling of Frozen Foods". Institut International du Froid, Paris (1972).
4. C. Ferrero, Doctoral Thesis, Universidad Nacional de La Plata (1992).
5. C. Ferrero, M. Martino and N. Zaritzky, Light microscopy measurements of ice recrystallization in frozen corn starch pastes using isothermal freeze fixation, Food Struct., 11(3):237-248 (1992).
6. D.S. Reid, M.S.B. Alviar and M.H. Lim. The rates of change of ice crystal size in model systems stored at different temperatures relevant to the storage of frozen food, in: Proceedings of the XVIIthInt. Con. of Refrigeration, Austria. Vol. C, pp. 397-401, Austrian Association ofRefrigeration and Air Conditioning (1987).
7. M.J. Miles, V.J. Morris, P.D. Orford and S.G. Ring, The roles of amylose and amylopectin in the gelation and retrogradation of starch. Carbohydrate Research 135:271-281 (1985).

FEATURES OF PROTEIN-STARCH GRANULE CONJUGATES

Koji Takahashi and Makoto Hattori

Applied Biological Science
Tokyo University of Agriculture and Technology
3-5-8 Saiwai-cho, Fuchu, Tokyo 183, Japan

INTRODUCTION

Starch is widely used as a very important food material, because starch is an excellent energy source and has useful physical properties. However, in order to cope with the recent increase in demands for the functionality and in application, it is insufficient to depend on only characteristics of native starches and modified starches from a viewpoint of material science. The authors are thus trying to improve the properties of starch by the conjugation of starch and protein with covalent bond. In this research, slightly modified carboxymethyl starch (CMS) was conjugated with whey protein isolate (WPI) by carbodiimide and the features of the WPI–CMS conjugate were characterized.

MATERIALS AND METHODS

Materials

Corn starch after being repeatedly washed with redistilled water at 4°C prior to air–drying was used (water content, 12.0%). WPI was supplied from Meiji Milk Products Co. 1-Ethyl-3-(3-dimethylaminopropyl)carbodiimide (EDC) hydrochloride was purchased from Dojindo. Other reagents were of special grade commercially available.

Carboxymethylation of Starch

Corn starch (2 g) was dispersed in 32 ml of methyl alcohol containing 0.6 g of monochloroacetic acid and 0.35 g of sodium hydroxide and shaken gently at 40°C for 6–16 hr. The reaction mixture was neutralized to pH 6.5 with acetic acid after cooling down to room temperature prior to filtration. After washing with methyl alcohol and air–drying, CMS was obtained. The degree of modification was determined by hydrochloride–methyl alcohol titration.[1]

Food Hydrocolloids: Structures, Properties, and Functions
Edited by K. Nishinari and E. Doi, Plenum Press, New York, 1994

Preparation of WPI-CMS Conjugate

Preparation of WPI-CMS conjugate was carried out by referring to the method of Hoare and Koshland.[2] CMS (0.2 g) was dispersed in 4 ml of 5% EDC solution maintaining the reaction medium at pH 7.0. 8 ml of WPI solution (25 mg/ml) was gradually added. The reaction mixture was incubated at 24°C for 5–300 min with gentle stirring and washed with distilled water by centrifuging five times at 5,000 rpm for 10 min after the reaction was stopped by adding 0.72 ml of acetic acid. The conjugates were obtained by air-drying.

Solubility

CMS or the conjugate was incubated in 5 ml of distilled water at 65°C for 1 hr with stirring. The solubility was estimated by determining saccharide concentration for the centrifuged supernatant (at 18,000 rpm for 10 min) by the phenol-sulfuric acid method.[3]

Differential Scanning Calorimetry (DSC)

DSC of CMS or the conjugate was performed using Seiko SSC-5020 DSC 100 as described previously.[4] From DSC curve, the gelatinization temperatures [onset temperature (To), peak temperature (tp) and final temperature (Tf)] and gelatinization enthalpy were estimated as the characteristics of gelatinization process.

Digestion with α-Amylase

The native corn starch, CMS or the conjugate (2 mg) dispersed in 1.8 ml of 0.02M sodium citrate buffer at pH 6.5 containing 0.1M sodium chloride was heated at 100°C for 10 min with stirring. After cooling down to room temperature, 0.2 units of α-amylase/0.2ml [Human salivary (500 units), Sigma] was added and the reaction mixture was incubated at 30°C for 2–20 hr. Digestibility was estimated by determining saccharide concentration of the filtrate at each digestion time by the phenol-sulfuric acid method.[3]

Treatment with Actinase

The conjugate was digested with Actinase (Kaken Pharmacy Co.) as follows. 0.1g of the conjugate was deipersed in 100 ml of 0.1M tris-hydrochloride buffer at pH 7.8 containing 5 mM calcium chloride and digested twice with 5 mg of Actinase at 30°C for 24 hr. After washing with fresh buffer by centrifugation at 5,000 rpm for 5 min five times, the Actinase-digested conjugate was incubated to eliminate calcium in 50 ml of 0.5M EDTA at pH 7.0 at room temperature for 2 hr and washed repeatedly with water and was recovered by air-drying.

Size exclusion chromatography (SEC)

The conjugate, CMS and native starch (10 mg each) were heated in 1 ml of 0.1M sodium borate buffer at pH 7.0 containing 0.1M potassium chloride at 110°C for 10 min. After cooling, each gelatinized sample was digested with 25 units of α-amylase (25u/ml, Sigma) at 30 °C for 24 hr. The enzymatic reaction was stopped by heating it at 100 °C for 10 min. The filtrate of the reaction mixture was applied onto a column of Toyopearl HW-65s and SEC was conducted as follows: column size, 1 x 90 cm; mobile phase, 0.01M sodium borate buffer at pH 7.0 containing 0.1M potassium chloride; flow rate, 18 ml/hr; temperature, 25 °C; fraction volume; 0.2 ml/tube; detection, absorbance at 280 nm and at 490 nm after coloring by phenol-sulfuric acid.[3]

RESULTS AND DISCUSSION

Features of CMS

The degree of modification (DM) of CMS preparations was clarified to be 0.6, 1.2, 2.1 and 3.5 (carboxymethyl residues/100 glucose residues) depending on the reaction time. It thus follows that slightly modified CMS could be prepared as compared with commercial CMS. Granular shapes and polarizing patterns of these CMSs in water at room temperature were essentially the same as those of the native starch. The solubility of these CMSs in water at 65°C for 1 hr showed about 34%, 50%, 64% and 66%, respectively, which were extremely higher than the native starch (about 5%). The gelatinization temperatures and enthalpies of CMSs decreased with the increase in DM (Table 1). Since hydroxy groups of CMS are partially substituted by carboxymethyl group, it is considered that the electrostatic repelling force by carboxyl groups results in the decrease in the intermolecular cohesion of CMS. For further experiments, we used CMS with DM 3.5 because it had the highest carboxyl group content and the micelle structure capable of gelatinizing.

Table 1. Gelatinization temperaure of CMSs with different degree of modification.

Degree[*] of modification	Gelatinization temp.(°C)		
	To	Tp	Tf
Native	63.0	68.0	73.2
0.6	54.7	59.8	67.6
1.2	48.7	54.3	63.7
2.1	47.1	52.3	62.2
3.5	44.1	47.6	56.1

[*]residue/100 glucose residues.
To, onset temp.; Tp, peak temp.; Tf, final temp.

Table 2. WPI contents of the conjugates.

Incubation time (min)	WPI content (%)[*]
5	1.96
15	3.26
30	5.21
60	7.08
300	7.46

[*]based on the conjugate weight.

Conjugation of CMS and WPI

The conjugated WPI amount was estimated by determining protein concentration for the centrifuged supernatant of the reaction mixture by the absorbance at 280 nm (Table 2). It increased from about 2.0% to 7.5% (based on the conjugate weight) with increase in the reaction time. The conjugate with the highest WPI content could be stained with Coomassie Brilliant Blue and little swelled by heating at 80°C for 10 min, while CMS extremely swelled and could not be stained (Fig. 1). SEC pattern for the conjugate (Fig. 2) indicated relatively small peak of proteinous components with the elution of starch chains at Fr. No. 253, which could not be observed in case of CMS and native starch. It is, therefore, considered that the conjugate with acid–amide bond could be prepared.

Changes in Solubility and Gelatinization Property

CMS and the conjugate were heated in distilled water at 65°C for 15 min. The solubility of CMS indicated about 66%. However, the conjugate indicated only 1.5%, which was lower than that of native starch. It is considered that bonding of WPI to the

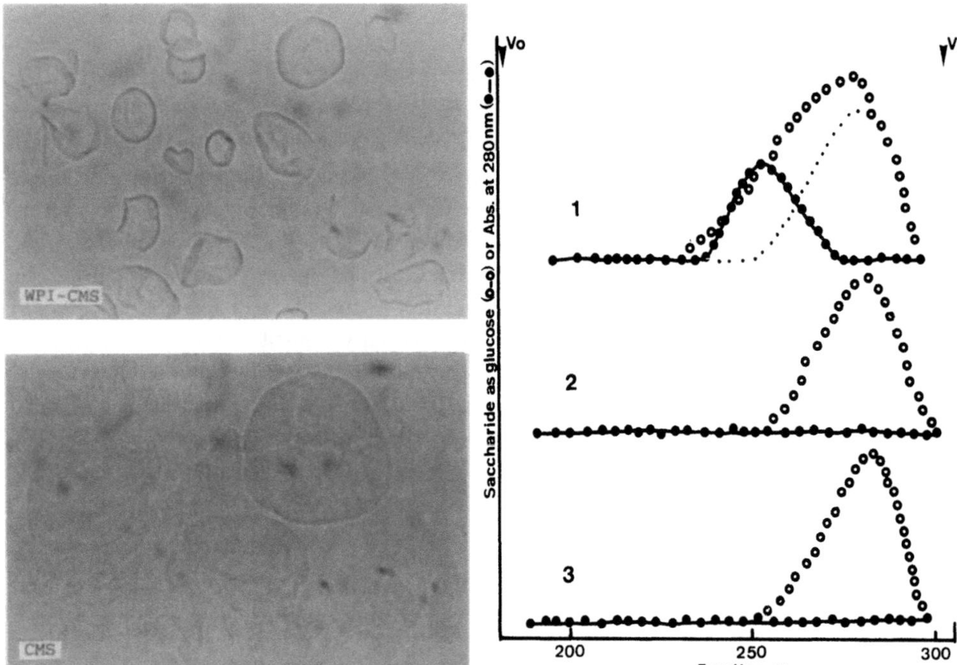

Fig. 1. Microphotograph of the conjugate heated at 80°C for 10 min compared with CMS.
Direct magnification, x400.

Fig. 2. SEC pattern for the conjugate compared with CMS and native starch. 1, the conjugate; 2, CMS; 3, native starch.

surface of CMS granules resulted in the marked decrease in solubility, meaning the capsulating effect of WPI.

DSC curves indicated that the conjugate gelatinized at 61°C (peak temperature, Tp), which was about 11 °C higher than Tp of CMS (Fig. 3). However, DSC curves for CMSs in 1.25–5.0% WPI solution were quite the same as that of CMS itself. When the conjugate was digested with Actinase, Tp showed about 54°C, which was close to that of CMS. These facts indicated that bonding of WPI to CMS resulted in the increase in the gelatinization temperature. The gelatinization enthalpy of CMS decreased from 11.3 mJ/mg to 9.5 mJ/mg by the conjugation. These mean that the gelatinization might be limited due to the suppression of swelling.

Retrogradation Behavior

The conjugate, CMS and native starch gelatinized on a DSC apparatus was preserved at 4°C for 7 days prior to analyzing again by DSC. The second DSC curve for native starch indicated relatively broad endothermic peak (Fig. 4) due to regelatinization of retrogradation as observed by Nakazawa et al.[5] Since the preserved CMS also showed a weak endothermic peak, it was thought that CMS could retrograde, but the degree of retrogrgadation for CMS estimated by the ratio of the second gelatinization enthalpy to the first was small as compared with that of native starch (Table 3). In case of the conjugate, no endothermic peak was observed. This indicates that the conjugate ceases retrogradation, because bonding of WPI to CMS might result in the inhibition of the rearrangement of starch chains.

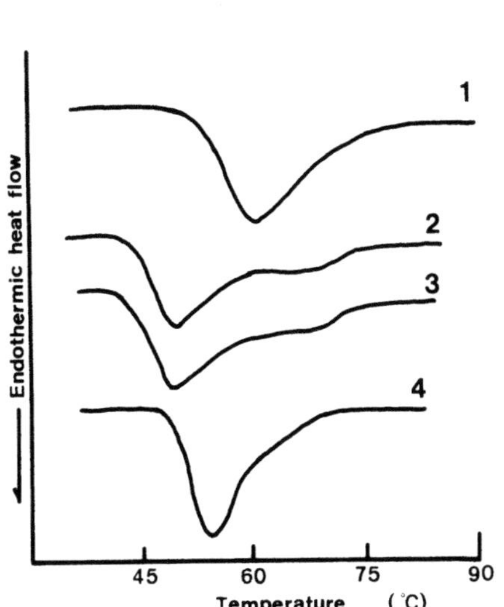

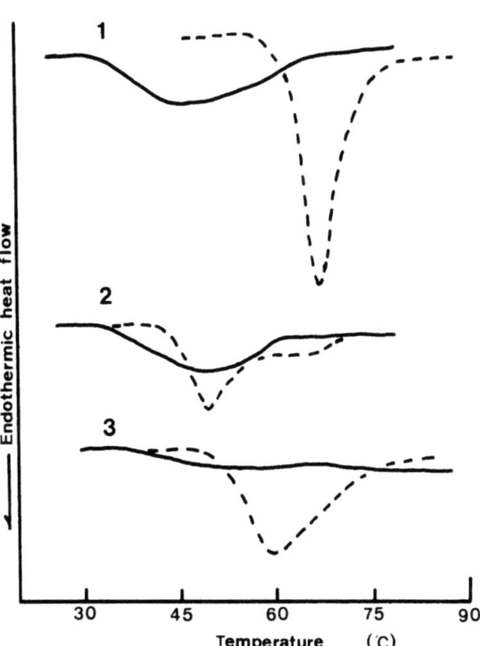

Fig. 3. DSC curves for the conjugate, CMS, and the Actinase–treated conjugate.
1, the conjugate; 2, CMS; 3, CMS in 5.0% WPI solution; 4, the Actinase–treated conjugate.

Fig. 4. DSC curve for the conjugate retrograded at 4 °C for 7 days compared with CMS and native starch.
----, first gelatinization;
——, second gelatinization.
1, native starch; 2, CMS; 3, the conjugate. compared with CMS and native starch.

Table 3. The degree of retrogradation of gelatinized native starch, CMS and the conjugate.

Sample	ΔHd (mJ/mg)		
	first[a]	second[b]	b/a(%)
Native	13.7	8.5	62
CMS	11.3	5.4	48
Conjugate	9.5	–	–

ΔHd, gelatinaization enthalpy; a, first gelatinization; b, second gelatinization.

Digestibility with α-Amylase

CMS was easily digested with α-amylase as compared with native starch (Fig. 5). This increase in digestibility was thought to be caused by high solubility and swelling power of CMS. On the other hand, the conjugate indicated great difficulty in digestion due to the capsulating effect of WPI.

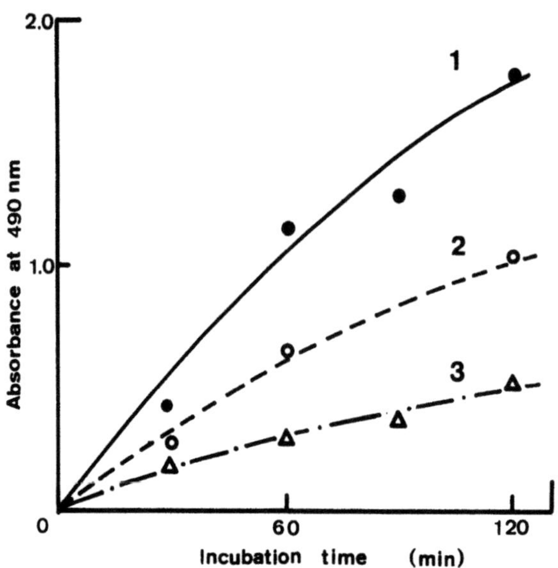

Fig. 5. Digestion of the conjugate with α-amylase compared with CMS and native starch. 1, CMS; 2, native starch; 3, the cojugate.

REFERENCES

1. H. Suzuki, Y. Tadokoro, and N. Taketomi, Carboxymethylation of starch by sodium monochloroacetate. *Denpunkougyougakkaishi.* 9:33(1961).
2. D.G. Hoare and D.E.Jr. Koshland, A method for the quantitative modification and estimation of carboxylic acid groups in protein. *J. Biol. Chem.* 242:2447 (1967).
3. M. Dubois, K.A. Gilles, J.K. Hamilton, P.A. Rebers, and F. Smith, Calorimetric method for determination of sugars and related substances. *Anal. Biochem.* 9:401 (1956).
4. K. Wada, K. Takahashi, K. Shirai, and A. Kawamura, Differential thermal analysis (DTA) applied to examining gelatinization of starches in foods. *J. Food Sci.* 44:1366 (1979).
5. F. Nakazawa, S. Noguchi, J. Takahashi, and M. Takada, Retrogradation of gelatinized potato starch studied by differential scanning calorimetry. *Agric. Biol. Chem.* 49:953 (1985).

RHEOLOGY AND DSC OF GELLAN-AGAROSE MIXED GELS

K. Nishinari[1,2], T. Takaya[2] and M. Watase[3]

[1]National Food Research Institute, Tsukuba 305, Japan
[2]Department of Food and Nutrition, Faculty of Science of Living
Osaka City University, Sumiyoshi, Osaka 558, Japan
[3]Chemical Research Laboratory, Faculty of Liberal Arts
Shizuoka University, Ohya, Shizuoka 422, Japan

ABSTRACT

The rheological and thermal properties of mixed gels of gellan and agarose were studied by dynamic viscoelastic measurement and differential scanning calorimetry (DSC). E' increased slightly up to a certain temperature T_0 and then began to decrease with increasing temperature. This temperature T_0 shifted to higher temperatures with increasing gellan content. The critical concentration for gelation C_0 becomes higher with increasing gellan content. Two endothermic peaks were observed in DSC curves for mixed gels. The higher temperature peak was attributed mainly to the melting of gellan gum network whilst the lower temperature peak was mainly attributed to the melting of agarose network. In the phase separated gels, gellan forms at first a network on cooling, and then agarose forms another network at lower temperatures.

INTRODUCTION

Gellan gum is a microbial polysaccharide produced by *Pseudomonas elodea*, and is widely used in food industry and biotechnology because it forms a transparent gel which is heat resistant and its gel strength is less dependent on pH with comparison to other polysaccharide gels.[1] Agarose has also been used in the similar area as gellan gum, and its gelling mechanism has been studied extensively by many researchers.[2] Rees, Morris and coworkers have proposed the gelling mechanism as the formation of double helices and their aggregation to form junction zones.[3] Much attention has been paid to the interaction between different polysaccharides, and as a result, considerable information has

Food Hydrocolloids: Structures, Properties, and Functions
Edited by K. Nishinari and E. Doi, Plenum Press, New York, 1994

been accumulated.[4] Differential scanning calorimetry (DSC) and viscoelastic measurements were carried out to study the gelation of gellan-agarose mixtures in the present work.

MATERIALS AND METHODS

Agarose was extracted from *Gelidium amansii* produced in Izu Suzaki region in July 1991, at 128℃ using an autoclave for one hour. The molecular weight determined by GPC was 2.6×10^4, the sulfate density was $0.0022 \text{mol}/C_6H_{10}O_5$, and the 3,6-anhydro-L-galactose content was 40.4%. Inorganic ion content in gellan gum used was as follows (μg/g):Na, 400; K, 30800; Mg, 300; Ca, 300. Gellan and agarose powders were mixed at various ratios (w/w) and were dissolved at 95℃, and then poured into cylindrical molds of 20mm inner diameter and 30mm length.

Storage Young's modulus E' and mechanical loss tangent $\tan\delta$ as a function of temperature at 2.5Hz were determined from the observation of longitudinal vibrations of cylindrically molded gels of 20mm diameter and 30mm length.[5]

DSC was carried out by use of a sensitive DSC apparatus DSC 120 (Seiko Electronics. Inc., Tokyo). About 50mg of the gellan-agarose mixed gel was sealed in the DSC pan hermetically. The temperature was kept at 2℃ for 30min and then raised at 2℃/min.

RESULTS AND DISCUSSION

Fig. 1 shows the temperature dependence of storage Young's modulus E' and mechanical loss tangent $\tan\delta$ for gellan-agarose mixed gels with

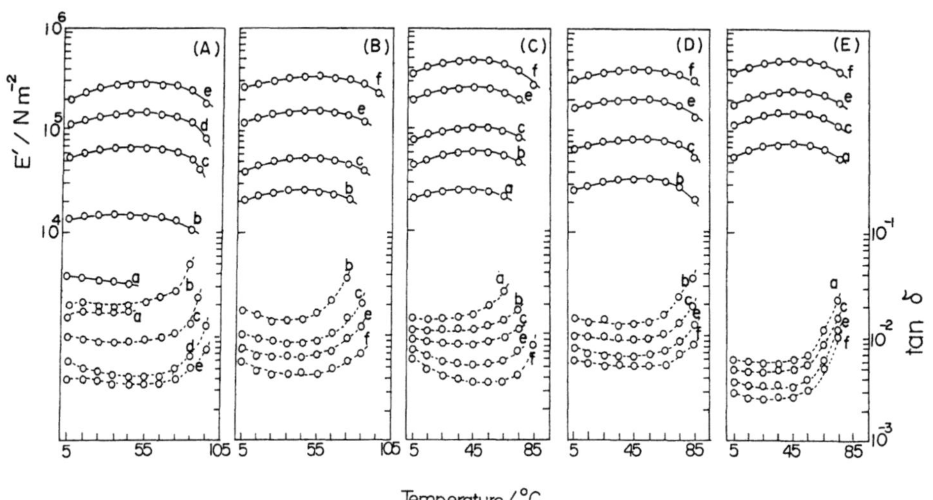

Fig. 1. Temperature dependence of storage Young's modulus E' and mechanical loss tangent $\tan\delta$ for gellan gum(G)-agarose(A) mixed gels. Total polymer concentration: a, 1 wt%; b, 1.5 wt%; c, 2 wt%; d, 2.5 wt%; e, 3 wt%; f, 4 wt%; g, 5 wt%; (A), G alone; (B), G:A=3:1; (C), G:A=1:1; (D), G:A=1:3; (E), A alone.

474

various mixing ratios. E' increased slightly up to a certain temperature T_0 and then began to decrease with increasing temperature. This temperature T_0 shifted to higher temperatures with increasing gellan content. This is explained by the fact that melting temperature T_m for a gel of gellan alone is far higher than that for a gel of agarose alone. Melting temperature T_m is defined later as an endothermic peak temperature in a heating DSC curve. Although T_0 is far lower than T_m, the molecular chains are considered to be released from junction zones above T_0 just like a thread is released from a reel with increasing temperature. [6]

The elastic modulus of agarose gels is larger than that of gellan gels at lower concentration region 1~2%, however, this situation is reversed at higher concentration region 3~4%. At higher concentration than ~2%, the additional agarose dose not contribute to the network structure effectively. E' of gels at 25°C as a function of polymer concentration C was expressed by $E' \sim C^n$ and the exponent n increased with increasing gellan gum content. According to modern theories of the concentration dependence of the elasticity for polymer gels proposed by Clark and Ross-Murphy[7], and by Oakenfull[8], the exponent n is larger in dilute polymer concentration region than in moderate concentration region where n is tending to two. The experimental findings that n increased with increasing gellan content should, therefore, be attributed to the fact that the critical concentration for gelation C_0 becomes higher with increasing gellan content.

Figs. 2(A)-(E) show the heating DSC curves for mixed gels of gellan gum and agarose with various mixing ratios. All the endothermic peaks were attributed to the transition from gel to sol. The endothermic peak temperature for gels of gellan alone and agarose alone shifted to higher temperatures with increasing polymer concentration. This shift is more pronounced in the former than in the latter, indicating that the heat

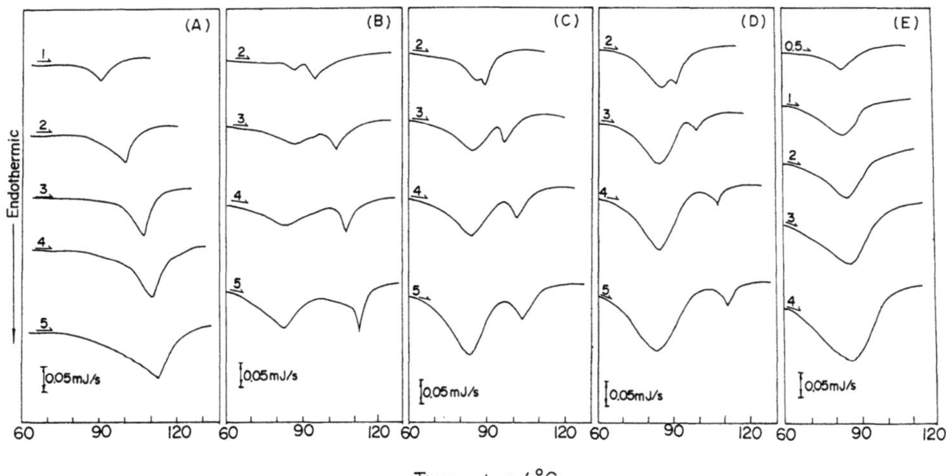

Fig. 2. Heating DSC curves of G-A mixed gels. Figures beside each curve represent the total polymer concentration in wt%. (A)~(E) represent the same meaning as in Fig. 1.

absorbed on forming one mole of junction zones $\triangle$Hm according to Eldridge
and Ferry[9] is larger in agarose gels than in gellan gum gels.   Two
endothermic peaks were observed in mixed gels. The higher temperature
peak was attributed mainly to the melting of gellan gum network whilst
the lower temperature peak was mainly attributed to the melting of
agarose network.   These mixed gels are considered to be phase separated
gels.   The higher temperature peak shifted to higher temperatures with
increasing polymer concentration whilst the lower temperature peak
shifted to lower temperatures in some cases suggesting that the formation
of junction zones in agarose is hindered by the presence of gellan gum.
In the phase separated gels, gellan forms at first a network on cooling,
and then agarose forms another network at lower temperatures.   The values
of $\triangle$Hm for higher temperature peak were 55kJ/mol for gellan alone,
29kJ/mol for the mixed gel with G:A=3:1, 24kJ/mol for the mixed gel with
G:A=1:3.   The fact that $\triangle$Hm decreased with increasing agarose content
suggests that the network formation of gellan is also inhibited by the
presence of agarose.

REFERENCES

1. G. R. Sanderson, Gellan gum, *in*: "Food Gels" , P. Harris Ed. , Elsevier
   Applied Sci. , London, pp. 201-232 (1990).
2. T. Matsuhashi, Agar, *in*: "Food Gels" , P. Harris Ed. , Elsevier Sci. ,
   London, pp. 1-50 (1990).
3. D. A. Rees, E. R. Morris, D. Thom & J. K. Madden, Shapes and interactions of
   carbohydrate chains, *in*: "The Polysaccharides" , G. O. Aspinall Ed. ,
   Academic Press, New York, Vol. 1. pp. 195-290 (1982).
4. P. A. Williams and G. O. Phillips, Interactions in mixed polysaccharide
   systems, *in*: "Food Polysaccharides", A. Stephens and I. C. M. Dea Eds. ,
   Marcel Dekker, in press.
5. M. Watase and K. Nishinari, Rheological and thermal properties of
   agarose and kappa-carrageenan gels containing urea, guanidine hydrochloride or formamide, Food Hydrocolloids, 1: 25 (1986).
6. K. Nishinari, S. Koide and K. Ogino, On the temperature dependence of
   elasticity of thermoreversible gels, J. Phys. (France), 46: 793 (1985).
7. A. H. Clark and S. B. Ross-Murphy, The concentration dependence of biopolymer gel modulus, Brit. Polymer J. , 17: 164 (1985).
8. D. Oakenfull, A method for using measurements of shear modulus to
   estimate the size and thermodynamic stability of junction zones in
   noncovalently cross-linked gels, J. Food Sci., 49: 1103 (1984).
9. J. E. Eldridge and J. D. Ferry, Studies of the cross-linking process in
   gelatin gels. III. Dependence of melting point on concentration and
   molecular weight, J. Phys. Chem. , 58:992(1954)

PHYSIOLOGICAL ASPECTS OF FOOD HYDROCOLLOIDS

David L. Topping

CSIRO Division of Human Nutrition
Glenthorne Laboratory
O'Halloran Hill, 5158
Australia

INTRODUCTION

Carbohydrate and protein hydrocolloids find extensive use in the food industry as gelling agents, emulsifiers and thickeners. These industrial applications are driven largely by technological or organoleptic considerations. The nutritional properties of individual hydrocolloids have attracted relatively little attention from food technologists which is understandable but in marked contrast to the perspective of nutritional scientists who are greatly interested in the physiological effects of complex carbohydrates, many of which are the same polysaccharides used by the food industry. Protein hydrocolloids attract relatively little attention (except possibly as fat substitutes) as in most industrialised countries the national diet is generally more than adequate in protein but is high in fat and low in complex carbohydrates (starches) and fibre. Health authorities in such countries recommend an increased consumption of the latter two and a reduction of total and saturated fat.[1] These coordinated changes have the express aim of reducing the incidence and/or improving the management of degenerative diseases such as coronary artery disease, diabetes, constipation and certain cancers eg of the colon. The evidence supporting these recommendations is generally strong. In the instance of simple constipation and diverticular disease, these may be alleviated by increasing fibre consumption to >20 g/day through the inclusion of high-fibre foods in the diet.[2] At the population level, increased fibre consumption is protective against cardiovascular disease[3] and colonic cancer[2] although for the latter, the greatest protection is with diets low in fat. In the instance of breast cancer, fibre density (ie g consumed/mJ of energy) seems to be one of the most protective dietary factors identified to date.[4]

NUTRITIONAL ATTRIBUTES OF POLYSACCHARIDES

Polysaccharides are an extremely diverse group of compounds and the vast majority are of non-animal origin. In the past, the latter have been simply classified as starch and non-starch polysaccharides (NSP) which is a very convenient subdivision as it also defines their most important nutritional property - small intestinal digestibility. For that reason, this review will compare the physiological effects of starches and NSP rather than carbohydrate hydrocolloids *per se.*

Humans secrete a very narrow range of enzymes capable of hydrolysing polysaccharides and essentially only α-(1-4) and α-(1-6) glycosidic linkages can be broken by endogenous enzymes in the small intestine which means that the only important digestible polysaccharides are starch and glycogen. All of the other (non-starch) polysaccharides are indigestible and conform to the definition of dietary fibre as *plant structural and exudative components indigestible to the enzymes of the human gastrointestinal tract.*[5] These polysaccharides when eaten pass into the colon of human and other monogastric omnivores essentially undigested. This passage of undigested polysaccharides is thought to explain the faecal bulking and relief of constipation seen in humans when fibre intakes are increased. However, it must be recognised that not all fibre sources are equally effective in increasing faecal bulk. For example, wheat bran[6] and rice bran[7] are very efficient in this regard but other products such as pectin and guar gum are not[6]. It will be quite obvious that the latter are typical of the carbohydrate hydrocolloids used by the food industry and are partially or wholly soluble in water. However, it is equally well-established that these water-soluble NSP are the ones with the most attractive physiological properties in the small intestine. A note of caution must be injected at this point as it is a general assumption that solubility *in vitro* is the same as solubility in the gut *in vivo.* Quite clearly this not be the case for NSP in unprocessed foods (such as fruit) or in foods such as vegetables subject to simple cooking although in processed foods where the NSP are fully hydrated, it is reasonable to assume that any physiological effect is due to the polysaccharide in solution.

PHYSIOLOGICAL EFFECTS OF NON-STARCH POLYSACCHARIDES

Blood Glucose Control

Much of the general interest in the potential benefits of water-soluble NSP has arisen from investigation of their potential benefit in diabetes which exists in two forms. Type 1 is characterised by partial or total lack of insulin and Type 2 by is a high degree of insulin resistance. In both instances, good control of blood glucose within the normal physiological range is at a premium and much effort has been expended in investigating the factors that could effect such improvements. Jenkins and co-workers have developed the concept of the *glycaemic index* whereby the rise in blood glucose for a given quantity of available carbohydrate (usually 50 g) in a food is expressed as a percentage of the rise due to the same amount of available carbohydrate in the form of a glucose solution or white bread. A number of difficulties have been identified with this approach including the fact that white bread is not the same all over the world and also that it is more relevant to calculate total area under the curve rather than the peak glucose response[1]. However, given that in Type 1 diabetes excursions of blood glucose outside the desirable range can produce coma and that in Type 2 diabetes normalisation of blood glucose by diet is most desirable, the index has much value. A number of factors have been identified as affecting the blood glucose response including chewing, speed of eating, the particle size of the food and the amylose/amylopectin ratio of the starch.[1] However, one consideration of direct relevance to this discussion is the impact of NSP and a considerable amount of work has been with guar gum and current knowledge indicates that this important hydrocolloid reduces the glycaemic index both in drinks and in foods[8]. The exact mechanism whereby this effect is achieved is still debatable but involves delayed gastric emptying, altered small intestinal transit and impaired glucose absorption. All of these effects of guar gum are thought to be a reflection of the high viscosity of guar gum in solution and much attention has been focussed on other plant NSP with similar physico-chemical characteristics *eg* the β-(1-3,1-4) glucans in oats and barley which are believed to mediate one of the most important effects of water-soluble NSP - lowering of plasma cholesterol.[9]

Reduction of plasma cholesterol

This is arguably the are of carbohydrate physiology that has attracted most attention and it is now established beyond doubt that water-soluble NSP lower plasma cholesterol quite substantially when fed either as isolates or in whole foods.[9-11] A number of water-soluble NSP have been investigated including pectin, guar gum and psyllium with most attention being focussed on guar gum. This product has been shown to produce reductions which are in the range achieved with drug therapies but without the adverse side effects (apart from some abdominal discomfort) seen with some drugs.[10] In fact, there is some evidence that guar gum may enhance the actions of existing therapies such as bezafibrate and lovostatin. Whether this potentiation is due to altered drug pharmacokinetics or to an independent action of the NSP on lipid metabolism remains to be established but is clearly a fruitful area for investigation. We have also shown a positive interaction between oat bran (which is high in β-glucan) and fish oils in the rat with a greater lowering of plasma lipids when the two were fed in combination.[12] Clearly this offers the opportunity for the development of novel food combinations targeted at specific consumer groups (*functional foods*) with greater potential for lipid-lowering. However, a note of caution must be injected as the quantities of NSP used to effect cholesterol reduction are rather higher than those used currently in food products. Nevertheless, there does seem to be scope for the increased use of NSP in the production of foods with nutritionally desirable characteristics.

MECHANISMS WHEREBY NSP LOWER PLASMA CHOLESTEROL

Clearly, this is a most pertinent question and there are currently three putative mechanisms: inhibition of hepatic cholesterol synthesis; enhanced bile acid and neutral sterol excretion; and, altered fat absorption. Of these mechanisms, the former seems to be improbable and the latter two are the most likely.

Inhibition of hepatic cholesterol synthesis

Given that animal digestive enzymes have such a restricted range of carbohydrate substrates, it is not surprising that the concept of fibre indigestibility in the human gastrointestinal tract has become almost an article of faith. However, it is now well-established that there is a very significant bacterial fermentation in the hind gut of humans and other omnivores. This fermentation resembles that in the rumen of obligate herbivores such as cattle and is effected by similar bacterial species with similar substrates and products.[13] Those products are the short-chain fatty acids (SCFA), principally acetate, propionate and butyrate. The breakdown of some water-soluble NSP is so complete that it is unlikely that they reach beyond the distal colon and one questions whether some of their effects (eg stimulation of electrolyte absorption) in the bowel are due to the NSP or SCFA.[10] This is a very important question as SCFA are now finding therapeutic use eg in the prevention of diarrhoea and the treatment of diversion colitis. In the case of cholesterol reduction, it has been proposed that one specific SCFA (propionate) is responsible.[14] This proposition is a very attractive one as propionate is a widely used (and generally regarded as reasonably safe) anti-mould agent in the bread industry. The suggestion that propionate may mediate effects of NSP on plasma lipids is based on animals studies in which lowering of plasma cholesterol has been seen with feeding of sodium or calcium propionate and these reductions are generally of the same order of magnitude as those seen with NSP isolates or foods. Even more compelling is the fact that propionate inhibits hepatic cholesterol synthesis in isolated liver cells.[14] However, this proposition has met some resistance for a number of good reasons.[10,15] Firstly, the concentrations required for suppression of cholesterogenesis are high and outside the normal physiological range. Secondly, the time-course of absorption of oral propionate is rather different to that absorbed from the colon. The administration of NSP in to the colon (the site of fermentation) does not lower plasma cholesterol - indicating that the site of action must be in the small intestine.[16] Finally, intra-rectal administration of propionate actually raises plasma cholesterol in experimental animals.[17]

Enhanced bile acid excretion

In general, NSP isolates and foods containing water-soluble NSP which lower plasma cholesterol appear to enhance faecal bile acid and neutral sterol excretion.[10] That this increase could contribute to reduction of plasma total and LDL cholesterol is not surprising as it is the basis of the therapeutic action of bile acid sequestrants such as cholestyramine which bind anions (including bile acids) in the small intestine. Binding of bile acids interrupts the enterohepatic circulation so that they are lost in the faeces. This drain leads to depletion of the body steroid pool which is made good by lowering of plasma cholesterol. Enhanced steroid excretion has been seen with foods such as oat bran[18] and NSP isolates including guar gum[19] and psyllium.[20] Under these conditions, hepatic cholesterol synthesis is greatly enhanced, consistent with the loss of steroids from the body.

The enhancement of bile acid excretion is not a general one but appears to be confined to specific molecular species which raises the possibility of specific binding by the NSP which implies that naturally-occurring NSP could replace bile acid sequestrants (which are generally tolerated rather poorly). However, two observations suggest that NSP are not binding bile acids selectively. Firstly, we have found recently that pectin (which lowers cholesterol in humans) was most effective in preventing diet-induced hypercholesterolaemia in the rat.[21] Pectin is an acidic polysaccharide and under gut conditions would be most likely to repel bile acids which is inconsistent with the increased bile acid secretion seen in pectin-fed rats. Secondly, Turley and co-workers[20] have shown that in hamsters the body pool of bile acids is expanded with both psyllium and cholestyramine but only the latter bound bile acids specifically. Thus, it appears that NSP increase steroid excretion but by a mechanism different to that of the bile acid sequestrants and it is possible that this is merely one further aspect of the effects of NSP on digestion.

Altered fat absorption

Viscosity is one of the most important physico-chemical properties of NSP which are believed to impact on gut physiology. As with carbohydrate absorption, there is some evidence to support its role in the reduction of plasma cholesterol. For example, Superko *et al*[22] showed that in mildly hypercholesterolaemic men the reduction in LDL cholesterol was greater with guar gum of higher viscosity than with guar gum of lower viscosity. It is known that viscosity of gut digesta is important in modifying the digestion of lipids.[23,24] In humans these changes in digestion are viewed as favourable as our concerns are with reducing the effects of overnutrition. However, in other applications such as broiler nutrition, relatively small changes in feed NSP content and composition can have quite marked antinutritive effects and greatly reduce efficiency.[25] Of course, these considerations add support to the view that water-soluble NSP act in the gut solely through modifications in digestive efficiency and are a powerful supporting argument. Nevertheless, there are some observations which conflict with this view. We have already noted that pectin (which forms viscous solutions only under very limited conditions) is extremely effective in lowering cholesterol. Moreover, animal studies have shown that viscosity is unrelated to plasma cholesterol both with synthetic and naturally occurring NSP.[26,27] On balance, it seems likely that viscosity is a contributor to the hypocholesterolaemic effects of NSP but that other effects such as altered transit and an expansion of the body bile acid pool is also a contributor.

STARCH AND RESISTANT STARCH

Of all of the materials used by food industry, starch is evidently one of the most significant. It is a major constituent of many important plant foods and mention has been made of the recommendations for increased starch consumption in societies where fat intakes are above the optimum. It has also been noted that starch is one of the few polysaccharides which are susceptible to hydrolysis by human digestive enzymes. The major consequence of starch digestion in the human small intestine is the provision of energy (as glucose) which enters the hepatic circulation *via* the portal vein. The gross energy yield of starch (4 kcal/g) is relatively modest level compared with fat (9 kcal/g) and ethanol (7 kcal/g) but one of the misperceptions of starch (at least in some cultures) is that it is "fattening". This is probably a consequence of the fact that the storage of glucose in the body requires the association of 4 mol of water/mol of glycogen glucose. In the gastrointestinal tract, starch and NSP also become hydrated and we have found in animals that there can be very substantial differences in the water held in the colon with various foods.[28,29]

The physiological effects of starches in humans have been examined almost exclusively using foods such as breads, cereals, vegetables and fruit. These studies have concentrated on the rate of starch digestion and glucose absorption and it is known that the rate of starch hydrolysis is known to be affected by a number of factors including the degree of gelatinisation, starch particle size, the degree of chewing (for whole foods), the presence of anti-nutritive factors and the ratio amylose/amylopectin.[1] In contrast to the interest in humans, the role of starch in the diets of experimental animals has been as a source of carbohydrate. In both instances it has been assumed that the digestibility of starch in the small intestine was complete. However, this is now known to be incorrect and, in addition to influencing glucose release, the rate of hydrolysis also affects the fraction of starch which escapes into the colon of humans and other omnivores. In that compartment this so-called resistant starch (RS) is fermented in the same way as other polysaccharides and it has been argued that if effects of fibre in the colon are due to the SCFA products of NSP fermentation then RS should be regraded as "fibre".[14] Indeed, it has been calculated that the contribution of RS to colonic fermentation may greatly exceed that of NSP.[30] Studies in experimental animals support this view and we have found in pigs that both the total pool of VFA in the colon and their distribution along the gut length cannot be predicted solely by the NSP content of foods and in both legumes and rice, RS augments their "fibre" content.[28,29]

Colonic fermentation has other implications which have yet to be fully evaluated. For example, it is assumed that NSP provide no energy and that starch has a gross energy of 4 kcal/g. However, SCFA are used by the body so that NSP effectively do provide energy and a yield of approximately 2 kcal/g has been derived.[31] A similar value applies to RS fermentation which is lower than that for small intestinal digestion because of the relative inefficiencies of bacterial fermentation. It follows that the true digestibility (and energy yield) of starchy foods in the body depends on the site of digestion.

CONCLUSIONS

The subjects of other plenary lectures at this Conference support the view that, up to the present, the interest of food technologists in the carbohydrate hydrocolloids they use as ingredients has been largely technological. Given the diverse benefits which have been shown for starch and non-starch polysaccharides it seems to be only a matter of time before their physiological attributes are exploited by food industry in new products.

REFERENCES

1. Complex Carbohydrates in Foods. **Report of the British Nutrition Foundation's Task Force.** Chapman and Hall, London (1990).
2. D.L.Topping, and S.H.Wong, Preventive and therapeutic aspects of dietary fibre, in:"Preventive Nutrition in Medical Practice," J.S.Vobecky and M.L.Wahlqvist, eds., Smith-Gordon, London (In the Press).
3. K.-T.Khaw, and E.T.Barrett-Connor, Dietary fiber and reduced ischaemic heart disease mortality rates in men and women:a 12-year prospective study, *Am. J. Epidemiol.* 133:24(1991).
4. P.A.Baghurst, and T.E.Rohan, High fibre diets and reduced breast cancer risk, Food and Cancer Prevention, *Porceedings*, In the press (1992).
5. H.C.Trowell, Ischaemic heart disease and dietary fiber, *Am. J. Clin. Nutr.* 5:926(1972).
6. B.O.Schneeman, Dietary fiber: physical and chemical properties,methods of analysis and physiological effects, *Food Technol.* 43:133(1989).
7. M.Kestin, R.Moss, P.M.Clifton, and P.J.Nestel, Comparative effects of three cereal brans on plasma lipids, blood pressure and glucose metabolism in mildly hypercholesterolemic men, *Am. J. Clin. Nutr.* 52:661(1990).
8. D.J.A.Jenkins, T.M.S.Wolever, and R.H.Taylor, Dietary fibre, fibre analogues and glucose transport: importance of viscosity, *Am. J. Clin. Nutr.* 25:926(1972).
9. P.J.Wood, J.W. Anderson, J.T.Braaten, N.A.Cave, F.W.Scott, and C.Vachon, Physiological effects of β-D-glucan rich fractions from oats, *Cereal Foods World* 34:878(1989).
10. P.A.Todd, P.Benfield, and K.L.Goa, Guar gum: a review of its pharmacological properties, and use as a dietary adjunct in hypercholesterolaemia, *Drugs* 39:917(1990).
11. D.L.Topping, Soluble fibre polysaccharides: effects on plasma cholesterol and colonic fermentation, *Nutr. Rev.* 49:195(1991).
12. P.D.Roach, K.Dowling, S.Balasubramaniam, R.J.Illman, A.Kambouris, P.J.Nestel and D.L.Topping, Fish oil and oat bran in combination effectively lower plasma cholesterol in the rat, *Atherosclerosis* In the press (1992).

13. S.E.Fleming, and D.S.Arce, Volatile fatty acids: their production, absorption and roles in human health, *Clin. Gastroenterol.* 15:787(1986).

14. W.-J.L.Chen, J.W.Anderson, and D.Jennings, Propionate may mediate the hypocholesterolemic effects of certain soluble plant fibers in cholesterol-fed rats, *Proc. Soc. Exp. Biol. Med.* 175:215(1984).

15. R.J.Illman, D.L.Topping, G.H.McIntosh, R.P.Trimble, G.B.Storer, M.N.Taylor, and B.-Q.Cheng, Hypocholesterolaemic effects of dietary propionate: studies in whole animals and perfused rat liver, *Ann. Nutr. Metab.* 32:97(1988).

16. F.Ahrens, H.Hagermeister, M.Pfeuffer, and C.A.Barth, Effects of oral and intracecal administration of pectin on blood lipids in minipigs, *J. Nutr.* 116:70(1986).

17. C.S.Venter, H.H.Vorster, and D.G.Van der Nest, D.G. Comparison between physiological effects of konjac-glucomannan and propionate in baboons fed "western" diets, *J. Nutr.* 120:1046(1991).

18. R.J.Illman, and D.L.Topping, Effects of dietary oat bran on faecal steroid excretion, plasma volatile fatty acids and lipid synthesis in the rat, *Nutr. Res.* 5:839(1985).

19. T.A.Miettenen, and S.Tarpila, Serum lipids and cholesterol metabolism during guar gum, plantago ovata and high fibre treatments, *Clin. Chim. Acta* 183:253(1989).

20. S.D.Turley, B.P.Daggy, and J.M.Dietschy, Cholesterol-lowering action of psyllium mucilloid in the hamster: sites and possible mechanisms of action, *Metabolism* 40:1063(1991).

21. D.L.Topping, M.Abbey and C.Triantafilidis, Effects of soluble non-starch polysaccharides on plasma cholesterol and liver lipoprotein receptor activity, *in*: "Gums and Stabilisers for the Food Industry" G.O.Phillips, P.A.Williams, and D.J.Wedlock, eds., IRL Press, Oxford (1992).

22. H.R.Superko, W.L.Haskell, L.Sawrey-Kubicek, and J.W.Farquhar, Effects of solid and liquid guar gum on plasma cholesterol and triglyceride concentrations in moderate hypercholesterolemia, *Am. J. Cardiol.* 62:51(1988).

23. I.Ikeda, Y.Tomari, and M. Sugano, M., Interrelated effects of dietary fiber and fat on lymphatic cholesterol and triglyceride absorption in rats, *J. Nutr.* 119:1383(1989).

24. K.Ebihara, B.O.Schneeman, Interaction of bile acids, phospholipids, cholesterol and triglyceride with dietary fibers in the small intestine of rats. *J. Nutr.* 119:1100(1989).

25. G.Annison, relationship between the levels of soluble nonstarch polysaccharides and the apparent metabolizable energy of wheats assayed in broiler chickens, *J. Ag. Food Chem.* 39:1252(1991).

26. D.L.Topping, D.Oakenfull, R.P.Trimble, and R.J.Illman, A viscous fibre (methyl cellulose) lowers blood glucose and plasma triacylglycerols and increases liver glycogen independently of volatile fatty acid production in the rat, *Br. J. Nutr.* 59:(1988)

27. A.J.Evans, R.L.Hood, D.G.Oakenfull, and G.S.Sidhu, Relationship between structure and function of dietary fibre: a comparative study on the effects of three galactomannans on cholesterol metabolism in the rat, *Br. J. Nutr.* In the press (1992).

28. D.L.Topping, R.J.Illman, J.M.Clarke, R.P.Trimble, K.A.Jackson, and Y.Marsono, Dietary fat and fiber alter large bowel and portal venous volatile fatty acids and plasma cholesterol but not biliary steroids in pigs, *J. Nutr.* In the press (1992).

29. Y.Marsono, R.J.Illman, J.M.Clarke, R.P.Trimble, and D.L.Topping, Plasma lipids and large bowel volatile fatty acids in pigs fed white rice, brown rice and rice bran, *Br. J. Nutr.* Accepted for publication (1992).

30. J.H.Cummings, G.R.Gibson, and G.T.MacFarlane, Quantitative estimates of fermentation in the hind gut of man, *Acta Vet. Scand (Suppl.)* 86:76(1989).

31. G.Livesey, Energy values of unavailable carbohydrate and diets: an inquiry and analysis, *Am. J. Clin. Nutr.* 51:617(1991).

EVALUATION OF SHORT-CHAIN FATTY ACID PRODUCTION
IN THE CECUM OF FIBER FED RATS

Hiroshi Hara, Yutaka Saito, Atsuko Idei, and Shuhachi Kiriyama

Department of Bioscience and Chemistry
Faculty of Agriculture
Hokkaido University
Sapporo 060, Japan

ABSTRACT

Short-chain fatty acid (SCFA) production rate in the cecum was estimated by the in vitro incubation system of the cecal content collected immediately after sacrifice, and compared with the SCFA content in the cecum under various conditions in rats. The cecal contents were anaerobically incubated at 37 °C for 1-4 hr. The initial concentration (SCFA content) and the increment of the each SCFA concentration after the incubation (SCFA production rate) were increased by fiber feeding. But, the proportions of the larger SCFA contents in the total SCFA were smaller than those of the production rates. We also observed that the SCFA content and production rates changed diurnally in rats fed guar gum, but the variation profiles of the SCFA content in the cecum were different from those of the production rates. Neomycin feeding clearly decreased the SCFA production, but not the contents in the fiber fed groups. The production rate reflects the in vivo digestible energy of dietary fiber sources measured by fecal excretion of fiber energy. These results reveal that the short time of the in vitro incubation of the cecal contents was a simple and useful method for the evaluation of the in vivo SCFA production rate from ingested dietary fibers, and that the contents of SCFA in the cecum did not usually reflect the SCFA production rate.

INTRODUCTION

Fermentation of dietary fibers and production of short-chain fatty acids (SCFAs) by gut flora had been estimated by the concentrations of SCFA in the colonic contents or feces [1,2]. The concentration of SCFA in the cecum or colon is determined by the production and absorption rates of SCFA, although the content of SCFA did not usually reflect the production rate of SCFAs. We evaluated the SCFA production rate in the cecum by using the in vitro incubation system of the cecal content. We used the short time (1-4 hr) incubation of the cecal content removed immediately after sacrifice in order to estimate the in vivo SCFA production rate in the cecum. The method was described by Carroll and Hungate [3]. The simple method was also applied, using rabbits [4] or porcupines [5].

The aim of the present study is the comparison of the SCFA production rate with the content of SCFA in the cecum under various conditions. This study consisted of three experiments. First, the proportion of the individual SCFA production rates was compared with those of the individual SCFA content under a meal feeding of a fiber-free or a corn husk diet. Secondly, diurnal variation of the SCFA production rate and cecal content were observed in the rats fed a diet containing guar gum. Finally, the effects of an antibiotic, neomycin, feeding on the SCFA production rate was compared with that of the SCFA content in the cecum.

Food Hydrocolloids: Structures, Properties, and Functions
Edited by K. Nishinari and E. Doi, Plenum Press, New York, 1994

MATERIALS AND METHODS

Animals and Diets

Experiment 1: The proportion of the individual SCFA production rate

Male Sprague-Dawley rats (Japan SLC, Hamamatsu, Japan), weighing 125-150g, were divided into 2 groups of 6 rats after a 6-hr (10:00-16:00) meal feeding of a fiber-free stock diet (Table 1)[6-9] for 12 days. The rats were fed a fiber-free or a diet containing10% corn husk for 22 days using a meal feeding procedure. On the last day, aortic blood was withdrawn under pentobarbital anesthesia (Nembutal, sodium pentobarbital: 50 mg/kg body weight, Abbott Co., North Chicago, IL) 5 hr after feedings of the test diets, and the cecum was removed immediately with the content.

Experiment 2: Diurnal variation of the SCFA production rate in the cecum

Male Sprague-Dawley rats (Japan SLC), weighing about 100 g, were divided into 6 groups of 6 rats after feeding a fiber-free stock diet for 7 days. The rats were fed a fiber-free or a diet containing 5% guar gum for 22 days. On the last day, aortic blood was withdrawn under pentobarbital anesthesia 4 , 12, and 20 hr after feedings of the test diets, and the cecum was removed with the content.

Experiment 3: Effects of neomycin on the SCFA production rate

Rats and feeding procedure were the same as Experiment 2. The test diets contained a 10% potato fiber or sugar beet fiber with or without 0.1% neomycin sulfate (Sigma Chemical Co., St. Louis, MO, U.S.A.). The feces were collected from the 19th to the 21th day after feeding of the test diets to measure gross energy excretion into feces. Rats were sacrificed under pentobarbital anesthesia 20 hr after feeding on the last day.

Throughout all the experiments, the rats were kept in a temperature controlled room at 23±2°C, maintaining a 12 h light-dark cycle (light period, 8:00-20:00).

Table 1 Composition of test diet

	Stock diet (fiber-free diet)[1]
	g / kg diet
Casein [2]	250
Corn oil [3]	50
Mineral mixture [4]	40
Vitamin mixture [5]	10
Granulated vitamin E [6]	1.0
Choline bitartrate	4.0
Sucrose	to make 1 kg

1. Dietary fiber sources were added to the stock diet up to 5% or 10% .Corn husk was obtained from Oji-Cornstarch Co. Ltd. (Ichihara, Japan), Guar gum was gifted from Asahi Kasei Co. (Tokyo, Japan). Potato fiber was prepared by treatment of bacterial amylases (Yakuruto Co. Tokyo) from potato pulp. Sugar beet fiber was gifted from Nihon Tensai Seito Co. Ltd. (Obihiro, Japan).
2. Casein (ALACID; New Zealand Dairy Board, Wellington, New Zealand).
3. Retinyl palmitate (7.66 μmol/kg diet) and ergocalciferol (0.0504 μmol/kg diet) were added to corn oil.
4. The mineral mixture is prepared based on the AIN-76 Workshop held in 1989[6]. It provided (mg/kg diet); Ca, 4491; P, 2997; K, 3746; Mg, 375; Fe, 100; I, 0.32; Mn, 10.0; Zn, 34.7; Cu, 6.00; Na, 4279; Cl, 6542; Se, 1.05; Mo, 1.00; Cr, 0.50; B, 0.50; V, 0.25; Sn, 2.00; As, 1.00; Si, 20.0; Ni, 1.00; F, 2.72; Co, 0.20.
5. The vitamin mixture was prepared in accordance with the AIN-76 mixture[7] except that menadione and L-ascorbic acid were added to make 5.81 μmol/kg[8] and 284 μmol/kg[9] diet, respectively.
6. Vitamin E granule (Juvela, Eisai Co., Tokyo) supplied 423 μmol all-rac-α-tocopheryl acetate per kg diet.

Analyses

The cecal contents were weighed, suspended in 10 ml of autoclaved deionized water in a sealed vial with a silicone septum filled with nitrogen gas, and incubated at 37°C. One ml of the suspension of the cecum content was collected by a long injection needle (18G x 90 mm, Terumo Corp. Tokyo, Japan) through a silicone septum, both before, and 1 or 4 hr after the incubation. We used a 4 hr incubation period in order to measure the minor SCFAs production rates correctly in experiment 1. The incubated cecal content was deproteinized by perchloric acid (final 5% (w/v)) cooled by ice, and a supernatant was added to the KOH solution to precipitate perchloric acid and to form potassium salts in the SCFAs. An individual SCFA was measured by gas-liquid chromatography (Shimadzu GC-14A with a glass column (1600 mm x 3 mm) packed 10% H_3PO_4 on 80-100 mesh chromosorb W-AW DMCS, Shimadzu Corporation) after adding phosphoric acid.

The feces were freeze-dried, weighed, and milled. Gross energy in fiber preparations and powdered feces were measured by bomb calorimetry (CA-3, Shimadzu Corporation, Kyoto, Japan).

<u>Calculation and Statistics</u>

Content and production rates of SCFAs by the cecal content were expressed per whole cecum. The production rate was calculated from the increase in each SCFA concentration of the suspension medium of the cecal content for the 1 or 4 hr incubation period. The cecal contents of SCFAs were evaluated from the initial concentrations of SCFAs in the suspension.

Digestible energy in dietary fiber sources in the test diet was evaluated by applying the following equation (1):

Digestible energy in dietary fiber (DF) (%)

$$= \frac{\text{Energy in DF consumed (KJ/ 3d)) - Fecal energy excretion derived from DF(KJ/3d)}}{\text{Energy in DF consumed (KJ/ 3d)}} \times 100 \quad (1)$$

Fecal energy excretion derived from DF was estimated by subtracting the average value of the fecal energy excretions of the fiber-free group from the fecal energy of each rat in the fiber groups. The subtracted values were corrected by the food intakes of individual rats of the fiber groups.

Data were analyzed by one-way or two-way (Figs.3 and 4) analysis of variance (ANOVA, p<0.05), and significant differences between diet groups were determined by a Student's *t*-test or the Duncan's multiple range test (p<0.05).

RESULTS

Figure1 presents SCFA contents and production rates in the cecum of rats fed a fiber-free or a corn husk diet. The contents and production rates of acetic and butyric acids were about increased 2-fold by the fiber feeding, respectively. The differences in the production rates between SCFA species were smaller than those of the cecal contents in both of the diet groups. Especially, the fractions of minor SCFAs, isovaleric and valeric acids contents in the cecum were very small in comparison with the proportions of these SCFA production rates, which were presented as 'Relative values' in Fig. 1.

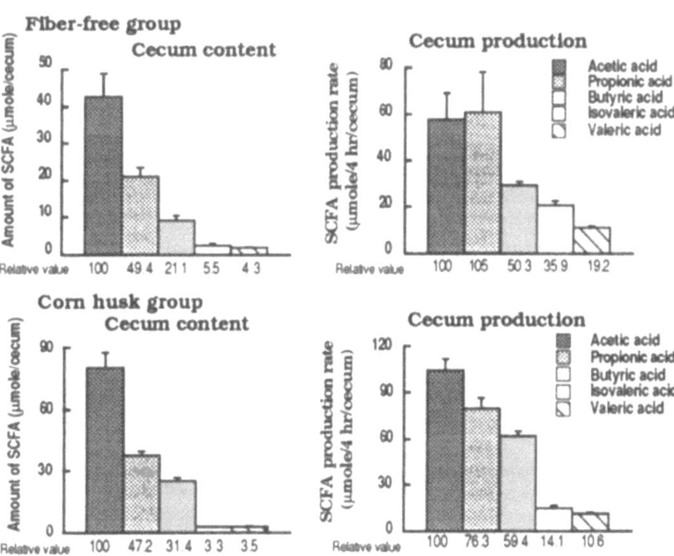

Fig.1 Short-chain fatty acid (SCFA) contents and production rates in the cecum of rats fed a fiber-free or 10% corn husk diet for 3 weeks. Values are mean ± SEM (n=6). Acetic and butyric acids contents and production rates in the fiber feeding groups are significantly higher than those in the fiber-free group respectively. Relative values were calculated, then were assigned a value of acetic acid as 100 in each graph.

The SCFA production rates changed diurnally in guar gum feeding rats, as shown in Fig. 2. The production rates 12 and 20 hr after feeding (putting out the light) were higher than that 4 hr after feeding in rats fed the fiber diet. In contrast, the content of the total SCFA was the highest 4 hr after the feeding. The variation profiles of the major SCFA, acetic, propionic and butyric acids were similar to that of the total SCFA (Data was not shown). The body weight gain and food intake for 3 weeks in the guar gum group were the same as those of the fiber-free group. The time of the sacrifice did not influence the weight of the cecal content in both of the diet groups (Data was not shown).

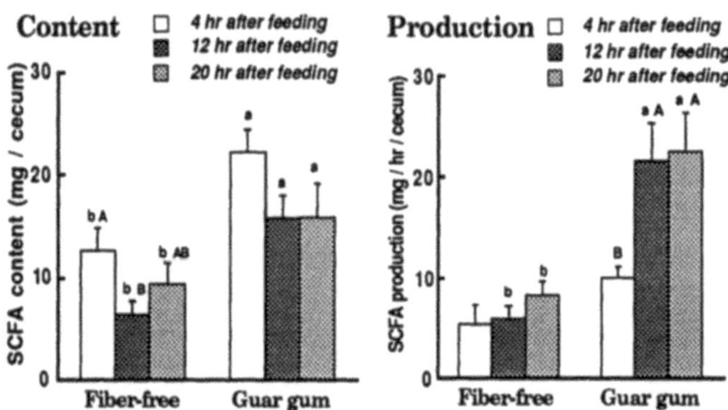

Fig. 2 Diurnal variations of short-chain fatty acid (SCFA) contents and production rates in the cecum of rats fed a fiber-free or 5% guar gum diet for 3 weeks. Values are mean ± SEM (n=6). The values not sharing a common capital letter are significantly different between times after feeding, and the values not sharing a common small letter are significantly different between the fiber-free and guar gum groups at the same time after feeding (p<0.05).

Figure 3 presents the effects of neomycin on the contents and production rates of the major SCFAs in the cecum. In rats fed 10% potato fiber (PF) diets, all the SCFA production rates were increased 2-3 fold by the fiber feeding but were clearly decreased by neomycin feedings. The increments of the SCFA production rates in sugar beet fiber (BF) groups were similar to those in the PF group, but the effects of neomycin was unclear except for butyric acid production. The degrees of the reduction of SCFA contents by neomycin were smaller than those of the production rates.

The contents and production rates of the sum of SCFAs were shown in Fig. 4. The production rates were clearly decreased by the neomycin feedings in both fiber groups, but the SCFA content in the cecum of the rats fed the BF diet was not influenced by the neomycin feeding. The changes in the digestible energy of the dietary fiber which were estimated by the fecal energy excretions of the dietary fiber sources were very similar to those of the total SCFA production rates, but not to the SCFA contents. The digestible energy in BF was higher than that in PF.

The body weight gains and food intakes were not influenced by the fiber and neomycin feedings, as shown in Table 2. The weight of cecal contents were markedly increased by neomycin feedings in both the PF and BF groups.

Table 2 Effects of neomycin feedings on the body weight gain, food intake and the weight of cecal content of rats fed the test diets for 3 weeks

	Body weight gain	Food intake	Cecum content
	g / 21 days		g / rat
Fiber-free diet	123 ± 5 b	377 ± 17 ab	0.39 ± 0.16 d
10% Potato fiber (PF) diet	145 ± 5 a	419 ± 14 ab	2.82 ± 0.19 c
PF + 0.1% neomycin diet	127 ± 3 ab	434 ± 33 a	4.13 ± 0.37 b
10% Sugar beet fiber (BF) diet	131 ± 7 ab	404 ± 20 ab	2.85 ± 0.17 c
BF + 0.1% neomycin diet	122 ± 8 b	364 ± 15 b	8.02 ± 1.07 a

All values are mean ± SEM (n=6). Means values in the same column with unlike alphabetical letters were significantly different (p<0.05).

Cecal contents of major SCFAs

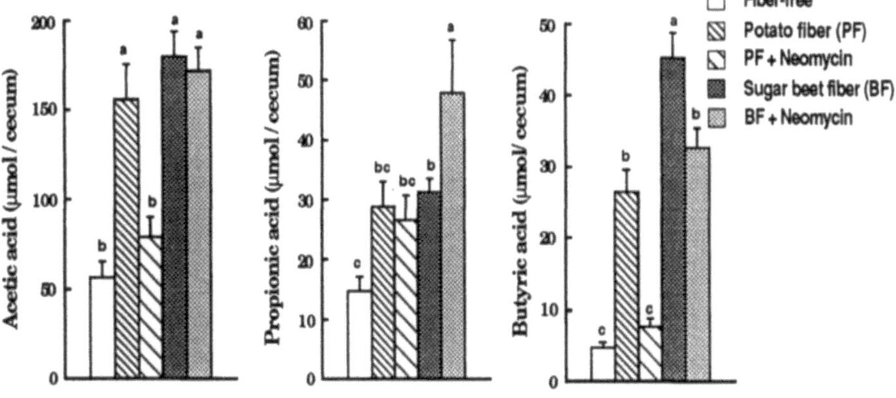

Cecal production of major SCFAs

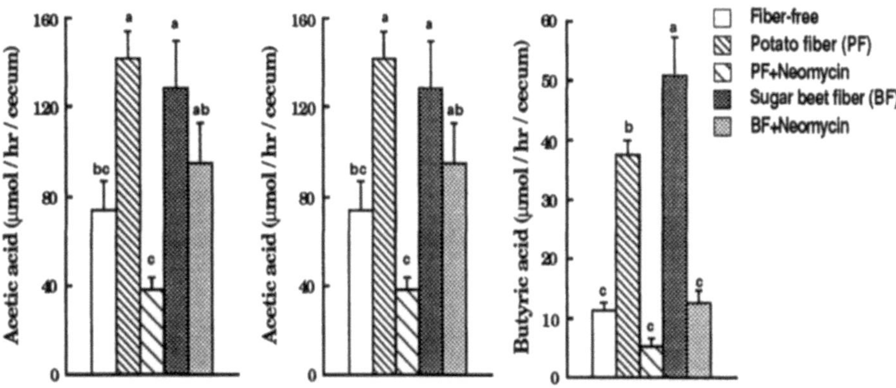

Fig. 3 The effects of neomycin feedings (0.1% in diets) on the short-chain fatty acid (SCFA) contents (upper panels) and production rates (lower panels) in the cecum of rats fed a 10% potato fiber or a 10% sugar beet fiber diet. Values are mean ± SEM (n=6). The values not sharing a common letter are significantly different (p<0.05).

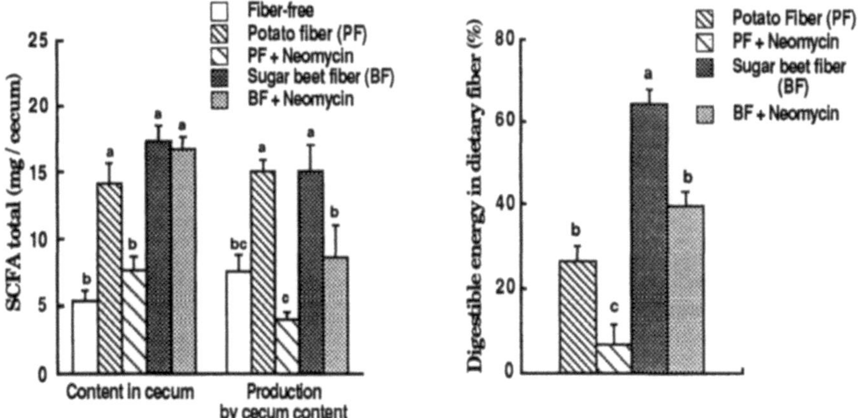

Fig. 4 Relationships between cecal contents and cecal production rates of short-chain fatty acids (SCFAs) and digestible energy values in dietary fiber sources in the rats fed the fiber diets with or without neomycin. Values are mean ± SEM (n=6). The values not sharing a common letter are significantly different (p<0.05).

DISCUSSION

We measured the SCFA contents in the cecum and the SCFA production rate during a short-period in vitro incubation of the cecal content of rats fed test diets containing various dietary fiber sources, and the changes in SCFA contents in the cecum did not usually coincide with those of the production rates of SCFAs.

Figure 2 demonstrates that the proportion of the production rate of individual SCFAs is different from that of the content in the cecum. That is, the proportion of the remaining larger SCFAs (contents) were smaller than those of the production rates of these SCFAs. The finding suggests that the larger SCFA is absorbed faster than the smaller one.

We observed that the profile of diurnal changes in the SCFA contents in the cecum was different from those of the production rates of SCFAs, especially, in the rats fed a guar gum containing diet (Fig. 2). The result demonstrates that the content of the SCFAs did not reflect those production rates at that time. Absorption rates of SCFA from the cecum may be slow 4 hr after feeding for some reason, for example, the mineral content and pH in the cecum. These factors are known to affect the SCFA absorption rate[10,11].

Figures 3 and 4 show that neomycin depresses the SCFA productions in the rats fed the potato fiber (PF) and sugar beet fiber (BF). Neomycin also reduced the digestible energy of both fiber sources in the same degree as the SCFA production rates. These results show that our estimation of the SCFA production rate reflects the in vivo energy supply from the dietary fibers. The digestible energy of BF was larger than that of PF, in spite of the fact that the SCFA production rate in the cecum of the BF group was the same as that of the PF group. The finding reveals that the BF was digested in a part other than the cecum, probably in the small intestine.

In the BF group, neomycin did not influence the SCFA content in the cecum (Fig. 4). Also in the PF group, the difference in the SCFA content between the groups with and without neomycin was smaller than the difference in the SCFA production rate (Fig. 4). These results suggest that the SCFA absorption rate was slower in the neomycin groups. Possibly, lower production rates of SCFAs affect the absorptive system for SCFAs in the cecum.

REFERENCES

1. J.C. Mathers and L.D. Dawson, Large bowl fermentation in rats eating processed potatoes., *Br. J. Nutr.* **66**: 313-329 (1991).
2. S.E.Fleming, M.D. Fitch and S. DeVries, The influence of dietary fiber on proliferation of intestinal mucosal cells in miniature swine may not be mediated primarily by fermentation., *J. Nutr.* **122**: 906-916 (1992).
3. E.J.Carroll and R.E.Hungate, The magnitude of the microbial fermentation in the bovine rumen., *Appl. Microbiol.* **2**: 205-214 (1954).
4. W.H. Hoover and R.N. Heitmann, Effects of dietary fiber levels on weight gain, cecal volume and volatile fatty acid production in rabbits., *J. Nutr.* **102**: 375-380 (1972).
5. J.L.Johnson and R.H.McBee, The porcupine cecal fermentation., *J. Nutr.* **91**: 540-546 (1967).
6. P.G.Reeves, AIN-76 diet: should we change the formulation?, *J. Nutr.* **119**: 1081-1082 (1989).
7. American Institute of Nutrition, Report of the American Institute of Nutrition, ad hoc Committee on Standards for Nutritional Studies., *J. Nutr.* **107**: 1340-1348 (1977).
8. A.E.Harpar, Amino acid balance and imbalance. 1. Dietary level of protein and amino acid imbalance., *J. Nutr.* **68**: 405-418 (1959).
9. American Institute of Nutrition, Second report of the ad hoc Committee on Standards for Nutritional Studies., *J. Nutr.* **110**: 1726 (1980).
10. D. Hollander, E.M. Gerard and C.A.R. Boyd, Transport of butyric acid in vascularly perfused anuran small intestine: importance of pH and anion transport., *Am. J. Physiol.*,**250**: G469-G474 (1986).
11. S.E. Fleming, S.Y. Choi and M.D. Fitch, Absorption of short-chain fatty acids from the rats cecum in vivo., *J. Nutr.*,**121**: 1787-1797 (1991).

DIETARY FIBER–INDUCED CHANGES IN BILE ACID CONJUGATION AND TAURINE METABOLISM IN RATS

Takashi Ide

Laboratory of Nutrition Biochemistry
National Food Research Institute
Ministry of Agriculture, Fisheries and Forestry
Tsukuba 305, Japan

ABSTRACT Effect of types of dietary fiber on biliary bile acid glycine/taurine ratio (G/T) and activities of hepatic enzymes in bile acid conjugation (bile acid–CoA:amino acid N–acyltransferase) and taurine synthesis (cysteine dioxygenase and cysteine sulfinate decarboxylase) was examined in the rats. Dietary pectin compared to cellulose greatly increased the G/T accompanying the decrease in the concentration of hepatic taurine. However, the N–acyltransferase reaction produced taurine– but little glycine–conjugated bile acids when both taurine and glycine were coexisted at the concentrations of physiological ranges in the assay media both in rats fed cellulose and pectin. Dietary pectin relative to cellulose greatly decreased the activity of cysteine dioxygenase but rather increased that of cysteine sulfinate decarboxylase. A taurine–lowering agent (guanidinoethyl sulfonate) severely decreased activities of these enzymes in taurine synthesis accompanying the decreases in hepatic taurine concentrations both in rats fed cellulose and pectin. This drug like as pectin profoundly increased the G/T irrespective of the dietary fiber sources. Although there was no consistent relationship between the G/T and hepatic concentration of taurine, an inverse relationship between the activity of cysteine dioxygenase and the G/T was observed among the groups of rats. It was suggested that the alteration in the activity of hepatic cysteine dioxygenase and thus the taurine synthesis is the factor to modify the G/T in rats fed different types of dietary fiber.

INTRODUCTION

Bile acids are conjugated either with glycine or taurine by the reaction catalyzed by bile acid–CoA:amino acid N–acyltransferase in the liver before being excreted in the bile. As physicochemical properties of glycine– and taurine–conjugated bile acids are considerably different each other, alterations in the bile acid glycine/taurine ratio may modify the various aspects of cholesterol and bile acid metabolism in the animals. There is a weight of comparable studies regarding the specificity toward the amino acid substrates of the bile acid conjugation enzyme among various animals species.[1-5] Considerable species to species difference in the bile acid glycine/taurine ratio[6,7] has been explained by the difference in the substrate specificity of the enzyme among the animal species. The taurine– and glycine–dependent activities in various animal species including rat were not separated during the purification process indicating that single enzyme is catalyzing bile acid conjugation in the liver.[2-4] The Km value of the purified rat enzyme for taurine (0.8

Food Hydrocolloids: Structures, Properties, and Functions
Edited by K. Nishinari and E. Doi, Plenum Press, New York, 1994

mM) was 40–50 times lower than that for glycine (40 mM).[3] Thus, the rat enzyme is exclusively specific for taurine, and the substrate specificity accounts for the predominance of taurine–conjugated bile acid in this animal species.[6-7]

However, we recently demonstrated that dietary fiber profoundly influenced the bile acid glycine/taurine ratio in rats. Thus, water–soluble dietary fibers (pectin, guar gum and konjak mannan)[8-11] elicited a large increase in the glycine/taurine ratio in biliary bile acid accompanying a decrease in the concentration of hepatic taurine. In contrast water–insoluble dietary fiber (cellulose, chitin and corn bran) rather decreased the bile acid glycine/taurine ratio and increased hepatic taurine concentration.[10] Also, dietary cholesterol increased the ratio irrespective of the dietary fiber sources, pectin or cellulose.[8] However, when hepatic glycine concentration was determined in rats fed a water–soluble dietary fiber (pectin), it was comparable to that of taurine.[9] Thus, in terms of the substrate specificity of a bile acid conjugation enzyme, the alteration in the hepatic taurine concentration could not be accounted for a huge increase in glycine conjugation at least in rats fed pectin. These results indicate the existence of an alternate unknown mechanism(s) to modify the bile acid glycine/taurine ratio in the liver. In the present study, we examined the effect of dietary fiber on bile acid conjugation and taurine metabolism in laboratory rats. The data obtained indicated a close relationship between bile acid conjugation reaction and hepatic taurine synthetic activity in the liver.

MATERIALS AND METHODS

Animals and diets Male Wistar rats obtained from the Imamichi Institute of Animal Reproduction, Ibaraki, Japan were used throughout in the present study. The animals were housed individually in a room with controlled temperature (20–22 °C), humidity (55–65%) and lighting (lights on from 0700 h to 1900 h), and fed purified experimental diets. The basal composition of the experimental diet was (in weight %): casein 20; corn starch, 15; corn oil, 5; mineral mixture,[12] 3.5; vitamin mixture,[12] 1.0; choline bitartrate, 0.2 and sucrose to 100. Dietary fiber sources were added to the experimental diets at the expense of sucrose.

Assays of enzyme activities Rats were bled from the inferior vena cava and livers were quickly excised. About 3 g of each liver was homogenized in 20 ml 0.25 M sucrose and the homogenate was centrifuged at 8,000 x*g* at 4 °C for 10 min. The supernatant fraction of the liver homogenate was employed for the assays of bile acid–CoA:amino acid *N*–acyltransferase,[13] cysteine dioxygenase[10] and cysteine sulfinate decarboxylase.[14]

Analyses of bile acids and amino acids Biliary bile acids were extracted, fractionated and determined enzymatically.[11, 15] Hepatic taurine and glycine concentrations were measured using an amino acid analyzer.[9]

RESULTS AND DISCUSSION

In the first experiment, rats were fed experimental diets either containing 10% cellulose or pectin for 26 days. As shown in Table 1, rats fed pectin relative to those fed cellulose showed increased bile flow. Pectin markedly increased biliary bile acid excretion, and the increase was essentially attributed to the glycine–conjugated bile acids, whereas it rather decreased the excretion of taurine–conjugates. As a consequence, bile acid glycine/taurine ratio in bile became greatly higher in rats fed pectin than in the animals fed cellulose.

Table 2 summarizes concentrations of taurine and glycine and the activity of cholyl-CoA:amino acid *N*–acyltransferase in the liver. Dietary pectin relative to cellulose significantly decreased the taurine concentration in the liver, whereas glycine concentration was comparable. The *N*–acyltransferase activity, both taurine– and glycine–dependent activities, in rats fed pectin was significantly higher than that in the animals fed cellulose. In spite of the observation that types of dietary fiber strikingly modified the bile acid glycine/taurine ratio (Table 1), the response of the enzyme reaction was rather in favor of the formation of taurine– than glycine–conjugated bile acids both in the animals fed

Table 1. Biliary bile acid secretion in rats fed cellulose and pectin

| | Dietary fiber | |
	Cellulose	Pectin
Bile flow (ml/hr)	0.81±0.02*	1.25±0.07
Biliary bile acid secretion (µmol/hr)		
Total	21.5±1.2*	45.2±3.3
Non–conjugated	0.56±0.10*	2.75±0.76
Glycine–conjugated	1.74±0.25*	28.5±2.8
Taurine–conjugated	19.2±0.9*	13.9±1.5
G/T ratio	0.09±0.01*	2.18±0.34

*Values are significantly different from those in rats fed pectin at p<0.05.

Table 2 Hepatic concentrations of taurine and glycine and activities of bile acid–CoA:amino acid N–acyltransferase in rats fed cellulose and pectin

| | Dietary fiber | |
	Cellulose	Pectin
Amino acids (µmol/g)		
Taurine	4.82±0.47*	1.88±0.10
Glycine	2.21±0.08	2.11±0.13
N–Acyltransferase (nmol/min/mg)		
20 mM Taurine	43.4±3.1*	52.9±3.4
100 mM Glycine	45.2±3.1*	71.7±3.5
4 mM Taurine+4 mM Glycine		
Glycine–conjugation	1.36±0.74	0.79±0.40
Taurine–conjugation	32.5±2.0*	41.7±2.8
2 mM Taurine+4 mM Glycine		
Glycine–conjugation	0.23±0.14	0.94±0.22
Taurine–conjugation	26.7±1.35*	36.5±1.7

*Values are significantly different from those in rats fed pectin at p<0.05.

cellulose and pectin when taurine and glycine were coexisted at concentrations of physiological ranges (2 or 4 mM) in the assay media. Moreover, the reduction in the concentration of taurine relative to glycine in the assay medium decreased taurine–conjugation, but did not increase the glycine–conjugation of bile acids. In accordance with this observation, the enzyme was found to be highly specific for taurine both in rats fed pectin and cellulose. Thus, Km values for taurine and glycine were 0.83 and 38 mM in rats fed cellulose and 1.02 and 40 mM in the animals fed pectin, respectively. The data obtained confirmed that hepatic N–acyltransferase is highly specific for taurine both in rats fed cellulose and pectin. Thus, the alterations in hepatic taurine concentration can not account for the modification by dietary fiber of bile acid glycine/taurine ratio as observed in the present study. In this context, Spaeth and Schnider[16] showed that the proportion of bile acid conjugated with taurine in rats, guinea pig and rabbits correlated well with the rate of taurine synthesis but not the taurine concentration in the liver. Thus, in the next trial, a relationship between the activities of enzymes in taurine synthesis and bile acid conjugation was examined using a taurine lowering agent (guanidinoethyl sulfonate, GES).[17] We recently demonstrated that this drug is a potent inhibitor for the enzymes in taurine synthesis in the liver (unpublished observation).

There were four groups of rats in this trial. Two groups of rats were fed the diet containing 10% pectin, and the other two groups of animals were fed the diet containing 10% cellulose for 26 days. The each one of the groups of rats fed cellulose and pectin was given GES in drinking water (1%, w/v). As shown in table 3, dietary pectin compared to cellulose increased total bile acid secretion both in rats treated with and without GES. The GES was totally ineffective in modulating this value. In consistent with the previous observation (Table 1), dietary pectin relative to cellulose greatly increased secretion of glycine–conjugated bile acid while decreased that of taurine–conjugate regardless of GES treatment. GES treatment exerted a similar effect on these parameters both in rats fed cellulose and pectin. Thus, dietary pectin relative to cellulose significantly increased bile acid glycine/taurine ratio both in rats treated with and without GES. Also, GES considerably increased this parameter both in rats fed cellulose and pectin.

Table 4 summarizes hepatic taurine and glycine concentrations and activities of enzymes in taurine synthesis. When GES was not included in drinking water, dietary pectin compared to cellulose greatly decreased hepatic taurine concentrations. Pectin also significantly decreased concentrations of this compound in serum, kidney and heart (data not shown). However, the extent of the reduction by dietary pectin was most prominent in the liver. GES treatment severely decreased hepatic taurine concentrations both in rats fed cellulose and pectin to a similar level. Types of dietary fiber were irrelevant in modifying hepatic glycine concentration. GES treatment significantly decreased hepatic concentrations of this amino acid both in rats fed cellulose and pectin. The extent of the decrease was more exalted in rat fed pectin. As a consequence, the hepatic glycine concentration in rats fed pectin became to be significantly lower than that in rats fed cellulose in the animals treated with GES.

Table 3 Effects of dietary fiber and guanidinoethyl sulfonate (GES) on biliary bile acid secretion in the rats

	Groups			
	Cellulose	Pectin	Cellulose +GES	Pectin +GES
Bile flow (ml/hr)	0.64±0.06*	1.30±0.11	0.83±0.05*	1.53±0.17
Biliary bile acid secretion (μmol/hr)				
Total	21.1±1.2*,**	52.7±4.3	37.3±3.5*	64.7±4.8
Non–conjugated	0.25±0.05	2.29±0.39**	1.00±0.21*	8.05±1.73
Glycine–conjugated	1.99±0.41*,**	39.8±3.2**	21.7±0.7*	50.8±4.1
Taurine–conjugated	18.9±1.2*	10.6±0.9	14.6±3.3*	5.83±0.79
G/T ratio	0.111±0.028*	3.79±0.20**	1.98±0.22*	9.32±0.93

*Values are significantly different from corresponding values in rats fed pectin at $p<0.05$.
**Values are significantly different from corresponding values in rats given guanidinoethyl sulfonate at $p<0.05$.

Taurine is synthesized from cysteine in animal tissues. Cysteine is converted to cysteine sulfinate by cysteine dioxygenase and in turn decarboxylated by cysteine sulfinate decarboxylase to form hypotaurine. Hypotaurine is finally converted to taurine presumably by the non–enzymatic process. Thus, the alterations in cysteine dioxygenase and cysteine sulfinate decarboxylase activities may modify the taurine synthesis in the tissues. When GES was not given, dietary pectin compared to cellulose severely decreased the activities of cysteine dioxygenase in the liver. In contrast, in the animals without receiving GES, hepatic cysteine sulfinate decarboxylase activities in rats fed pectin were significantly higher than those in the animals fed cellulose. GES treatment severely decreased cysteine sulfinate decarboxylase activities irrespective of the dietary fiber sources. Also, this drug significantly reduced cysteine dioxygenase activities in rats fed cellulose and tended to decrease this parameter in the animals fed pectin. The differences in the activities of these enzymes between rats fed cellulose and pectin became to be statistically insignificant in rats treated with GES. However, dietary pectin compared to cellulose still tended to decrease the activities of cysteine dioxygenase and to increase those of cysteine sulfinate decarboxylase in the liver even in rats treated with GES.

Table 4 Effects of dietary fiber and guanidinoethyl sulfonate (GES) on the concentrations of taurine and glycine and activities of the enzymes in taurine synthesis in rat liver

	Groups			
	Cellulose	Pectin	Cellulose +GES	Pectin +GES
Amino acids (μmol/g)				
Taurine	5.27±0.69*,**	1.46±0.08	0.816±0.048	0.866±0.051
Glycine	2.63±0.11**	2.87±0.14**	2.06±0.14*	1.29±0.11
Enzyme activities (nmol/min/mg)				
Cysteine dioxygenase	8.63±2.04*,**	1.61±0.15	2.78±0.31	0.939±0.226
Cysteine sulfinate decarboxylase	27.6±2.1*,**	49.9±2.1**	6.50±0.32	10.3±0.7

*Values are significantly different from corresponding values in rats fed pectin at p<0.05.
**Values are significantly different from corresponding values in rats given guanidinoethyl sulfonate at p<0.05.

We could demonstrate that GES severely inhibits the activities of cysteine dioxygenase and cysteine sulfinate decarboxylase, the enzymes in taurine synthesis, in the liver. This drug is previously shown to be a competitive inhibitor for taurine transport in the tissue.[17] Present observation suggests that the decrease in hepatic taurine synthesis is also the one of the mechanism to account for the taurine–lowering effect of GES. We confirmed previous observation[10] that dietary pectin severely decreased the activities of cysteine dioxygenase in the liver. Thus, it was suggested that dietary pectin decreases hepatic taurine concentration through the impairment in hepatic taurine synthesis. However, in the present study, it was found that the dietary fiber rather increased activities of cysteine sulfinate decarboxylase in the liver. It is uncertain that which of these two enzymes, cysteine dioxygenase or cysteine sulfinate decarboxylase, is a rate–limiting enzyme in taurine synthesis in the liver. However, available evidences[18-21] strongly indicate that the dioxygenase rather than the decarboxylase plays a crucial role in regulating hepatic taurine synthesis in various nutritional status in the rats. Thus, it is plausible that activities of the dioxygenase rather than the decarboxylase reflect the rate of taurine synthesis in the liver of rats fed different types of dietary fiber, as well.

This study showed that neither the substrate specificity of bile acid–CoA:amino acid *N*–

acyltransferase nor the alterations in hepatic concentrations of taurine did not account for the modifications by dietary fiber of bile acid glycine/taurine ratio in the rat. Apparently there was no consistent relationship between hepatic taurine concentration and biliary bile acid glycine/taurine ratio among the groups (compare Tables 3 and 4). However, there was a consistent inverse correlationship between the activities of cysteine dioxygenase and bile acid glycine/taurine ratio among the groups. This observation possibly indicates that the alteration in the activity of hepatic cysteine dioxygenase and thus in the hepatic taurine synthesis is the crucial factor in modulating bile acid glycine/taurine ratio in the rat.

REFERENCES

1. D.A.Vessey, The biochemical basis for the conjugation of bile acid either with glycine or taurine, *Biochem. J.* 174:621 (1978).
2. P.G.Killenberg and J.T. Jordan, Purification and characterization of bile acid–CoA:amino acid N–acyltransferase from rat liver, *J. Biol. Chem.* 253:1005 (1978).
3. D.A.Vessey, The co–purification and common identity of cholyl CoA:glycine and cholyl CoA:taurine–N–acyltransferase activities from bovine liver, *J. Biol. Chem.* 254:2059 (1979).
4. M.R.Johnson, S.Barnes, J.B.Kwakye and R.B.Diasio, Purification and characterization of bile acid–CoA:amino acid N–acyltransferase from human liver, *J. Biol. Chem.* 266:10227 (1991).
5. J.B.Kwakye, M.R.Johnson, S.Barnes and R.B.Diasio, A comparative study of bile acid CoA:amino acid:N–acyltransferase (BAT) form four mammalian species, *Comp. Biochem. Physiol.* 100B:131 (1991).
6. J.G.Jacobsen and L.H.Smith, Jr., Biochemistry and physiology of taurine and taurine derivatives, *Physiol. Rev.* 48:424(1968).
7. G.A.D.Haslewood and V.Wooton, Comparative studies of 'bile salts'. 1. Preliminary survey, *Biochem. J.* 47: 584(1950).
8. T.Ide and M.Horii Predominant conjugation with glycine of biliary and lumen bile acids in rats fed on pectin, *Br. J. Nutr.* 61:545 (1989).
9. T.Ide, M.Horii, K.Kawashima and T.Yamamoto, Bile acid conjugation and hepatic taurine concentration in rats fed on pectin, *Br. J. Nutr.* 62:539 (1989).
10. T.Ide, M.Horii, T.Yamamoto, and K.Kawashima, Contrasting effects of water–soluble and water–insoluble dietary fibers on bile acid conjugation and taurine metabolism in the rat, *Lipids* 25:335(1990).
11. T.Ide and M.Sugano, Interaction of dietary protein differing in sulfur amino acid content and pectin on bile acid conjugation in immature and mature rats, *J. Nutr.* 121:985 (1991).
12. American Institute of Nutrition,Report of the American Institute of Nutrition Ad Hoc Committee on standards for nutritional studies, *J. Nutr.*107:1340 (1977).
13. B.F.Kase and I.Björkhem, Peroxisomal bile acid–CoA:amino acid N–acyltransferase in rat liver, *J. Biol. Chem.* 264:9220 (1989).
14. C.A.Weinstein and O.W.Griffith, Cysteinesulfinate decarboxylase, *Method Enzymol.*143:404(1987).
15. T.Ide and M.Horii, A simple method for the extraction and determination of non–conjugated and conjugated luminal bile acids in rats, *Agric.Biol.Chem.* 51: 3155 (1987).
16. D.G.Spaeth and D.L.Schneider, Taurine synthesis, concentration, and bile salt conjugation in rat, guinea pig, and rabbit, *Proc.Soc.Expt.Biol.Med.* 147:855 (1974).
17. R.J.Huxtable, H.E.Laird II and S.E.Lippincott, The transport of taurine in the heart and the rapid depletion of tissue taurine content by guanidinoethyl sulfonate, *J.Pharmacol.Expt.Ther.* 211:465 (1979).
18. N.Kohashi, K.Yamaguchi, Y.Hosokawa, Y.Kori, O.Fujii and I.Ueda, Dietary control of cysteine dioxygenase in rat liver, *J.Biochem.*84:159 (1978)
19. Y.Hosokawa, S.Nizeki, H.Tojo, I.Sato and K.Yamaguchi, Hepatic cysteine dioxygenase activity and sulfur amino acid metabolism in rats:possible indicator in the evaluation of protein quality, *J.Nutr.*118:456 (1988).
20. A.A.Jerkins and R.D.Steele, Dietary sulfur amino acid modulation of cysteine sulfinic acid decarboxylase, *Am.J.Physiol.* 261:E551 (1991).
21. A.A.Jerkins, L.E.Bobroff and R.D.Steele, Hepatic cysteine sulfinic acid decarboxylase activity in rats fed various levels of dietary casein, *J.Nutr.* 119:1593 (1989).

ANTI-HYPERTENSIVE SUBSTANCES IN VISCOUS MATERIAL OF FERMENTED SOYBEAN (NATTO)

Akiko Okamoto[1], Hiroshi Hanagata[1], Eiko Matsumoto[1], Yukio Kawamura[2] and Fujiharu Yanagida[1]

[1] Department of Brewing and Fermentation, Tokyo University of Agriculture, Sakuragaoka, Setagaya, Tokyo 156, Japan
[2] National Food Research Institute, Kannondai, Tsukuba, Ibaraki 305, Japan

ABSTRACT   The angiotensin converting enzyme (ACE) inhibitor in the viscous material of natto was investigated. The relatively strong inhibitory activities were detected both in the water soluble and the water insoluble but ethanol soluble fraction in the viscous material. The former contained the inhibitory substance with the high molecular weight and considered to be a proteineous substance. The latter had the inhibitor of the small molecular weight (*ca* 600), which was stable over the wide range of pH and thermostable. It showed competitive inhibition against the hydrolysis of Bz-Gly-His-Leu by ACE.

INTRODUCTION

The close relationship between food and health has been well known. Foods supply not only various basic nutrients that maintain the functions of living body, but also some pleasure of eating. Recently, new aspects of food functions, other than nutrition or taste have been drawing considerable attention. Among them are the physiological function of some food components against certain ailments. One of the representatives is the epidemiological relationship between foods and hypertension. Some foods are known to be effective to decrease blood pressure,[1] which suggests the existence of some components having medicine-like activities that effect to the regulation systems of blood pressure.

The renin-angiotensin system is considered to be a blood pressure regulation system which is apt to be effected by food components. The system starts by the conversion of angiotesinogen to angiotensin I (DRVYIHPFHL) by the action of renin which is secreted from kidney. The angiotensin I is further converted to angiotensin II (DRVYIHPF), the active form of the hormone, by the action of angiotensin converting enzyme (ACE). Angiotensin II rises blood pressure by acting directly to vein, sympathetic nerves and adrenal grands. Therefore, food components which inhibit ACE have the possibility to prevent hypertension by inhibiting the formation of angiotensin II.

Food Hydrocolloids: Structures, Properties, and Functions
Edited by K. Nishinari and E. Doi, Plenum Press, New York, 1994

From this point of view, we investigated natto, a Japanese traditional fermented food. Natto is produced by the fermentation of boiled soy beans with *Bacillus natto*, and is characterized by the viscous material covering the beans. The relationship between natto and blood pressure is studied by Hayashi *et al.*,[2] and suppress of blood pressure of rats by feeding natto is reported. In this report, we deals with the anti-hypertensive substances with ACE inhibitory activity in the viscous material of natto.

MATERIALS AND METHODS

Preparation of ACE

ACE was prepared from rabbit and porcine lung acetone powder by following procedure. Fresh lung was cut into small pieces after removing bronchus and tracheas, then roughly homogenized with 5-10 volumes of cold acetone using Wahling blender. The homogenate was filtered with suction and the resultant residue was used for the extraction of ACE after removing acetone completely. Thus prepared acetone powder was stable when stored at -20°C.

Extraction of ACE from the acetone powder was achieved as follows: The acetone powder was suspended in 10 volumes of 125mM of Tris-HCl buffer (pH 8.3) containing 1M NaCl and homogenized with polytron. The homogenate was stirred overnight at 4°C, and subjected to centrifugation (16,000rpm x 20min). The resultant supernatant showed the high ACE activity. This activity was inhibited by EDTA completely, therefore, the extract was used for the whole experiment as a ACE solution.

Assay Procedure of ACE Inhibitory Activity

The ACE inhibitory activity was measured principally according to the method of Cushmen *et al.*[3] ACE solution (50μl) was preincubated with each sample (50μl) for 5min at 37°C, then the enzyme reaction was started by adding 8.33mM of the substrate (Bz-Gly-His-Leu, Peptide Institute Co.) to the mixture. After incubation for 30min at 37°C, the reaction was stopped by adding 250μl of 1N HCl. Hippuric acid produced by this reaction was extracted with 1.5ml ethyl acetate, and exactly 1ml of the ethyl acetate layer after separating two layers by centrifuging was collected in another test tube to remove ethyl acetate by evaporation. The extracted hippuric acid was redissolved in distilled water, and its amount was measured from the absorbance at 228nm. Comparing with the amount of hippuric acid produced by the ACE preincubated with distilled water instead of samples, the ACE inhibitory activities of samples were calculated. The degree of inhibition was expressed with IC_{50}, the sample concentration that inhibit 50% of ACE activity.

Gel Filtration

Sephadex G-25 Column Chromatography. This chromatography was achieved according to the following condition: column size, Ø1.5cm x 17cm; flow rate, 19.5ml/hr; fraction volume, 1ml/tube. The column was eluted with 0.1M NaCl.

Bio-gel P-2 column chromatography. This chromatography was achieved with the following condition: column size, Ø1.0cm x 70cm; flow rate, 5.6ml/hr; fraction volume, 1ml/tube. The column was eluted with distilled water.

Determination of Molecular Weight

The molecular weight of the low molecular weight inhibitor was estimated using Bio-gel P-2 with bacitracin (MW 1,400), glutathione, disulfide (MW 600) and glutathione, reduced (MW 300) as standard molecular markers.

RESULTS

ACE Inhibitory Activity of Natto

Natto was homogenized with 10 volumes of distilled water followed by centrifugation. The supernatant showed the IC_{50} of 0.4 mg/ml, indicating that natto is possessed of relatively strong ACE inhibitory activity.

Fractionation of Natto

Two kinds of natto was then separated to viscous material and beans by washing with 10 volumes of distilled water. The residual beans were homogenized with distilled water followed by centrifugation. Table 1 shows the inhibitory activities of both viscous material and bean extract. Both parts had considerably strong activities, and viscous material showed stronger activity when compared these two portions. Therefore, the ACE inhibitory substance of this viscous material were investigated in more detail.

Table 1. ACE inhibitory activities of natto

Kind of natto	IC_{50} (mg/ml)	
	viscous materials	beans
General	0.32	0.33
High fermented	0.27	0.51

Dialysis Test of the Viscous Material

The viscous material was then extracted with ethanol after lyophilized. The water extract and the ethanol extract were subjected to dialysis (Table 2). IC_{50} of the water extract decreased after dialysis, while the value of the ethanol extract increased. This fact suggests that the viscous material contains at least two kinds of ACE inhibitory substances, namely, the inhibitory substance of high molecular weight which is contained in the water extract and of low molecular weight which is in the ethanol extract.

Fractionation of the Viscous Material

For the detailed study of these inhibitory substances, the viscous material separated from beans was further fractionated. Because the fraction of viscous material was turbid, it was centrifuged (10,000rpm x 15min) to obtain the precipitate. The supernatant and the

Table 2. Effect of dialysis to the viscous substance [1]

Sample	IC$_{50}$ (mg/ml)	
	before dialysis	after dialysis
General		
water extract	0.32	0.15
EtOH extract	0.33	0.74
High fermented		
water extract	0.27	0.14
EtOH extract	0.54	1.82

[1]Dialysis was achieved with the dialysis tube of MW cutoff 3,500.

precipitate were then lyophilized, and the both preparations were extracted with hot 80% ethanol. The supernatant, hot 80% ethanol extract of it and hot 80% ethanol extract of the precipitate were named fraction 1, fraction 2 and fraction 3, respectively. ACE inhibition of fraction 1 and fraction 3 showed relatively strong compared to fraction 2, and further investigations were carried out with these samples.

Characterization of Fraction 1

Fraction 1, which contains the substances with high viscosity, was subjected to the following examination.

Effect of Temperature. The inhibitory activity was stable up to 40°C. However, it almost disappeared when incubated above 50°C, indicating that this substance is unstable to heating. This fact suggests that the inhibitory substance in fraction 1 is a proteineous one.

Effect of Proteases. To certain the above suggestion, fraction 1 was subjected to treatment with digestive proteases (Table 3). As shown in the table, treatment with pepsin, trypsin and α-chymotrypsin caused increase of the ACE inhibitory activities, while the combined digestion both with trypsin and α-chymotrypsin decrease the activity. This fact that the ACE inhibitory activity changes by the treatment with proteases indicates that the substance is proteineous, that is, increasing of the inhibitory activity suggests the generation of ACE inhibitory peptides and decrease of the activity is considered to due to degradation of the peptides.

Table 3. ACE inhibitory activities of viscous substance hydrolysate Prepared by treatment with digestive proteases [1]

Protease	ACE inhibitory activity (%)	
	Before hydrolysis	after hydrolysis
Pepsin	10.6	61.8
Trypsin	35.4	52.4
α-Chymotrypsin	41.3	44.9
Trypsin and α-Chymotrypsin	49.8	19.4

[1]The reaction mixture was incubate at 37°C for 24hr with shaking.

Effect of a Protease Inhibitor. When fraction 1 was incubated with 1mM phenylmethanesulfonyl fluoride (PMSF), 63.5% of the ACE inhibitory activity was decreased, however, 36.5% of the inhibitory activity still remained.

Characterization and Partial Purification of Fraction 3

Effect of pH. The pH stability of the substance was studied against pHs from 2 to 10. The inhibitory activity was almost the same for any given pH, indicating that the inhibitory substance in fraction 3 is stable under pH conditions.

Effect of Temperature. At any temperatures, the inhibitory activities were almost constant values, indicating that the substance is stable at any temperature conditions.

Partial Purification of Fraction 3. Fraction 3 was partially purified with Sephadex G-25 column chromatography followed by Bio-gel P-2 column chromatography. After Sephadex G-25 column chromatography, the ACE inhibitory activity was detected as a major peak at the elution volume of 27ml. The elution pattern of Biogel P-2 showed a major activity peak at the elution volume of about 42ml, which differ from the peak with absorbance at 210 and 280 nm. The active fraction gave IC_{50} of 0.50 mg/ml when measured after lyophilization.

Molecular Weight. The molecular weight of this partial purified inhibitor was determined by Bio-gel P-2 gel filtration. The molecular weight was estimated to be about 600 from its elution volume. At this stage, the IC_{50} of this preparation was calculated to be 0.83mM based on the molecular weight.

Mode of Inhibition. The mode of inhibition of the partial purified inhibitor was determined by the lineweaber-burk plot. The V_{max} value was not altered with the change of the inhibitor concentration, while the K_m value was changed. This result indicates that the inhibitor causes competitive inhibition.

DISCUSSION

Hayashi mentions in his report using spontaneous hypertensive rat (SHR) that the blood pressures of the rats decreased when fed with natto, while that of the rat which was fed with boiled soy bean rose above 200mmHg. This fact suggests that something which suppresses the blood pressure is produced in natto preparation through the fermentation and/or during digestive process. Our results indicated that ACE inhibitory substances both with high and low molecular weight are present in natto.

As mentioned in INTRODUCTION, natto is made of boiled soybeans with *B.natto* strewed on the surface of the beans. This bacteria produces many substances mainly on the surface of the beans throughout the fermentation. The most characteristic substance is the viscous material. This material is reported to mainly consist of polymer of D-glutamate and fractan[4,5] in addition to many substances produced by the bacteria. The results of present experiments reveal that the ACE inhibitory substances exist in these substances. Among these, the fraction 3 is considered to contain *B. natto* itself, therefore, the inhibitory substance might be arising from the bacterial bodies. This consideration shows a good agreement with the fact reported by Hayashi *et al.*, that the substance effective to decrease the blood pressure exists in ethanol extract of dead *B. natto*. Further purification of this substance for the structural analysis is in progress.

The inhibitory substance of high molecular weight obtained from the fraction 1 was considered to be a proteineous substance. This fraction is very viscous and probably including many substances. Especially, many kinds of enzymes relating to the fermentation are considered to be included in it. Therefore, one possibility is that the ACE inhibition of this fraction substance may be caused by the degradation of ACE by proteases in this fraction. Actually, the inhibitory activity decreased by treatment of the sample with PMSF, indicating that proteases are responsible for the inhibitory action to a certain extent. However, the fact that about 35% of the inhibitory activity remains after the treatment with PMSF suggests that the real inhibitory substance other than proteases.

In addition, treatment of this fraction with various proteases caused increase of ACE inhibitory activity (Table 3). As mentioned in RESULTS, this is probably to due to the formation of inhibitory peptides from the proteineous components of this viscous material. There are many reports about the ACE inhibitory peptide sequences.[6] Those peptides are liberated by digestive proteases in intestine. From our results, the viscous material of natto is suggested to possesses such peptides latently. The detailed study of these peptides is currently under way.

ACKNOWLEDGEMENT

We wish to thank Takano Foods Co. for the supply of natto used in this experiment.

REFERENCES

1. K.Ikeda, Y Nora and Y. Yamori, Effect of Pork and Krill-Proteins on Blood Pressure in Stroke-Prone SHR (SHRSP), *Jpn. Heart J.* 29 : 583 (1988).

2. U. Hayashi, K. Nagao, Y. Tosa and Y. Yoshioka, Natto no Eiyouka ni Kansuru Jikkenteki Kenkyu, *Natto Kagaku Kenkyu Kaishi* 1 : 85 (1977).

3. D. W. Cushman and H. S. Cheung, Spectrophotometric Assay and Properties of the Angiotensin-Converting Enzyme of Rabbit Lung, *Biochemical Pharmacol.* 20 : 1637 (1971).

4. H. Fujii, On the Formation of Mucilage by Bacillus natto. Part III. Chemical Constituents of Mucilage in Natto, *Nippon Nogei Kagaku Kaishi* 37 : 407 (1963).

5. H. Fujii, On the Formation of Mucilage by Bacillus natto. Part IV. Chemical Constituentsof Mucilage in Natto, *Nippon Nogei Kagaku Kaishi* 37 : 474 (1963).

6. Y. Kawamura, T. Sugimoto, S. Takane and M. Satake, Biologically Active Peptides Derived from Food Proteins, *Biryo Eiyouso Kenkyu* 6 : 117 (1989).

BENEFICIAL EFFECT OF DIETARY FIBER
ON IRON BIOAVAILABILITY IN RATS

Taro Kishida, Yota Shimizu and Shuhachi Kiriyama

Laboratory of Nutritional Biochemistry, Faculty of Agriculture
Hokkaido University, Sapporo 060, Japan

ABSTRACT

We examined the effect of demineralized wheat bran (DWB), demineralized onion dietary fiber (DODF) and other dietary fibers (DF) (experiment 1) and the effect of polystyrene foams (PSF) of various foaming degrees (reciprocal of density)(PSF30, PSF50, PSF90) (experiment 2) on Fe-II or Fe-III bioavailability. Fe bioavailability was evaluated by Hb regeneration. Experiment 1: Both Fe-II and Fe-III were utilized for Hb regeneration at the same rate. The concurrent addition of DODF, DWB or PSF90 significantly improved the regeneration rate of Hb compared to that in rats fed Fe-II alone (Fe-II control). DODF and DWB showed higher values of settling volume in water (SV) than those of other DFs. All DFs did not improve Hb regeneration in rats fed Fe-III. Experiment 2: 5% of PSF50 and PSF90 significantly improved the regeneration rate of Hb compared to that of Fe-II control. The larger foaming degree PSF had, the better improved Fe-II bioavailability. DODF and DWB of high SV, also improved it. Then we suggest that the promoting effect of DF on Fe bioavailability is closely related to the volume of DF exhibited in the intestinal lumen.

INTRODUCTION

DF would decrease mineral bioavailability because it may have cation exchange capacity, bind mineral ion and form complexes difficult to absorb in intestinal lumen, as phytin does[1]. But Fairweather-Tait et al. [2] reported sugar-beet fiber increased absorption of ^{59}Fe by about 13% in rats. In a previous study, we found that dietary addition of DWB accelerated Fe-II bioavailability, but had no effect on Fe-III bioavailability as determined by the rate of Hb regeneration in rats[3]. The present study was designed to examine the effect of DWB, DODF and other DFs and the effect of PSFs of various foaming degrees on Fe bioavailability.

MATERIALS AND METHODS

Preparation of demineralized wheat bran (DWB): After Wheat bran (WB) was demineralized by washing with 1N HCl 20 times, HCl was removed by washing with demineralized water. Resultant wet DWB was dehydrated with ethanol and dried in air. Different wheat brans were used in experiments 1a and 1b.
PSF: Every PSF was ground to pass 0.5 mm sieve, demineralized by washing with 6N HCl, and treated in a similar manner adopted in DWB preparation.
DODF: Onion was roughly ground with a disposer, boiled for 30 minutes, dehydrated with ethanol, dried and ground to pass 0.5 mm sieve. This crude onion DF was

Food Hydrocolloids: Structures, Properties, and Functions
Edited by K. Nishinari and E. Doi, Plenum Press, New York, 1994

demineralized with 1N HCl 3 times and treated in a similar manner adopted in DWB preparation.

In all experiments, the protein source of Fe-deficient diet was egg white powder (15%), mineral mixture was Fe-free mineral mixture 2 (MM2)[4] and vitamin mixture was without ascorbic acid. All test diets contained 20 ppm Fe in the form of $FeSO_4$ or $Fe_2(SO_4)_3$. Rats were preliminarily fed a standard diet contained 40 ppm Fe for 3 days in experiment 1a and a Fe-deficient diet for 1 wk in other experiments. 5% of DFs were added to the diets except for the case of 10% PSF diet used in experiment 2.

Experiment 1a: Rats were fed the Fe-deficient diet for 4 wk and randomly allocated to groups of 6 rats. One group continued to be fed the Fe-deficient diet and other groups were fed the diet with added Fe-II or Fe-III alone (Fe-II or Fe-III control) and the diets added Fe-II together with PSF90, DWB or cellulose and the diets added Fe-III together with DWB for 2 wk. During experimental period, one more group continued to be fed the standard diet. Body weight, food intake and Hb concentration were measured. Hb regeneration efficiency (HRE) was estimated by the following equation:

$$\text{Total Hb of whole rat (g)} = \frac{-1.803\log(\text{body wt (g)}) + 10.203}{100} \times (\text{body wt}) \times \text{Hb conc (g/dL)}$$

$$HRE = \frac{\text{Hb-Fe gain (mg)}}{\text{Fe intake (mg)}} \qquad (\text{Hb-Fe} = 3.35 \text{ mg/g Hb})$$

Experiment 1b: After feeding the Fe-deficient diet for 3 wk, rats were divided into groups and fed the diet with added Fe-II or Fe-III alone and the diets added either Fe-II or Fe-III together with PSF90, DWB or cellulose for 2 wk.

Experiment 1c: The diets added either Fe-II or Fe-III together with DODF or PSF90 were fed to rats for 2 wk.

Experiment 2: The diets containing Fe-II with or without added 5% of PSF30, PSF50, PSF90 or 10% of PSF90 were fed to rats for 2 wk.

RESULTS

In all experiments, Hb concentration continued to decrease after Fe-deficient diet feeding and it was 4.5-5.5 g/dL before the test feeding began.

Experiment 1a: Figure 1 shows changes in Hb concentration during the Fe depletion and repletion periods. Both Fe-II and Fe-III control diets showed almost the same changes in Hb concentration. The concurrent addition of DWB or PSF90 significantly improved the regeneration rate of Hb concentration compared to that in rats fed the Fe-II control diet. However, DWB did not improve Hb regeneration rate in rats fed Fe-III .

Experiment 1b: Figure 2 shows changes in Hb concentration during the Fe depletion and repletion periods. As experiment 1a, both Fe-II and Fe-III controls regenerated Hb at the same rate. PSF90 significantly improved the regeneration rate of Hb, but DWB had no effect in rats fed Fe-II. All DFs did not improved Hb regeneration rate in rats fed Fe-III .

Experiment 1c: Figure 3 shows changes in Hb concentration during the Fe depletion and repletion periods. DODF and PSF90 significantly improved the recovery rate of Hb concentration compared to that of rats fed Fe-II control diet. DODF also significantly improved the recovery rate of Hb concentration in rats fed Fe-III, but this improvement was smaller than that in rats fed Fe-II. PSF90 had no effect in rats fed Fe-III. DODF and DWB used in experiment 1a had higher values of settling volume in water (SV) than SV of DWB used in experiment 1b.

Experiment 2: 5% of PSF30 did not improve the regeneration rate of Hb. 5% of PSF50, PSF90 and 10% of PSF90 significantly improved the regeneration rate of Hb compared to that of Fe-II control group. Figure 4 is HRE in the repletion period. 5% of PSF30 added to the Fe-II control diet had no effect. 5% of PSF50, PSF90 improved Hb regeneration rate though not significant. 10% of PSF90 produced significantly higher value of Hb concentration compared to that of Fe-II control group.

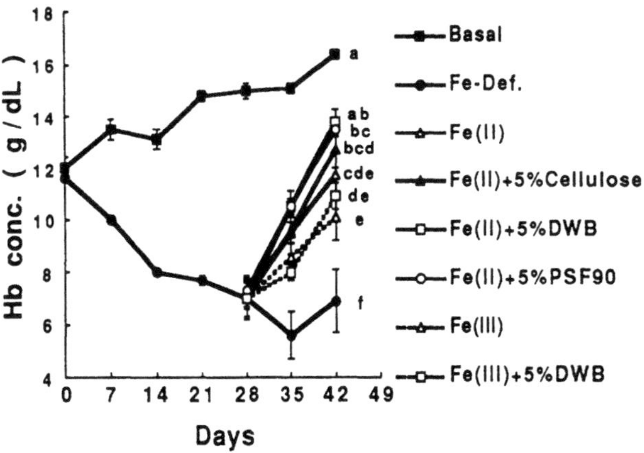

Figure 1 Changes in Hb concentration (experiment 1a). Points represent the mean Hb concentration of six rats fed the Fe-II or Fe-III control diets, with or without added PSF90, DWB and cellulose. Vertical lines indicate standard error. Points with different letter are significantly different (p<0.05) by ANOVA and Fisher's LSD test.

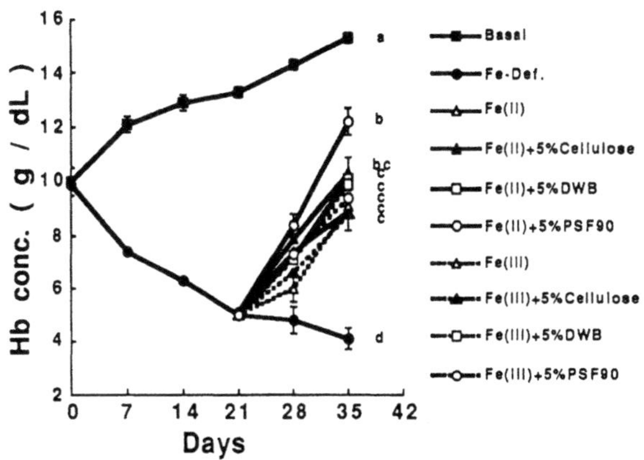

Figure 2 Changes in Hb concentration (experiment 1b). Points represent the mean Hb concentration of six rats fed the Fe-II or Fe-III control diets, with or without added PSF90, DWB and cellulose. Vertical lines indicate standard error. Points with different letter are significantly different (p<0.05) by ANOVA and Fisher's LSD test.

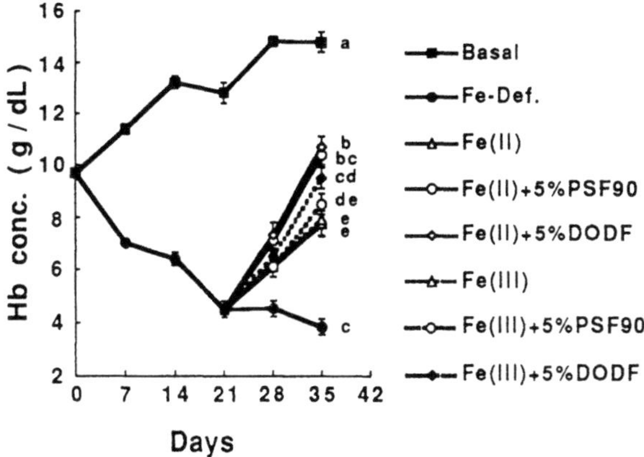

Figure 3 Changes in Hb concentration (experiment 1c). Points represent the mean Hb concentration of six rats fed the Fe-II or Fe-III control diets, with or without added PSF90 and DODF. Vertical lines indicate standard error. Points with different letter are significantly different (p<0.05) by ANOVA and Fisher's LSD test.

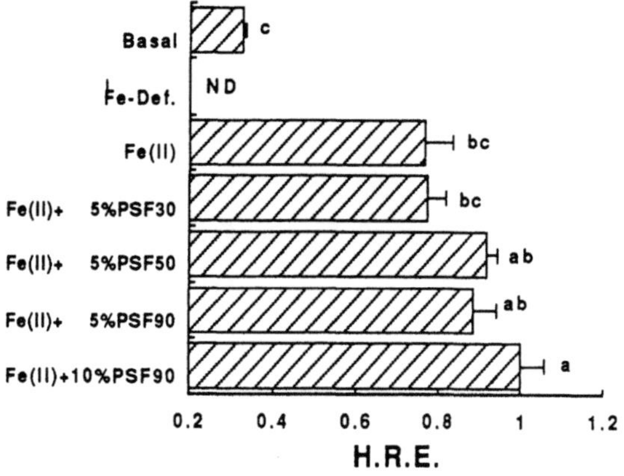

Figure 4 Hb regeneration efficiency (HRE) in the repletion period of experiment 2. The diets containing Fe-II with or without added 5% of PSF30, PSF50, PSF90 or 10% of PSF90 were fed for 2 wk. Horizontal lines indicate standard error. Values with different letter are significantly different (p<0.05) by ANOVA and Fisher's LSD test.

DISCUSSION

In the present study, we observed that DWB had promoting effect on Fe-II bioavailability and DODF and PSF90 also exerted the same effect, but all DF had no or a little effect on Fe-III bioavailability. Table 1 summarizes HRE in rats fed various DFs and SV of DFs added to Fe-II control diet. DODF and DWB of high SV also improved Fe-II bioavailability. Cellulose of low SV has no effect. The larger foaming degree PSF had or the more PSF added, the better improved Fe-II bioavailability. The promoting effect of DF on Fe bioavailability may be closely related to the volume of DF exhibited in the intestinal lumen. DF of high SV may delay intestinal transit time and improve nutritient absorption[5]. Fe absorption may be improved by the similar mechanism.

Table 1 HRE in rats fed various DFs and SV of DF
added to Fe-II control diet

DF added to Fe-II control diet	SV[1] of dietary fiber	HRE[2]
	ml/g dry sample	
Cellulose	4.9	0.66
DWB used in experiment 1a	20.5	0.94
DWB used in experiment 1b	15.9	0.79
DODF	23.9	0.83

1 SV=settling volume in water.
2 HRE=Hb regeneration efficiency.

REFERENCE

1. B. Lonnerdal, A. Sandberg, B. Sandstrom and C. Kunz. "Inhibitory effects of phytic acid and other inositol phosphates on zinc and calcium absorption in suckling rats," *J. Nutr.*, **119**: 211 (1989).
2. S. Fairweather-Tait and A. J. A. Wright. "The effects of sugar-beet fibre and wheat bran on iron and zinc absorption in rats," *Br. J. Nutr.*, **64**: 547 (1990).
3. S. Saeki, A. Sugihara, S. Kiriyama, "Beneficial effect of dietary fiber on iron bioavailability in rats," *Nippon Nogeikagaku Kaishi* abstract, **65**: 513 (1991).
4. K. Ebihara,Y. Imamura and S. Kiriyama, "Effect of dietary mineral composition on nutritional equivalency of amino acid mixture and casein in rats," *J. Nutr.*, **109** : 2106(1979).
5. T. Arizuka, H. Takeda and S. Kiriyama, "Correlation between anti-toxic activity and settling volume in water of beet dietary fiber in rats fed a toxic dose of amaranth," *Nippon Nogeikagaku Kaishi* , **66**: 719(1992).

W/O/W emulsion, 399
Wall effect, 295
Water, 281
Water binding, 368
Water-holding ability, 371
Water permeability, 401
Welan, 109
Whey protein isolate, 467

Xanthan, 131, 141, 247, 461

Yoghurt drinks, 151

Zeta potential
  of egg PC, 270
  of casein, 342
Zipper model, 107